MODERN REACTIONS IN ORGANIC SYNTHESIS

edited by

C. J. TIMMONS

Department of Chemistry
University of Nottingham

VAN NOSTRAND REINHOLD COMPANY
LONDON
NEW YORK CINCINNATI TORONTO MELBOURNE

VAN NOSTRAND REINHOLD COMPANY LTD
Windsor House, 46 Victoria Street, London SW1

INTERNATIONAL OFFICES
New York Cincinnati Toronto Melbourne

Library of Congress Catalog Card No. 78-96108

First Published 1970

PRINTED IN GREAT BRITAIN BY J. W. ARROWSMITH LTD., BRISTOL

MODERN REACTIONS IN ORGANIC SYNTHESIS

Contents

1

Introduction

C. J. TIMMONS

This book contains seven chapters that deal with some of the more widely useful methods in organic chemistry and highlights the developments and applications of these methods that have occurred in the 1960s.

The choice of such topics is necessarily a personal one. Many readers would undoubtedly have made a different choice. In some ways there is not a completely free choice, because it is always easier to think of a suitable topic than to find a contributor who has both the ability and the time to write a chapter. To the present contributors who have so willingly reviewed the various fields in which they are all actively engaged, the editor is most grateful.

It is hoped that the reviews will be of use to the practising organic chemist. It is also felt that the undergraduate honours chemist, say in his final year, will find much of interest in the various chapters, because within the fields chosen he can find the more important information collected together, rather than having to ferret it out from original papers or to consult larger more specialized volumes for each topic.

The average chemist is well aware how the development and application of new physical techniques has changed the face of chemistry in the last ten years. One has only to think back on how organic chemistry was conducted in the 1950s, before the everyday use of nuclear magnetic resonance spectra and mass spectra, to realize the difference.

The developments in synthetic chemistry and the applications of new methods and reagents have been less spectacular because they have ranged more widely. However, one has only to glance at any major contribution to synthetic organic chemistry today to see that it employs numerous reagents and methods that were not available before 1960.

In this book reduction and oxidation methods have been chosen because they represent two of the most fundamental types of reaction. Gone are the days when the main oxidizing agents were chromic acid and permanganate. Organic chemists frequently think that free radicals are only the concern of physical chemists and forget that many useful organic reactions proceed by a free-radical mechanism. This applies particularly to many reactions carried out industrially on a vast and ever increasing scale. Not all chemists will immediately consider an electrochemical step in a synthetic scheme, but

Chapter 5 on electrochemical methods illustrates the versatility of this approach.

From 1850 to 1950 chemists were mainly interested in reactions that took place with molecules in their ground electronic state, albeit perturbed largely by the approach of the reagent. More recently there has been great interest in the reactions of molecules in electronically excited states brought about by the absorption of light. What is now unfolding is a huge panorama of the chemistry of the excited state. Chapter 6 on photochemical methods shows that as many possible molecules as resources will allow are now being studied photochemically.

Aromatic compounds and heterocyclic compounds have long been the staple components of much organic study, and here too there have been numerous advances which are detailed in the last two chapters, which deal with new approaches to the synthesis of these molecules.

The intensive study of organic chemistry has now passed into its second century. It is in no way a dead or worn-out pursuit. The applications of physical techniques are allowing chemists to unravel and elucidate the structures of extremely complicated molecules. It then remains to the organic chemist to synthesize these and related structures. How well he succeeds in doing this is shown by the fact that not only does he achieve an academic synthesis, using the modern synthetic methods now available, but in a number of cases he goes on to manufacture the material industrially.

This book reviews, within chosen fields, some of the synthetic methods recently developed or applied that seem important in the views of the contributors.

2

Reduction Methods

R. F. EVANS, *University of Queensland, Brisbane*

Reduction, so far as this short review is concerned, is regarded as both the addition of hydrogen to an unsaturated centre in an organic molecule and the replacement of another atom or group in that molecule by hydrogen, i.e., it includes both hydrogenation and hydrogenolysis. Reductions effected with deuterium, i.e., with hydrogen of atomic mass 2, are also included, but those involving tritium, the isotope of atomic mass 3, will not be mentioned here on account of the complications arising from its radioactivity [1].

Reduction methods can be conveniently classified according to the active entities involved [2]. Hydrogen atoms are implicated or thought to be implicated in (1) catalytic reduction; (2) electrochemical reduction at a low-overvoltage electrode; (3) photochemical reduction; (4) reduction involving complex metal hydrides derived from Group IVB metals; (5) *some* reductions involving hydrazine (di-imide reductions); (6) homogeneous reduction brought about by dihydrido derivatives of Group VIII metals. Hydrogen is transferred as hydride ion in (7) reduction involving complex metal hydrides containing boron or aluminium; (8) homogeneous hydrogenation involving certain monohydrido complexes of Group VIII metals; (9) hydrogen transfer from one substrate to another. Finally, hydrogen is transferred as protons in (10) reduction brought about by dissolving metals; (11) electrochemical reduction at a high-overvoltage electrode; and (12) reduction whose first stage is attack by an anion.

In discussing reduction methods under these headings, no attempt (other than to furnish references to more detailed texts) will be made to give details of the experimental conditions or to list each and every functional group that has been reduced by a particular technique. Under each heading will be included something about the mechanism, together with supporting evidence. Examples of the application of the particular technique will be culled from the recent literature (end of 1967) to illustrate, *inter alia*, sophisticated applications of the technique, improvements in the efficiency or selectivity of the technique, or the reputed advantages of a new technique over the more traditional ones.

For convenience, the above scheme will be modified slightly. All reductions involving hydrazine, whatever their mechanism, will be discussed together with the di-imide reductions. Similarly, the section on homogeneous hydrogenations will include examples where hydrogenation is probably not effected by the transference of hydrogen atoms, i.e., both classifications (6) and (8) above.

CATALYTIC REDUCTION [3, 4, 5]

This is a widespread and convenient method which consists of shaking the substrate, alone or dissolved in a solvent, with a source of hydrogen in the presence of a finely divided metal. The hydrogen source is usually the gas but sometimes sodium borohydride, hydrazine or cyclohexene. The metal is usually a Group VIII metal, which is sometimes supported on a finely divided powder like carbon or barium sulphate.

In the reduction process [6, 7, 8], a hydrogen molecule is assumed to dissociate into atoms which are chemisorbed on the surface of the catalyst

$$H_2 + 2* \rightleftharpoons 2\underset{\displaystyle *}{\overset{}{\underset{|}{H}}} \qquad (2.1)$$

where * represents a surface site of the catalyst which may consist of one or more than one metal atom.

The molecule to be reduced is also chemisorbed, and an olefin, for example, may be chemisorbed in one of three ways:

(a) as a σ-diadsorbed species which involves two active catalyst sites

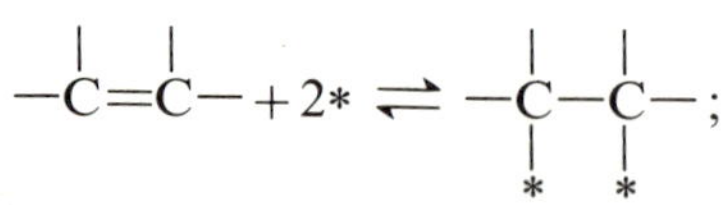

(2.2)

(b) as a π-adsorbed entity in which a π-donor bond has been formed with one surface metal atom 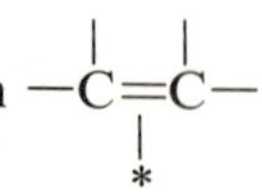 ; or

(c) as a π-allyl complex with one surface metal atom by loss of an α-methylene hydrogen atom.

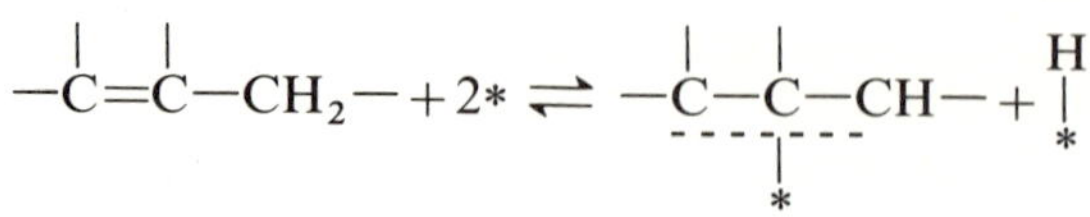

A variant of (c) has recently been suggested [9]. The next step in the reduction is the union of one adsorbed hydrogen atom with the adsorbed species of the substrate to form a 'half-hydrogenated state' which is reminiscent of a metal alkyl. Thus adsorbed 1,2-dimethylcyclopentene in equation (2.2) leads to equation (2.3).

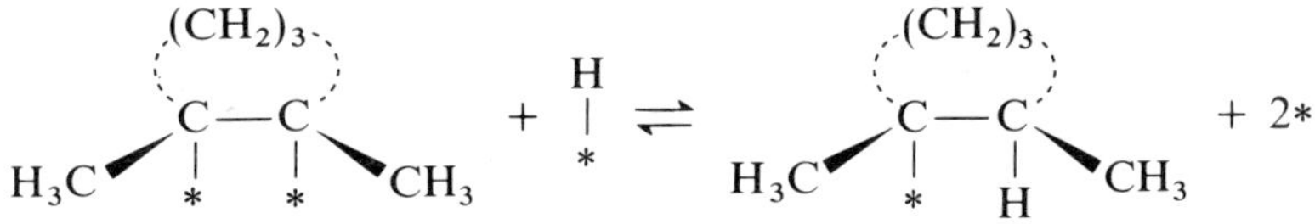

(2.3)

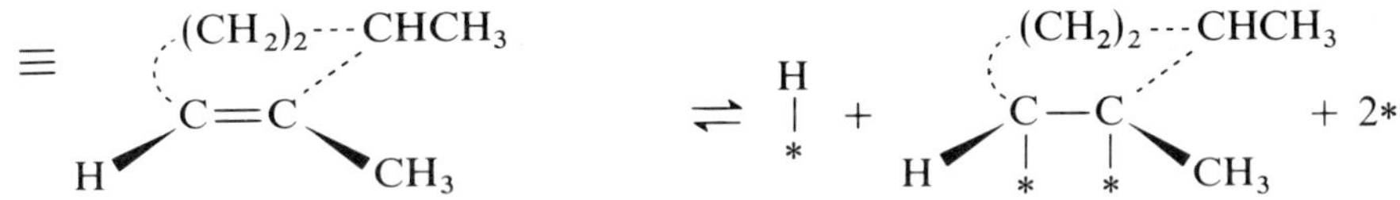

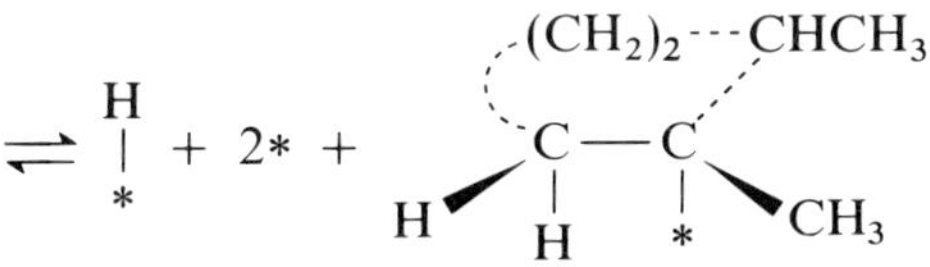

(2.3a)

Reaction of the 'half-hydrogenated state' with a second chemisorbed hydrogen atom affords the saturated compound.

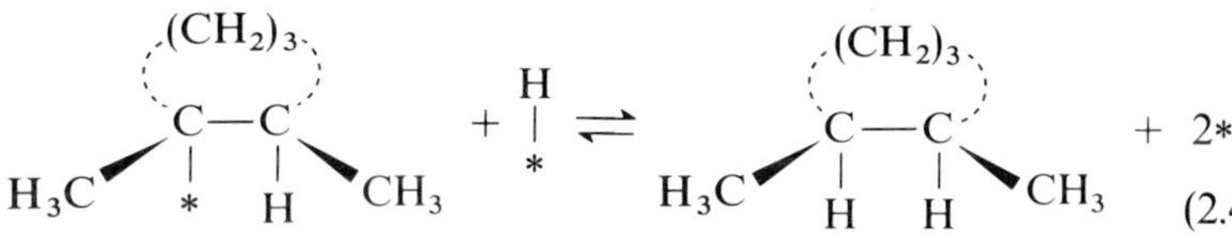

(2.4)

This mechanism results in the net *cis* addition of two hydrogen atoms, retention of configuration at the carbon atoms forming the original double bond, and in the example being discussed, the production of *cis*-2,3-dimethylcyclopentane. In practice, the *trans* form is also obtained. The reason for this lies in the reversible nature of reactions (2.1–2.4). In the back reaction of equilibrium (2.3), the hydrogen atom lost may be different from the one originally added and may lead eventually to an isomeric olefin (2.3a). Indeed, 2,3-dimethylcyclopentene can be isolated during the reduction of the 1,2-isomer [10]. Chemisorption of 2,3-dimethylcyclopentene through its double bond to the catalyst surface can give two entities, one of which also leads to the *cis*-cyclopentane compound. The other, although adding two hydrogen atoms in *cis* fashion, leads (2.4a) to *trans*-1,2-dimethylcyclopentane.

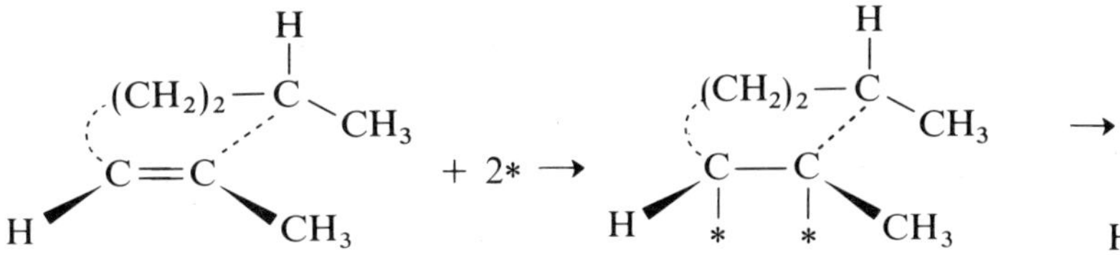

(2.4a)

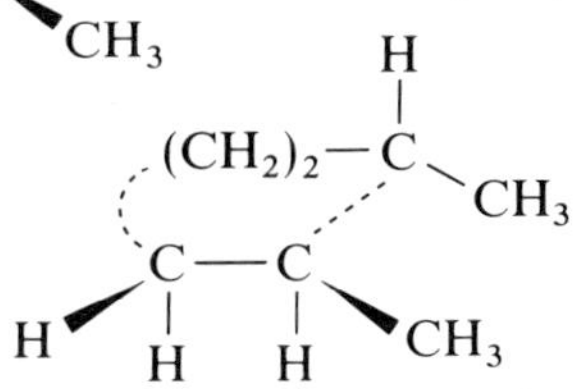

Another result of such equilibria is that catalytic deuteriation of an olefin causes scrambling, i.e., it gives a mixture of hydrocarbons, some of which contain more and some less than two deuterium atoms per molecule and at positions other than at the ends of the original double bond.

Intermediates akin to the 'half-hydrogenated state' have been postulated for the hydrogenolysis of single bonds, especially for benzyl derivatives [11, 12]. The choice of metal catalyst is important, since in the case of optically active benzyl derivatives, retention of configuration is observed with nickel, but inversion with palladium. Thus various derivatives of atrolactic ester (1; X = OH, OAc, $OCOCF_3$) hydrogenolyse with in-

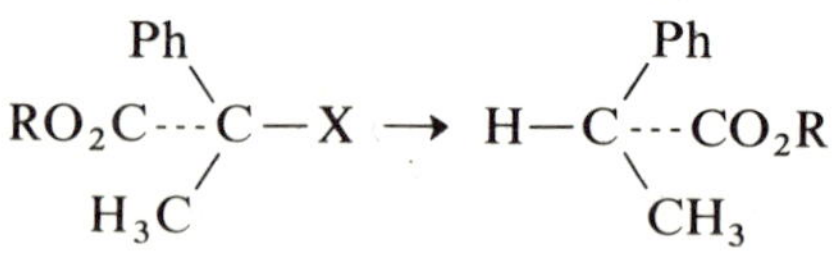

(1)

version at a rate which increases in the order given, i.e., with the electronegativity of the displaced group X [11]. Hydrogenolysis of *cis*- and *trans*-homobenzylic derivatives (2; X = OH, OAc), however, proceeds by reduction of a common olefinic intermediate formed by prior β-elimination [14].

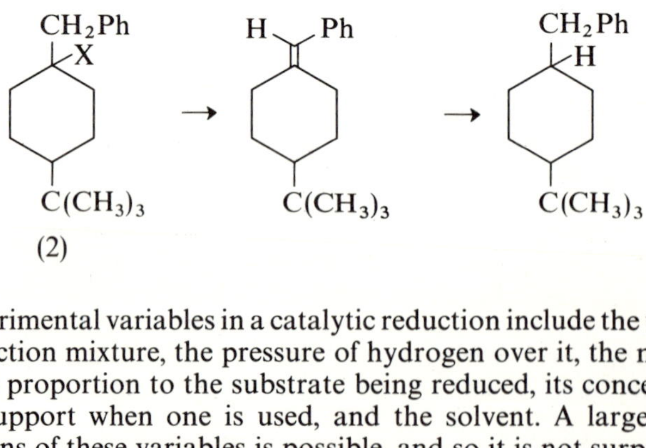

(2)

The experimental variables in a catalytic reduction include the temperature of the reduction mixture, the pressure of hydrogen over it, the nature of the catalyst, its proportion to the substrate being reduced, its concentration on the inert support when one is used, and the solvent. A large number of combinations of these variables is possible, and so it is not surprising that a wide variety of functional groups in organic molecules is reduced by catalytic hydrogenation. Lists may be drawn up which give the approximate order in which functional groups are reduced [4(a)]. However, these lists are only a guide, since structural features in the molecule to be reduced or changes in the reduction conditions can cause a reversal in the order, particularly of functional groups close together in such a list. The catalytic reduction of aromatic or heterocyclic nitro groups to amino groups is usually a much more facile process than hydrogenation of the ring, and control of the accompanying vigorous heat evolution may be a problem. However, in the dioxan derivative (3) the pyridine ring is reduced first [15].

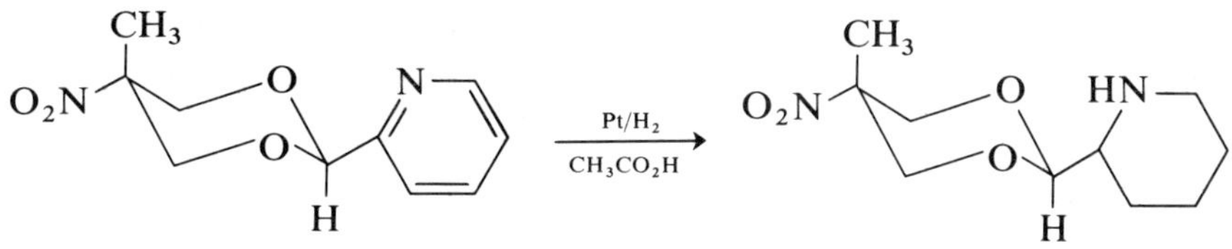

(3)

Ketonic carbonyl groups are readily reduced to secondary alcohols, but hydrogenolysis of the C—C bond occurs first in the strained ketone (4) [16]. Such hydrogenolysis is rare outside small ring systems [17].

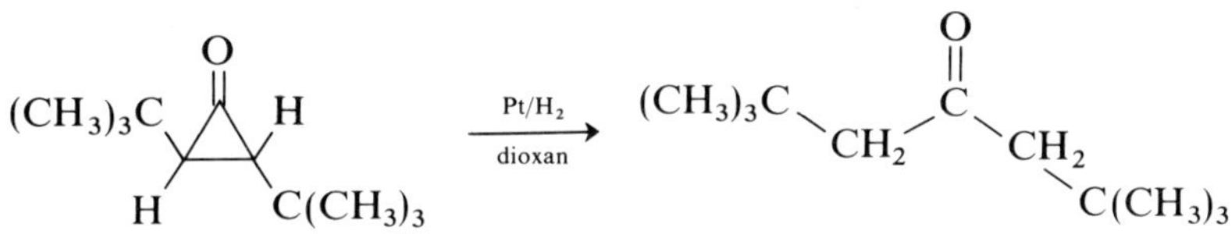

(4)

Generally, a molecule approaches the catalyst so that its less hindered side is adsorbed and the hydrogen is added *cis* to that. However, other groups in the molecule help decide which side is adsorbed. Reduction of 2-cyclopentylidene cyclopentanol (5) affords the *trans* alcohol.

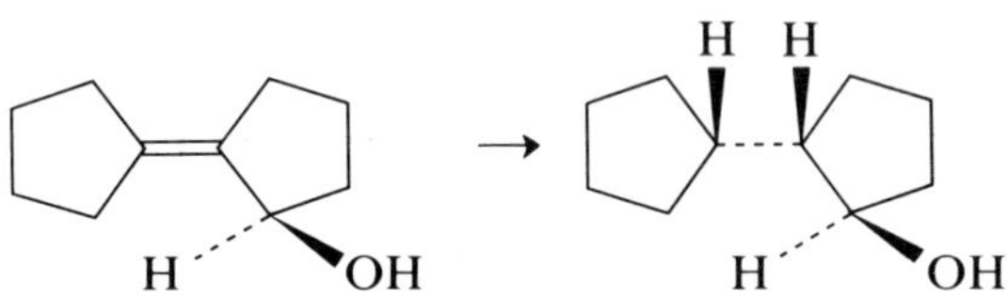

(5)

The molecule is adsorbed so that both the π-electrons of the double bond and one of the lone pairs on the oxygen atom can interact with the catalyst surface [18]. The large alkyl group in 1-*t*-butylnaphthalene (6) hinders the adsorption of the substituted benzene ring on a palladized charcoal catalyst surface, but it is this ring which is preferentially reduced to the tetrahydro derivative [19].

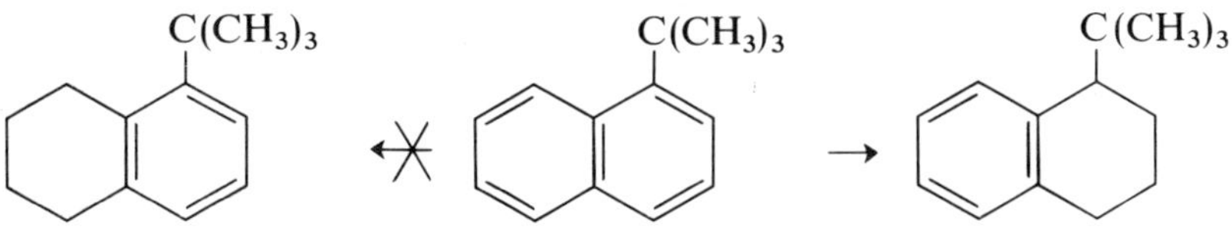

(6)

The steric strain in the transition state leading to a neopentylbenzene-type system must be less than that in the transition state leading to the alternate 2-*t*-butyltoluene-type system.

Hydrogenation of the C=N bond in the Schiff base (7) obtained by condensing an α-keto acid with one enantiomer of phenylglycine must be favoured from one side of the molecule. Continuance of the reduction so that hydrogenolysis occurs affords a new amino acid of 40–60% optical purity. This process has been termed hydrogenolytic asymmetric transamination [20].

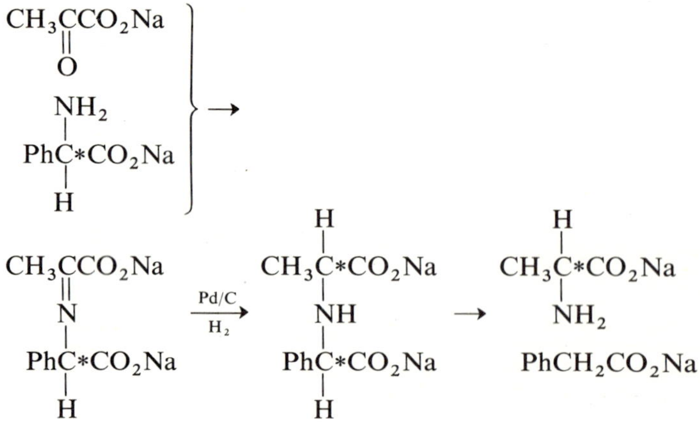

(7)

(* signifies asymmetric carbon atom responsible for optical activity.)

Recent attempts to improve the catalytic method of reduction include the production of more active specimens of the conventional metal catalysts, the introduction of metals outside Group VIII for selective reductions, and the delineation of factors influencing stereoselective reductions.

Platinum metal obtained by sodium borohydride reduction [21] of chloroplatinic acid is twice as active a catalyst as that usually obtained by shaking platinic oxide with hydrogen. Thus reduction of olefins (8) occurs much more quickly

$$CH_3(CH_2)_5CH{=}CH_2 \rightarrow CH_3(CH_2)_6CH_3$$

(8)

More active forms of palladium and nickel catalysts can be similarly produced. The nickel catalyst is probably the boride Ni_2B, which has a low tendency to isomerize olefins during the hydrogenation and converts internal acetylenes like (9) into *cis* olefins of high purity.

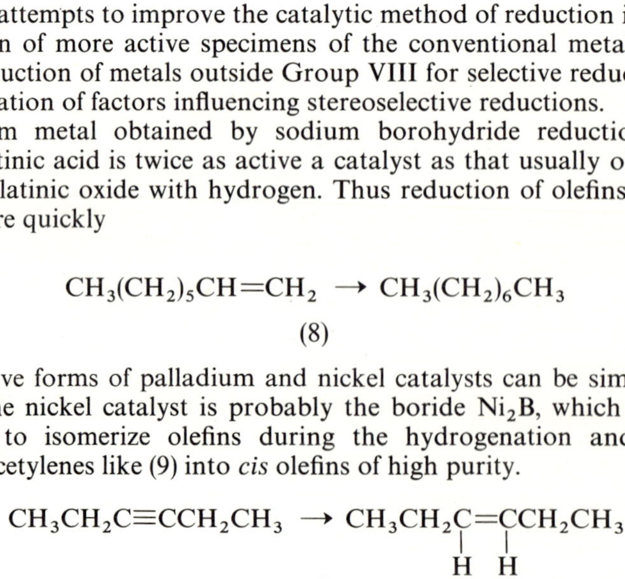

(9)

The activity of platinum catalysts, conventionally prepared from the oxide by shaking with hydrogen, can also be increased by mixing with another metallic oxide prior to reduction. The active catalyst is suspected to be an alloy or solid solution [22].

Rhenium oxides, from borohydride reduction, from direct hydrogenation or from alkali metal in liquid ammonia reduction of other rhenium compounds, catalyse the reduction, *inter alia*, of aromatic carboxylic acids to primary alcohols without affecting the benzene ring [23]. The reverse is true for metallic rhodium [24], which selectively reduces benzene-1,2- and 1,4-dicarboxylic acids to the corresponding *cis* cyclohexane acids. The sulphides of the platinum metals and of cobalt have been recently introduced for the selective reduction of aromatic halogenonitro compounds to the halogenoamines [25]. The solvent in catalytic reductions plays a number of roles. It obviously acts as a diluent to moderate the reaction; it also can displace adsorbed species from the catalyst surface (some more effectively than others) and so influence the extent of the reaction. Thus it has long been recognized that during the reduction of unsaturated hydrocarbons, the extent of the hydrogenation increases with the polarity of the solvent, so that it is least in cyclohexane, greater in ethanol and greatest in acetic acid [26].

Acidic solvents are also important because they can change the nature of the species being reduced. Pyridine (10) can be reduced relatively easily under acidic conditions, because it is the protonated form and not the neutral molecule that is being reduced [27].

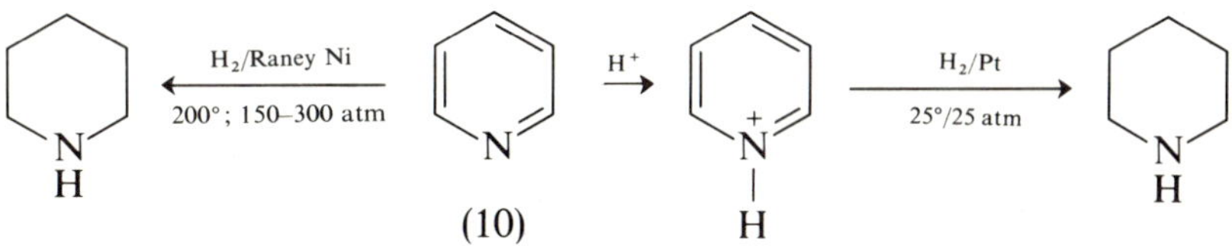

(10)

Adsorption of the pyridinium ion on the catalyst surface can only be flatwise via the π-electron cloud, as with benzene [28]. With the neutral molecule, adsorption is through the unshared electrons on the nitrogen atom, so that the molecule is oriented at an angle to the catalyst surface, which renders transference of chemisorbed hydrogen difficult. If the product of a reduction is a strong base, it will be protonated by acid which will stop the strong chemisorption on the catalyst surface that causes the complete cessation of activity—the so-called poisoning of catalysts. Solvents can also react with intermediate reduction products and remove them from the sphere of reduction. Catalytic reduction of pyrimidine (11) and derivatives in aqueous acidic solutions under ambient conditions, using palladized charcoal, generally leads to salts of the 1,4,5,6-tetrahydro compounds after the absorption of two moles of hydrogen [29].

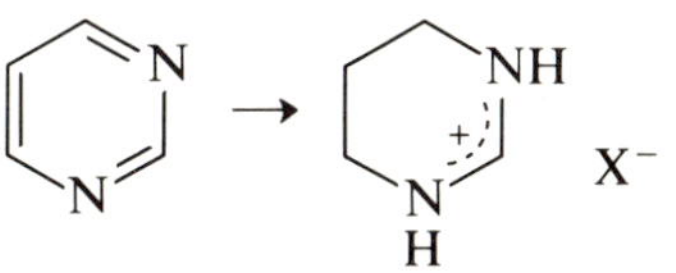

(11)

($X = Cl, CH_3CO_2$)

However, 5-aminopyrimidine (12) adsorbs one mole of hydrogen only. The NH_2 group of the intermediate enamine is hydrolysed to OH and the keto tautomer is hydrated to the stable gem diol [30]. The same gem diol is obtained from the reduction of 5-hydroxypyrimidine (13).

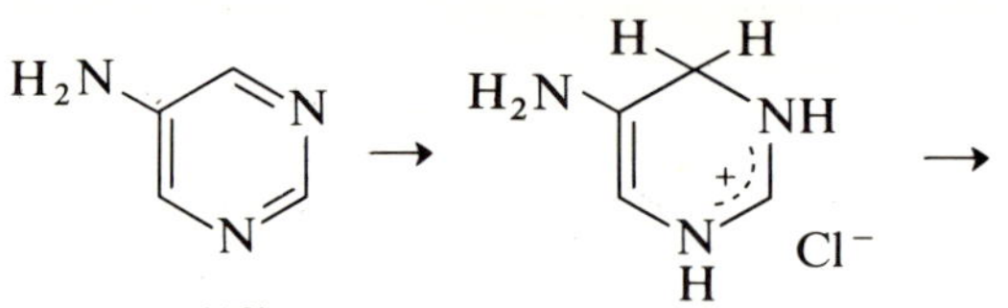

(12)

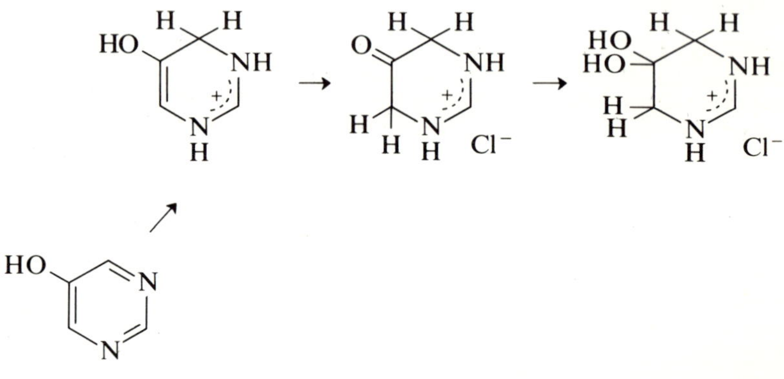

(13)

Catalytic reduction of 5-acetamidopyrimidine (14) in hot acetic anhydride affords a different, non-ionic tetrahydro derivative, in which all the nitrogen atoms are acetylated [31].

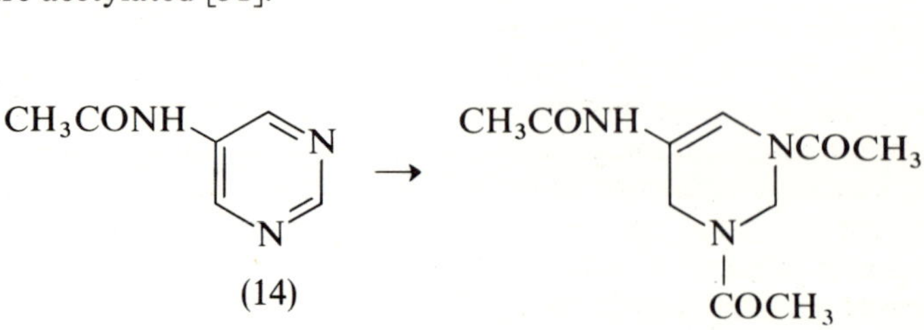

(14)

Certain azacyclic ketones (15) which are normally reduced to the secondary alcohols (16), are reduced to the azahydrocarbons under acidic conditions [32, 33, 34]. In this latter reduction, the alcohol cannot be an intermediate, since it is not hydrogenolysed when subjected to the same conditions [32].

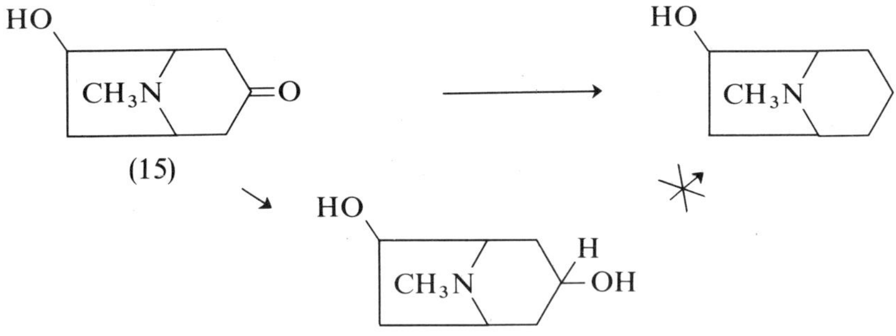

(15)

(16)

Alkali is known to have a catalytic effect upon the reduction of some enone systems, and also increases the proportion of *trans* olefinic compound formed by catalytic reduction of acetylenes [34].

ELECTROCHEMICAL REDUCTION AT A LOW-OVERVOLTAGE ELECTRODE [36]

Reducing conditions exist at the cathode of an electrolytic cell. If the cathode is a metal of low overvoltage, the discharge of protons $H^+ + e^- \rightarrow H\cdot$ is catalysed and the resulting hydrogen atoms are adsorbed as a film on the metal surface. The situation is now reminiscent of that existing on the metal surface during catalytic reductions [37] and similar functional groups ought to be reduced by both techniques. Thus acetylene dicarboxylic acid (17) has been reduced only to the *cis* olefinic compound (maleic acid) in sulphuric acid solution inside a helical cathode of silver-palladium alloy. The hydrogen diffused through the metal, and under these conditions selective and stereospecific reduction of the acetylenic bond occurred. Other unsaturated groups like C=C, C=O, C=N and C≡N are not attacked [38].

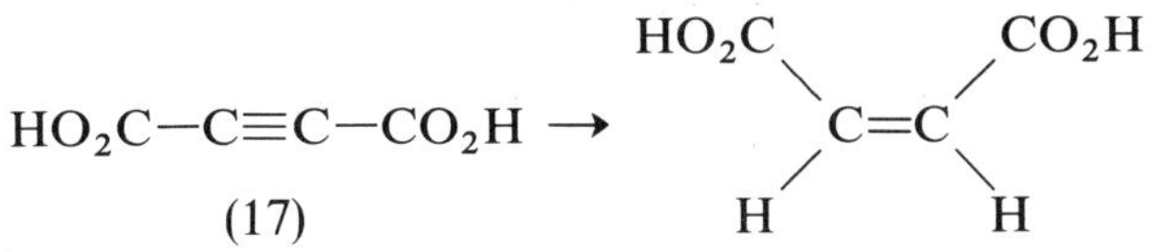

(17)

PHOTOCHEMICAL REDUCTION

In this method the substrate A, dissolved in a solvent such as ethanol, absorbs energy from a source of ultraviolet radiation, either a lamp or the sun. The activated molecules A*

$$A + h\nu \rightarrow A^*$$

then abstract hydrogen atoms from the solvent

$$A^* + C_2H_5OH \rightarrow AH\cdot + C_2H_5O\cdot$$

or any other hydrogen-containing substance present.

The new radical AH· may now acquire a second hydrogen atom from a solvent molecule

$$AH\cdot + C_2H_5OH \rightarrow AH_2 + C_2H_5O\cdot$$

or from a solvent radical

$$AH\cdot + C_2H_5O\cdot \rightarrow AH_2 + CH_3CHO$$

or it may dimerize

$$AH\cdot + AH\cdot \rightarrow HA{-}AH$$

or it may react with solvent radicals

$$AH\cdot + C_2H_5O\cdot \rightarrow H{-}A{-}OC_2H_5$$

so that in effect a solvent molecule has been added to the substrate molecule. Thus photoreduction of acridine

A = 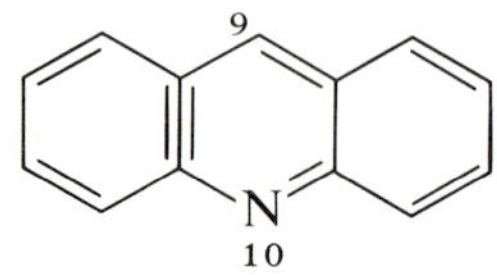

gives mainly the 9,10-dihydro derivative in dilute solution, but the tetrahydroderivative of the 9,9′-dimer in concentrated solutions [39].

Ketones usually undergo reductive dimerization to give the pinacols [40].

$$R_2C{=}O \rightarrow R_2C{=}O^* \rightarrow R_2\dot{C}{-}OH \rightarrow \begin{array}{l} R_2C{-}OH \\ \quad | \\ R_2C{-}OH \end{array}$$

Certain conjugated diacetylenes, *inter alia*, add a solvent molecule across one of the triple bonds [41].

$$CH_3(CH_2)_3C{\equiv}C{-}C{\equiv}C{-}(CH_2)_3CH_3 \xrightarrow[\text{pentane}]{h\nu}$$

$$\textit{cis/trans}\ CH_3(CH_2)_3C{\equiv}C{-}C(R){=}CH(CH_2)_3CH_3$$

$$R = \text{2- and 3-pentyl}$$

Some selectivity is apparent in the case of the heterocyclic molecule (18), where the nitro group is reduced to hydroxylamino [42] and the *N*-oxide group is untouched.

(18) 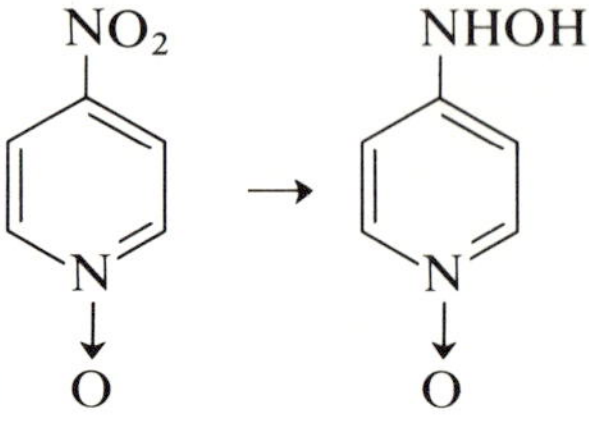

Hydrogen sources include aliphatic alcohols—even tertiary ones [43]—

ethers [44], and aliphatic hydrocarbons [41] used also as solvents, the substrate itself [40] and tri-*n*-butylstannane [45]. Ketone molecules are usually excited to the triplet state [43, 45, 46] before photoreduction, while both singlet and triplet excited states are involved in the case of acridine [47]. Ultraviolet light can also homolytically cleave a molecule like benzoyl bromide into two radicals, which can then abstract hydrogen atoms from the solvent molecules H—R [44].

$$\mathrm{PhCOBr} \longrightarrow \mathrm{PhCO}\cdot + \mathrm{Br}\cdot$$

$$\mathrm{PhCO}\cdot + \mathrm{H{-}R} \longrightarrow \mathrm{PhCHO} + \mathrm{R}\cdot$$

Usually, only small quantities of substance can be photoreduced at a time, and frequently the product may be obtained more conveniently by other methods.

METAL HYDRIDE REDUCTIONS

Certain metal hydrides, particularly those of Group IVB elements, also effect reduction through free-radical intermediates, and the reduction may be catalysed by ultraviolet light or radical initiators [48]. Tin hydrides in particular have attracted attention for their stereoselective reduction of various alkyl halides. Triphenyl tin hydride, Ph_3SnH, reduces the 7-chloro derivative of *endo*-7-phenylnorcarane [49] (19; X = Cl) to the parent hydrocarbon (19; X = H).

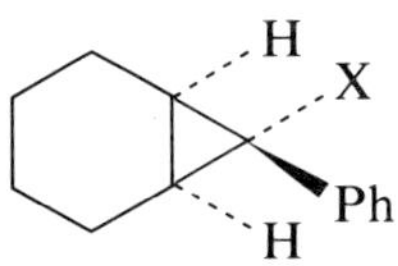

(19)

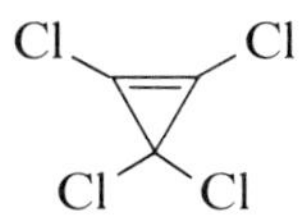

(20)

The tri-*n*-butyl analogue reduces the halogen atoms in the cyclopropene (20) in steps and without affecting the double bond [50].

HYDRAZINE REDUCTIONS

A reduction method of long standing involving hydrazine is the high-temperature Wolff–Kishner decomposition of the hydrazones (21) of aldehydes and ketones in alkaline media to form hydrocarbons [51]. The process involves anion formation, isomerization to an azo compound, and loss of nitrogen to a carbanion intermediate.

$$\underset{(21)}{\mathrm{R_2C{=}N{-}NH_2}} \xrightarrow{\mathrm{OH^-}} [\mathrm{R_2C{=}N{-}\bar{N}H} \longleftrightarrow \mathrm{R_2\bar{C}{-}N{=}NH}] \longrightarrow$$

$$\mathrm{R_2CH{-}N{=}\bar{N}} \xrightarrow{-\mathrm{N_2}} \mathrm{R_2\bar{C}H} \xrightarrow{\mathrm{R'OH}} \mathrm{R_2CH_2}$$

If potassium *t*-butoxide be used as the alkaline catalyst in dimethyl sulphoxide, the reduction can be effected at room temperature [52].

Another novel application of hydrazine chemistry lies in its utility for the indirect replacement of certain functional groups by hydrogen. The functional group (e.g., Cl, OCH_3) is nucleophilically displaced by hydrazine, and the hydrazino group oxidized by silver oxide, cupric ion, or even air and aqueous alkali [53]. Thus 5-nitropyrimidine (22; R = H) was obtained from its 4,6-dihydrazino derivative (22; R = $NHNH_2$) unaccompanied by reduction of the nitro group or of the heterocyclic ring [54].

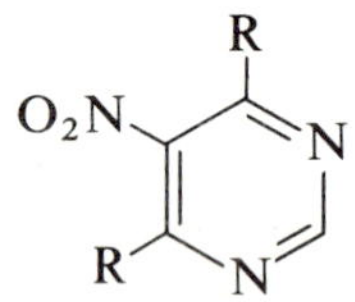

(22)

(23)

However, the big advance in hydrazine reductions has stemmed from the realization that under certain conditions the active reducing species is *cis*-di-imide (23). This is produced by using hydrazine in conjunction with oxidizing agents such as hydrogen peroxide, atmospheric oxygen, and potassium ferricyanide [55, 56]. Di-imide transfers its two hydrogen atoms via a non-polar transition state (24) to a symmetrical double (N=N, C=C) or triple bond (C≡C).

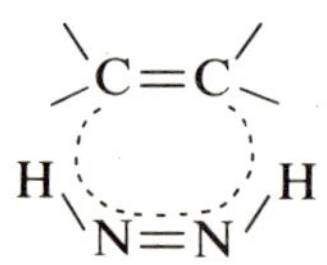

(24)

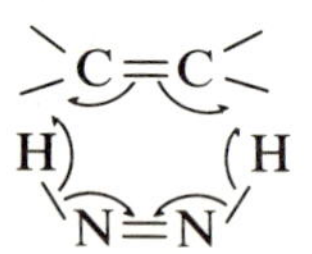

(25)

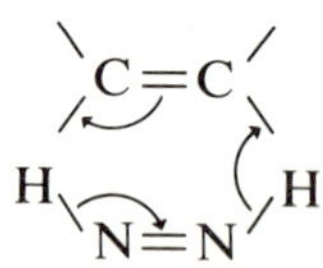

(26)

In general di-imide does not normally transfer its hydrogen atoms to an unsymmetrical double or triple bond (N=O, C=N, C=O, C≡N) or cause hydrogenolysis (even of a labile S—S bond [55(g)]), so that the hydrogen atoms are probably transferred as atoms (25) and not as protons and hydride ions (26).

When the polarity of the unsymmetrical unsaturated bond is modified by suitable substituents, reduction is possible. Thus benzaldehyde [57] and some derivatives (27; Z = CHO, Y = H, *p*-CH_3O, Cl, CH_3) afforded the corresponding benzyl alcohols (27; Z = CH_2OH), while in the seven-membered ring ketone (28; Y = C=O) the carbonyl group was reduced (28; Y = CH·OH) and not the olefinic double bond [58].

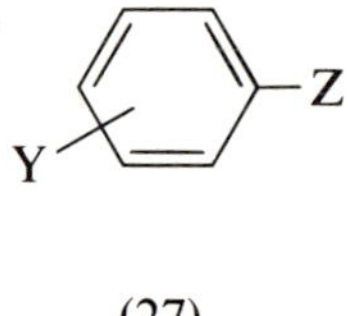

(27)

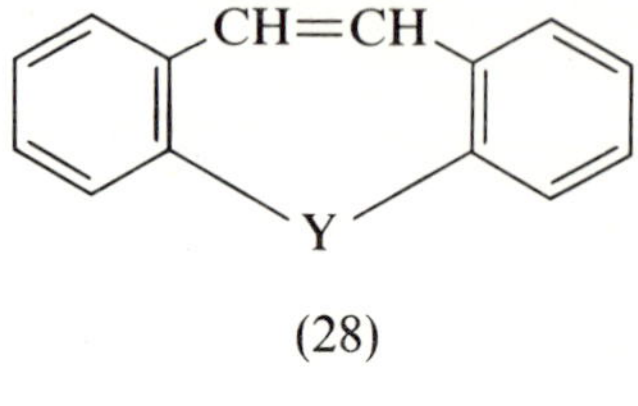

(28)

Other sources of di-imide have been discovered, such as the decarboxylation of azodicarboxylic acid [55(a,b,d)], the thermal decomposition of anthracene-9,10-di-imine [55(f)] or arene sulphonyl hydrazines [55(c)] and the alkaline decomposition of hydroxylamine-O-sulphonic acid [59].

$$HO_2C{-}N{=}N{-}CO_2H \rightarrow 2CO_2 + N_2H_2 \qquad (2.5)$$

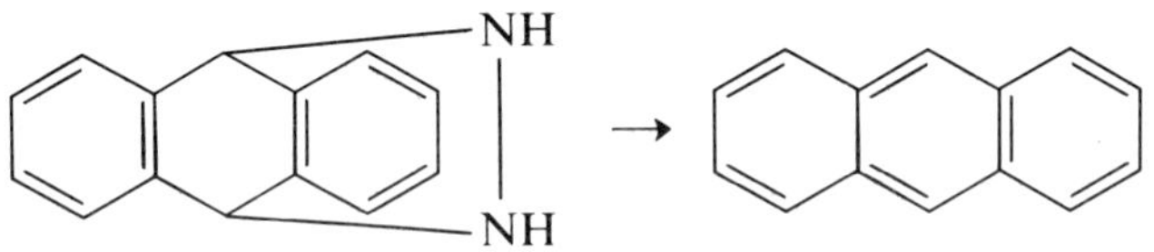

$+ N_2H_2$ (2.6)

$$ArSO_2NHNH_2 \rightarrow ArSO_2H + N_2H_2 \qquad (2.7)$$

$$2H_2N{-}O{-}SO_3^- \rightarrow 2SO_4^{2-} + N_2H_2 + 2H^+ \qquad (2.8)$$

Di-imide's importance as a reducing agent lies in the stereospecific way in which the two hydrogen atoms are added *cis* to the unsaturated bond, without migration of the latter. The use of deuterated di-imide thus leads to the incorporation of deuterium in the molecule, two deuterium atoms per double bond, and without scrambling. Hence *cis*-but-2-en-1,4-diol affords meso-2,3-dideuteriobutan-1,4-diol [60] (2.9).

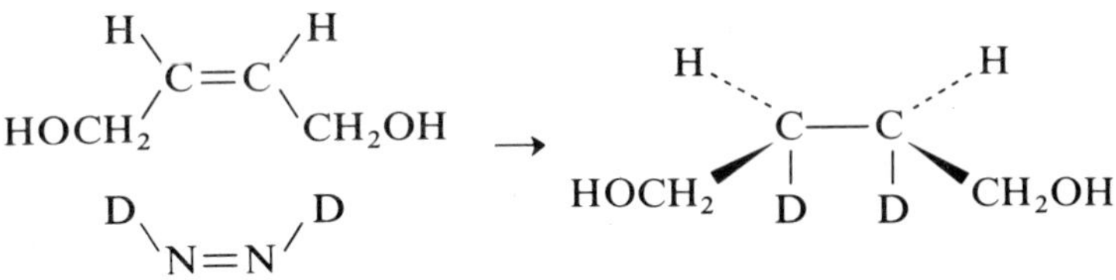

(2.9)

Even the double bonds of transient species may be reduced by di-imide, as in the cyclodecapentaene momentarily produced during the photochemical isomerization of *trans*-9,10-dihydronaphthalene [61].

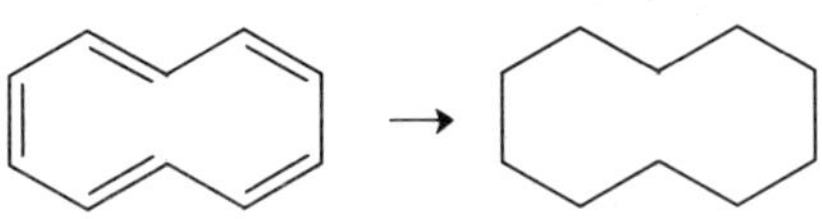

(2.10)

Reactivity differences between different olefins have been explained in terms of torsional strain, bond-angle bending strain, and α-alkyl substituent effects [62]. In many cases, di-imide attacks the olefin molecule from its least hindered side, which may lead to the less thermodynamically stable hydrogenation product. Thus α-pinene gives mostly *cis*-pinane containing an axial methyl group [55(e)] (2.11).

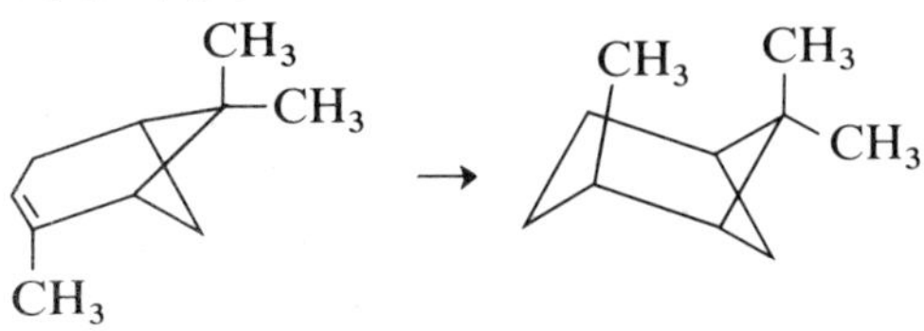

(2.11)

However, some 7-substituted norbornadienes react with dideuterodi-imide in such a way that the hindered double bond is attacked, and that from the hindered *exo* side [63] (2.12).

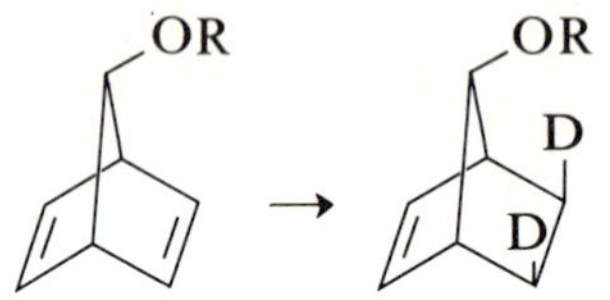

$R = H, COCH_3, C(CH_3)_3$ (2.12)

The electron-donating effect of the oxygen atom of the 7-substituent can play a part in stabilizing the N=N bond of the transition state involving an *exo* approach, but not an *endo*. Hydrazine in the presence of metal catalysts (palladium, nickel) will bring about the reduction of groups which di-imide alone does not [64]. Aromatic nitro compounds, $ArNO_2$, afford the amino compound, $ArNH_2$, or the intermediate azoxy, ArN(O)=NAr, azo, ArN=NAr, or hydrazo compounds, ArNHNHAr. It is not known whether the hydrazine functions as a source of hydrogen which becomes chemisorbed on the metal surface and then transferred to the substrate (as in more orthodox catalytic reductions), whether chemisorbed di-imide is the actual reducing species, or whether both mechanisms play a part.

Sodium hydrazide has been used as a reducing agent for olefinic, polynuclear aromatic hydrocarbons, and nitrogen heteroaromatic compounds [65]. The reducing process involves anionic attack by $\bar{N}HNH_2$, subsequent loss of a proton to give a dianion, followed by the expulsion of di-imide, and the acquisition of two protons. Thus quinaldine gives the 1,4-dihydro compound (2.13).

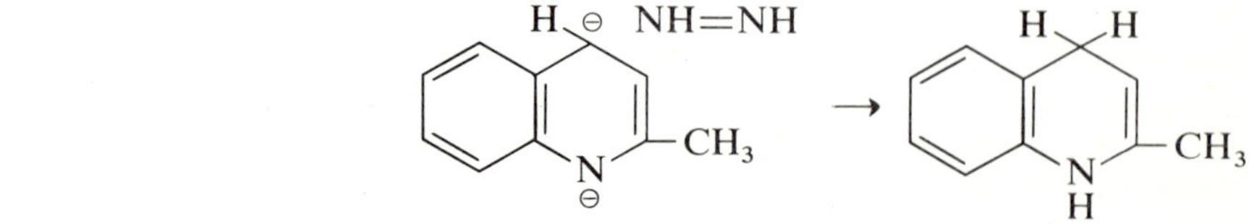

(2.13)

HOMOGENEOUS HYDROGENATION [66]

This novel technique is still in its infancy as regards organic chemical applications but seems to have a great future. Hydrogen is activated, not by chemisorption on the surface of a metal catalyst, but by incorporation into the co-ordination sphere around a metal atom, usually but not exclusively belonging to a Group VIII metal. Three types of hydrogen activation have been recognized [67]:

(a) heterolytic splitting, in which a ligand is split out of the co-ordination complex and replaced by a hydride derived from hydrogen; there is no change in the formal oxidation number of the metal under these circumstances

$$Ru^{III}Cl_6^{3-} + H_2 \rightleftharpoons Ru^{III}HCl_5^{3-} + H^+ + Cl^- \qquad (2.14)$$

(b) homolytic splitting, in which complex metal hydride formation is accompanied by formal oxidation of the metal

$$2Co^{II}(CN)_5^{3-} + H_2 \rightleftharpoons 2Co^{III}H(CN)_5^{3-} \qquad (2.15)$$

(c) insertion of a hydrogen molecule and formation of a dihydrido species

$$Ir^{I}Cl(CO)(PPh_3)_2 + H_2 \rightleftharpoons Ir^{III}H_2Cl(CO)(PPh_3)_2 \qquad (2.16)$$

The system of current interest and probable greatest utility is the *cis* dihydrido rhodium species $RhCl(H_2)(PPh_3)_2$, formed reversibly when a solution of tris(triphenyl phosphine)chlororhodium^{I} is exposed to hydrogen gas [68(a)]. The active species (29) contains two metal–hydrogen bonds, with a solvent molecule (S) occupying the last octahedral position. The unsaturated compound to be reduced displaces the solvent molecule. A stereospecific *cis* transfer of both hydrogen atoms is then postulated to occur via a transition state similar to that involved in di-imide reductions [68(b)]. If this is so, the hydrogens are probably transferred as atoms.

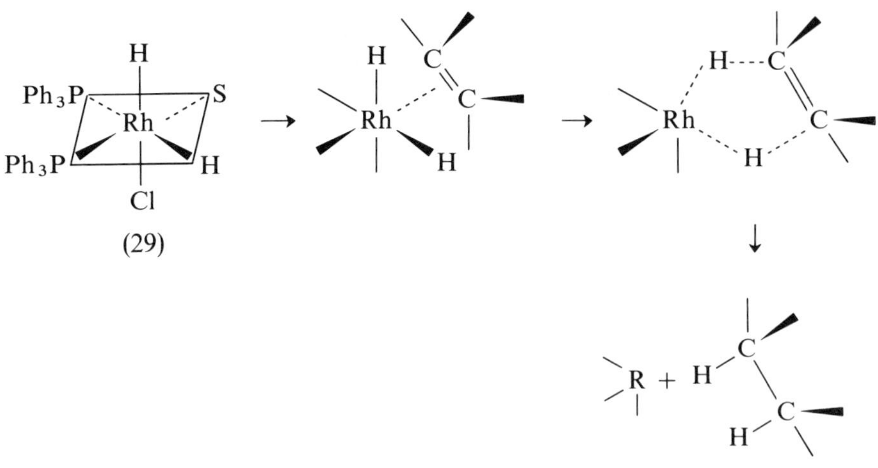

(29)

This soluble rhodium catalyst has been used to hydrogenate acetylenes to the *cis*-olefin, and olefins to the saturated hydrocarbons. Deuterium has been added *cis* to the olefinic double bonds in cyclohexene, and in unsaturated carboxylic acids and their derivatives, without scrambling occurring [69(a,d)]. Olefinic double bonds are preferentially reduced in

$\alpha\beta$-unsaturated aldehydes [68(c)], and even in *p*-quinones [69(c)]. This reduction of an olefinic double bond is subject to steric influences [68(d)], so that an unsymmetrically disubstituted double bond is reduced before a trisubstituted one [70–72]. Another advantage of the soluble rhodium catalyst over heterogeneous catalysts is that its activity is not drastically diminished by the presence of sulphur compounds [69(b)]. The active ruthenium species $[Ph_3P]_3RuHCl$ also brings about the *cis* homogeneous hydrogenation of acetylenes and olefins [73] under mild conditions. A cobalt complex, $(Ph_3P)_3CoH_2$, catalyses reduction of ethylene to ethane and also of carbon tetrachloride to chloroform [74].

Hydridoiridium complexes have attracted attention because they catalyse reduction of an unsymmetrical unsaturated bond—the carbonyl bond—either by molecular hydrogen [75] or by hydrogen transference from a protonic solvent [75, 76]. Thus n-butyraldehyde in formic acid afforded n-butanol [75],

$$CH_3(CH_2)_2CH{=}O \rightarrow CH_3(CH_2)_3OH, \qquad (2.17)$$

while 4-*t*-butylcyclohexanone in isopropyl alcohol–water–trimethyl phosphite mixtures preferentially formed the axial alcohol [76].

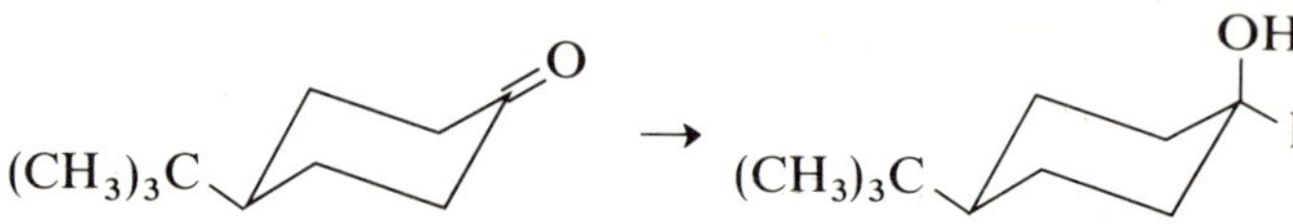

(2.18)

A whole series of complexes of the general formula $(R_3Q)_2MX_2$, where R = alkyl or aryl, Q = P, As or Sb, M = Ni, Pd or Pt and X = halogen or pseudo-halogen, has been investigated for the homogeneous hydrogenation of the olefinic double bonds in esters of long-chain unsaturated carboxylic acids [77]. However, more vigorous conditions of temperature and pressure are required, and hydrogenation is preceded by *cis–trans* isomerization, and migration of double bonds to give a conjugated system. Hydrogenation of the latter occurs to give a stable monoene, so that methyl linoleate (30) gives a mixture of methyl oleate (31; $w = v = 7$) and isomeric octadecenoates (31).

$$CH_3(CH_2)_4CH{=}CH{-}CH_2{-}CH{=}CH(CH_2)_7COOCH_3 \qquad (30)$$
$$\downarrow$$
$$CH_3(CH_2)_y{-}CH{=}CH{-}CH{=}CH{-}(CH_2)_xCOOCH_3$$
$$\downarrow$$
$$CH_3(CH_2)_wCH{=}CH{-}(CH_2)_v{-}COOCH_3$$

(31)

Other soluble catalysts used for this and analogous hydrogenations of unsaturated fats include the metal acetylacetonates [78] and the carbonyls [79, 80]. Cobalt octacarbonyl brings about reduction under hydroformylation conditions [81]. It is first converted into cobalt hydrocarbonyl, $HCo(CO)_4$, which transfers its hydrogen to the electrophilic carbon atom of an aldehydic carbonyl group. The resulting cobalt complex is hydrogenolysed to the primary alcohol and cobalt heptacarbonyl which is reconverted to the octacarbonyl by the carbon monoxide.

$$Co_2(CO)_8 + H_2 \rightarrow 2HCo(CO)_4$$

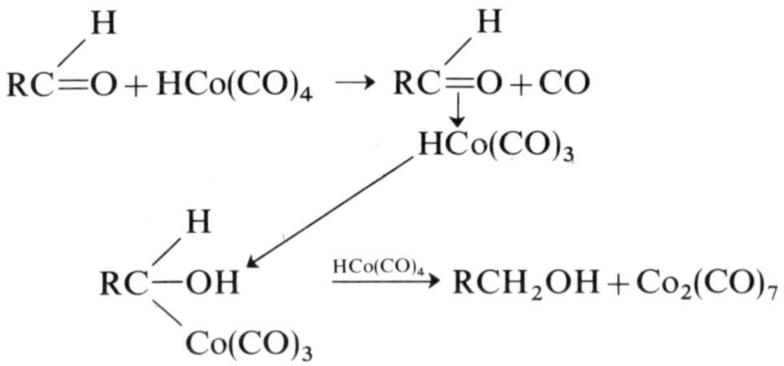

$$Co_2(CO)_7 + CO \rightarrow Co_2(CO)_8$$

Hydrogenolysis of a secondary alcohol to a hydrocarbon has also been known to occur [82].

The pentacyanocobaltate system is also an active hydrogenation catalyst. It can be used only in water and will reduce conjugated but not isolated olefinic double bonds [83]. At high cyanide concentrations, the hydridocobaltate anion (2.15) adds, for example, to the conjugated system of butadiene

$$HCo(CN)_5^{3-} + CH_2{=}CH{-}CH{=}CH_2 \rightarrow CH_3CH{=}CH{-}CH_2Co(CN)_5^{3-} \rightarrow CH_3CH_2{-}CH{=}CH_2$$

to give a σ-allyl intermediate subsequently transformed by more hydrido species to but-1-ene. At low cyanide concentrations the σ-allyl intermediate equilibrates with a π-allyl intermediate which subsequently gives *trans*-but-2-ene [67].

$$CH_3CH{=}CHCH_2Co(CN)_5^{3-} \xrightarrow{HCo(CN)_5^{3-}} CH_3CH_2CH{=}CH_2 + 2Co(CN)_5^{3-}$$

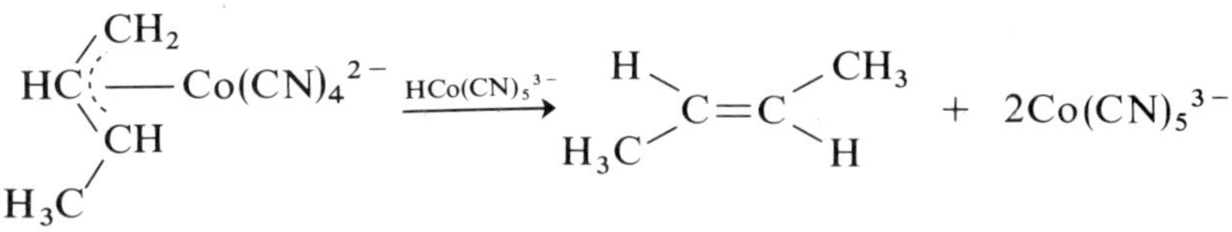

Aromatic nitro compounds $ArNO_2$, can be reduced to the hydroxylamino, ArNHOH, hydrazo, ArNHNHAr, or azoxy compounds, ArN(O)=NAr, the olefinic double bond of $\alpha\beta$-unsaturated ketones being preferentially hydrogenated [84].

One may speculate that the transference of hydrogen atoms is responsible for hydrogenation of symmetrical unsaturated bonds (C=C, C≡C), while hydride ion is involved for the others.

REDUCTION INVOLVING BORON AND ALUMINIUM HYDRIDES [4(b), 85]

Although this method of reduction is an example of the general phenomenon of hydride transfer, it is considered separately on account of the numerous examples and widespread applicability of the method. The reagents used are of two types. They include the nucleophilic substances lithium aluminium hydride, $LiAlH_4$ (LAH), the deuteride, $LiAlD_4$ (LAD), the borohydrides (MBH_4) of sodium, lithium and potassium, and sodium borodeuteride, $NaBD_4$, as well as the electrophilic species diborane, B_2H_6, alane, AlH_3, and derivatives thereof.

Hydrogenation involving the nucleophilic reagents proceeds by the transference of a hydride ion from the anion to the positively polarized end of a multiple bond such as a carbonyl group in aldehydes, or ketones. Thus

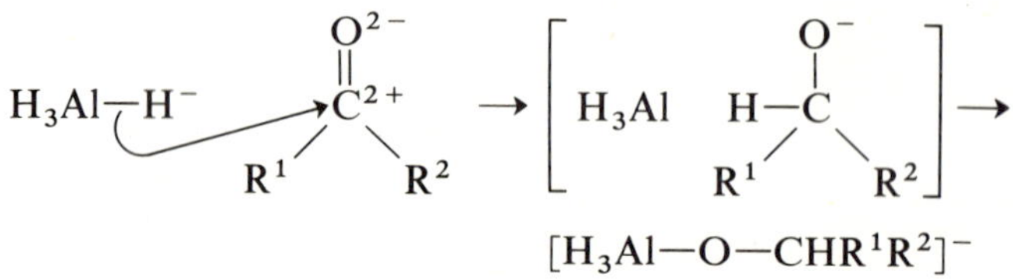

with LAH the hydride transference is quickly followed by formation of an alkoxy aluminohydride anion. The remaining three hydrogen atoms attached to the aluminium atom in this anion are in turn available for transference, but not so readily, since the electron-withdrawing inductive effect of the alkoxyl group opposes loss of the hydride ion. The hydride ion may be transferred intermolecularly to a fresh molecule of substrate to be reduced, or it may be transferred intramolecularly to some other site within the same molecule. A spectacular example of this occurred during the synthesis of strychnine [86] when the C-20 amide carbonyl group in a precursor (32) was reduced to CH_2 and the intermediate aluminohydride complex transferred hydrogen to C-8 from the more highly hindered side of the molecule.

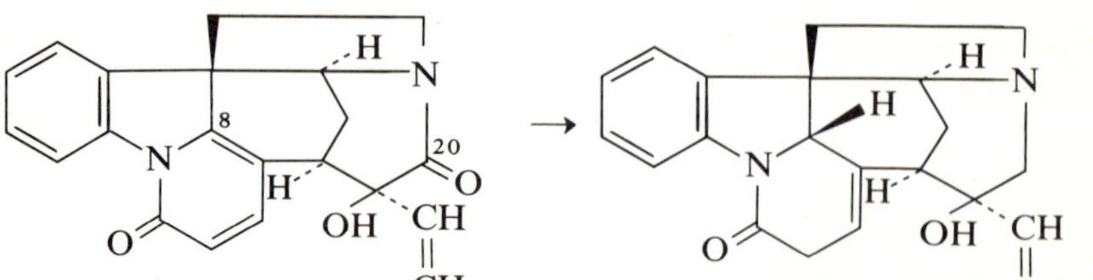

(32)

Since alkoxy groups weaken the reducing power of any hydrogen atoms remaining in the AlH_4^- anion, less powerful and more selective reducing agents can be formed from LAH by reaction with 1–3 equivalents of alcohols prior to reduction. In particular, lithium tri-t-butoxy aluminohydride (obtained by reacting LAH with three equivalents of t-butanol) has established itself as a useful reagent for the preparation of aldehydes starting with acid chlorides [87] (2.19; X = Cl) or phenol esters [88] (2.19; X = OPh).

$$\mathrm{R{-}\underset{\underset{\displaystyle O}{\|}}{C}{-}X + Li\left[\begin{matrix}H \\ AlOC(CH_3)_3\end{matrix}\right] \rightarrow R{-}\underset{\underset{\displaystyle O}{\|}}{C}{-}H} \qquad (2.19)$$

A novel approach has been to pretreat LAH with optically active alcohols and use the product to effect asymmetric reductions. LAH pretreated with (+)-menthol reduced Δ^2-pyridine salts (33) to optically active piperidines [89(a)].

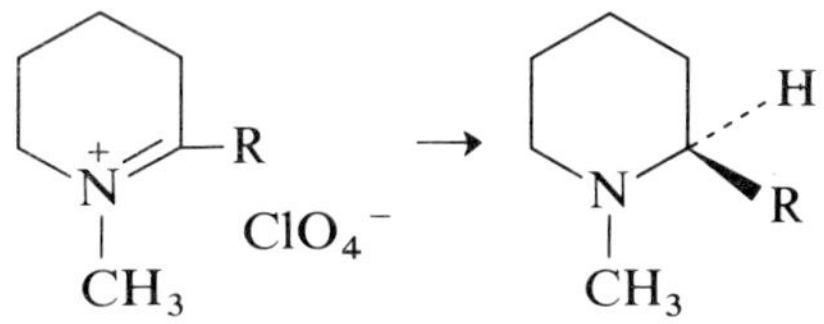

(33)

Asymmetric reductions of ketones to secondary alcohols have been brought about with LAH pretreated with amino alcohols, glycols and sugars [89(b,c,d)]. LAH pretreated with pyridine also has its reducing power drastically reduced to the point where carbonyl groups of diaryl ketones only are reduced. The actual reducing species seems to be a complex anion (34) containing 1,2- and 1,4-dihydropyridine species in the approximate ratio of 1 to 3 [89(e)].

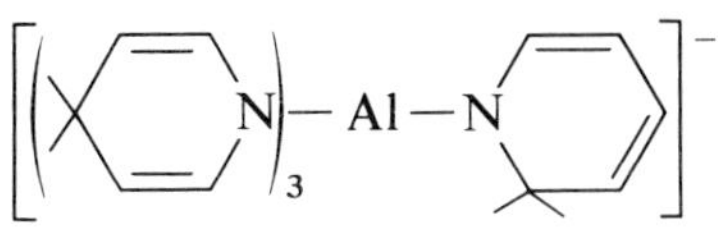

(34)

LAH is the most powerful reducing agent of its class, but it cannot be used in protic solvents. Substances containing active hydrogen (—OH, —NH) usually lose it as molecular hydrogen, and the substance becomes part of a complex aluminohydride anion which hinders the approach and hence attack by the similarly charged hydride or aluminohydride ion. The presence of surplus electrons in a substrate usually means that no reduction occurs (e.g., in the nuclei of π-excessive nitrogenous heteroaromatic compounds), or it may take an unusual course. LAH reduces smoothly the ester groups of diethyl cyclohexylmalonate to the expected and usual primary alcohol stage (2.20; $R^1 = R^2 = CH_2OH$). However, in the case of the enolate anion, conversion to methyl and aldehyde groups occurs [90] (2.20; $R^1 = CH_3$, $R^2 = CHO$; $R^1 = CH_3$, $R^2 = CH_2OH$).

$$\mathrm{C_6H_{11}.CH(CO_2C_2H_5)_2 \rightarrow C_6H_{11}.CHR^1R^2}$$
$$\searrow \mathrm{C_6H_{11}.\overset{\ominus}{C}(CO_2C_2H_5)_2} \nearrow \qquad (2.20)$$

Hydrogenolysis was brought about in cyclopropylamine derivatives because of initial replacement of active hydrogen [91].

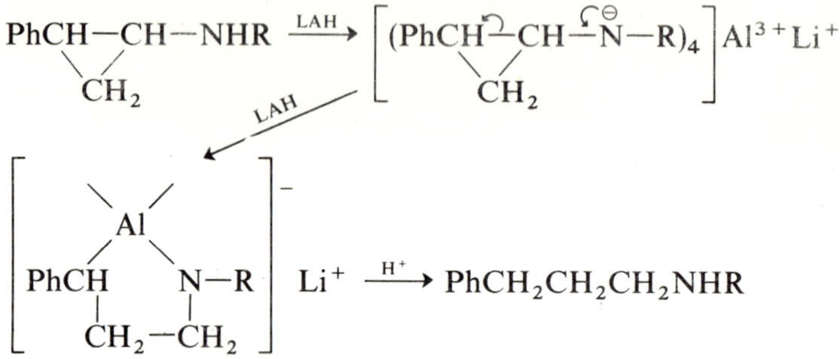

(2.21)

Most hydrogenolyses of single bonds effected by LAH involve a nucleophilic displacement and production of a new C—H bond. This occurs with the halogen atoms in certain halogenoarenes [92], especially when the halogen atoms are activated, and with sulphonate esters. Neighbouring group participation can play a part with the latter and cause formation of isomers. Thus 4-methoxy-1-pentyl-*p*-bromobenzenesulphonate affords 4-methoxypentane by nucleophilic displacement at C-1 as well as the isomeric 1-methoxypentane obtained through displacement at C-2 in the intermediate ion-pair (35) produced by anchimeric assistance of the methoxy group [93].

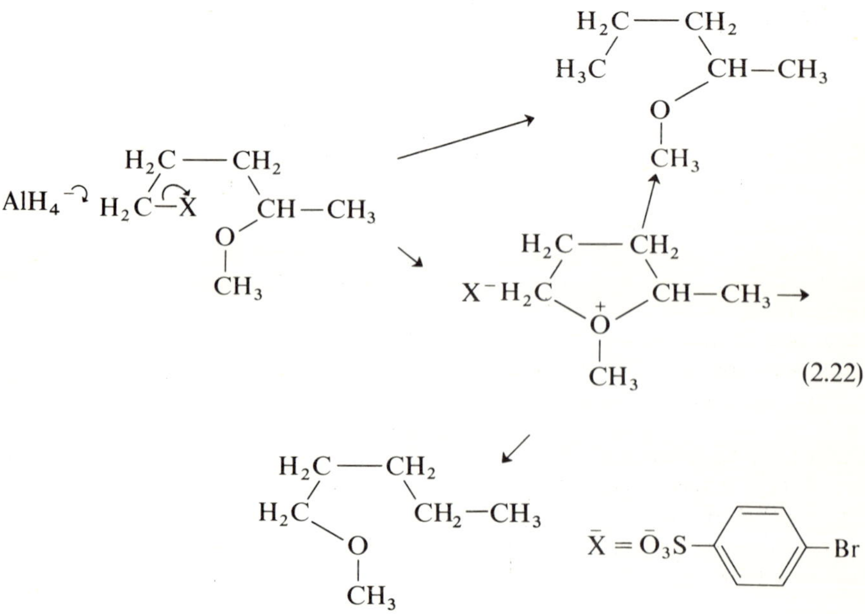

(2.22)

(35)

Hydrogenolytic splitting of C—N bonds occurs in strained amides [94] (2.23), of C—O bonds in epoxides, by attack at the least hindered carbon atom [95], while both C—O and C—C bonds are split when the epoxide is unusually strained [96] (2.25).

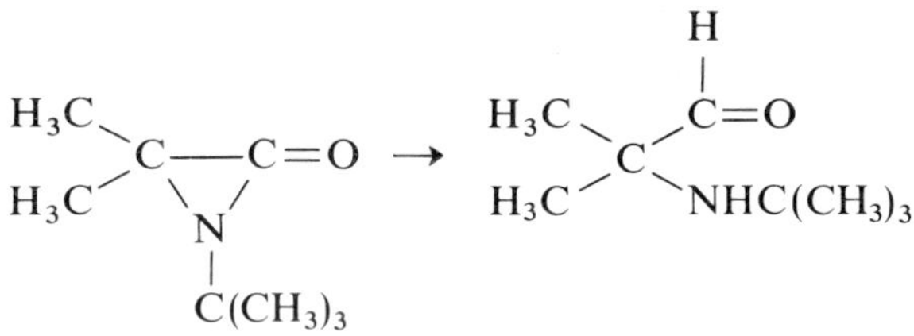

(2.23)

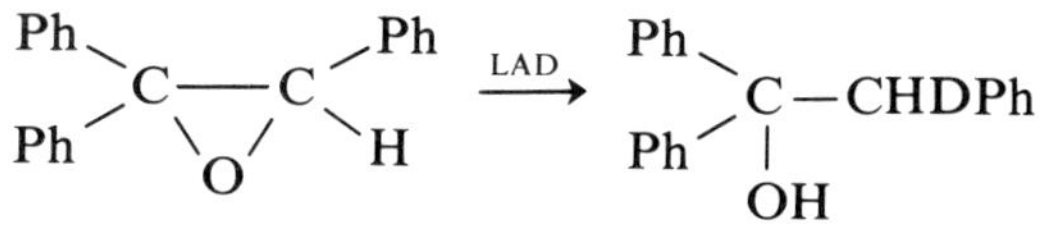

(2.24)

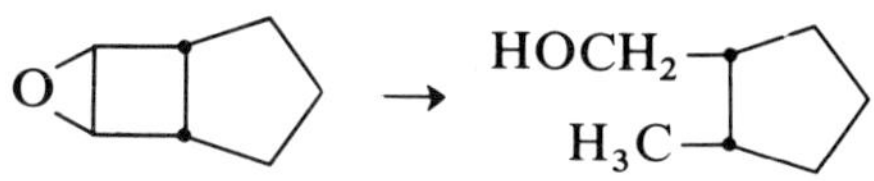

(2.25)

It has been generally believed that carbon–carbon double bonds are inert to LAH unless they are activated by both aryl and keto groups. Thus both groups in methyl cinnamate (36) are reduced, and if LAD be used, the product is a 50/50 mixture of the *erythro* and *threo* alcohols [97].

$$\underset{(36)}{PhCH{=}CH{-}CO_2CH_3} \rightarrow PhCHDCHDCD_2OH$$

However, the solvent plays an important role. There is no olefinic reduction in hydrocarbon solvents, while in ether and tetrahydrofuran prolonged reduction gives phenylcyclopropane.

However, it has recently been found that LAH can function as a homogeneous hydrogenation catalyst for the reduction of olefins and acetylenes. LAH adds to the unsaturated system to form a metal alkyl which subsequently is hydrogenolysed. In tetrahydrofuran, the acetylene (37) underwent *trans* addition leading to *trans*-pent-2-ene [98].

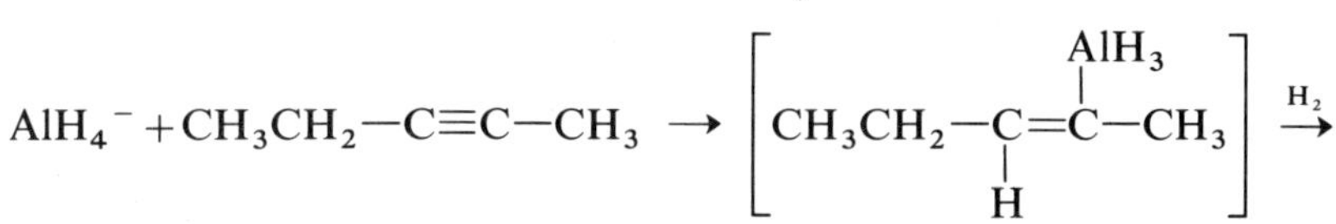

(37)

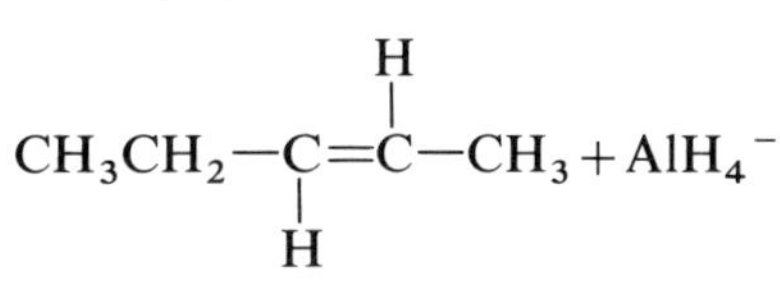

Sodium borohydride is a less powerful reducing agent than LAH but has the advantage that it can be used in protonic solvents. These frequently protonate intermediate reduction products to give positively charged species, which in turn are reducible by borohydride. It is thus possible for the weaker reducing agent to produce a more highly reduced compound. Reduction of pyridinium salts (38; $R^1 = R^4 = CH_3$; $R^6 = Ph$) in aprotic solvents usually stops at the 1,2-dihydro stage (39) when the C=N bond has been saturated. Protic solvents (water, alcohol) can transfer a proton to the $\alpha\beta$-enamine system in (39), and the new quaternary cation is reduced by

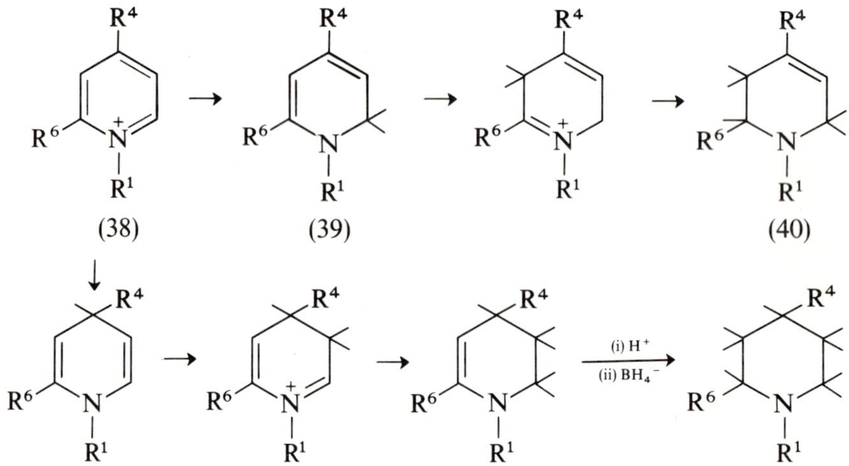

(38) (39) (40) (41) (42)

borohydride to the final 1,2,5,6-tetrahydro product (40) containing a stable isolated olefinic double bond. If the alkyl group on the nitrogen is large enough (38; $R^1 = (CH_3)_2CH$; $R^4 = R^6 = H$), it will hinder the transference of hydride ion to C-2, direct some to C-4 and form the 1,4-dihydro derivative (41). A double sequence of proton transfer and hydride attack converts (41) into the fully reduced heterocycle (42; $R^1 = (CH_3)_2CH$; $R^4 = R^6 = H$) [99]. Proton transfer occurs because the nitrogen atom makes the necessary electrons available to the β-carbon atom by an electromeric shift.

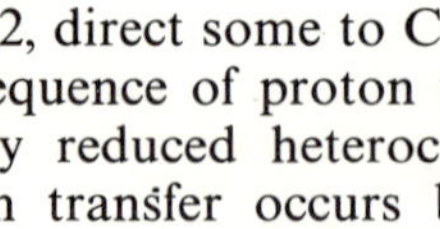

It is more effective than a sulphur atom in this respect, so that thiazolium salts (43) give tetrahydro compounds (44), in which H_R signifies hydrogen transferred by borohydride and H_S that transferred by solvent [100].

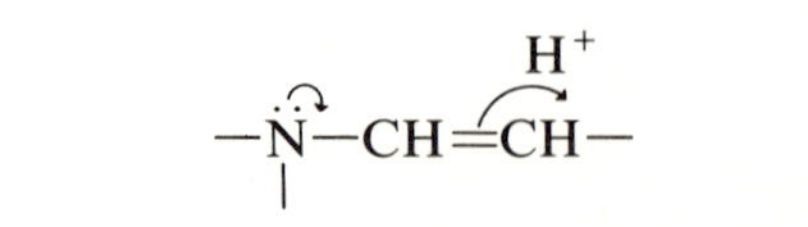

(43) (44)

A similar electromeric shift is responsible for the formation and later reduction of immonium cations when acetic acid is added to a solution of an $\alpha\beta$-unsaturated steroidal tertiary amine containing borohydride [101].

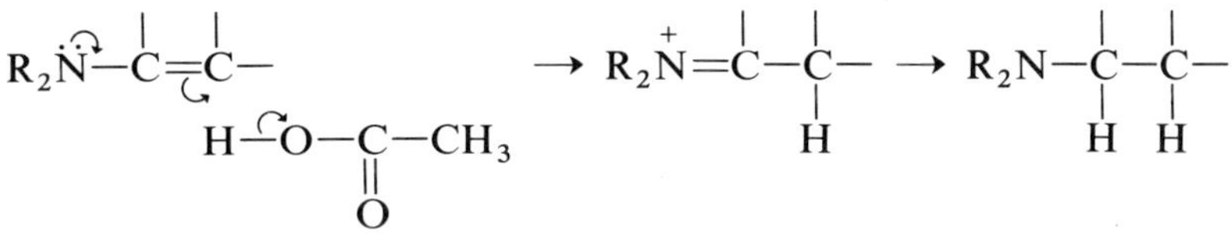

Again, the interaction of protic solvents with hexahydropyrimidine to give similar cations lies behind the reductive cleavage of hexahydropyrimidine and derivatives [102] by sodium borohydride.

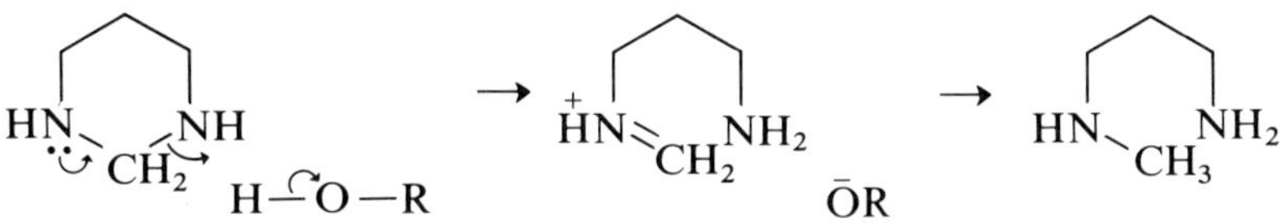

An unexplained solvent effect was observed with sodium borohydride in pyridine, which preferred to reduce an olefinic double bond activated by conjugation with nitrile or ethoxycarbonyl, rather than the carbonyl group itself [103].

$$(CH_3)_2C{=}C(CO_2C_2H_5)_2 \rightarrow (CH_3)_2CHCH(CO_2C_2H_5)_2$$

Yet in another example a large excess of methanolic sodium borohydride reduced both types of group [104].

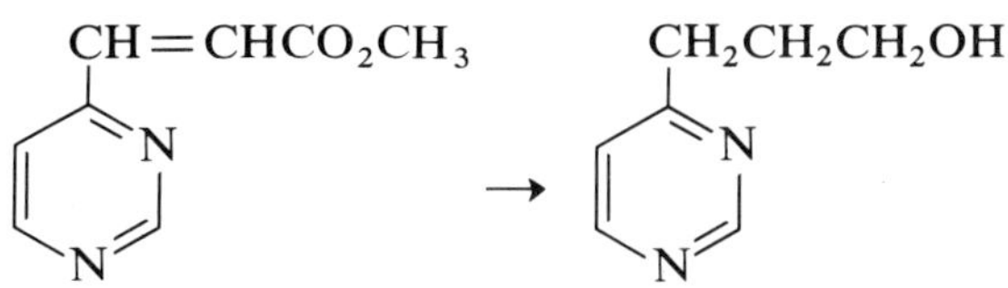

(2.26)

The nucleophilic properties of borohydride are exemplified in the hydrogenolysis of easily ionized alkyl halides to the parent hydrocarbon [105], by trapping the intermediate carbonium ions (2.27).

$$\begin{array}{c} R_3CCl \rightleftharpoons R_3C^+ + Cl^- \\ \quad\quad\quad \downarrow BH_4^- \\ \quad\quad\quad R_3CH \end{array} \tag{2.27}$$

Nucleophilic displacement of alkoxy anions by hydride occurs in the reduction of alkoxy sulphonium salts [106]

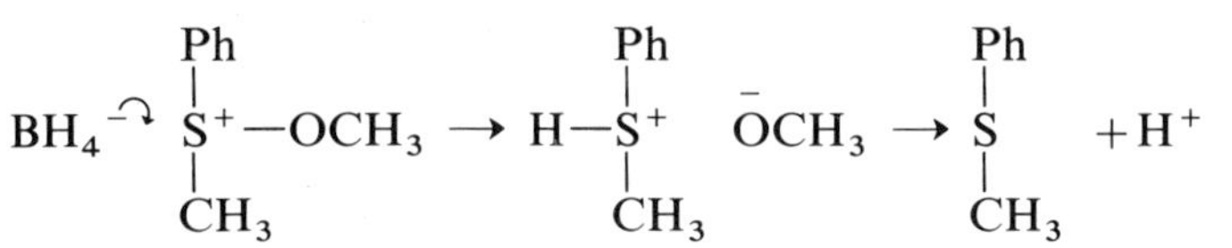

(2.28)

and of halogen in certain crowded polynitrohalogenoarenes [107]. Aryl diazonium borofluorides are converted to the parent arene, possibly by trapping of the aryl carbonium ion too [108].

$$ArN_2^+ \xrightarrow{-N_2} Ar^+ \xrightarrow{BH_4^-} ArH$$

Borohydride cannot be expected to attack aromatic rings unless the nucleophilic properties of the π-electron cloud are reduced by electron-withdrawing substituents like NO_2. Thus the ring in 2,4-dinitrochlorobenzene is reduced to a cyclohexene derivative [112].

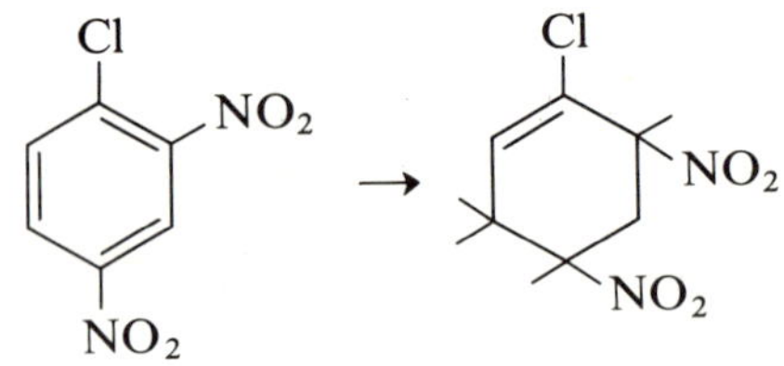

With more nitro groups, the C—Cl bond is hydrogenolysed as well.

The benzene nucleus in the cations derived from certain 'piano-stool' organometallic compounds is also reduced [113].

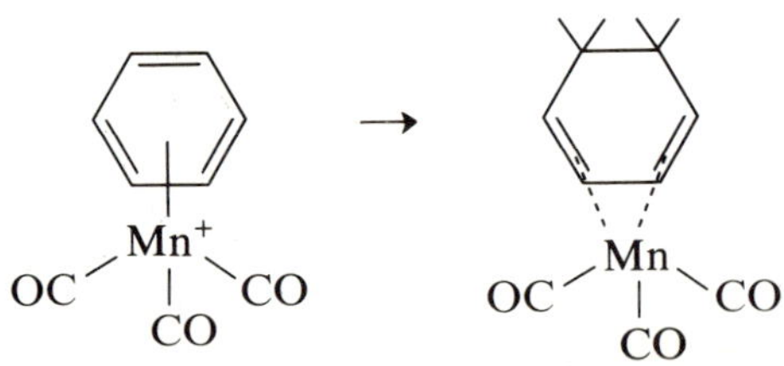

π-Excessive heteroaromatics are not reduced, and even π-deficient heteroaromatics like pyridine require a positive charge as above or a 3-nitro or cyano group before the ring is reduced [114] to the tetra- or hexahydro stage. With these groups in the 2- or 4-position, the substituents are preferentially reduced ($NO_2 \rightarrow$ —N=N—; CN $\rightarrow$ CH_2NH_2). This is one of the few examples where borohydride *alone* will reduce a nitro group. This can be reduced if an electron-transfer catalyst like a cobalt bipyridyl complex is also present [115], or if palladized charcoal catalyst is used. Thus certain *o*-nitro aromatic compounds furnish heterocyclic compounds [116], presumably by the trapping of an intermediate hydroxylamine (2.29).

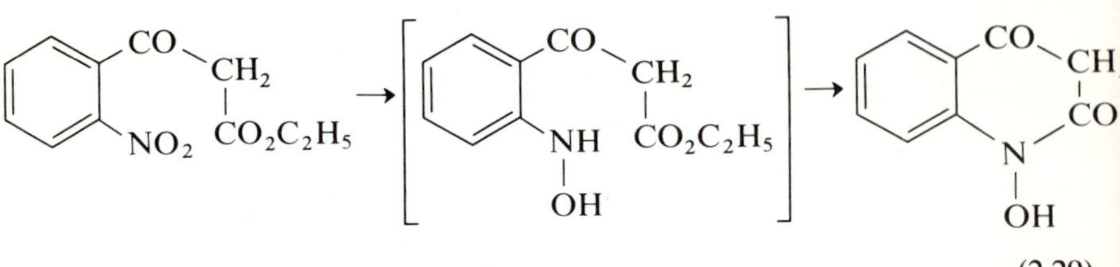

(2.29)

A recent unusual and promising variant on these reductions has been to irradiate the solution containing the substrate and sodium borohydride with ultraviolet light. The borohydride then reduces an *activated* molecule of the substrate. Among the unusual reductions that then can be so effected is the reduction of the aromatic ring [117, 118] of tryptophane (2.30).

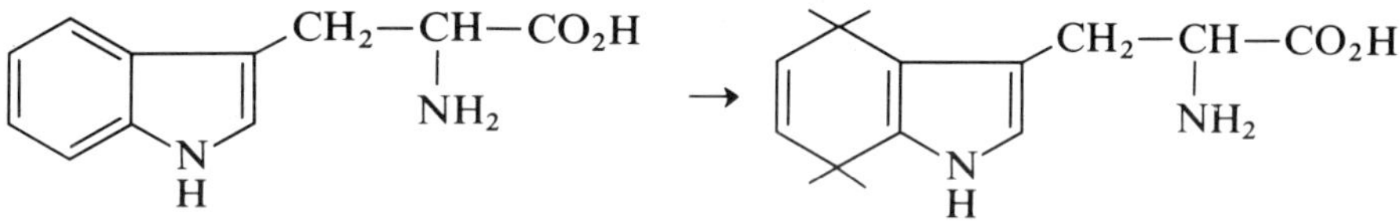

(2.30)

Photoreduction of heterocyclic systems and of the benzene ring of phenols has also been effected. But in spite of these widespread investigations borohydride is still widely used for the reduction of the carbonyl group in aldehydes and ketones to primary and secondary alcohols respectively. If the borohydride anion be associated with an organic cation such as triphenylphosphorane [119] or long-chain quaternary ammonium [120], greater selectivity is possible and only an aldehyde carbonyl is effectively reduced.

Sodium borohydride reduction of simple acyclic and monocyclic ketones usually affords the more thermodynamically stable alcohol—the product-development control factor. In rigid bicyclic ketones, steric hindrance to the approach of the borohydride ion to the carbonyl function becomes the predominant factor—the steric-approach control factor [121]. So 2-methylcyclohexanone affords 69 % of the *trans* diequatorial alcohol

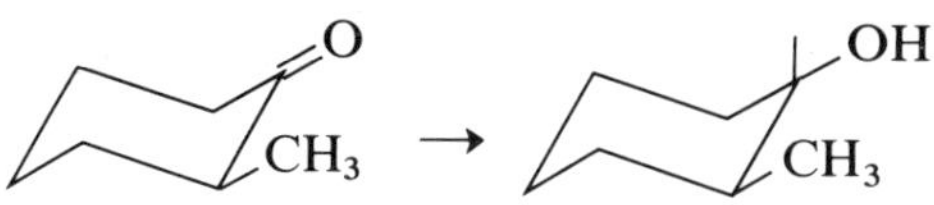

(2.31)

while norbornan-2-one produces 86 % of the less stable *endo* alcohol by attack from the less hindered *exo* direction.

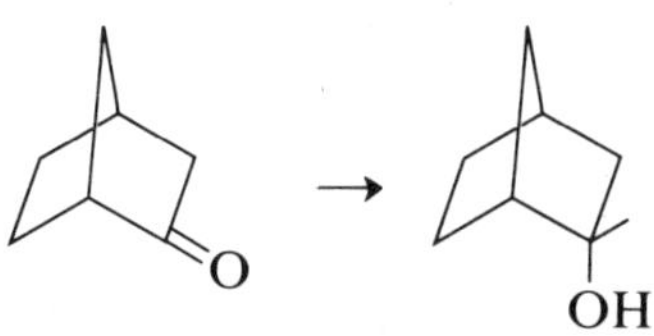

(2.32)

If the ketone be converted into its tosylhydrazone, the parent hydrocarbon is obtained on borohydride reduction [122].

$$R_2C{=}N.NH.SO_2.C_6H_4.CH_3(p) \rightarrow R_2CH_2 \quad (2.33)$$

The electrophilic boron hydride reducing agents include diborane, B_2H_6, which is the gaseous dimer of borine, BH_3; borine stabilized by co-ordination with dimethylamine, trimethylamine or pyridine

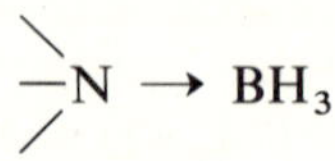

and some dialkyl derivatives of borine.

Diborane may be prepared in large quantities from the reaction of sodium borohydride with iodine or mercurous chloride [123] and attacks electron-rich sites in the substrate molecule in order of electron density. In progesterone (45) the order of attack is (1) carbonyl at C-3; (2) carbonyl at C-20; (3) the Δ^4 double bond [124].

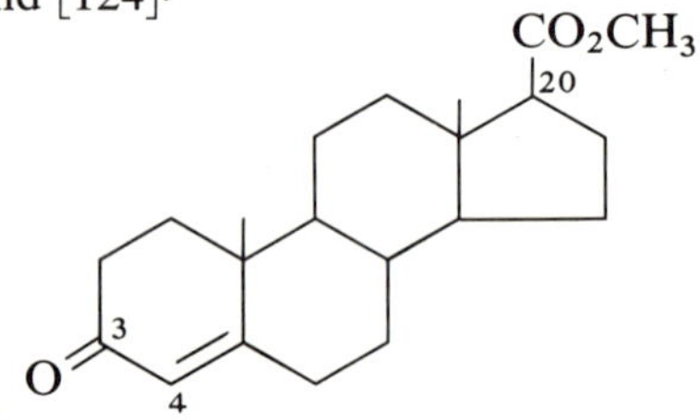

(45)

The carbonyl group of a ketone is usually reduced to the secondary alcohol. Co-ordination is first assumed to occur between the carbonyl oxygen atom and the monomeric borine assumed to be in equilibrium with its dimer.

$$B_2H_6 \rightleftharpoons 2BH_3$$

$$R_2C{=}O + BH_3 \rightarrow \left[R_2C{=}O \rightarrow BH_3 \leftrightarrow R_2C{=}\overset{+}{O}{-}\overset{-}{B}H_3 \leftrightarrow R_2\overset{+}{C}{-}O{-}\overset{-}{B}H_2(H) \right] \rightarrow R_2CH{-}OBH_2$$

(46)

Intramolecular hydride transfer is possible in the complex, and the intermediate (46) still contains hydrogen atoms directly attached to boron; the process of co-ordination with a fresh molecule of ketone, followed by hydride transfer, may be repeated twice over until the trialkyl borate $(R_2CHO)_3B$ is formed. This will then be hydrolysed to the secondary alcohol and boric acid during the working up. However, borate anion can be a good leaving group, and heterolysis in certain cases affords a carbonium ion which is reduced to the hydrocarbon. This seems to be the case for aryl ketones. Thus xanthone (47) afforded xanthene [125].

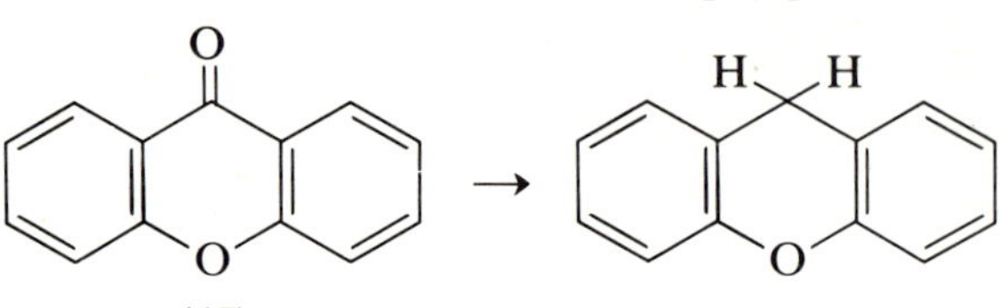

(47)

Ionization of the borate intermediate is aided by the presence of boron trifluoride [126] during the reduction of cyclopropyl aryl ketones to the hydrocarbon. Diborane is a selective reducing agent for the rapid reduction of a carboxylic group to the primary alcohol, even in the presence of an acid chloride function.

Diborane is remarkable for the ease with which its monomer adds to isolated acetylenic or olefinic bonds—called the hydroboration reaction. Olefins give in turn mono-, di- and trialkyl boranes, all of which may be subsequently hydrolysed to the reduced hydrocarbon.

$$RCH{=}CH_2 \rightarrow RCH_2CH_2BH_2 \rightarrow (RCH_2CH_2)_2BH \rightarrow (RCH_2CH_2)_3B \xrightarrow{C_2H_5CO_2H} 3RCH_2CH_3$$

Borine adds *cis* to a multiple bond, and from the less hindered side, and there is no hydrogen–deuterium scrambling if B_2D_6 is used, so that the C_{10} cyclic acetylene was converted into a *cis* deuterated olefin [127].

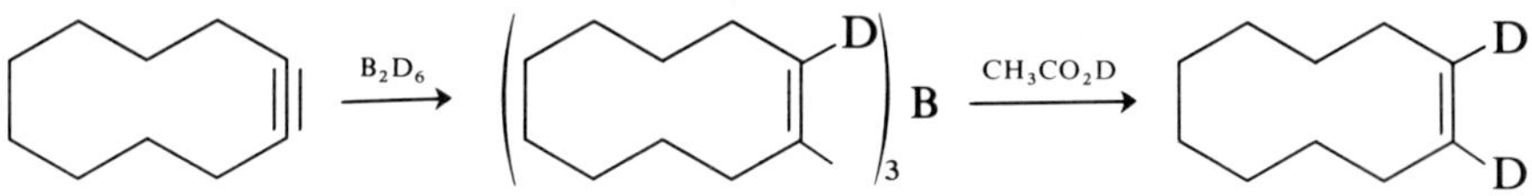

The intermediate dialkyl boranes are less reactive and more selective reagents than diborane itself. Di-isopino-campheyl borane (produced from α-pinene and diborane) slowly reduces camphor completely to the less stable of the two possible alcohols—isoborneol—again by approaching from the less hindered direction [128].

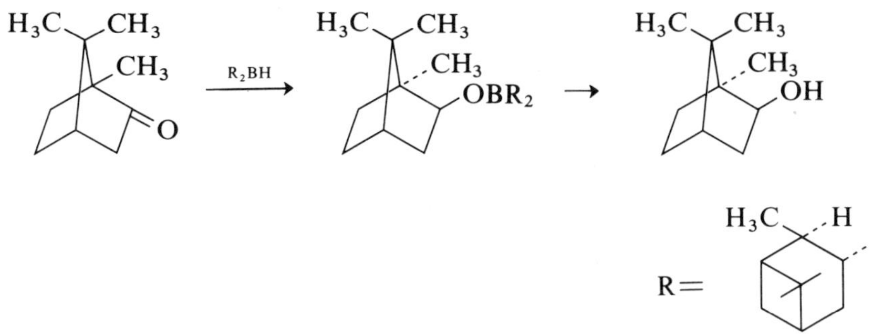

The last (reductive) step in the cyanohydrin synthesis of sugars has been simplified by reacting, e.g., D-erythronolactone with bis(3-methyl-2-butyl)-borane which protects the hydroxyl groups by conversion into esters and preferentially reduces the carbonyl to CHOH [129].

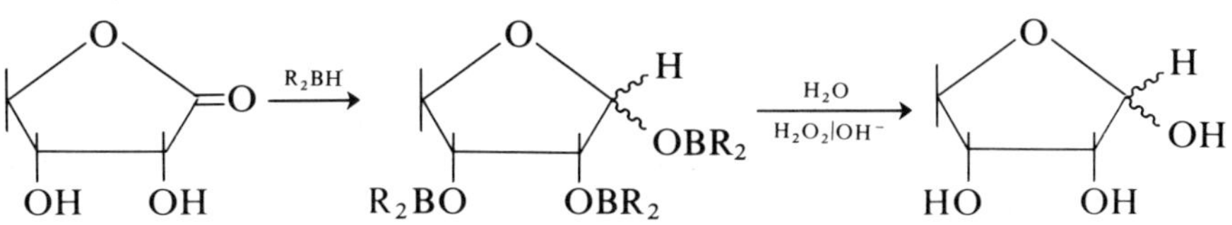

$$R = (CH_3)_2CHCH(CH_3)-$$

Hydroboration of aldoximes and ketoximes by diborane affords the hydroxylamine [130], while the acetate goes further to the acetylated amine.

$$R^1R^2C{=}NOH \rightarrow R^1R^2CHNHOH$$

$$R^1R^2C{=}NOC(=O)CH_3 \rightarrow R^1R^2CHNHC(=O)CH_3$$

Direct reduction of a carbon–carbon double bond by diborane has been noted when the bond has polar character. Thus dibenzofulvenes afford the corresponding fluorenes [131].

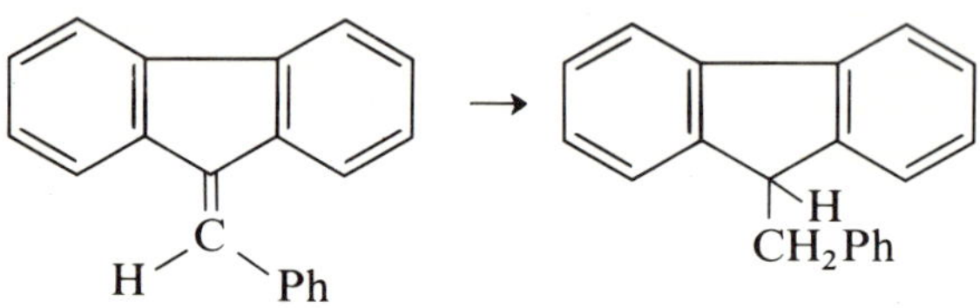

Addition complexes of borine with nitrogenous bases have been used for the hydrogenation of the C=N bond in Schiff bases and also for reductive deuteriations [132].

Alane may be conveniently prepared from LAH and concentrated sulphuric acid in tetrahydrofuran [133]. Alane complements LAH in that it selectively attacks different functions and sometimes gives a more highly reduced product. Thus with a cinnamic ester, only the ester moiety is attacked [134].

$$PhCH{=}CHCO_2C_2H_5 \rightarrow PhCH{=}CHCH_2OH$$

Alane reduces pyridine bases to the tetrahydro stage [135], compared with the dihydro level attained with LAH,

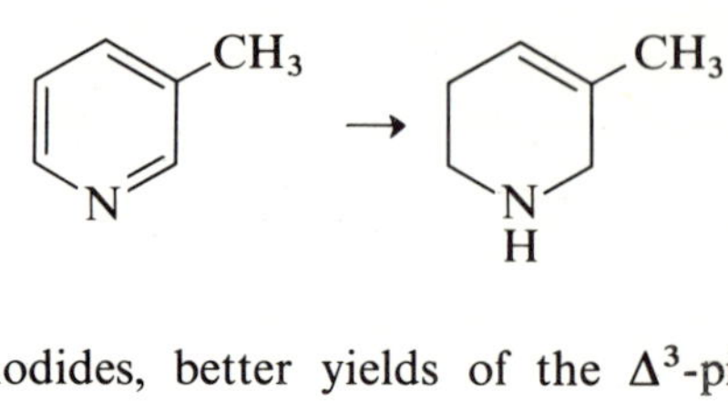

and with the methiodides, better yields of the Δ^3-piperideines are obtained [136].

Alane or its deuterium analogue opens unsymmetrical epoxide rings differently from LAH.

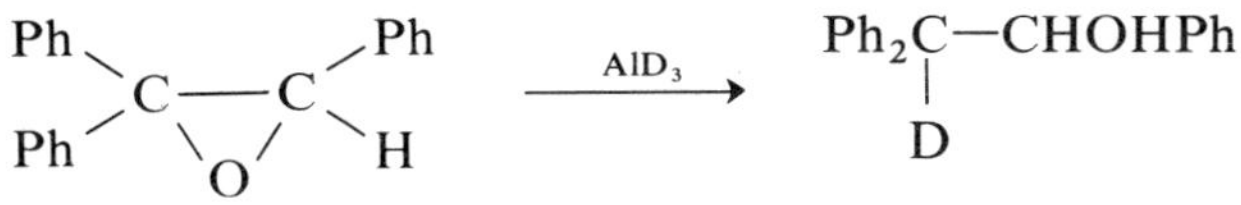

With excess alane, it seems that the epoxide alane complex is the species reduced [95] by a bimolecular mechanism.

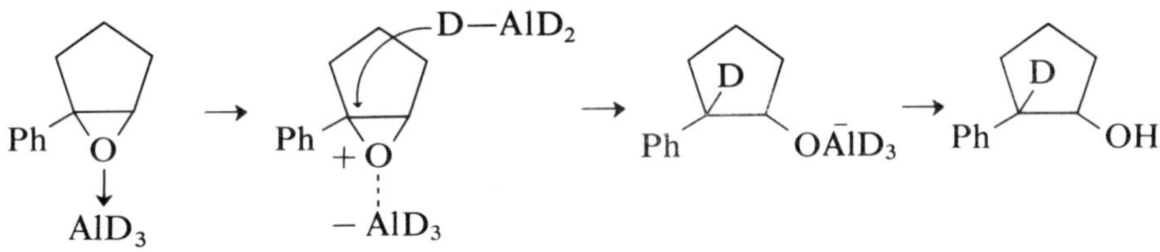

Inverse addition of insufficient alane led to 1,2-dihydride shifts prior to reduction.

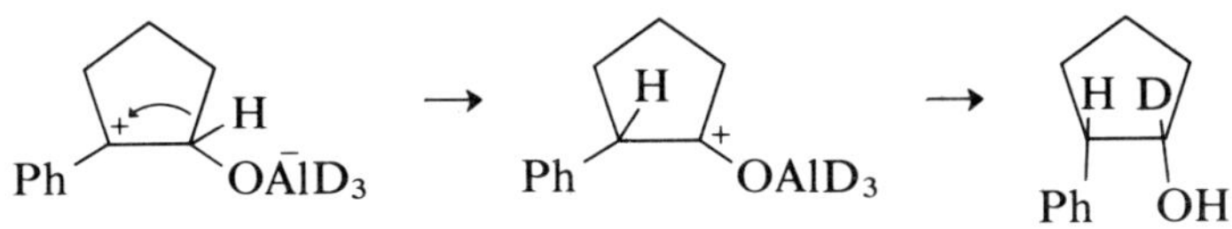

Opening of the epoxide ring in cyclohexene oxides is subject to the conformational control by groups in the six-membered ring. *cis*-4-t-Butylcyclohexene oxide affords the *cis*-3-t-butylcyclohexanol [137].

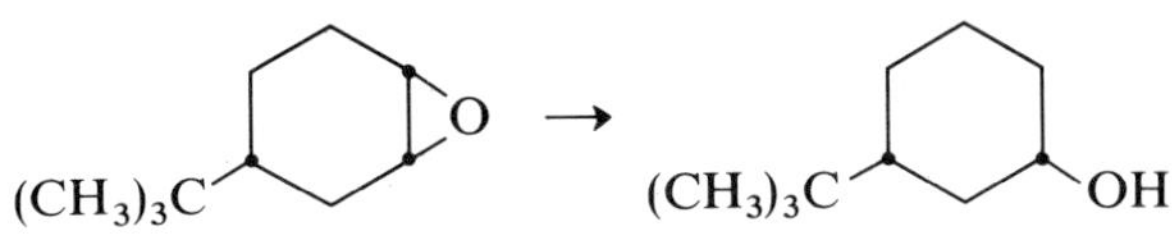

Chlorinated derivatives of alane exist in LAH solutions containing varying amounts of aluminium chloride [138, 139], and these have been used for the reduction of acetals and ketals [140], and for the reduction to hydrocarbons of alcohols which can form reasonably stable carbonium ions [141]. Of the alkyl derivatives of alane, di-isobutylaluminium hydride has been used to reduce quinoline to its 1,2-dihydro derivative.

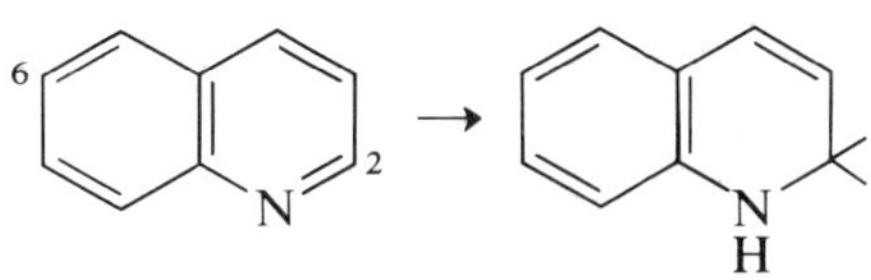

This it does more effectively than LAH because the lipophilic alkyl groups render the reagent readily soluble in ether. It is also more susceptible to steric effects than LAH. It will not reduce 2,6-dimethyl quinoline, on account of the 2-methyl group [142]. It is therefore a more selective reducing agent and has been used to reduce heterocyclic nitrile groups to aldimino [143] groups hydrolysed to aldehydes during the work-up of the reduction mixture. LAH usually reduces —CN to $—CH_2NH_2$.

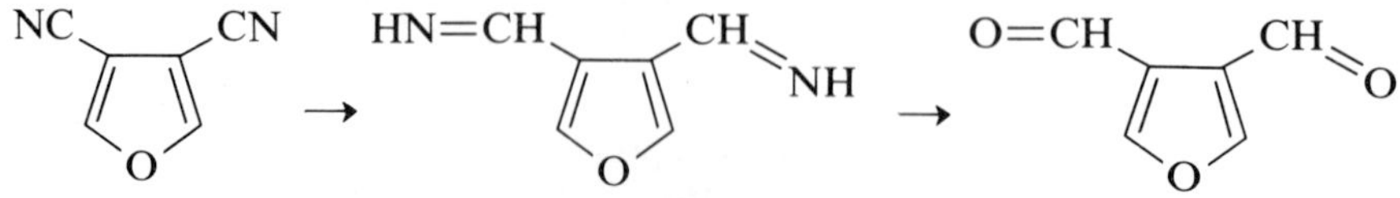

HYDROGEN TRANSFER REDUCTION [144]

In this method, a hydrogen atom accompanied by two electrons is transferred, usually from one carbon atom to another, without exchanging with labile protons in the reaction medium or reacting with them to form molecular hydrogen. Thus in biological systems, reduction involves the transference of such a hydride ion from the 4-position of a 1,4-dihydropyridine derivative to the substrate in question [145]. Such reduction has been duplicated in model systems in the laboratory, as for example in the deuterogenation of the indolenine salt [146] (2.34).

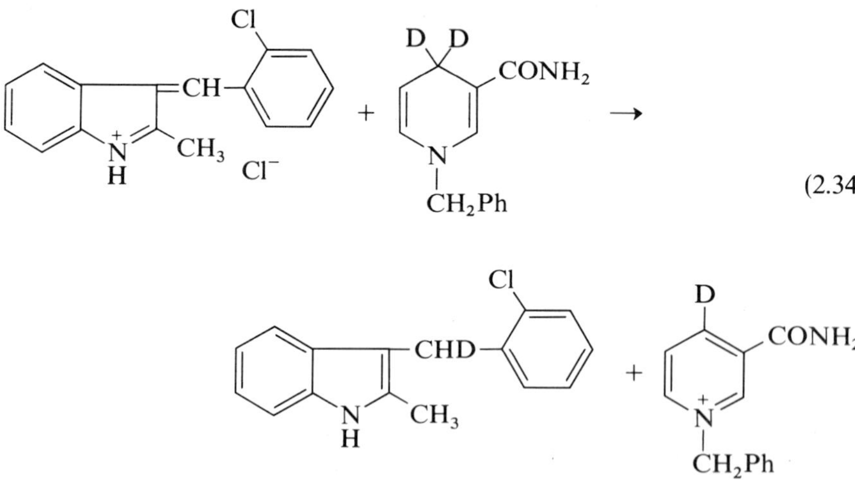

(2.34)

The Hantzsch ester (2.35) reduces other *N*-heteroaromatic substances to dihydro derivatives [147].

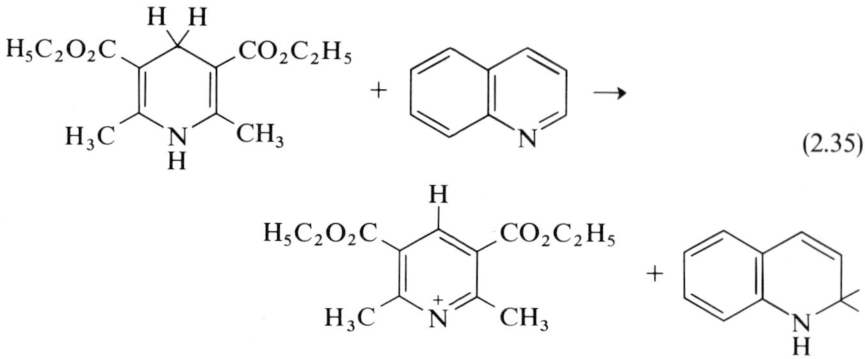

(2.35)

Formic acid is also a hydride transfer agent. Formic acid-*d* reduces pyridinium salts [148] to the tetrahydro and hexahydro derivatives (2.36)

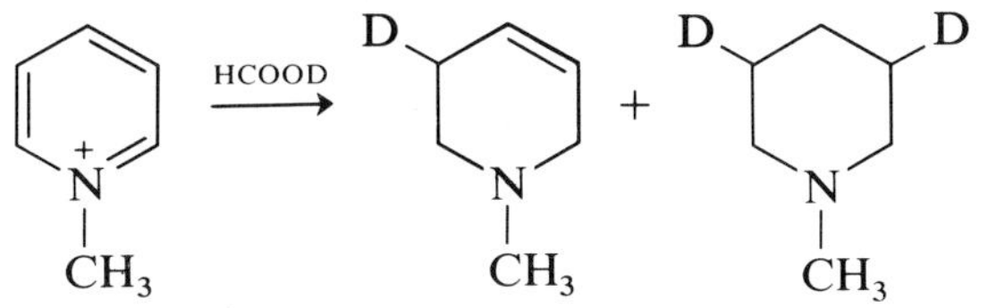

(2.36)

by a mechanism analogous to that of borohydride in protonic solvents. Its use in homogeneous hydrogenation has been mentioned earlier. The alkoxide ion is also capable of transferring a hydride ion to a suitable substrate by a mechanism of the type shown in (2.37).

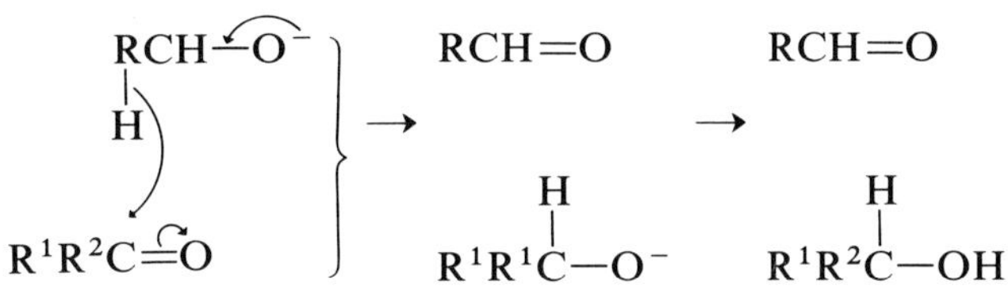

(2.37)

Hot alkaline ethylene glycol reduces the norbornanone derivative [48] to the *endo* secondary alcohol.

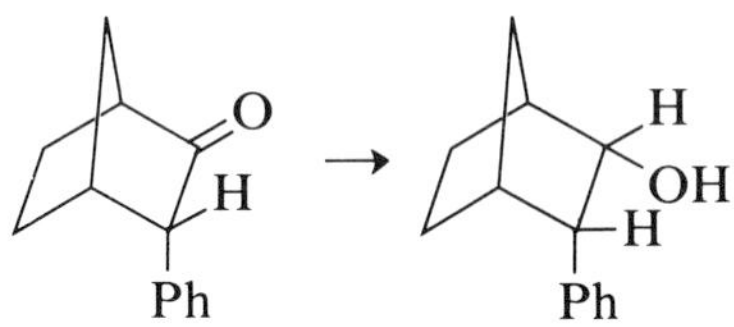

(48)

However, the *exo* isomer afforded both *exo* and *endo* alcohols, so that the phenyl group in (48) hindered approach of the anion from the *endo* direction [149]. Copper catalysis of the alkoxide hydrogenolysis of aryl halides has also been noted [150].

However, reduction of ketones by the alkoxides of aluminium, in particular the isopropoxide—the Meerwein–Ponndorf–Verley reaction—is a well-known technique of long standing [151] which involves hydride transfer to the carbonyl groups of aldehydes and ketones probably by a cyclic intermediate (2.38).

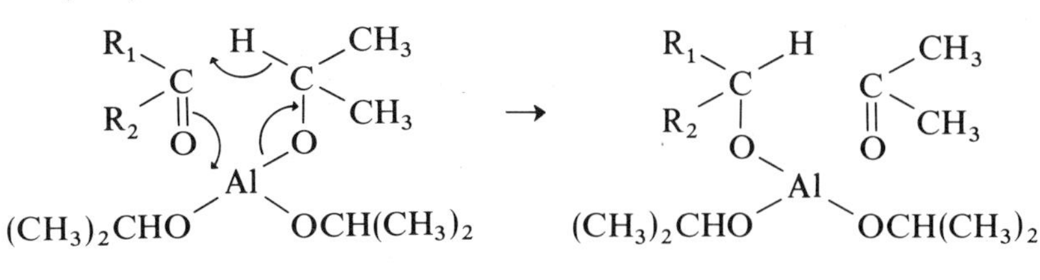

(2.38)

By using the potassium derivatives of optically active alcohols it is possible to bring about asymmetric reduction of ketones [152], likewise the transference of deuteride from the magnesium halide salt of an optically active deuterio alcohol (isoborneol (2.39)) to acetaldehyde leads to optically active ethanol-1-*d* [153].

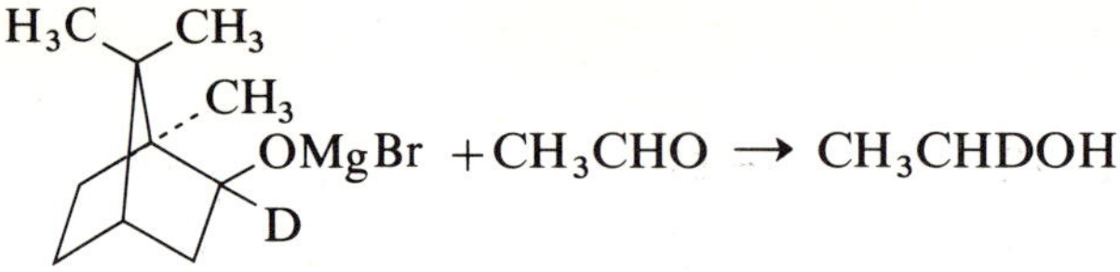

(2.39)

Reduction of cyclohexa-1,2-dione by aluminium isopropoxide afforded *cis* and *trans* diols. The *trans* diol arose from a complex containing two aluminium atoms co-ordinated on opposite sides of the molecule, the *cis* diol from a complex containing only one aluminium atom [154].

Hypophosphorus acid [155] and phosphonium iodide may also be sources of hydride ion. The latter smoothly reduces the immonium double bond in the protonated species derived from pyrrole-2-carboxamide to the 2,5-dihydro derivative (2.40) and is the reagent recommended for this task [156].

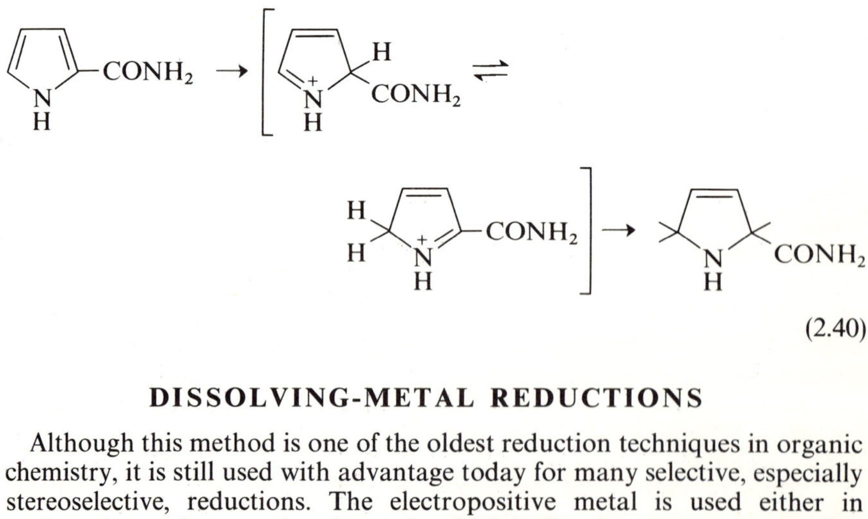

(2.40)

DISSOLVING-METAL REDUCTIONS

Although this method is one of the oldest reduction techniques in organic chemistry, it is still used with advantage today for many selective, especially stereoselective, reductions. The electropositive metal is used either in conjunction with protonic solvents in which it dissolves, or as a suspension in an inert solvent to which a source of protons is later added. Reduction effected by alkali metals dissolved in liquid ammonia has a special name—the Birch reduction [157].

The metal functions as a source of electrons, one or two of which are transferred to the substrate [158]. A one-electron transfer to two atoms X and Y linked by a single bond causes fission into a free radical and an anion

$$\begin{aligned} X{-}Y + e^- &\rightarrow X^{\cdot} + Y{:}^- \\ &\text{or } X{:}^- + Y\cdot \end{aligned} \qquad (2.41)$$

whereas a double bond is converted to a single bond of a resonance-stabilized radical anion.

$$X{=}Y + e^- \rightarrow [X^-{-}Y^\cdot \leftrightarrow X^\cdot{-}Y^-] \tag{2.42}$$

If the structure of XY is such that extensive delocalization can occur, a second electron can be added to give a dianion.

$$[X^-{-}Y^\cdot \leftrightarrow X^\cdot{-}Y^-] + e^- \rightarrow X^-{-}Y^- \tag{2.43}$$

Usually the radical anion accepts a proton from a proton source

$$(X{-}Y)^{\dot{-}} \rightarrow \dot{X}{-}Y{-}H \xrightarrow[H^+]{e^-} H{-}X{-}Y{-}H \tag{2.44}$$

and the new radical receives a second electron followed by a second proton, affording the completely reduced product, rather than the dianion's being first formed and acquiring two protons.

$$\bar{X}{-}\bar{Y} \xrightarrow{2H^+} H{-}X{-}Y{-}H \tag{2.45}$$

In an aprotic solvent, the radical anion may dimerize, and then acquire two protons when a proton source is added

$$(X{-}Y)^{\dot{-}} \rightarrow \begin{bmatrix} X{-}X \\ | \quad | \\ Y^- \; Y^- \end{bmatrix} \xrightarrow{2H^+} \begin{matrix} X{-}{-}X \\ | \qquad | \\ YH \quad YH \end{matrix} \tag{2.46}$$

or it may expel a substituent which can form a good leaving group.

The radical ion with its odd electron can be detected by its electron spin resonance spectrum [159]. If the substrate initially was a univalently charged cation, addition of one electron should give a free radical. This has been observed during the treatment of a quaternary salt of isonicotinic ester with zinc under acetonitrile. The free radical was sufficiently stabilized by resonance so that it could be distilled in high vacuum [160].

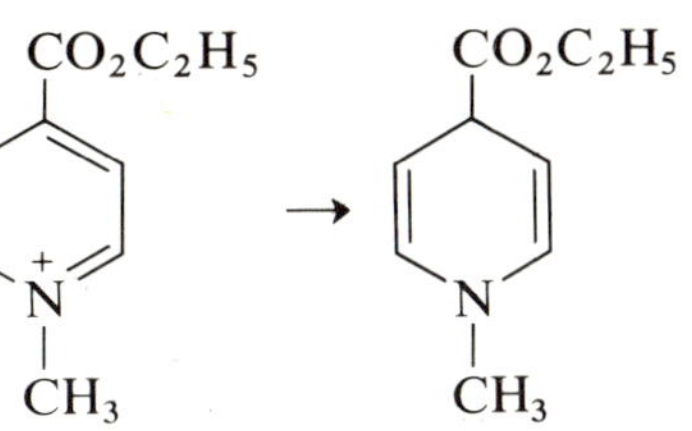

(2.47)

A dark green anion radical is produced when naphthalene reacts with sodium suspended in ether solvents.

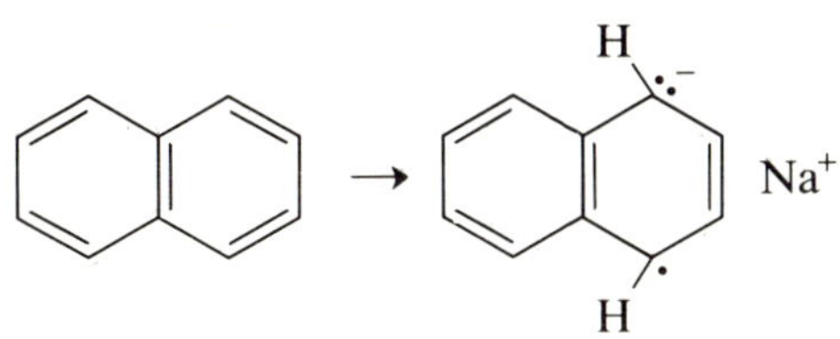

(2.48)

Cyclo-octatetraene forms a dianion with potassium suspended in tetrahydrofuran, the two extra electrons making the anion

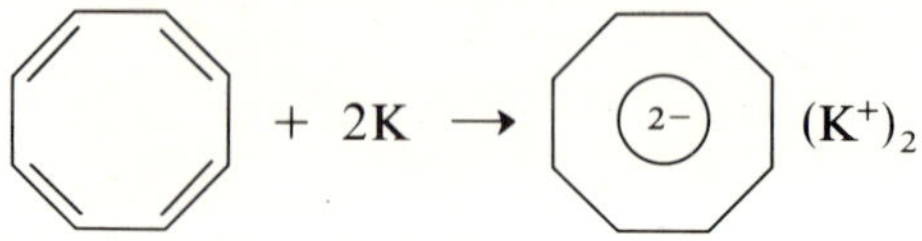

a system obeying Hückel's rule [161].

Both systems are resonance-stabilized, can be obtained in relatively high concentration in ether solvents, and have been used in turn to effect reduction processes in other systems. The naphthalene anion radical cleaves the S—O bond in toluene sulphonates [162],

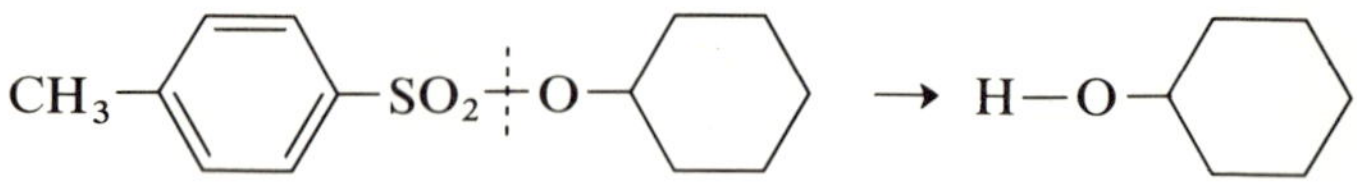

while the dianion can transfer its two electrons to two tropylium cations to afford two free radicals which subsequently unite [163].

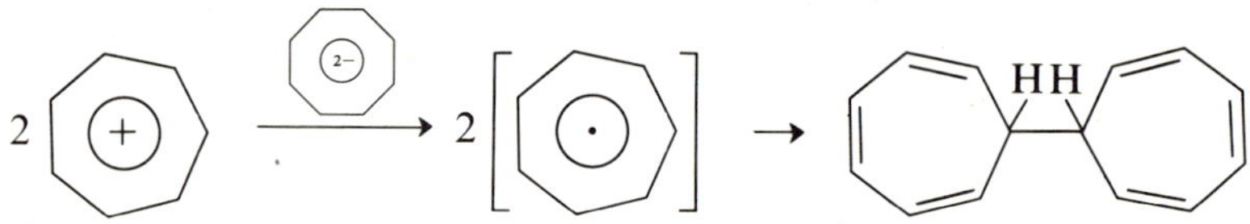

Ketones, however, are the star substrates in these reductions. Pinacols result when one electron is transferred from magnesium metal under anhydrous aprotic conditions [164]. The anion radical dimerizes to a dianion,

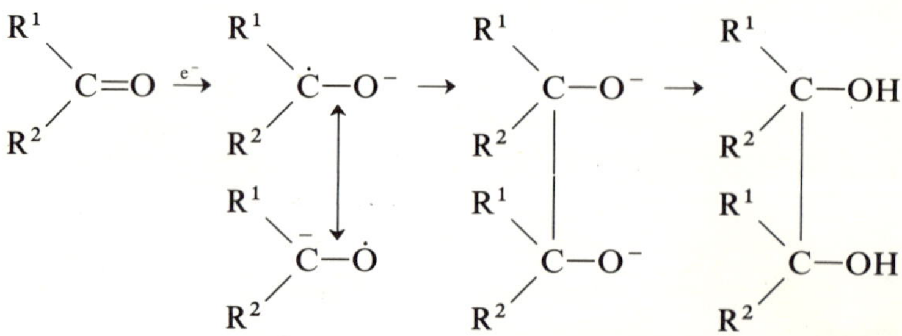

which subsequently acquires two protons during the hydrolytic working-up. The process is termed bimolecular reduction. With sodium and an alcohol, the anion radical (presumably at its less hindered face) accepts a proton from the solvent. Acceptance of a second electron from the metal affords the metal alkoxide.

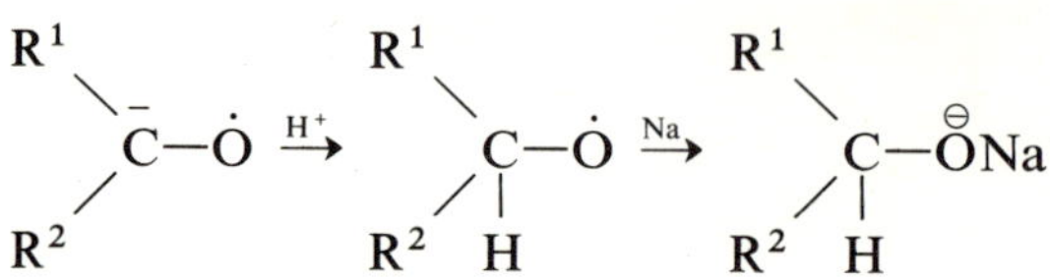

This can sometimes equilibrate the starting ketone by a cyclic mechanism analogous to the Meerwein–Ponndorf–Verley reduction to produce an isomeric, and usually more stable, alkoxide [165].

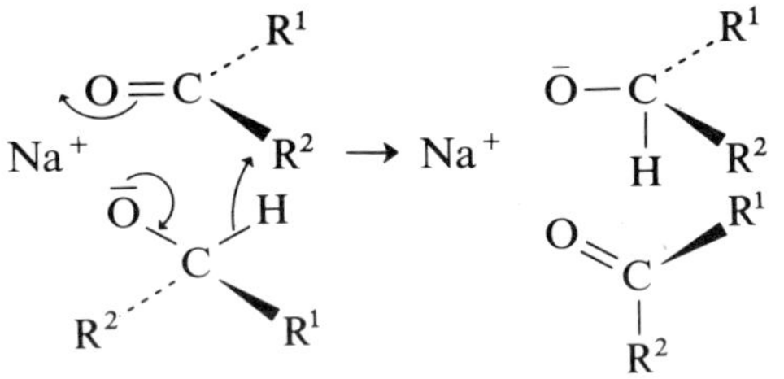

An unusual two-electron reduction of a bicyclic ketone led to transannular reduction [166].

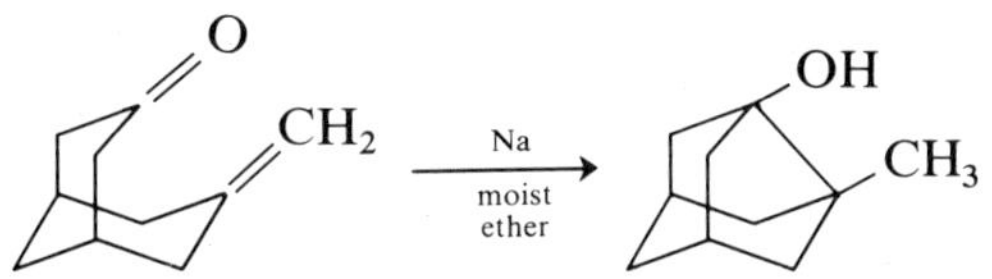

Under acidic conditions, a metal such as zinc (in the form of amalgam) can pass no less than four electrons successively to one molecule of protonated species, leading to the hydrocarbon.

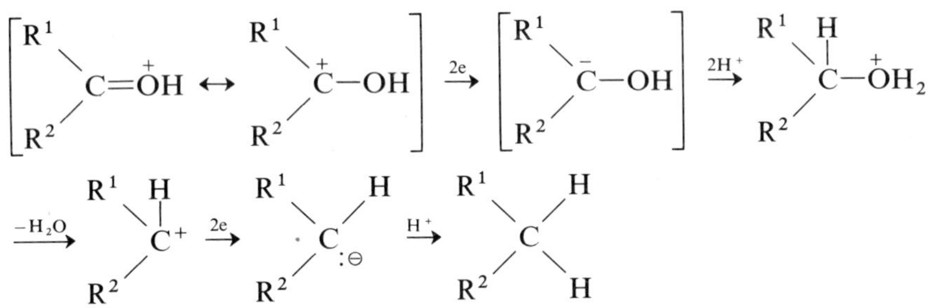

This well-known Clemmensen reduction [168] has recently been used for deuterium-labelling [169] and investigated in 1,3- and 1,4-diketones [170].

Reductions under acid conditions thus frequently involve a species other than the neutral molecule of substrate. The reduction of pyrrole (49) with zinc dust and hydrochloric acid is another example [171].

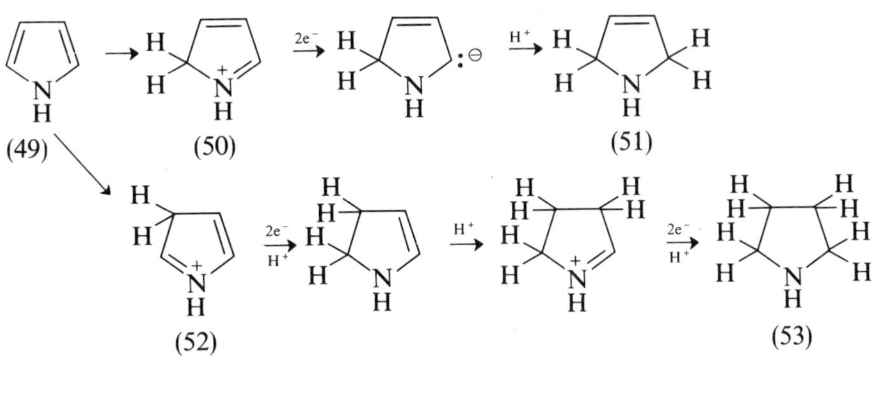

(49) (50) (51) (52) (53)

Protonation may occur at the α-carbon atom, and the C=N bond of the cation (50) is reduced after acquiring two electrons and one proton to the Δ^3-pyrroline (51), whose isolated olefinic double bond is stable to further reduction. Protonation may also occur at a β-carbon atom (52) and this can lead to the formation of the fully reduced pyrrolidine (53). In practice both (51) and (53) were isolated. The ability of alkali metals dissolved in liquid ammonia to partially reduce aromatic systems is a valuable contribution to organic synthesis. Thus lithium and liquid ammonia reduce benzene to 1,4-dihydrobenzene in the presence of ethyl alcohol, which protonates the intermediate radical ion (54) and anion (55) [172].

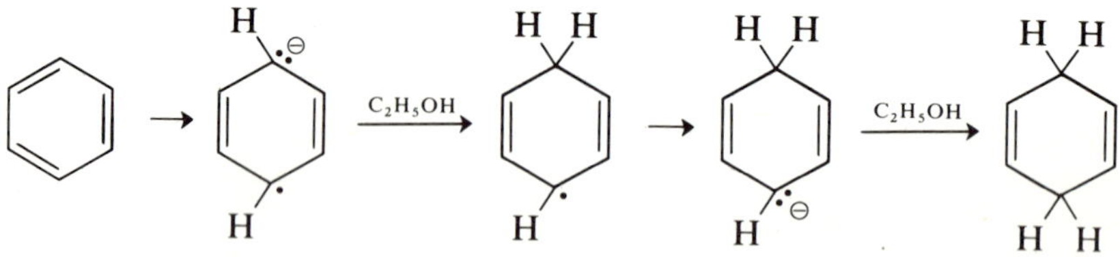

(54) (55)

However, by working at elevated temperatures (60–130°) it is possible to effect reduction without addition of an alcohol or other proton source [173]. Presumably the anion radical abstracts the proton required from the ammonia. The amide anion so produced, however, can isomerize the 1,4-dihydrobenzene to its more stable conjugated isomer, the 1,2-dihydrobenzene (57). The conjugated system of double bonds in this hydrocarbon can again

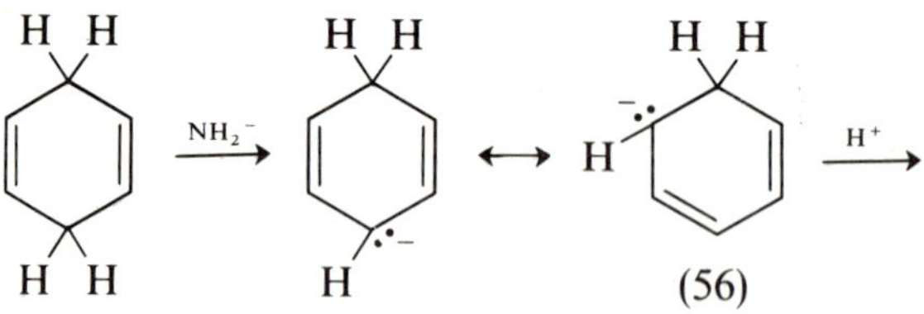

(56)

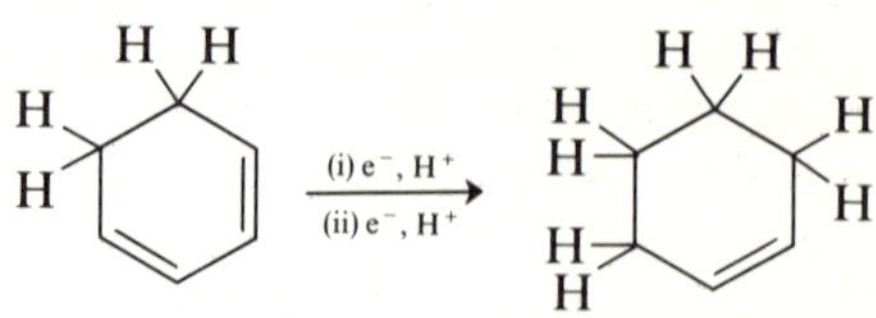

(57) (58)

go through the procedure of alternate electron- and proton-accepting, so that the final product is the tetrahydro derivative cyclohexene (58). All traces of iron compounds must be removed from the liquid ammonia used, especially if the reducing metal used is sodium, since they catalyse unwanted side reactions between the metal and proton source. Lithium metal is less susceptible in this respect [174]. The presence of a little water can be beneficial in some instances. In the methiodide system the presence of 0·5% water ensured that C—N hydrogenolysis involved the phenyl–nitrogen bond only

(2.49), unlike the anhydrous medium, where *N*-methyl fission also occurred.

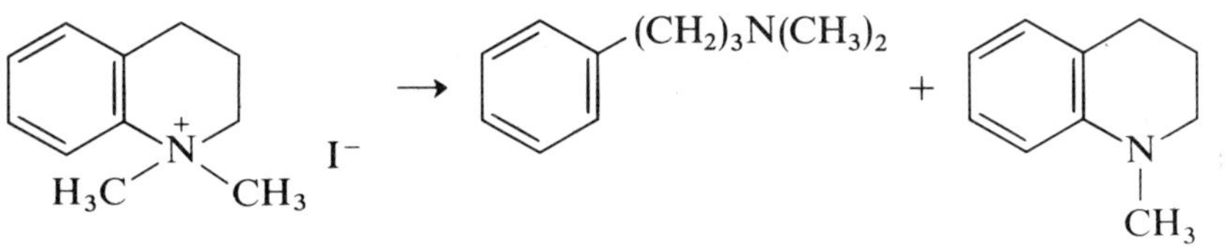

(2.49)

Ammonia can be replaced by its organic derivatives. Thus methylamine together with alkali metals at elevated temperatures, and without a proton source, reduces benzene and alkyl benzenes to the tetrahydro stage [176]. Ethylenediamine and lithium bring about a similar reduction of 1,4-di-t-butylbenzene [177]. These examples illustrate the value of alkali metal in nitrogenous base solutions for effective partial reduction of the nucleus in benzene derivatives (catalytic methods, for example, invariably give the fully reduced cyclohexane compound).

In the formation of dihydro derivatives of benzene, an alkyl group and other electron-donating substituents should hinder the acceptance of an electron to form an anion radical and so retard the reduction. Calculations by the molecular orbital method of the electron density in the anion radical reveal that it is greatest at the positions *ortho* or *meta* to the substituent [178]. If the radical anion is protonated initially at the site of highest electron density, further electron transfer and protonation should lead to 1-substituted-2,5-dihydrobenzenes.

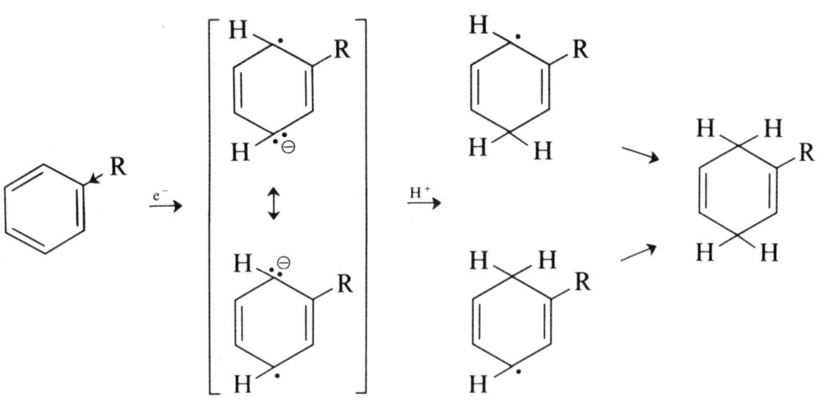

(2.50)

Electron-withdrawing substituents, on the other hand, facilitate acceptance of an electron by the nucleus and speed up the reaction. The *para* position of the anion radical turns out to have the greatest electron density, and an argument analogous to that above confirms the experimental observation that a 1-substituted-1,4-dihydrobenzene should result.

Though π-excessive heteroaromatic compounds would be expected to be stable, being reluctant to accept the first electron, reduction can occur if the heterocyclic ring is annelated with a benzene ring. Thus lithium in liquid ammonia alone converts 5-methoxy-1-methylindole via the radical

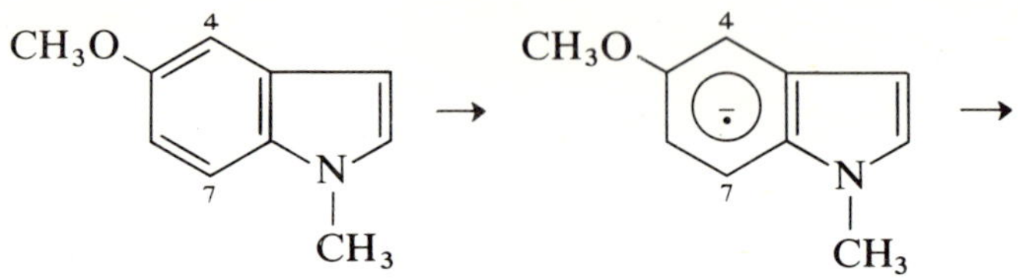

(59) (60)

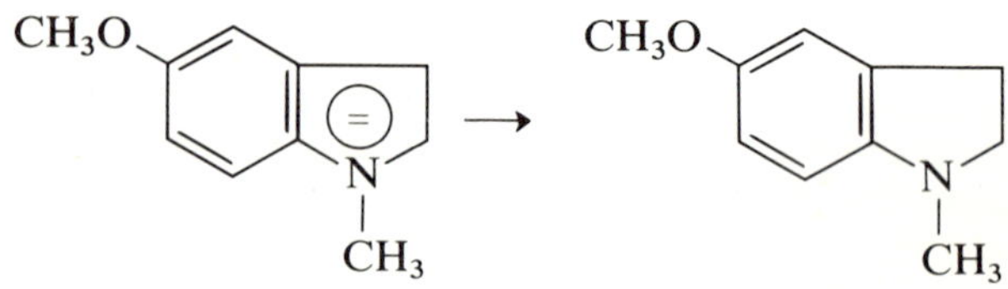

(61) (62)

anion (60) and dianion (61) to the indoline (62). Ammonia is regarded as the source of protons. If methanol be present at the start of the reduction, it intercepts the radical anion (60) and leads to the formation of the 4,7-dihydro derivative. Similar results were obtained for 6-methoxyquinoline [179]. The indole compound (59: NH for NCH_3) gave the 4,7-dihydro derivative whether methanol was present or not, presumably because the acidic NH group was converted into the anion, and this negative charge ensured that electron addition occurred in the benzene ring. Anion formation hinders the Birch reduction of phenols, but it has been proved possible, provided that a larger than usual concentration of lithium is used [180].

Dialkyl acetylenes can be reduced to *trans* olefins (64) by sodium in liquid ammonia, whereas catalytic hydrogenation tends to yield *cis*. Since the anions from terminal acetylenes cannot be reduced by this technique, this means in this case that a second electron does not add to a triple bond already carrying one negative charge (i.e., the acetylide anion). Therefore the reduction proceeds through one-electron addition and proton abstraction to yield the *trans* radical (63), which acquires a second electron and proton before inversion of any intermediate occurs.

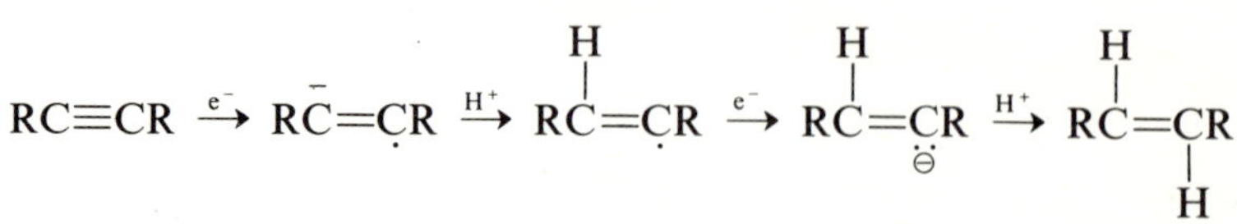

(63) (64)

Cyclic acetylenes, from cyclononyne to cyclotridecyne, afford a mixture of *cis* and *trans* olefins, the *cis* arising from reduction of an allene produced by isomerization of the original cyclic acetylene [181]. The product from diacetylenes with isolated triple bonds depends upon the molecular geometry. In the bispropargylamine (65), the t-butyl group ensures that the two unsaturated centres are near each other, so that reduction affords a cyclic product as well as the expected bis allyl compound [182].

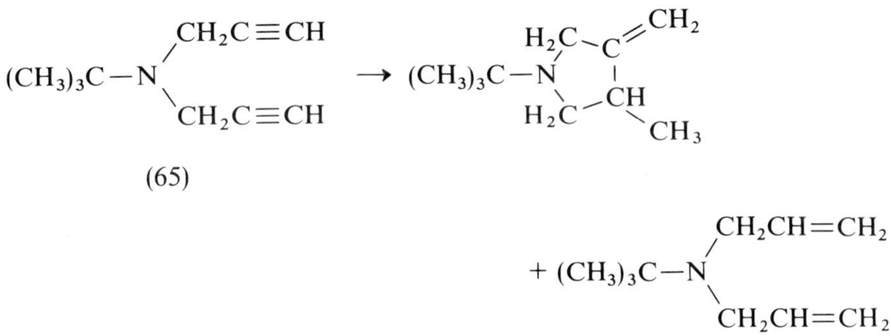

(65)

Nonconjugated olefinic double bonds are expected to be stable to solutions of alkali metals in liquid ammonia, but the strain in alkylated cyclopropene and cyclobutenes presumably ensures reduction of the double bond [183]. Only one of the double bonds in norbornadiene (66) is reduced, because of

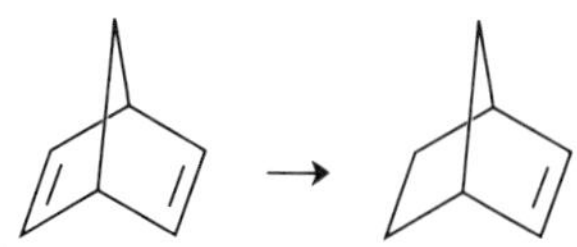

(66)

interaction (66) between the two π-electron systems [183]. However, olefinic bonds conjugated to aromatic systems or to carbonyl groups are readily reduced, and the latter system has been extensively investigated in recent years [184, 185]. The initially formed radical anion (67) abstracts a proton from the liquid ammonia to produce, once a second electron has been added,

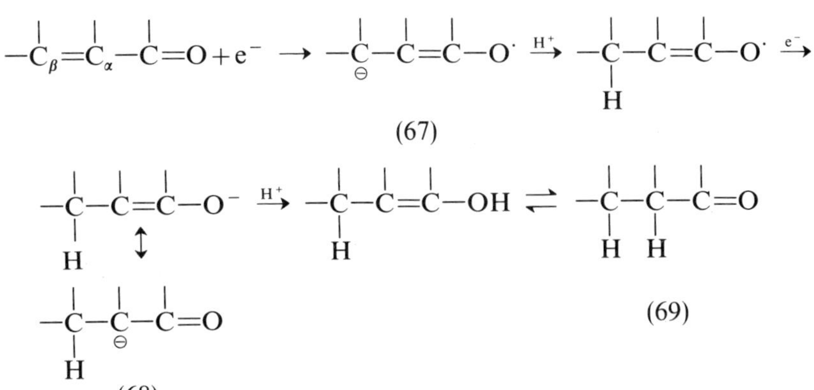

(67)

(68)

(69)

the enolate anion (68). This anion is usually too weak a base to abstract a second proton from ammonia, and further reduction involving addition of a second electron is also resisted. Intermediates like (68) have nucleophilic character and so react with carbon dioxide or methyl iodide. Thus $\alpha\beta$-unsaturated ketones like cyclohexenone (70) can be reductively methylated when methyl iodide is added to a solution of the ketone and lithium in liquid ammonia.

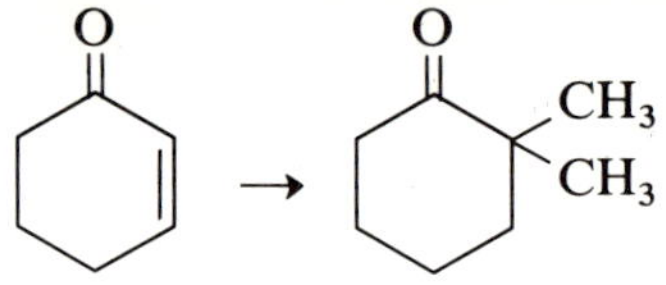

(70)

The stereochemical consequences of furnishing a more acidic proton source than ammonia (e.g., an alcohol) to the solution containing (68) have been much investigated. The product is the saturated ketone (69). In the case of the system (71), reduction proceeds through the transition state of lowest energy, which was one where the four *p*-orbitals involved overlapped. Protonation afforded (73), which was the less stable of two possible products.

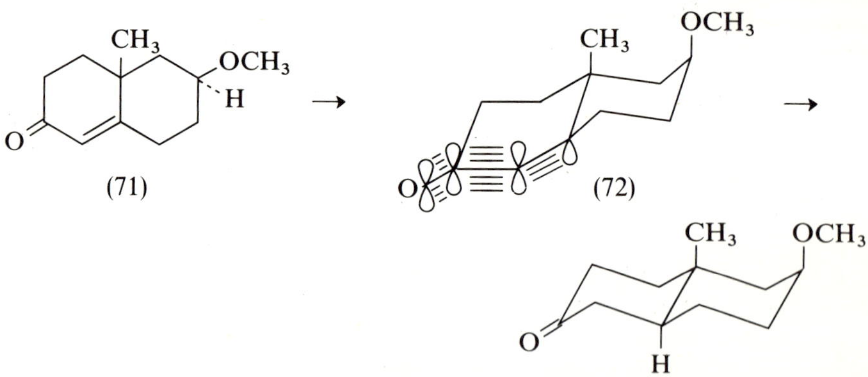

(71) (72) (73)

In the lithium ammonia reduction of 3,4-disubstituted cyclohexen-2-ones the *cis*/*trans* ratio in the product was bound up with the steric requirements of the substituents [186].

Hydrogenolysis of C—N bonds has been mentioned earlier. Other examples include both C—S and S—N cleavage [187] in tosylaminopeptides (74) and C—C bonds both in tertiary nitriles [188] (75) and in a cyclopropane

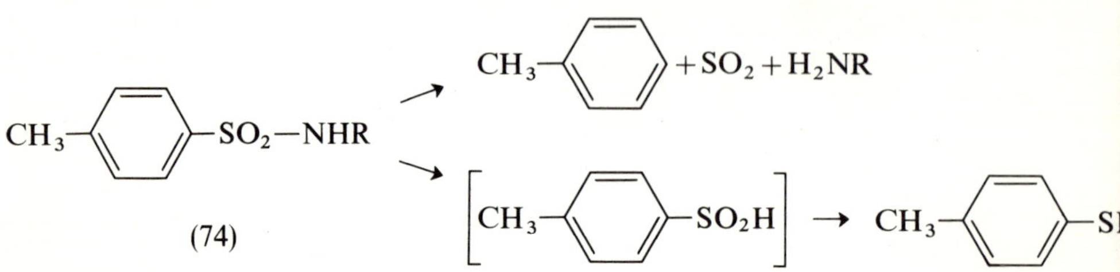

(74)

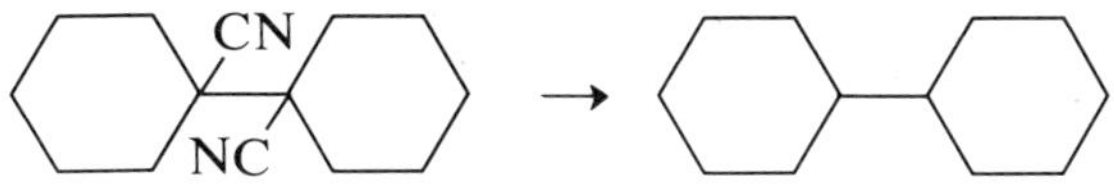

(75)

compound (76) [189], where the C—C bond having maximum overlap with

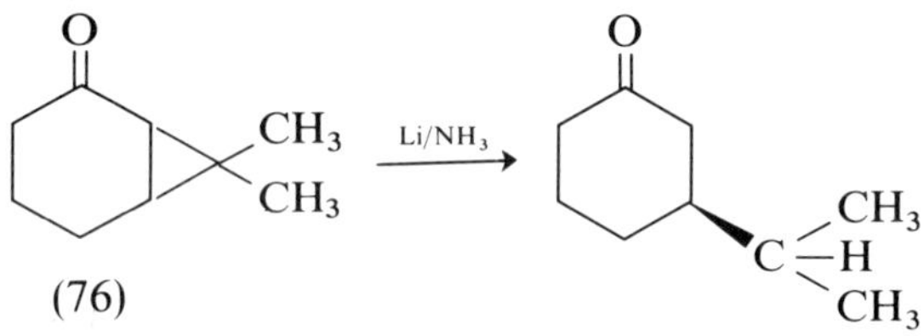

(76)

the π-orbital system of the carbonyl group was split.

Related to dissolving-metal reductions, in which the metal valency changes from zero to some positive integer, is that involving the conversion of a metallic ion such as chromium from a low valency state (2) to a higher (3). Chromous chloride reduction of 9-α-bromo-$\Delta^{1,4}$-3-ketosteroids (partial formula 77) involved the transference of one electron to the substrate, which caused the splitting off of the stable bromide ion. This left a carbon radical which formed a carbanion after acquiring a second electron from another chromous ion. The negative charge in the anion must have been delocalized, since protonation caused the formation of a cyclopropane ring [190].

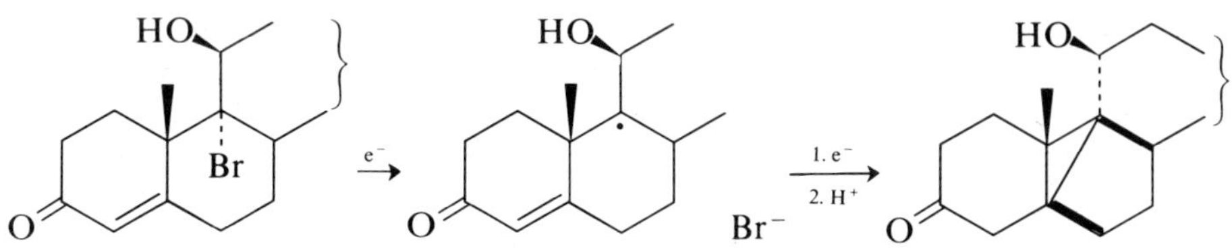

(77)

Chromous chloride has been used to reduce nitro steroids to oximino sterols [191] and acetylenes to *trans* olefins via a complex in which a hydrated chromous ion is co-ordinated on each side of the triple bond [192]. Reduction of alkyl halides involves dimerization of the intermediate free radical [193]. Magnesium in conjunction with propan-2-ol has been used for the specific replacement of halogen in aryl and alkyl halides by hydrogen. It will even reduce carbon tetrachloride to methane [194].

Cyanohydrin anions have been recently cited as one-electron reducing agents and not hydride transferring agents for the reduction of aromatic nitro groups to hydroxylamino (NHOH) groups [195].

ELECTROLYTIC REDUCTION AT A CATHODE OF HIGH OVERVOLTAGE POTENTIAL

The substrate, in solution containing a supporting electrolyte, is electrolysed, using a metal cathode of lead, mercury or some other metal of high overvoltage potential. The cathode transfers electrons to the substrate molecule adsorbed upon it and the new species goes on to acquire protons and more electrons just as with dissolving-metal reductions. The advantage of electrolytic reduction is the control that can be exercised over the reducing power of the system. This can be varied within wide limits by variation in the size and material of the electrodes, the current density and the pH of the solution. Such a feature has been realized from the early days of the method, a classical example being the electrolytic reduction of aromatic nitro compounds to amino compounds or to a variety of intermediate compounds like azo and azoxy derivatives, as pre-decided [196]. The electrolytic reduction of aliphatic nitro compounds is at present the subject of intense investigation [197]. Ketones, too, are ideal substrates for this method of reduction and the product is the pinacol, the secondary alcohol or the hydrocarbon, depending upon whether one, two or four electrons per molecule of substrate are transferred.

It has been found that at a mercury cathode non-conjugated steroidal ketones are reduced stereospecifically to the thermodynamically more stable equatorial alcohol [198], a 1,3-diketone may give the diol or pinacol [199],

$$PhCOCH_2COPh \rightarrow PhCHOHCH_2CHOHPh$$

or

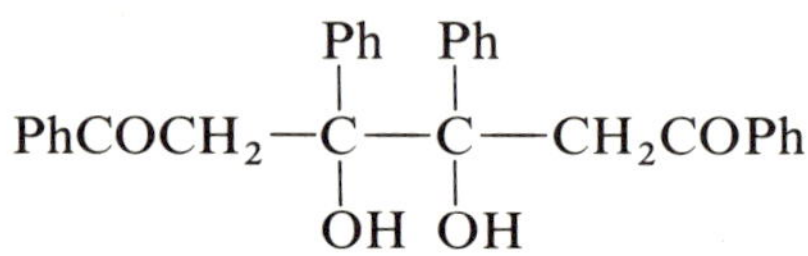

while reduction of some steroid ketones to hydrocarbons at a lead electrode has been used to introduce deuterium into the molecule [200]. Reduction of the oximes of camphor (78; R = CH_3) or norcamphor (78; R = H) at a mercury cathode [201] was explained by

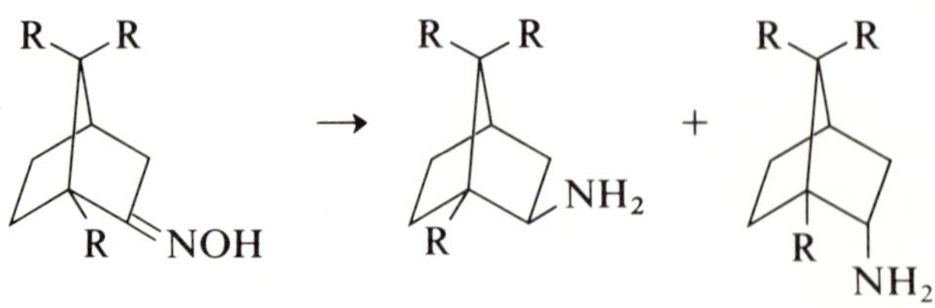

(78) (79) (80)

postulating that the less hindered side of each oxime is adsorbed on the cathode surface (*endo* for camphor; *exo* for norcamphor). Electrons are transferred and something akin to an organomercurial is formed. When this protonolyses, retention of configuration occurs, so that the *exo* amine (79; R = CH_3) is obtained with camphoroxime, but the *endo* amine (80; R = H) with norcamphoroxime. It was pointed out that sodium reductions of each system gave mixtures of the *exo* and *endo* amino compounds, so that the claim that the results of electrolytic reduction at a high-overvoltage electrode parallel those obtained with dissolving metals [37] must be modified in one case at least.

Compounds containing olefinic bonds activated by carbonyl or other groups can also accept electrons during electrolysis at a mercury cathode in a catholyte containing a quaternary ammonium *p*-toluenesulphonate. With acrylic ester (81) the radical anion (82) dimerizes and accepts two

$$\underset{(81)}{CH_2{=}CHCO_2C_2H_5} \xrightarrow{e^-} \underset{(82)}{\dot{C}H_2{-}\bar{C}HCO_2C_2H_5} \rightarrow \begin{array}{l} CH_2{-}\bar{C}HCO_2C_2H_5 \\ | \\ CH_2{-}\bar{C}HCO_2C_2H_5 \end{array} \rightarrow \underset{(83)}{\begin{array}{l} CH_2{-}CH_2CO_2C_2H_5 \\ | \\ CH_2{-}CH_2CO_2C_2H_5 \end{array}}$$

protons [202] to form adipic ester (83), while allyl cyanide (84) gives 3,4-dimethyl adiponitrile (85) by prior tautomerization [203].

$$\underset{(84)}{CH_2{=}CHCH_2CN} \rightarrow CH_3CH{=}CHCN \rightarrow \underset{(85)}{\begin{array}{l} CH_3CHCH_2CN \\ \quad\ | \\ CH_3CHCH_2CN \end{array}}$$

If the two activated double bonds are in the same molecule, cyclization [204] may occur.

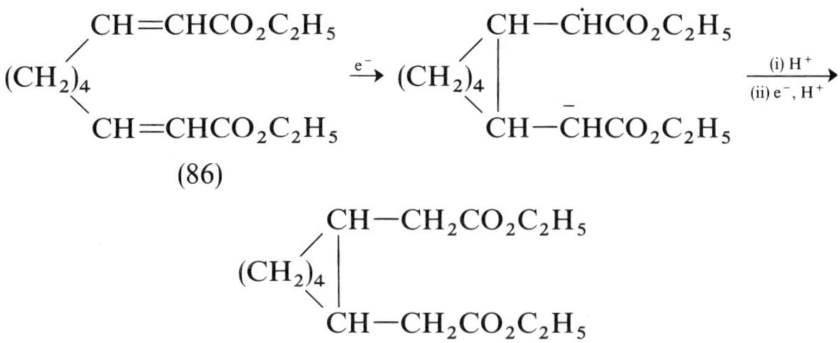

(86) (87)

Processes (81 ⟶ 83) and (84 ⟶ 85) are examples of 'intermolecular' reductive coupling, while process (86 ⟶ 87) illustrates an electrohydrocyclization technique.

Amine solvents have also been used in electrolytic cells. Electrolysis of methylamine solutions of lithium chloride containing benzene hydrocarbons between platinum electrodes leads to dihydro derivatives (88 ⟶ 89).

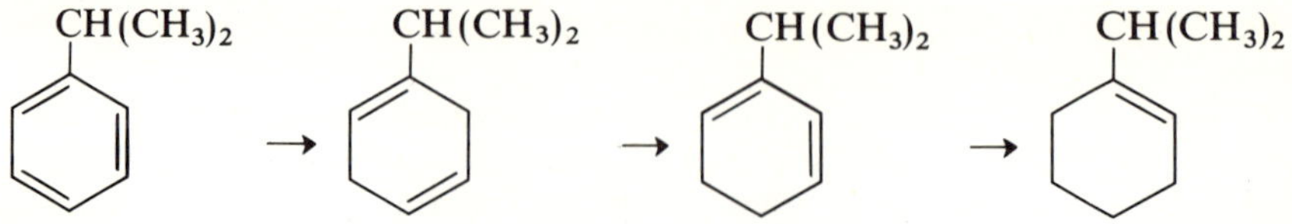

(88) (89) (90)

If an asbestos divider is placed in the cell, the lithium methylamide formed in the cathode department is kept apart from the methylamine hydrochloride formed in the anode department. It can therefore catalyse the isomerization of the dihydro species to a conjugated dihydro compound, which can be reduced further to the tetrahydro compound (90) [205]. Ethylenediamine may be substituted for methylamine [206] in reductions of a similar nature. Hexamethylphosphoramide has attracted attention for its ability to accept electrons from an aluminium cathode and release them to the benzene ring of tetralin (91) and bring about reduction to the hexa-, octa- and decahydro

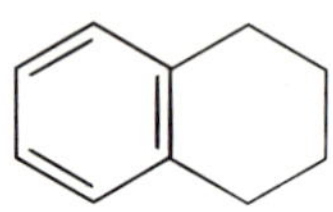

(91)

stage [207]. Hydrogenolyses have also been carried out electrochemically. The selective reduction (2.49) involving phenyl-nitrogen fission can also be effected by electrolysis of the methiodide in a liquid ammonia solution of sodium iodide between platinum electrodes [208]. Hydrogenolytic fission of carbon–halogen bonds has been observed during electrolysis of aliphatic trichloromethyl compounds [209] and optically active halogenocyclopropyl carboxylic acids [210]. In the first example, whether one or two chlorine atoms were replaced by hydrogen depended upon the supporting electrolyte. In the second, the stereochemical consequences of partial retention or inversion depended upon whether the carboxyl group was ionized.

REDUCTION INVOLVING ANIONIC ATTACK

In this mode of reduction the substrate is subject to a nucleophilic attack by an anion which is subsequently replaced by hydrogen, usually in the form of a proton. Sodium hydrazide has been discussed earlier, and this section will concentrate on the salts of sulphur oxyacids.

Sodium hydrosulphite, $Na_2S_2O_4$, is a powerful reducing agent, converting pyridinium salts into their 1,4-dihydro derivatives. The reduction proceeds through a charge-transfer complex [211] of the pyridinium cation and a sulphoxylate anion which rearranges to a 1,4-addition product (either a sulphinate or sulphoxylate) [212], subsequently protonolysed to the 1,4-dihydro compound (93).

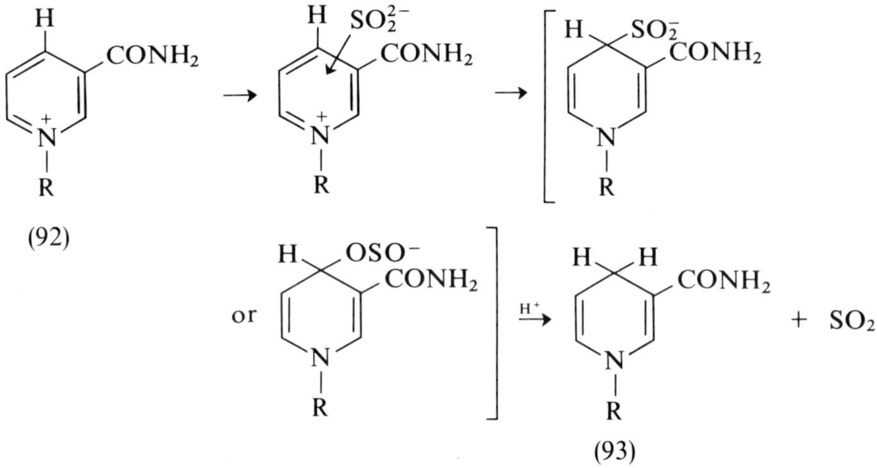

(92)

(93)

Hydrosulphite reduces an NO_2 group to NH_2 when it is attached to either aromatic [213] or heteroaromatic rings [214], although frequently a sulphonic

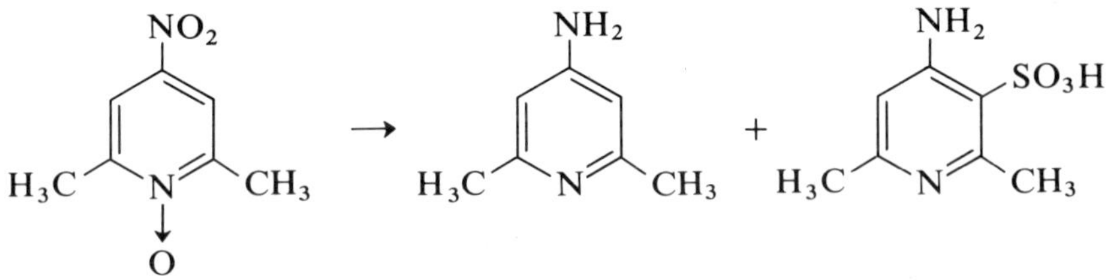

acid grouping is introduced elsewhere in the molecule. In common with sodium sulphite, sodium bisulphite and sodium metabisulphite, sodium hydrosulphite will completely reduce an aromatic ring (in oestradiol (94)) when the latter is simultaneously activated by ultraviolet light [215].

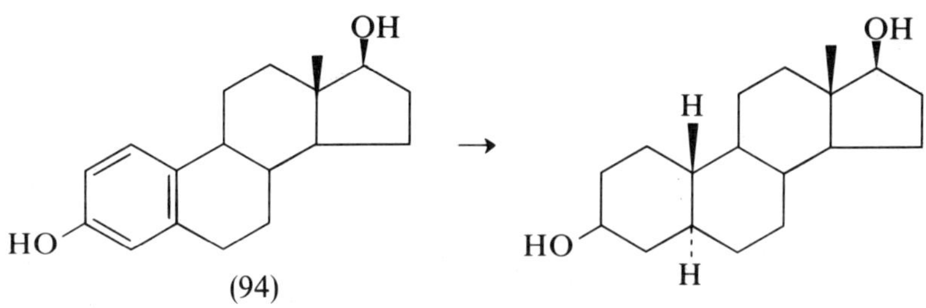

(94)

REFERENCES

[1] Evans, E. A. *Tritium and its Compounds*, pp. 99–216, Butterworths, London (1966).

[2] Evans, R. F. *Rev. Pure Appl. Chem.* **15**, 23 (1965).

[3] McQuillin, F. J. in *Techniques of Organic Chemistry* (ed. A. Weissberger), Vol. II, Part 1, pp. 497–580, Interscience, New York (1963); Augustine, R. L. *Catalytic Hydrogenation*, Arnold, London (1965).
[4] House, H. O. in *Modern Synthetic Reactions*: (a) pp. 1–22; (b) pp. 23–49; (c) pp. 50–77, Benjamin, New York (1965).
[5] Freifelder, M. *Adv. Catalysis* **14**, 203 (1963).
[6] Bond, G. C. and Wells, P. B. *Adv. Catalysis* **15**, 92 (1964).
[7] Siegel, S. *Adv. Catalysis* **16**, 123 (1966).
[8] Burwell, R. L. *Chem. Engng News* 56 (1966).
[9] Smith, G. V. and Swoap, J. R. *J. org. Chem.* **31**, 3904 (1966).
[10] Siegel, S., Thomas, P. A. and Holt, J. T. *J. Catalysis* **4**, 73 (1965).
[11] Khan, A. M., McQuillin, F. J. and Jardine, I. *J. chem. Soc.* (*C*), 136 (1967).
[12] Garbisch, E. W., Schreader, L. and Frankel, J. J. *J. Am. chem. Soc.* **89**, 4233 (1967).
[13] Mitsui, S. and Kudo, Y. *Chemy Ind.* 381 (1965); Mitsui, S., Sevda, Y. and Konno, K. *Chemy Ind.* 1354 (1963).
[14] Garbisch, E. W. *Chem. Commun.* 806 (1967).
[15] Aeberli, P. and Houlihan, W. J. *J. org. Chem.* **32**, 3211 (1967).
[16] Pazos, J. F. and Greene, F. D. *J. Am. chem. Soc.* **89**, 1030 (1967).
[17] Newham, J. *Chem. Rev.* **63**, 123 (1963).
[18] Howard, T. J. and Morley, B. *Chemy Ind.* 73 (1967).
[19] van Bekkum, H., Nieuwstad, T. J., van Barneveld, J. and Wepster, B. M. *Tetrahedron Lett.* 2269 (1967).
[20] Harada, K. *J. org. Chem.* **32**, 1790 (1967).
[21] Brown, H. C. and Brown, C. A. *J. Am. chem. Soc.* **84**, 1493, 1495, 2827, 2828 (1962); **85**, 1003, 1005 (1963).
[22] Bond, G. C. and Welster, D. E. *Proc. chem. Soc.* 398 (1964); *Chemy Ind.* 878 (1967).
[23] Broadbent, H. S. and Johnson, J. H. *J. org. Chem.* **27**, 4400 (1962); Broadbent, H. S. and Seegmiller, D. W. *J. org. Chem.* **28**, 2347 (1963).
[24] Freifelder, M., Dunningan, D. A. and Baker, E. J. *J. org. Chem.* **31**, 3438 (1966).
[25] Greenfield, H. and Dovell, F. S. *J. org. Chem.* **32**, 3670 (1967).
[26] Ruzicka, L. and van Veen, A. G. *Annln Chem.* (*Liebigs*) **468**, 135, 141 (1929).
[27] Skomoroski, R. M. and Shriesheim, A. *J. phys. Chem.* **65**, 1340 (1961); Jones, J. I. *J. chem. Soc.* 1392 (1950).
[28] Siegel, S., Smith, G. V., Dmuchovsky, B., Dubbell, D. and Halpern, W. *J. Am. chem. Soc.* **84**, 3136 (1962); Rooney, J. J. *J. Catalysis* **2**, 53 (1963).
[29] Brown, D. J. and Evans, R. F. *J. chem. Soc.* 527 (1962).
[30] Evans, R. F. *J. chem. Soc.* 2450 (1964).
[31] Evans, R. F. and Shannon, J. S. *J. chem. Soc.* 1406 (1965).
[32] Reiff, L. P. and Aaron, H. S. *Tetrahedron Lett.* 2329 (1967).
[33] Davies, W. A. M., Pinder, A. R. and Morris, I. G. *Tetrahedron Lett.* 405 (1962).
[34] Davies, W. A. M., Jones, J. B. and Pinder, A. R. *J. chem. Soc.* 3504 (1960).
[35] McQuillin, F. J. and Ord, W. O. *J. chem. Soc.* 2902 (1959).
[36] Popp, F. D. and Schultz, H. P. *Chem. Rev.* **62**, 19 (1962).
[37] Brewster, J. H. *J. Am. chem. Soc.* **76**, 6361 (1954).
[38] Lee, J. B. and Lashmore, P. *Chemy Ind.* 1758 (1966).
[39] Zanker, V. and Schnith, H. *Chem. Ber.* **92**, 2210 (1959); Zanker, V., Mader, F. and Körber, W. *Angew. Chem. Int. Edn* **3**, 388 (1964).
[40] Becker, H.-D. *J. org. Chem.* **32**, 2140 (1967).
[41] Ben-Efraim, D. A. *Tetrahedron Lett.* 957 (1967).
[42] Kaneko, C., Yokoe, I. and Yamada, S. *Tetrahedron Lett.* 775 (1967).
[43] Kendall, D. S. and Leermakers, P. A. *J. Am. Chem. Soc.* **88**, 2766 (1966).

[44] SCHMIDT, V. *Angew. Chem. int. Edn* **4**, 146 (1965).
[45] WAGNER, P. J. *J. Am. chem. Soc.* **89**, 2503 (1967).
[46] YANG, N. C. and MUROV, S. *J. Am. chem. Soc.* **88**, 2766 (1966).
[47] KIRA, A., IKEDA, Y. and KOIZUMI, M. *Bull. chem. Soc. Japan* **39**, 1673 (1966).
[48] GREENE, F. D. and LOWRY, N. N. *J. org. Chem.* **32**, 882 (1967).
[49] JENSEN, F. R. and PATTERSON, D. B. *Tetrahedron Lett.* 3837 (1966).
[50] BRESLOW, R., GROVES, J. T. and RYAN, G. *J. Am. chem. Soc.* **89**, 3073, 5048 (1967).
[51] TODD, D. *Org. React.* **4**, 378 (1948).
[52] CRAM, D. J., SAHYUN, M. R. V. and KNOX, G. R. *J. Am. chem. Soc.* **84**, 1734 (1962); GRUNDON, M. F., HENBEST, H. B. and SCOTT, M. D. *J. chem. Soc.* 1855 (1963).
[53] ALBERT, A. and CATTERALL, G. *J. chem. Soc.* (*C*) 1533 (1967).
[54] BIFFIN, M. E. C., BROWN, D. J. and LEE, T. C. *J. chem. Soc.* (*C*) 573 (1967).
[55] (a) COREY, E. J., MOCK, W. L. and PASTO, D. J. *Tetrahedron Lett.* 347 (1961).
(b) HÜNIG, S., MÜLLER, H. R. and THIER, W. *Tetrahedron Lett.* 353 (1961).
(c) VAN TAMELEN, E. E., DEWEY, R. S. and TIMMONS, R. J. *J. Am. chem. Soc.* **83**, 3725 (1961).
(d) DEWEY, R. S. and VAN TAMELEN, E. E. *J. Am. chem. Soc.* **83**, 3729 (1961).
(e) VAN TAMELEN, E. E. and TIMMONS, R. J. *J. Am. chem. Soc.* **84**, 1067 (1962).
(f) COREY, E. J. and MOCK, W. L. *J. Am. chem. Soc.* **84**, 685 (1962).
(g) VAN TAMELEN, E. E., DEWEY, R. S., LEASE, M. F. and PIRKLE, W. H. *J. Am. Chem. Soc.* **83**, 4302 (1961).
[56] HÜNIG, S., MÜLLER, H. R. and THIER W. *Angew. Chem. int. Edn* **4**, 271 (1965); MILLER, C. E. *J. chem. Educ.* **42**, 254 (1965).
[57] VAN TAMELEN, E. E., DAVIS, M. and DEEM, M. F. *Chem. Commun.* 71 (1965); CURRY, D. C., UFF, B. C. and WARD, N. D. *J. chem. Soc.* (*C*) 1120 (1967).
[58] LOOKER, J. J. *J. org. Chem.* **32**, 472 (1967).
[59] APPEL, R. and BUECHNER, W. *Annln Chem.* (*Liebigs*) **654**, 1 (1962).
[60] COREY, E. J., MOCK, W. L. and PASTO, D. J. *J. Am. chem. Soc.* **83**, 2957 (1961).
[61] VAN TAMELEN, E. E. and BURKOTH, T. L. *J. Am. chem. Soc.* **89**, 151 (1967).
[62] GARBISCH, E. W., SCHILDCROUT, S. M., PATTERSON, D. B. and SPHRECHER, C. M. *J. Am. chem. Soc.* **87**, 2932 (1965).
[63] BAIRD, W. C., FRANZUS, B. and SURRIDGE, J. H. *J. Am. chem. Soc.* **89**, 410 (1967).
[64] FURST, A., BERLO, R. C. and HOOTON, S. *Chem. Rev.* **65**, 51 (1965); EVANS, R. F. *Rev. pure appl. Chem.* **12**, 146 (1962).
[65] KAUFFMAN, T. *Angew. Chem. int. Edn* **3**, 342 (1964).
[66] LAUER, J. *Annls Chim.* **10**, 301 (1965); COLLMAN, J. P. in *Transition Metal Chemistry, A Series of Advances* (ed. R. L. Carlin), Vol. 2, pp. 79–89, Dekker, New York (1966).
[67] HALPERN, J. *Chem. Engng News* 68 (1966).
[68] (a) YOUNG, J. F., OSBORN, J. A., JARDINE, F. H. and WILKINSON, G. *Chem. Commun.* 131 (1965).
(b) OSBORN, J. A., JARDINE, F. H., YOUNG, J. F. and WILKINSON, G. *J. chem. Soc.* (*A*) 1711 (1966).
(c) JARDINE, F. H. and WILKINSON, G. *J. chem. Soc.* (*A*) 270 (1967).
(d) JARDINE, F. H., OSBORN, J. A. and WILKINSON, G. *J. chem. Soc.* (*A*) 1574 (1967).
[69] BIRCH, A. J. and WALKER, K. A. M. *Tetrahedron Lett.*: (a) 4939 (1966); (b) 1935 (1967); (c) 3457 (1967); (d) *J. chem. Soc.* (*C*) 1894 (1966).
[70] BIELLMANN, J.-F. and LIESENFELT, H. (a) *Compt. rend. Acad. Sci., Paris* **263**, 251 (1966); (b) *Bull. Soc. chim. Fr.* 4029 (1966).
[71] BROWN, M. and PIZKIEWICZ, L. W. *J. org. Chem.* **32**, 2013 (1967).
[72] DJERASSI, C. and GUTZWILLER, J. *J. Am. chem. Soc.* **88**, 4537 (1966).
[73] JARDINE, I. and MCQUILLIN, F. J. *Tetrahedron Lett.* 4871 (1966).

[74] MISONO, A., UCHIDA, Y., SAITO, T. and SONG, K. M. *Chem. Commun.* 419 (1967).
[75] COFFEY, R. S. *Chem. Commun.* 923 (1967).
[76] HADDAD, Y. M. Y., HENBEST, H. B., HUSBANDS, J. and MITCHELL, T. R. B. *Proc. chem. Soc.* 361 (1964); TROCHA-GRIMSHAW, T. and HENBEST, H. B. *Chem. Commun.* 544 (1967).
[77] BAILAR, J. C. and ITATANI, H. *J. Am. chem. Soc.*, **89**, 1592, 1600 (1967).
[78] EMKEN, E. A., FRANKEL, E. N. and BUTTERFIELD, R. O. *J. Am. Oil Chem. Soc.* **43**, 14 (1966).
[79] FRANKEL, E. N., JONES, E. P., DAVISON, V. L., EMKEN, E. and DUTTON, H. J. *J. Am. Oil Chem. Soc.* **42**, 130 (1965).
[80] FRANKEL, E. N., EMKEN, E. A. and DAVISON, V. L. *J. org. Chem.* **30**, 2739 (1965).
[81] ALDRIDGE, C. L. and JONASSEN, H. B., *J. Am. chem. Soc.* **85**, 886 (1963).
[82] FU, Y. C., GREENFIELD, H., METTIN, S. J. and WENDER, I. *J. org. Chem.* **32**, 2837 (1967).
[83] KWIATEK, J. and SEYLER, J. K. *J. organometall. Chem.* **3**, 421 (1965); KWIATEK, J., MADOR, I. L. and SEYLER, J. K. *J. Am. chem. Soc.* **84**, 304 (1962).
[84] U.S. 3,205,217 and U.S. 3,205,266.
[85] GAYLORD, N. G. *Reduction with Complex Metal Hydrides*, Wiley-Interscience, New York (1956); BROWN, W. G. *Org. React.* **6**, 469 (1951); BROWN, H. C. and ZWEIFEL, G. *Org. React.* **13**, 1 (1963); BROWN, H. C. *Hydroboration*, Benjamin, New York (1962); LYLE, R. E. and ANDERSON, P. S. *Advances in Heterocyclic Chemistry*, Academic Press, New York, **6**, 46 (1966).
[86] WOODWARD, R. B., CAVA, M. P., OLLIS, W. D., HUNGER, A., DAENIKER, H. U. and SCHENKER, K. *Tetrahedron* **19**, 247 (1963).
[87] BROWN, H. C. and SUBBA RAO, B. C. *J. Am. chem. Soc.* **80**, 5377 (1958).
[88] WEISSMAN, P. M. and BROWN, H. C. *J. org. Chem.* **31**, 283 (1966).
[89] ČERVINKA, O., *Colln. Czech. chem. Commun.:* (a) **26**, 673 (1961); (b) **30**, 1684 (1965); (c) *Tetrahedron Lett.* 1179 (1967); (d) LANDOR, S. R. and REGAN, J. P. *J. chem. Soc.* (C) 1159 (1967); (e) LANSBURY, P. T. and PETERSON, J. O. *J. Am. chem. Soc.* **85**, 2236 (1963).
[90] MARSHALL, J. A., ANDERSON, N. H. and HOCHSTETTER, A. R. *J. org. Chem.* **32**, 113 (1967).
[91] KAISER, C., BURGER, A., ZIRNGIBL, L., DAVIS, C. S. and ZIRKLE, C. L. *J. org. Chem.* **27**, 768 (1962); ZAUGG, H. E., DE NET, R. W. and MICHAELS, R. J. *J. org. Chem.* **28**, 1795 (1963).
[92] KARABATSOS, G. J., SHONE, R. L. and SCHEPPLE, S. E. *Tetrahedron Lett.* 2113 (1964); CORBETT, J. F. and HOLT, P. F. *J. chem. Soc.*, 2385 (1963).
[93] ALLRED, E. L. and WINSTEIN, S. *J. Am. chem. Soc.* **89**, 4008 (1967).
[94] SHEEHAN, J. C. and LENGYEL, I. *J. org. Chem.* **31**, 4244 (1966).
[95] LANSBURY, P. T., SHARF, D. J. and PATTISON, V. A. *J. org. Chem.* **32**, 1748 (1967).
[96] PACQUETTE, L. A. YOUSSEF, A. A. and WISE, M. L. *J. Am. chem. Soc.* **89**, 5246 (1967).
[97] SNYDER, E. I. *J. org. Chem.* **32**, 3531 (1967).
[98] SLAUGH, L. H. *J. org. Chem.* **32**, 109 (1967); *Tetrahedron* **22**, 1741 (1966); MAGOON, E. F. and SLAUGH, L. H. *Tetrahedron* **23**, 4509 (1967).
[99] ANDERSON, P. S. and LYLE, R. E. *Tetrahedron Lett.* 153 (1964).
[100] CLARKE, G. M. and SYKES, P. *J. chem. Soc.* (C) 1411 (1967).
[101] MARSHAL, J. A. and JOHNSON, W. S. *J. org. Chem.* **28**, 421 (1963).
[102] EVANS, R. F. *Austr. J. Chem.* **20**, 1643 (1967); WILSON, E. M. *Chemy Ind.* 471 (1965).
[103] KADIN, S. B. *J. org. Chem.* **31**, 620 (1966).
[104] BROWN, M. S. and RAPOPORT, H. *J. org. Chem.* **28**, 3261 (1963).
[105] BROWN, H. C. and BELL, H. M. *J. org. Chem.* **27**, 1928 (1962); *J. Am. chem. Soc.* **88**, 1473 (1966).

[106] JOHNSON, C. R. and PHILLIPS, W. G. *J. org. Chem.* **32**, 3233 (1967).
[107] KAPLAN, L. A. *J. Am. chem. Soc.* **86**, 740 (1964).
[108] HUFFMAN, J. W. *J. org. Chem.* **28**, 601 (1963); HENDRICKSON, J. B. *J. Am. chem. Soc.* **83**, 1251 (1961).
[109] MATSUMURA, S., NAGAI, T. and TOKURA, N. *Tetrahedron Lett.* 3929 (1966).
[110] KUNIEDA, T. and WITKOP, B. *J. Am. chem. Soc.* **89**, 4243 (1967).
[111] ELIEL, E. L., DELLA, E. W. and ROGIĆ, M. M. *J. org. Chem.* **27**, 4712 (1962).
[112] SEVERIN, T. and ADAM, M. *Chem. Ber.* **96**, 448 (1963).
[113] WINKHAUS, G. *Z. anorg. Chem.* **319**, 404 (1963).
[114] YAMADA, SHUN-ICHI and KIKUGAWA, Y. *Chemy Ind.* 2169 (1966); 1325 (1967).
[115] VLCEK, A. A. and RUSINA, A. *Proc. chem. Soc.* 162 (1961).
[116] COUTTS, R. T. and WIBBERLEY, D. G. *J. chem. Soc.* 4610 (1963).
[117] YONEMITSU, O., CERUTTI, P. and WITKOP, B., *J. Am. chem. Soc.* **88**, 3941 (1966).
[118] WATERS, J. A. and WITKOP, B. *J. Am. chem. Soc.* **89**, 1022 (1967).
[119] DRISCOLL, J. S. and MATTHEWS, C. N. *Chemy Ind.* 1282 (1963).
[120] SULLIVAN, E. A. and HINCKLEY, A. A. *J. org. Chem.* **27**, 3731 (1962).
[121] BROWN, H. C. and MUZZIO, J. *J. Am. chem. Soc.* **88**, 2811 (1966).
[122] CAGLIOTI, L. and GRASSELLI, P. *Chemy Ind.* 153 (1964).
[123] FREEGUARD, G. F. and LONG, L. H. *Chemy Ind.* 471 (1965).
[124] STEFANOVIĆ, M. and LAJŠIC, S. *Tetrahedron Lett.* 1777 (1967).
[125] WECHTER, W. J. *J. org. Chem.* **28**, 2935 (1963).
[126] BREUER, E. *Tetrahedron Lett.* 1849 (1967).
[127] COPE, A. C., BERCHTOLD, G. A., PETERSON, P. E. and SHARMAN, S. H. *J. Am. Chem. Soc.* **82**, 6370 (1960).
[128] BROWN, H. C. and VARMA, V. *J. Am. chem. Soc.* **88**, 2871 (1966).
[129] GINDICI, T. A. and FLUHARTY, A. L. *J. org. Chem.* **32**, 2043 (1967).
[130] FEUER, H., VINCENT, B. F. and BARTLETT, R. S. *J. org. Chem.* **30**, 2877 (1965); HASSNER, A. and CATSOULACOS, P. *Chem. Commun.* 590 (1967).
[131] RABINOVITZ, M., SALEMNIK, G. and BERGMAN, E. D. *Tetrahedron Lett.* **34**, 3271 (1967).
[132] DAVIES, R. E., BROWN, A. E., HOPMANN, R. and KIBBY, C. L. *J. Am. chem. Soc.* **85**, 487, (1963).
[133] BROWN, H. C. and YOON, N. M. *J. Am. chem. Soc.* **88**, 1464 (1966).
[134] JORGENSEN, M. J. *Tetrahedron Lett.* 559 (1962).
[135] FERLES, M., HAVEL, M. and TESAŘOVÁ, A. *Colln. Czech. chem. Commun.* **31**, 4121 (1966).
[136] HOLIK, M. and FERLES, M. *Colln. Czech. chem. Commun.* **32**, 3067 (1967).
[137] RICKBORN, B. and LAMKE, W. E. *J. org. Chem.* **32**, 537 (1967); RICKBORN, B. and QUARTUCCI, J. *J. org. Chem.* **29**, 3185 (1964).
[138] ELIEL, E. L. *Rec. chem Prog.* **22**, 129 (1961).
[139] ASHBY, E. C. and PRATHER, J. *J. Am. chem. Soc.* **88**, 729 (1966).
[140] ELIEL, E. L., BADDING, V. G. and RERICK, M. N. *J. Am. chem. Soc.* **84**, 2371 (1962).
[141] BREWSTER, J. H. and BAYER, H. O. *J. org. Chem.* **29**, 105 (1964).
[142] BADGER, G. M., CROCKER, H. P., ENNIS, B. C., GAYLER, J. A., MATTHEWS, W. E., RAPER, W. G. C., SAMUEL, E. L. and SPOTSWOOD, T. M. *Austr. J. Chem.* **16**, 814 (1963).
[143] TROFIMENKO, S. *J. Am. chem. Soc.* **85**, 1357 (1963).
[144] DENO, N. C., PETERSON, H. J. and SAINES, G. S. *Chem. Rev.* **60**, 7 (1960).
[145] WALLENFELS, K. and SCHULZ, H. *Annln Chem.* (*Liebigs*) **621**, 178 (1959).
[146] SCHELLENBERG, K. A. and MCLEAN, G. W. *J. Am. chem. Soc.* **88**, 1077 (1966).
[147] BRAUDE, E. A., HANNAH, J. and LINSTEAD, R. *J. chem. Soc.* 3249 (1960).
[148] ČERVINKA, O. and KŘÍŽ, O. *Colln. Czech. chem. Commun.* **30**, 1700 (1965).
[149] KLEINFELTER, D. C. *J. org. Chem.* **32**, 840 (1967).

[150] BACON, R. G. R. and HILL, H. A. O. *Q. Rev. chem. Soc.* **19**, 95 (1965).
[151] WILDS, A. L. *Org. React.* **2**, 178 (1944).
[152] ČERVINKA, O., BĚLOVSKY, O., FÁBRYOVÁ, A., DUDEK, V. and GROHMAN, K. *Colln. Czech. chem. Commun.* **32**, 2618 (1967).
[153] STREITWIESER, A., Jr. and GRANGER, M. R. *J. org. Chem.* **32**, 1528 (1967).
[154] SNYDER, C. H. *J. org. Chem.* **31**, 4220 (1966).
[155] GÜNTHER, W. H. H. *J. org. Chem.* **31**, 1202 (1966).
[156] ROBERTSON, A. V. and WITKOP, B. *J. Am. chem. Soc.* **84**, 1697 (1962).
[157] BIRCH, A. J., *Q. Rev. chem. Soc.* **4**, 69 (1950); BIRCH, A. J. and SMITH, H. *Q. Rev. chem. Soc.* **12**, 17 (1958); SMITH, H. *Organic Reactions in Liquid Ammonia*, Vol. I, part 2, *Chemistry in Non-Aqueous Ionizing Solvents*, Wiley, New York (1963).
[158] KRAPCHO, A. P. and BOTHNER-BY, A. A. *J. Am. chem. Soc.* **81**, 3658 (1959).
[159] CARRINGTON, A. *Q. Rev. chem. Soc.* **17**, 67 (1963); MCCLELLAND, B. J. *Chem. Rev.* **64**, 301 (1964).
[160] KOSOWER, E. M. and POZIOMEK, E. J. *J. Am. chem. Soc.* **85**, 2035 (1963).
[161] KATZ, T. J. *J. Am. chem. Soc.* **82**, 3784 (1960).
[162] CLOSSON, W. D., WRIEDE, P. and BANK, S. *J. Am. chem. Soc.* **88**, 1581 (1966).
[163] MURRAY, R. W. and KAPLAN, M. L. *J. org. Chem.* **31**, 962 (1966).
[164] *Org. Synth.*, Coll. Vol. **I**, 459 (1941).
[165] HOUSE, H. O., MÜLLER, H. C., PITT, C. G. and WICKHAM, P. P. *J. org. Chem.* **28**, 2407 (1963).
[166] EAKIN, M., MARTIN, J. and PARKER, W. *Chem. Commun.* 206 (1965).
[167] BREWSTER, J. H. *J. Am. chem. Soc.* **76**, 6364 (1954).
[168] MARTIN, E. L. *Org. React.* **1**, 155 (1942).
[169] ENZEL, C. R. *Tetrahedron Lett.* 1285 (1966).
[170] KAPLAN, M. L. *J. org. Chem.* **32**, 2346 (1967); ST. C. BUCHANAN, J. G. and DAVIS, B. R. *J. chem. Soc.* (C) 1340 (1967).
[171] HUDSON, C. B. and ROBERTSON, A. V. *Tetrahedron Lett.* 4015 (1967).
[172] KRAPCHO, A. P. and BOTHNER-BY, A. A. *J. Am. chem. Soc.* **82**, 751 (1960).
[173] SLAUGH, L. H. and RALEY, J. H. *J. org. Chem.* **32**, 369 (1967).
[174] HARVEY, R. G. *J. org. Chem.* **32**, 238 (1967).
[175] WRÓBEL, J. T., BIEN, A. S. and PAZDRO, K. M. *Chemy Ind.* 1759 (1966).
[176] SLAUGH, L. H. and RALEY, J. H. *J. org. Chem.* **32**, 2861 (1967).
[177] STOLOW, R. D. and WARD, J. A. *J. org. Chem.* **31**, 965 (1966).
[178] ZIMMERMAN, H. E. *Tetrahedron* **16**, 169 (1961); STREITWIESER, A., Jr. *Molecular Orbital Theory for Organic Chemists*, pp. 425–431, Wiley, New York (1962).
[179] REMERS, W. A., GIBS, G. J., PIDACKS, C. and WEISS, M. J. *J. Am. chem. Soc.* **89**, 5513 (1967).
[180] FRIED, J., ABRAHAM, N. A. and SANTHANAKRISHNAN, T. S. *J. Am. chem. Soc.* **89**, 1044 (1967).
[181] SVOBODA, M., ZÁVADA, J. and SICHER, J. *Colln. Czech. chem. Commun.* **30**, 413 (1965).
[182] HENNION, G. P. and ODE, R. H. *J. org. Chem.* **31**, 1975 (1966).
[183] DE MONTELLANO, O., LOVING, B. A., SHIELDS, T. C. and GARDNER, P. D. *J. Am. chem. Soc.* **89**, 3365 (1967).
[184] SMITH, H. A., HUFF, B. J. L., POWERS, W. J. and CAINE, D. *J. org. Chem.* **32**, 2851 (1967); STORK, G., GOLDMAN, N., COOMBS, R. V. and TSUJI, J. *J. Am. chem. Soc.* **87**, 278 (1965).
[185] STORK, G. and DARLING, S. D. *J. Am. chem. Soc.* **86**, 1761 (1964).
[186] MALHOTRA, S. K., MOAKLEY, D. F. and JOHNSON, F. *Tetrahedron Lett.* 1089 (1967).
[187] KOVACS, J. and GHATAK, U. R. *J. org. Chem.* **31**, 119 (1966).
[188] ARAPAKOS, P. G. *J. Am. chem. Soc.* **89**, 6795 (1967).

[189] DAUBEN, W. G. and DEVINY, E. J. *J. org. Chem.* **31**, 3794 (1966).
[190] ROBINSON, C. H., GNOJ, O., OLIVETO, E. P. and BARTON, D. H. R. *J. org. Chem.* **31**, 2749 (1966).
[191] HANSON, J. R. and PREMUZIC, E. *Tetrahedron Lett.* 5441 (1966).
[192] CASTRO, C. E. and STEPHENS, R. D. *J. Am. chem. Soc.* **86**, 4359 (1964).
[193] CASTRO, C. E. and KRAY, W. C. *J. Am. chem. Soc.* **85**, 2768 (1963).
[194] BRYCE-SMITH, D., WAKEFIELD, B. J. and BLUES, E. T. *Proc. chem. Soc.* 219 (1963).
[195] KRAMER, D. N., GUILBAULT, G. G. and MILLER, F. M. *J. org. Chem.* **32**, 1163 (1967).
[196] HABER, F. *Z. phys. Chem.* **32**, 193 (1900); quoted in SIDGWICK, N. V. *The Organic Chemistry of Nitrogen*, 3rd ed, by I. T. Millar and H. D. Springall, pp. 388–389, Clarendon Press, Oxford (1966).
[197] INVERSEN, P. E. and LUND, H. *Tetrahedron Lett.* 4027 (1967); MASUI, M., SAYO, H. and KISHI, K. *Tetrahedron* **21**, 2831 (1965).
[198] KABASAKALIAN, P., MCGLOTTEN, J., BASCH, A. and YUDS, M. D. *J. org. Chem.* **26**, 1738 (1961).
[199] EVANS, D. H. and WOODBURY, E. C. *J. org. Chem.* **32**, 2158 (1967).
[200] THROOP, L. and TÖKEŚ, L. *J. Am. chem. Soc.* **89**, 4789 (1967).
[201] FRY, A. J. and NEWBERG, J. H. *J. Am. chem. Soc.* **89**, 6374 (1967).
[202] BAISER, M. M. *Tetrahedron Lett.* 973 (1963).
[203] ORT, M. R. and BAISER, M. M. *J. org. Chem.* **31**, 1646 (1966).
[204] ANDERSON, J. D., BAISER, M. M. and PETROVICH, J. P. *J. org. Chem.* **31**, 3890 (1966).
[205] BENKESER, R. A. and KAISER, E. M. *J. Am. chem. Soc.* **85**, 2858 (1963); **86**, 5272 (1964).
[206] STERNBERG, H. W., MARKBY, R. E. and WENDER, I. *J. electrochem. Soc.* **110**, 425 (1963).
[207] STERNBERG, H. W., MARKBY, R. E., WENDER, I. and MOHILNER, D. M. *J. Am. chem. Soc.* **89**, 186 (1967).
[208] WRÓBEL, J. T., BIEŃ, A. S. and PAZDRO, K. M. *Chemy Ind.* 1760 (1966).
[209] NAGAO, M., SATO, N., AKASHI, T. and YOSHIDA, T. *J. Am. chem. Soc.* **88**, 3447 (1966).
[210] ANNINO, R., ERICKSON, R. E., MICHALOVIC, J. and MCKAY, B. *J. Am. chem. Soc.* **88**, 4424 (1966).
[211] KOSOWER, E. M. and BAUER, S. W. *J. Am. chem. Soc.* **82**, 2191 (1960).
[212] BIELLMANN, J.-F. and CALLOT, H. *Tetrahedron Lett.* 3991 (1966); CAUGHEY, W. S. and SCHELLENBERG, K. A. *J. org. Chem.* **31**, 1978 (1966).
[213] FISCHER, F. G., EILINGSFELD, H. and NEUMANN, D. P. *Annln Chem.* (*Liebigs*) **651**, 49 (1962).
[214] EVANS, R. F. and BROWN, H. C. *J. org. Chem.* **27**, 1665 (1962).
[215] WATERS, J. A. and WITKOP, B. *J. Am. chem. Soc.* **89**, 1022 (1967).

3

Oxidation Methods

R. G. R. Bacon, *Queen's University, Belfast*

The practice and theory of oxidation processes have been developing over almost two centuries, and the scope of the subject is now extremely wide. Organic oxidation is briefly definable as loss of hydrogen or gain of oxygen by the substrate. It is applicable to virtually every type of carbon compound, and it variously takes the form of an addition, elimination, substitution, fission, coupling, or rearrangement process. Similarly, there is great diversity in the nature and behaviour of the oxidants employed in organic chemistry. They include some organic species as well as a very wide range of inorganic species. One-electron or two-electron transfer steps may be discerned in the oxidation–reduction process, and it often results in gain of hydrogen or loss of oxygen by the oxidant.

In cases under consideration, oxidation–reduction most commonly occurs between an organic and an inorganic compound in a solvent, and often in the presence of additional reagents or catalysts. An adequate understanding of such reactions requires knowledge of the parts played by all components of the system. Accordingly, oxidation provides one of the main areas in chemistry where the traditional separation of organic from inorganic topics is most rapidly disappearing.

Studies of oxidation mechanisms, particularly during the last two decades, have brought substantial progress in the formulation of co-ordinating principles. Consequently, in the period under review, which is roughly since 1960, various authors have been stimulated to review oxidation comprehensively in monographs [1]. Also, the need for detailed and systematic accounts of applications and modes of action of the commoner oxidants used in organic chemistry is being met by the publication of a work with multiple authorship [2]; otherwise, information has to be sought in the original literature or in reviews of varying date. A large amount of data concerning oxidants is included in the monumental survey of organic reagents recently produced by L. F. and M. Fieser [3]. Selected references to current investigations of oxidation processes appear in some annual compilations [4].†

† In view of the very wide scope of this chapter, review articles are given prominence among the references. Accordingly, numerous important papers have not been directly cited.

The range of organic oxidation methods is continually enlarged by the introduction, often for specific types of reaction, of compounds not previously exploited for the purpose. Whereas, for example, the use of permanganate and chromic acid was established during the 19th century, and the appearance of selenium dioxide (introduced in 1932) [5] and periodate (introduced in 1928) [6] occurred in recent decades, the present review includes reference to ruthenium tetroxide, thallium(III) salts, palladium(II) salts, carbonium ions, and other oxidants, which have much more recently attracted attention. The scope of organic oxidation methods is also extendable by the use of particular techniques, exemplified in this review by photoactivation, heterogeneous and homogeneous catalysis, and electrolytic procedures.

In the present survey of oxidation methods, subdivisions have been made, as follows, in accordance with the different types of oxidants employed.

(1) Oxygen, and related reagents containing oxygen–oxygen bonds, i.e., ozone, and inorganic and organic peroxidic compounds.

(2) Some non-metallic elements, oxides, and oxyanions, from positions neighbouring oxygen (Groups V, VI, and VII) in the Periodic Table.

(3) Salts, complexes, oxides, and oxyanions of transition and post-transition metals in their higher oxidation states. In the same section mention is made of some catalytic and electrolytic processes of oxidative type, which occur on metal surfaces.

(4) Organic compounds capable of acting oxidatively, either by hydrogen-abstraction or by oxygen-donation.

3.1 OXYGEN, OZONE, AND PEROXIDIC COMPOUNDS

(a) Oxygen

Since the 1940s one of the leading contributions to knowledge of radicals and homolytic processes in chemistry has come from studies of autoxidation, i.e., absorption of oxygen by organic substrates at moderate temperatures, frequently leading to hydroperoxides or their breakdown products [7]. Catalysis is of paramount importance, both in autoxidations and in reactions performed with molecular oxygen under more drastic conditions. Various types of catalysis are illustrated in the following examples of recent investigations.

(i) Numerous organic compounds are produced industrially by gaseous-phase reactions of oxygen with alkanes, alkenes, and aromatic hydrocarbons, typically at temperatures in the range 250–450°, under the influence of specific heterogeneous catalysts [8]. Silver, for example, has been established (since 1931) as a unique catalyst for the conversion of ethylene into ethylene oxide.

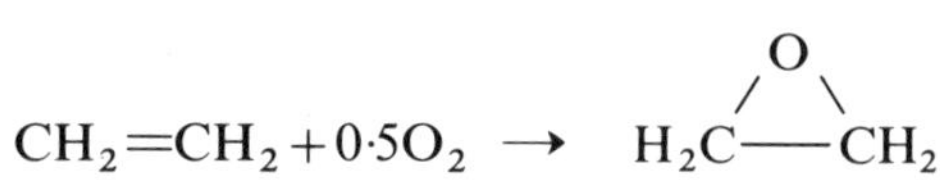

In the case of propene, however, recent investigations have shown that it may be efficiently converted into acrolein with the aid of catalysts based either on cuprous oxide or on certain mixed oxides, e.g., those of bismuth and molybdenum:

$$CH_2{=}CH{-}CH_3 + O_2 \rightarrow CH_2{=}CH{-}CHO + H_2O$$

R. S. Mann and D. J. Rouleau [9] have discussed the analogous conversion of isobutene into methacrolein.

$$CH_2{=}C(CH_3)_2 + O_2 \rightarrow CH_2{=}C(CH_3){-}CHO + H_2O$$

If ammonia is present during the oxidation of propene, an important synthesis of acrylonitrile results.

$$CH_2{=}CH{-}CH_3 + NH_3 + 1{\cdot}5O_2 \rightarrow CH_2{=}CH{-}CN + 3H_2O$$

(ii) Heterogeneous catalysis is also effective in promoting some liquid-phase oxidations by oxygen at relatively low temperatures. An example given in *Organic Syntheses* [10], again demonstrating the utility of metals of Group IB, is the conversion of furfuraldehyde into furoic acid, at 55°, in aqueous alkali containing a suspension of mixed silver(I) and copper(I) oxides.

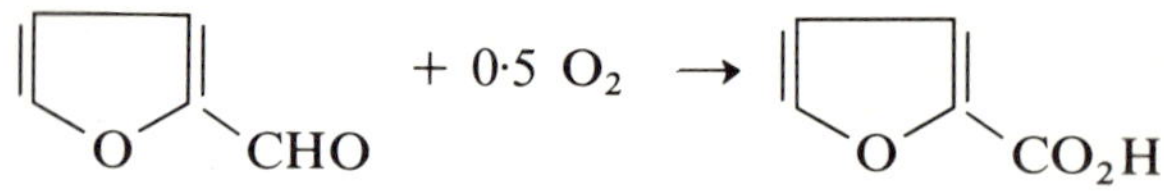

A technique which has found considerable use in recent years [11] consists of treating a compound with oxygen at 20–60° in water or an organic solvent containing a stirred suspension of a platinum catalyst. Historically, this method stems from H. Wieland's experiments in the 1920s, on which he based his advocacy of organic oxidations as hydrogen-transfer processes, promotable by suitable catalysts. Primary or secondary alcohols are thus dehydrogenated, respectively, to aldehydes (further oxidizable to carboxylic acids) or to ketones, molecular oxygen serving as the hydrogen-acceptor.

$$RR'CH{-}OH + 0{\cdot}5O_2 \rightarrow RR'CO + H_2O$$

This procedure has found special application in carbohydrate chemistry, where its selectivity is useful; aldoses are converted into aldonic acids, while glycosides (by oxidation of the primary hydroxyl group) give the corresponding uronides. Cyclic polyhydric alcohols, e.g., the stereoisomeric inositols, undergo preferential oxidation at axial hydroxyl groups.

(iii) It is well known that various transition-metal ions in solution function as homogeneous catalysts for reactions of molecular oxygen, e.g., for the oxygenation of alkyl side chains in aromatic hydrocarbons. A case which has been much studied in industrial laboratories in recent years is the conversion of *p*-xylene, by a sequence of steps, into terephthalic acid, with the aid of cobalt and bromide ions; a complicated series of oxidation-reduction reactions, involving metal ions, peroxides, organic radicals, and bromine atoms, has been postulated.

Catalysis by copper salts has recently acquired special prominence in relation to liquid-phase oxidation by oxygen. W. Brackman and his co-workers [12] have described cases such as the following: the conversion of aldehydes, mixed with ammonia, into cyanides:

$$PhCHO + NH_3 + 0{\cdot}5O_2 \rightarrow PhCN + 2H_2O;$$

oxidation of unsaturated carbonyl compounds in the allylic position:

$$CH_3{-}CH{=}CH{-}CHO + O_2 \rightarrow CHO{-}CH{=}CH{-}CHO + H_2O$$
$$\downarrow$$
$$CHO{-}CH(OCH_3){-}CH_2{-}CHO;$$

oxidative degradation of aldehydes:

$$CH_3{-}CH_2{-}CHO + O_2 \rightarrow CH_3{-}CH(O_2H){-}CHO$$
$$\downarrow$$
$$CH_3{-}CHO$$

These reactions were carried out at 20–40°, under basic conditions, in methanol; copper(II) chloride and sodium methoxide were present in the first case, copper(II) acetate, pyridine, and triethylamine in the second and third cases. The reactions summarized in the above equations are considered to be homolytic, as is exemplified by the following sequence of steps suggested for the cyanide synthesis (complexing ligands being omitted from the copper species).

$$RCHO + NH_3 \rightleftharpoons RCH{=}NH + H_2O$$
$$RCH{=}NH + Cu(II) \rightarrow RCH{=}N\cdot + Cu(I) + H^+$$
$$RCH{=}N\cdot + O_2 \rightarrow RC{\equiv}N + HO_2\cdot$$
$$HO_2\cdot + Cu(I) \rightarrow HO_2^- + Cu(II)$$

In aromatic chemistry, copper species are effective in promoting reactions of molecular oxygen with phenols and amines; the result may be oxygenation, giving quinonoid compounds, or dehydrogenation, giving dimers or larger polymers [13], e.g.,

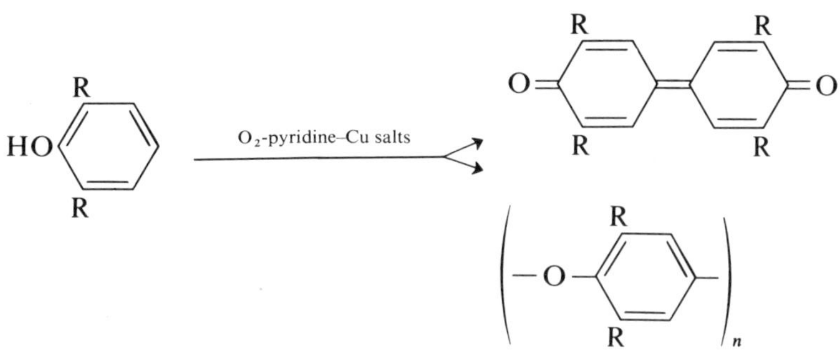

Catalysis by transition metals in such oxidations could involve intermediate metal complexes which are too labile for characterization. However, there are particular cases of catalysed oxidations in which metal–oxygen complexes show sufficient stability to be recognizable as intermediates.

One example [14] is catalysis of the reaction

$$PhCH_2CH_3 + O_2 \rightarrow PhCOCH_3 + H_2O$$

by the rhodium complex $RhCl(PPh_3)_3$, which in the process forms the oxygen-containing complex $RhCl(O_2)(PPh_3)$. Another example [15] is the oxidation of t-butyl isocyanide to the isocyanate by way of a nickel-oxygen complex.

$$Ni(1,5\text{-}C_8H_{12})_2 + 4Bu^tNC \rightarrow Ni(Bu^tNC)_4$$

$$Ni(Bu^tNC)_4 + O_2 \rightarrow Ni(O_2)(Bu^tNC)_2$$

$$Ni(O_2)(Bu^tNC)_2 + 4Bu^tNC \rightarrow Ni(Bu^tNC)_4 + 2Bu^tNCO$$

In view of current interest in metal complexes containing simple inorganic molecules as ligands, further developments in this type of oxidation may be expected.

(iv) Because oxidations tend to occur particularly readily when the substrate is an anion, reaction is often favoured by use of a basic medium. In the case of oxidations by molecular oxygen, the vulnerability of phenols in alkaline solution is a well-known example. The reactivity of anions can be exploited by carrying out oxidations in the presence of strong bases and in dipolar aprotic solvents which solvate the cation. Primary aromatic amines, for example, are oxidized to azo compounds in the presence of potassium t-butoxide [16].

$$2ArNH_2 + O_2 \rightarrow ArN{=}NAr + 2H_2O$$

Likewise, thiols give disulphides and desulphurization products by reaction with oxygen in dimethyl sulphoxide containing potassium t-butoxide [17].

This method of oxidation is particularly useful in the case of hydrocarbons, provided that they possess a suitable degree of acidity. Examples of conversions of hydrocarbons into oxygenated and dehydrogenated products by oxygen in the presence of alkali have in fact long been known; e.g., *p*-nitrotoluene gives *pp'*-dinitro-bibenzyl and -stilbene. Recently such reactions have been extensively studied by G. A. Russell and his co-workers [9], who found that the process could be made highly selective, depending on the acidity of the hydrocarbon, and on the base and solvent chosen. A typical case was the efficient conversion of fluorene into fluorenone by oxygen in the presence of potassium t-butoxide; preferred solvents were mixtures of t-butanol with pyridine, hexamethylphosphoramide, dimethylformamide, or dimethyl sulphoxide. Nitrobenzene and related compounds were useful catalysts and in such cases electron spin resonance spectra provided evidence of the production of nitro-aromatic radical-anions. A homolytic mechanism was therefore proposed:

$$R_2CH_2 + B^- \rightleftharpoons R_2CH:^- + BH$$

$$R_2CH:^- + O_2 \rightarrow R_2CH\cdot + O_2\cdot^-$$

$$(\text{catalysis step:}\quad R_2CH:^- + ArNO_2 \rightarrow R_2CH\cdot + ArNO_2\cdot^-)$$

$$R_2CH\cdot + O_2 \rightarrow R_2CH{-}O{-}O\cdot$$

$$R_2CH{-}O{-}O\cdot + R_2CH:^- \rightarrow R_2CH{-}O{-}O^- + R_2CH\cdot$$

(and other steps, including fission of the hydroperoxide:

$$R_2CH{-}O_2H \rightarrow R_2CO + H_2O).$$

(v) The use of irradiation to stimulate oxidations by molecular oxygen is an aspect of organic photochemistry (cf. Chapter 6). One type of reaction involves oxygen acting as a dienophile in a 1,4-addition to a conjugated diene or aromatic compound [7, 18].

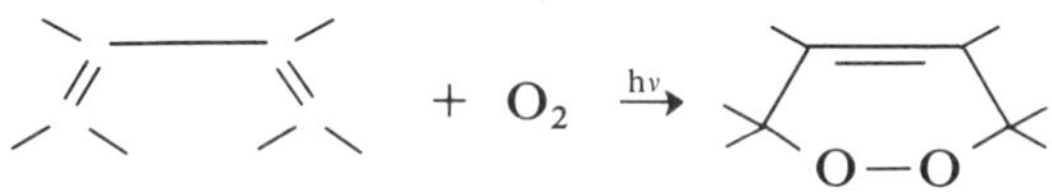

Recent investigators favour the view that this type of reaction occurs between the substrate molecule in its ground state and an oxygen molecule in an excited singlet state; the same result may be achieved by using electrically or chemically generated singlet-state oxygen [19].

Photo-initiated oxidations of a dehydrogenative type are exemplified by the nuclear coupling of stilbene derivatives to phenanthrene derivatives.

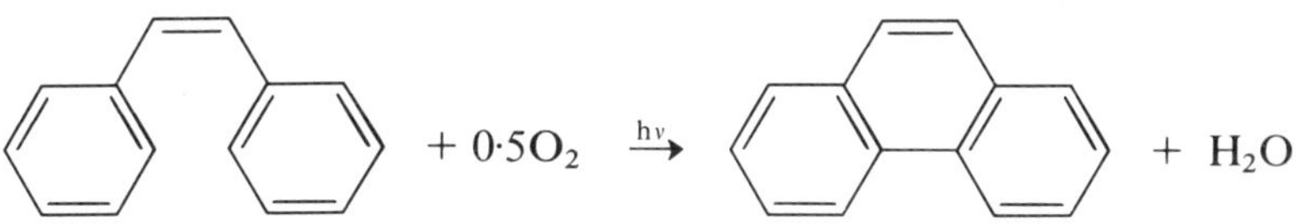

Mechanistic interpretations of such reactions may be complicated by the fact that inclusion of minor or major amounts of supplementary oxidants, such as iodine and/or copper(II) salts, has been found advantageous [20].

(b) Ozone

Both the theory and the practice of ozonolysis have undergone developments in recent years. R. Huisgen removed the reaction from its somewhat isolated position in organic chemistry by classifying it within the large category of 1,3-dipolar cycloaddition processes which his research school has studied so extensively [21]. The ozone molecule is seen as a 1,3-dipolar structure when represented in the form $^{+}{:}\ddot{\underset{..}{O}}{-}\ddot{\underset{..}{O}}{-}\ddot{\underset{..}{O}}{:}^{-}$. The only dipolarophiles to which it is known to add are C=C double-bonded compounds.

Criegee's studies in the 1950s led him to the view that an initial cyclic adduct between an olefin and ozone undergoes fragmentation and recombination to give an isomeric adduct containing a peroxidic bond [7]; this is subjected to reduction, hydrolysis, etc., in subsequent treatment.

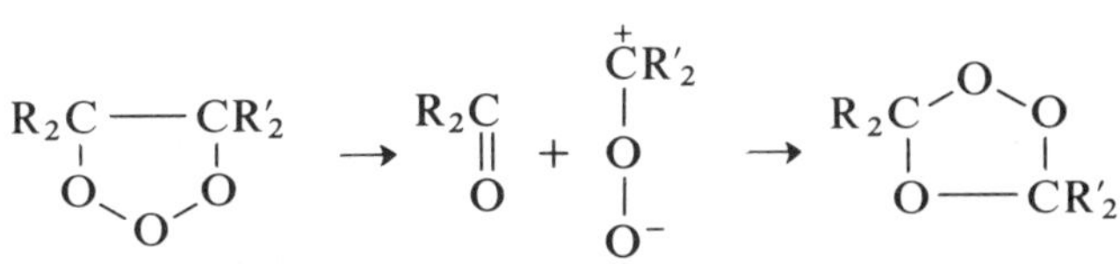

(or other formulation of initial adduct) (and other products)

P. S. Bailey's comprehensive review [22] summarizes knowledge of ozonolysis up to about 1958. More recently [23] investigators have compared results obtained in the ozonolysis of various unsymmetrical *cis* and *trans* olefins with what might be predicted from Criegee's fragmentation hypothesis; it is concluded that the hypothesis does not account for all the observed results. A useful development has been the isolation of ozonides in a sufficiently pure state for spectroscopic studies and even for X-ray crystallographic investigation.

The value of the technique for preparative purposes has been increased by improvements in ozonolysis procedures and in subsequent treatments of the product. For example, in polycyclic aromatic chemistry, the ring-opening of phenanthrene and acenaphthylene leads respectively to 2,2′-biphenyl derivatives and 1,8-naphthalene derivatives [24]. Some typical results, exemplified with acenaphthylene (X representing the 1,8-naphthalene residue) are as follows.

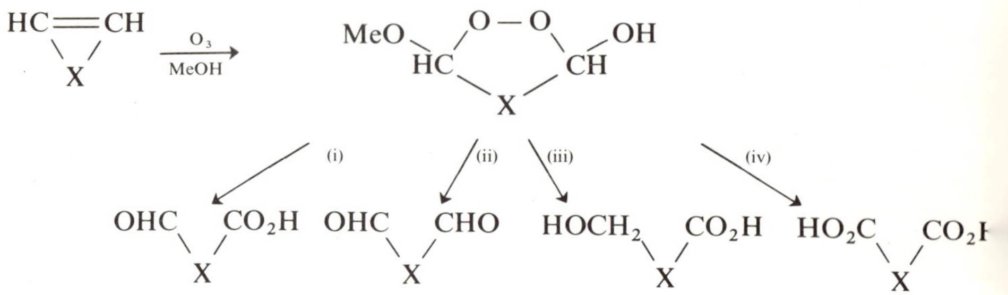

In this case, the ozonolysis was conducted in methanol at $-30°$, yielding the hemiperacetal, which gave (i) the aldehydoacid by treatment with aqueous bicarbonate, (ii) the dialdehyde by reduction with iodide, (iii) the hydroxyacid (as lactone) by treatment with formaldehyde, (iv) the dicarboxylic acid by oxidation with hydrogen peroxide.

For steps of type (ii), giving two carbonyl groups, various reductants have been recommended. Typically they are oxygen-acceptors, such as triphenylphosphine (converted into its oxide) and dimethyl sulphide (converted into the sulphoxide). Ozone may itself act as a simple oxygen-donor to organic acceptors, e.g., converting tertiary amines into their oxides.

The transformations of acenaphthylene shown in the preceding example demonstrate variations in ozonolysis procedure leading to different end products. More elaborate procedural variation is exemplified by 'cyan-ozonolysis', described by E. K. Fields [9], the result of which is not only double-bond fission, but also the incorporation of two additional carbon atoms. For example, if a cyclic olefin reacts with ozone in a cold aqueous emulsion containing sodium cyanide and hydrochloric acid, the end product is a dihydroxy-dicarboxylic acid.

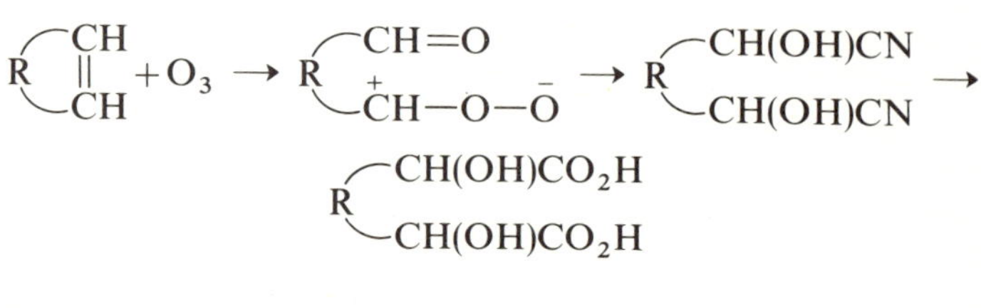

Besides being a useful technique for organic syntheses, ozonolysis retains its value as a chemical aid in the determination of molecular structures; this traditional application has been summarized by K. W. Bentley [25].

(c) Hydrogen Peroxide

This versatile oxidant [1, 7] is handled as a stabilized aqueous solution of varying concentration, including the 90% material which, with suitable precautions, is increasingly used for organic oxidations. The versatility of the reagent is largely due to the variations in its mode of reaction which result from the use of catalysts or particular experimental conditions. Variations in procedure may be summarized as: (i) use of alkaline media, in which the nucleophilic ion HOO^- is present; (ii) use of strongly acidic media, giving protonated reagent species, or use of carboxylic acids, giving per-acids as reagents; (iii) inclusion of transition-metal ions with reducing properties, which generate hydroxyl radicals; (iv) inclusion of other catalytic species, particularly certain transition-metal ions which may acquire peroxidic ligands; (v) use of ultraviolet irradiation.

In its reactions, hydrogen peroxide (like peroxides in general) sometimes acts homolytically and sometimes heterolytically. In the latter case, bond fission may occur in the manner $H\big)-O_2H$, whereby the reagent displays nucleophilic character, or in the manner $HO-\big)OH$, whereby it displays electrophilic character, as in the formation of an amine oxide.

$$R_3N{:} + HO-OH \rightarrow R_3\overset{+}{N}-OH + OH^-$$

In a recent review [26] numerous oxidative and rearrangement reactions of peroxidic compounds have been co-ordinated on the basis of electrophilic character. Both heterolytic and homolytic modes of reaction appear in the following examples, illustrating oxidations of types (i)–(v).

(i) Nucleophilic reaction of alkaline hydrogen peroxide with a suitably activated olefinic bond ($\alpha\beta$ to a carbonyl group) is a useful source of epoxy-compounds. A related addition reaction with a cyano-group (at $\sim$pH 8) has been found by G. B. Payne and his co-workers [27] to give an unstable peroxycarboximidic acid, which reacts with excess of hydrogen peroxide to give oxygen and an amide (an effect discovered by B. Radziszewski in 1884), whilst the inclusion of an olefin in the system results in the formation of an epoxide and an amide.

$$R-C\equiv N + H_2O_2 \rightarrow R-C(O_2H)=NH$$

$$R-C(O_2H)=NH + H_2O_2 \rightarrow R-CO-NH_2 + H_2O + O_2$$

$$R-C(O_2H)=NH + \;\gt C=C\lt\; \rightarrow R-CO-NH_2 + \;\gt C\underset{O}{\diagdown\!\diagup}C\lt$$

An organic cyanide (e.g., $R = CH_3$ or C_6H_5) may thus be used to catalyse the epoxidation of an unactivated olefin by hydrogen peroxide. If an olefinic bond and a cyano group are contained in the same molecule, e.g., acrylonitrile, then an epoxy-amide results. Similarly, hydrogen peroxide and an organic cyanide convert aniline into azoxybenzene, and pyridine into its *N*-oxide.

(ii) When mixed, e.g., with formic or acetic acids, hydrogen peroxide gives the corresponding peracids, and such solutions are well-known reagents [28] for converting an olefinic bond (by *cis* addition) into an epoxide, which then (by *trans* ring-opening) gives a mono-ester of the diol.

$$\mathrm{RCO{-}OH + H_2O_2 \rightleftharpoons R{-}CO{-}O{-}OH + H_2O}$$

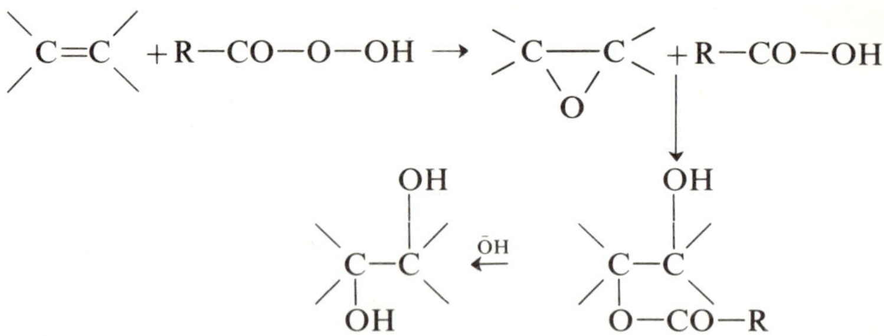

In strong protonic acids, however, hydrogen peroxide is protonated to give the species $\mathrm{H_2\overset{+}{O}{-}OH}$, i.e., the hydrated form of the hydroxyl cation, which D. H. Derbyshire and W. A. Waters showed (in 1950) to be electrophilic towards an aromatic ring; e.g., mesitylene was converted into mesitol. It has more recently been shown [29] that a mixture of 85% hydrogen peroxide, hydrofluoric acid, and boron trifluoride, in sulpholane as solvent, is an extremely potent source of $\mathrm{H_2\overset{+}{O}{-}OH}$, and oxidatively degrades benzene or cyclohexane.

More useful transformations, occurring at alkyl-substituted olefinic bonds, may be carried out with hydrogen peroxide in aqueous sulphuric acid [7]. The olefin is presumably preferentially protonated in such cases and the hydrogen peroxide converts it into a hydroperoxide by nucleophilic attack. In the following example [30] the hydroperoxide is subsequently subjected to an acid-catalysed rearrangement which is similar to that used [7] in the manufacture of phenol from cumene hydroperoxide.

$$\mathrm{Me_3CCH_2CMe{=}CH_2 + H_2O_2 \xrightarrow{H^+} Me_3CCH_2CMe_2OOH}$$
$$\downarrow$$
$$\mathrm{Me_3CCH_2OH + COMe_2}$$

(iii) The classical case of catalysis by a reducing metal ion is oxidation by Fenton's reagent, in which the effective oxidant is the hydroxyl radical:

$$\mathrm{Fe^{2+} + HO{-}OH \rightarrow HO\cdot + Fe(OH)^{2+}}$$

This method has been extensively discussed over the last twenty years, and most recently by R. O. C. Norman and his co-workers [31], who used $\mathrm{H_2O_2{-}Fe(II)}$ or $\mathrm{H_2O_2{-}Ti(III)}$ as oxidant systems for aromatic compounds and alcohols, and detected intermediate radicals by the electron spin resonance technique. A typical reaction is the production of biphenyl and phenol from benzene, which may be formulated as follows; the electron transfer symbolized by $(-e)$ might involve Fe(III), O_2, or HO· as oxidant.

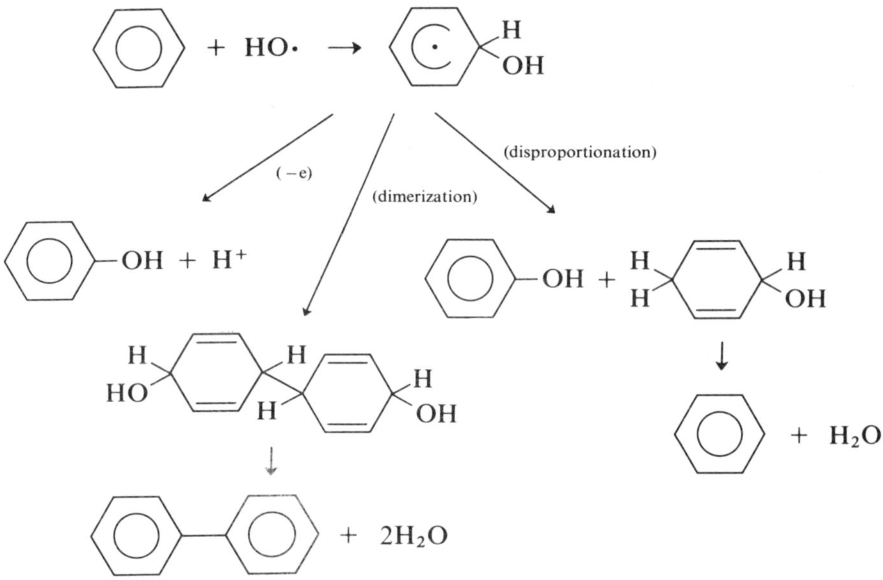

In the case of oxidation of alcohols, electron spin resonance studies by various investigators have shown the vulnerability of both α- and β-hydrogen atoms, giving intermediate radicals, such as $(CH_3)_2\dot{C}{-}OH$ from 2-propanol and $\cdot CH_2{-}C(CH_3)_2{-}OH$ from t-butanol.

Although prominent in studies of reaction mechanisms, Fenton's reagent has more limited interest for syntheses, since similar effects are often more satisfactorily achieved with other reagents. Examples of its utility may be seen in the publications of Coffmann and his co-workers, who showed that a variety of aliphatic compounds, e.g., alcohols, carboxylic acids, nitriles, could be fairly efficiently dimerized by the reagent. In the following example [32], in which the oxidant system is $FeSO_4{-}H_2O_2{-}$aq. H_2SO_4, a diol results by coupling of the radical $\cdot CH_2C(CH_3)_2OH$.

$$2HOC(CH_3)_3 + 2HO\cdot \longrightarrow HOC(CH_3)_2CH_2CH_2C(CH_3)_2OH + 2H_2O$$

(iv) Catalysis of oxidations by hydrogen peroxide, not involving reducing species, may be achieved with various transition-metal salts, oxides, and oxyanions, and also with selenium dioxide. This method has been used particularly for the hydroxylation of olefinic bonds [28]. If osmium tetroxide is thus used with hydrogen peroxide (the Milas reagent), the tetroxide may itself be the effective oxidant. Unlike the other catalysts, it gives a *cis* diol, which is considered to be formed via a cyclic osmic ester; the function of the peroxide may be to regenerate the tetroxide from its reduction products. A study of the oxidation of propene and other olefins by the Milas reagent has been reported by C. J. Norton and R. E. White [9]. No evidence was found for a stable intermediate complex between the olefin and osmium tetroxide; the resulting diols underwent a good deal of further oxidation.

Effects of these catalysed oxidations by hydrogen peroxide are not confined to hydroxylation of olefinic bonds. Tungstate and molybdate ions are good catalysts for amine oxidations, e.g., for the conversion of cyclohexylamine into cyclohexanone oxime [33].

$$R_2CHNH_2 + 2H_2O_2 \rightarrow R_2C{=}NOH + 3H_2O$$

With the same catalysts, or with titanium(IV) salts, hydrogen peroxide converts phenols into quinones and cyclohexadienone derivatives [34], whilst selenium dioxide catalyses a Baeyer–Villiger type of reaction between hydrogen peroxide and steroidal ketones [35]. The mechanism of these reactions is not known, but it is significant that the catalysts are, in general, of types which react with hydrogen peroxide to give metal complexes containing peroxidic ligands [36], e.g., of types $L_xM{-}O{-}OH$, $L_xM{-}O{-}O{-}ML_x$, and 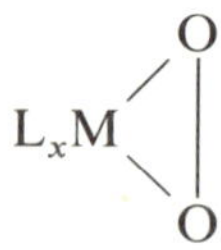 (in which M represents a metal and L_x other ligands); these species may be the effective oxidants.

A special type of catalyst is the iron-containing enzyme peroxidase, which in conjunction with hydrogen peroxide, has practical value for oxidations of phenols and aromatic amines. The varied products of these reactions sometimes resemble and sometimes differ from those given by analogous oxidant systems, such as Fenton's reagent. Peroxidase-catalysed reactions have been extensively reported by B. C. Saunders and his co-workers [37].

(v) Photolytically promoted oxidations of organic compounds by hydrogen peroxide have not so far shown much practical value. The hydroxylation of olefins is slow and inefficient [28]. Phenol has been reported to give catechol and other products in low yield [38].

(d) Persulphates and other Inorganic Peroxides

Peroxydisulphuric acid (used in the form of potassium, sodium, or ammonium persulphate) and peroxymonosulphuric acid (Caro's acid; generated by treating persulphates with sulphuric acid) are related to hydrogen peroxide in the following hydrolytic sequence, occurring in acidic media.

$$HO_3S{-}O{-}O{-}SO_3H + H_2O \rightarrow H{-}O{-}O{-}SO_3H + H_2SO_4$$
$$H{-}O{-}O{-}SO_3H + H_2O \rightarrow H{-}O{-}O{-}H + H_2SO_4$$

Like organic peracids, Caro's acid reacts by heterolytic cleavage of the O—O bonds, e.g., converting ketones into lactones (the Baeyer–Villiger reaction), converting aromatic amines into nitroso compounds and hydroxylating olefinic bonds. Persulphates also react heterolytically in their best-known application, i.e., the *para*- (or *ortho*-)sulphation of phenols, in aqueous alkaline solution (Elbs reaction). Study of this oxidation suggests that it involves electrophilic attack on the aromatic nucleus [39].

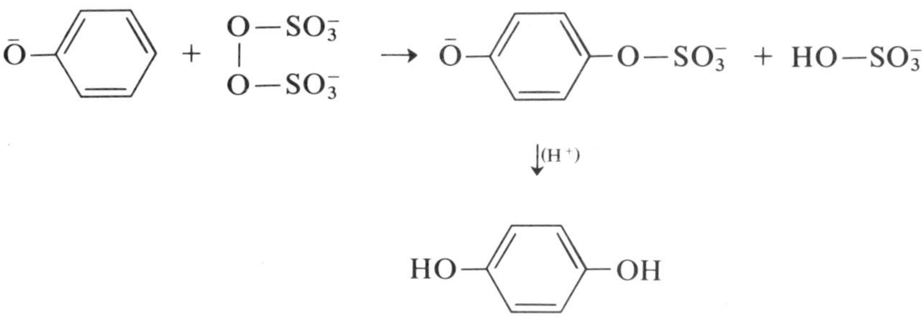

Apart from the Elbs reaction, organic oxidations by persulphate, in the absence of catalysts, find limited applications; a recent example [40] is the conversion of lactams into imides.

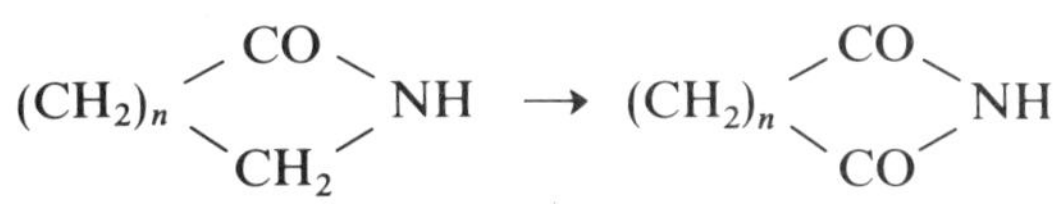

In the presence of various metal ions, such as Fe^{3+}, Cu^{2+}, and particularly Ag^{+}, homolytic decomposition of persulphate ensues, and it becomes a powerful and fairly useful oxidant in preparative applications, such as the following transformations [41], which may be achieved in aqueous media with the aid of catalytic amounts of silver nitrate.

$$PhCH_2{-}CH_3 \rightarrow PhCO{-}CH_3$$
$$PhCH_2{-}CO_2H \rightarrow PhCHO + PhCH_2{-}CH_2Ph$$
$$p\text{-cresol} \rightarrow \text{Pummerer's ketone, etc.}$$
$$2RCH_2{-}NH_2 \rightarrow RCH{=}N{-}CH_2R \xrightarrow{(H^+)} RCHO + RCH_2{-}NH_2$$
$$R_2CH{-}CO_2H \rightarrow R_2CH{-}SO_4^{-} \xrightarrow{(H^+)} R_2CH{-}OH$$

Catalysis is considered to involve steps such as the following.

$$S_2O_8^{2-} + Ag^+ \rightarrow SO_4^{\cdot -} + SO_4^{2-} + Ag^{2+}$$

Hence, oxidation is likely to be due to silver(II) and possibly silver(III) species (cf. p. 80), as well as sulphate radical-ions. There has been much interest in the kinetics of persulphate decompositions and oxidative reactions since the 1940s, partly because of their relevance for vinyl polymerization processes; the subject has been recently reviewed [42].

Occasional use is made of peroxidic salts of inorganic acids other than sulphuric, such as perborates and perphosphates. An example is the conversion of primary aromatic amines into azo compounds by sodium perborate, as in the following reaction [43]

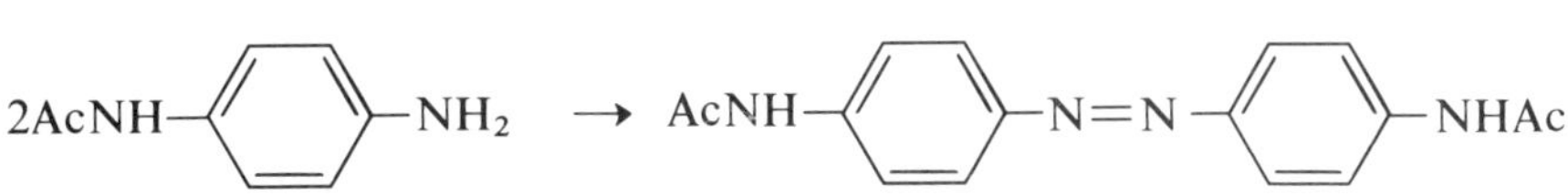

(e) Organic Peroxides

This class of oxidants provides a wide choice of reagents, essentially within the following five subdivisions: alkyl hydroperoxides, RO—OH; dialkyl peroxides, RO—OR; peroxyacids, RCO—O—OH; peroxyesters, RCO—O—OR′; diacyl peroxides, RCO—O—O—COR [7]. As in the cases of other peroxidic compounds, their reactions may be heterolytic or homolytic. There is a marked tendency for those of homolytic type to produce mixtures, resulting, e.g., from alternative reaction paths, degradation processes, and polymerizations. Hence, although organic peroxides have considerable value for preparative purposes [1, 3], the copious literature [7] is more commonly concentrated on mechanistic aspects of their reactivity. Reference is confined in this review to two features in which there is much current interest [44]: heterolytic oxygen-donating reactions of peroxyacids, and certain applications of peroxyesters in conjunction with copper salts.

The heterolytic reactions of peracids represent the major application of peroxidic organic compounds in preparative chemistry. One aspect, the hydroxylation of olefins via epoxidation, was referred to in connection with oxidations by hydrogen peroxide. The transition state of the epoxidation reaction has been given a representation such as (A).

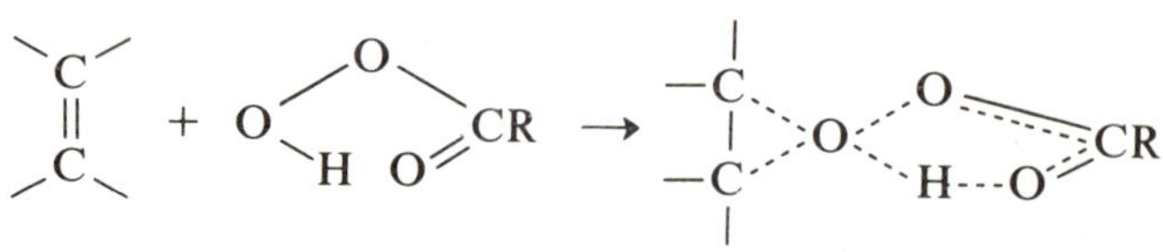

(A)

A recently advocated alternative view [45] is that the peracid reacts as a 1,3-dipolar species, giving a cyclic peroxidic adduct (B) which then undergoes fission.

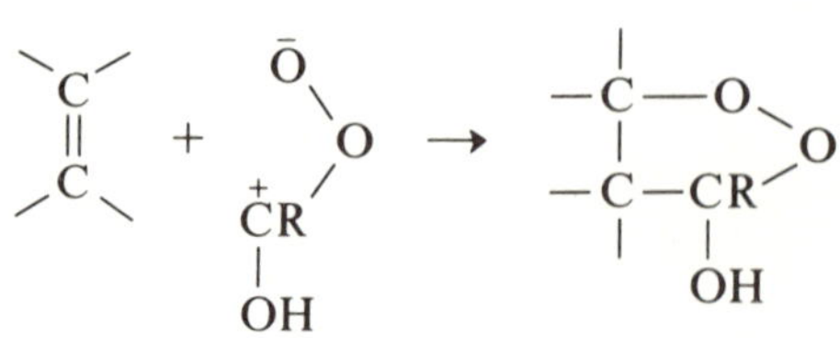

(B)

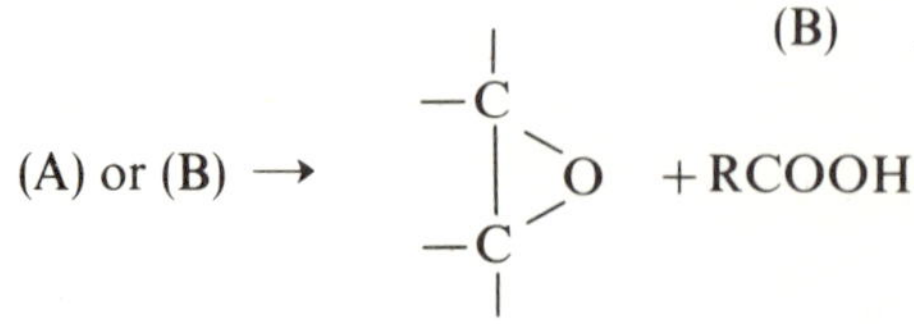

For the preparation of epoxides, preferred reagents are, e.g., perbenzoic, permonophthalic (o-$CO_2H{\cdot}C_6H_4{\cdot}CO_3H$), or perlauric acids, which may be obtained as pure crystalline compounds and employed in aprotic solvents, such as chloroform or ether. For stereochemical studies, optically active species, such as (+)-percamphoric acid, have been extensively used; asymmetric syntheses of epoxides and sulphoxides may be thus achieved [46]. *m*-Chloroperbenzoic acid (m-$C_6H_4Cl{\cdot}CO_3H$) is finding an increasing number of applications. Its reaction with imines, in chloroform or methylene chloride, provides an improved route to oxaziridines [47].

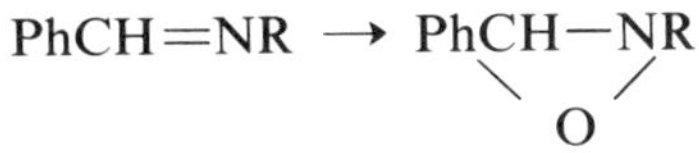

An example of its utility is seen in the last stage of a recent synthesis of cyclopenin, a mould metabolite containing an epoxy grouping attached to a benzodiazepine ring system [48]; there was no precedent for the successful epoxidation of an olefinic bond thus substituted.

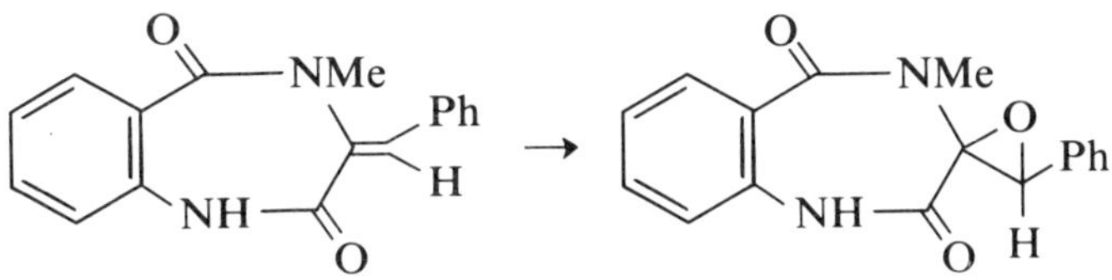

It converts primary amines into nitro compounds, a transformation which exemplifies the use of peracids to donate oxygen to an atom with an unshared pair of electrons. Nitroso, nitro, and azoxy compounds, and amine *N*-oxides are thus prepared from the appropriate nitrogenous compounds. Likewise, in organosulphur chemistry, there is the well-known conversion of sulphides by peracids into sulphoxides and sulphones, and less familiar oxidations, such as the following examples [49], which respectively represent the conversion of disulphides into thiolsulphinates, thiones into sulphines, and dithiocarboxylic esters into their oxides.

$$RS{-}SR' \rightarrow RS{-}SR'{=}O$$
$$RR'C{=}S \rightarrow RR'C{=}S{=}O$$
$$RC(SMe){=}S \rightarrow RC(SMe){=}S{=}O$$

New applications continue to be found for trifluoroperacetic acid, the most powerful oxidant among the peracids [44]. Thus, H. Hart and his co-workers [9] have used it in conjunction with boron trifluoride, which may be regarded as a Lewis acid for the trifluoroacetate moiety, making the electrophilic hydroxy-cation available as an oxygen-donor to aromatic hydrocarbons and olefins, e.g.,

$$CF_3{-}CO{-}O{-}OH + BF_3 \rightleftharpoons CF_3{-}COO{-}\overset{-}{B}F_3 + HO^+$$

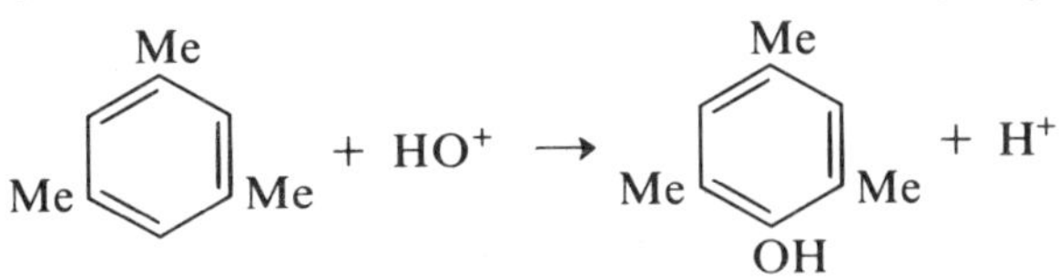

$$CMe_2{=}CMe_2 + HO^+ \rightarrow \overset{+}{C}Me_2{-}CMe_2{-}OH \rightarrow CMe_3{-}COMe + H^+$$

$$CF_3{-}COO{-}\overset{-}{B}F_3 + H^+ \rightleftharpoons CF_3{-}COOH + BF_3$$

Copper-catalysed reactions of peroxyesters, such as t-butyl peracetate or perbenzoate [44], provide further examples of oxidative effects due to the interaction of transition-metal ions and peroxidic bonds (cf. oxidations by hydrogen peroxide). The mechanism of the peroxyester reaction has been extensively discussed by J. K. Kochi [50]. It is considered to involve an oxidation-reduction cycle and it is commonly used to effect allylic substitution of an olefin by an ester group.

$$\mathrm{Bu^tO{-}OCOMe + Cu(I) \rightarrow Bu^tO\cdot + Cu(II)OCOMe}$$

$$\mathrm{{-}CH_2{-}CH{=}CH{-} + Bu^tO\cdot \rightarrow {-}\dot{C}H{-}CH{=}CH{-} + Bu^tOH}$$

$$\mathrm{{-}\dot{C}H{-}CH{=}CH{-} + Cu(II)OCOMe \rightarrow {-}CH(OCOMe){-}CH{=}CH{-} + Cu(I)}$$

Since the allylic radical shown in the second step is a resonance-stabilized species, the allylic ester produced in the third step (oxidation of a radical by a metal ion) may possess a rearranged olefinic bond. The reaction may take more complicated forms, e.g., the sequence of steps occurring when norbornadiene reacts with t-butyl perbenzoate and cuprous bromide, results in substitution by a t-butoxy-group [51].

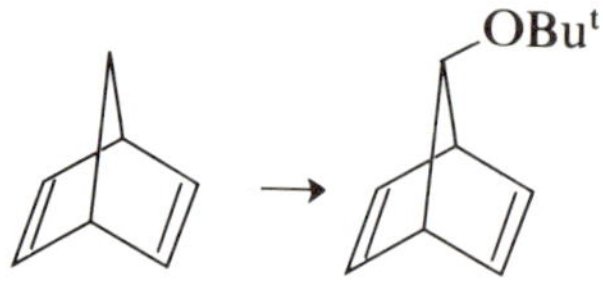

3.2 NON-METALLIC ELEMENTS, THEIR OXIDES AND OXYANIONS

Oxidative properties arise in Groups V, VI, and VII through hydrogen- or electron-accepting properties of elements, or oxygen-donating properties of oxides or oxyanions.

(a) Group V

The main oxidants in this Group are compounds of nitrogen, i.e., nitric and nitrous acids, their salts and esters, and nitrogen oxides. Nitrous oxide will effect certain oxidations with the aid of heat and pressure. Dinitrogen tetroxide is reactive under quite mild conditions towards readily oxidized organic compounds, such as alcohols and sulphoxides. Nitrogen dioxide oxidizes even aliphatic hydrocarbons at 300–400°. Nitric acid, one of the oldest oxidants used in organic chemistry, remains a recommended reagent for preparations of some types of compounds, e.g., the conversion of D-glucose into D-glucaric acid [52]. There have been a number of recent kinetic studies; for instance, oxidations of benzyl alcohols and ethers, and of benzaldehyde, with aqueous nitric acid are considered to be homolytic reactions, with NO_2 or HNO_2^+ as the oxidizing entity [53].

An odd-electron structural character is a feature of some other nitrogen-containing oxidants. Frémy's salt, nitrosodisulphonate, $ON(SO_3K)_2$, has been widely used for the conversion of phenols into quinones since its introduction for this purpose by H. J. Teuber in 1951. Similar oxidations may be effected, though rather less successfully, by nitroxide radicals, RR′NO· [54]. Another example of homolytic reactivity is intramolecular oxidation arising from the photolysis of nitrite esters, exploited by D. H. R. Barton and his co-workers for transformations of steroids [55]. This reaction requires particular spatial relationships of the groups concerned and is thought to involve the following sequence.

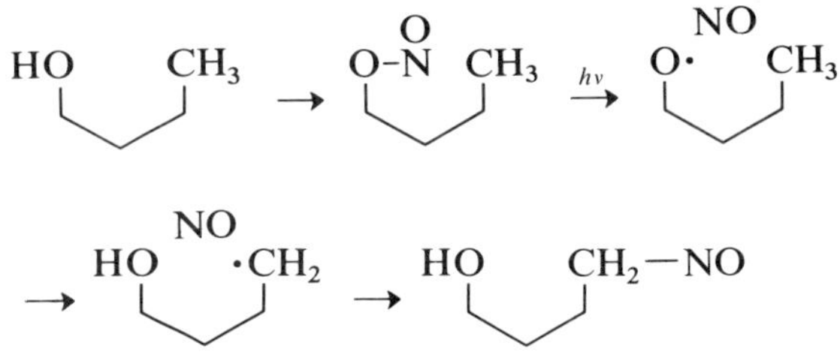

The nitroso compound which ends the photolysis sequence is isomeric with an oxime, and the latter is hydrolysable to an aldehyde ($-CH_2-NO \rightarrow -CH{=}N-OH \rightarrow -CHO + NH_2OH$); the overall effect is therefore the oxygenation of the methyl group.

Compounds of the other non-metals in Group V furnish few useful examples of organic oxidation. Arsenate is occasionally used, but applications of phosphorus compounds are dominated by the powerful reductive effects of trivalent species (see p. 84). An exceptional case of oxidation is the formation of salts of the tropylium cation by reaction between cycloheptatriene and phosphorus pentachloride [56].

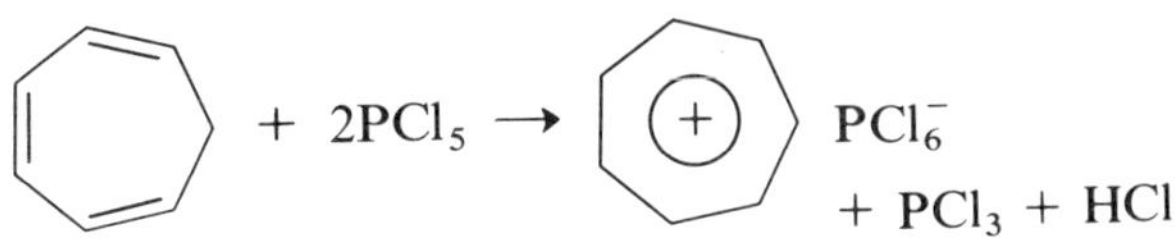

(b) Group VI

(i) Molecular oxygen and compounds with O—O bonds are the subject of Section 3.1, but that category did not include one simple oxygen-containing species, hydroxyl ion, which is capable of reacting oxidatively. It has been known for well over a century that oxidation may occur when organic compounds are treated with fused sodium or potassium hydroxide; the process provided a classical method of chemical degradation. Aromatic and heterocyclic compounds may undergo fission or hydroxylation; the latter effect is often applied for oxidations of polycyclic quinones. Olefinic carboxylic acids are degraded to dicarboxylic acids, whilst alcohols react in a complicated manner, yielding carboxylic acids and other products. Hydrogen evolution may occur in these oxidations. Progress has recently been made in the study of some of these reactions [57] but further elucidation is needed. The literature has been reviewed by B. C. L. Weedon [25].

(ii) Until superseded by selenium and then by metal catalysts, sulphur was the standard reagent for dehydrogenating alicyclic to aromatic compounds

[58]. It may be effective for other high-temperature dehydrogenations; e.g., at 500° amines are converted into imines and thence into thiazoles [59].

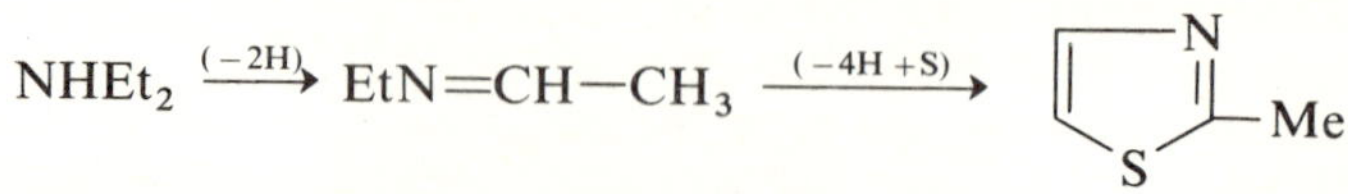

The use of elemental sulphur or polysulphide ion, e.g., at 200–300° in the presence of water or aqueous alkali, is a means of oxygenating organic compounds, as in the Willgerodt reaction, or in the conversion of alkyl into carboxylic groups on aromatic nuclei. The literature relating to these complicated reactions has been reviewed up to about 1960 [60].

Oxygen derivatives of sulphur may be effective oxidants. Sulphuric acid is an obvious example, and, if its other effects on organic molecules can be avoided, it may be useful in preparations; e.g., adamantane gives 1-hydroxyadamantane and adamantone [61]. Sulphur dioxide has been used as an oxidant, at 250-300° and under 300 atm pressure, by A. J. Shipman [9]; the method was applied mainly to the conversion of alkyl-substituted aromatic compounds into the corresponding carboxylic acids, according to the overall equation

$$RCH_3 + 1{\cdot}5SO_2 \rightarrow RCO_2H + 1{\cdot}5S + H_2O$$

(iii) The dehydrogenating properties of elemental selenium are well known [58], and its dioxide [5] shows continued value, e.g., for the conversion of ketones into α-diketones, allylic methylene groups into carbonyl groups, side-chain methyl into aldehydo or carboxylic groups, and 1,4-diketones into dehydrogenated derivatives. R. Stewart [1] has summarized recent progress in the elucidation of mechanisms of oxidation by selenium dioxide; reaction with the organic substrate converts the hydrated oxide, H_2SeO_3, into the divalent selenium derivative, H_2SeO_2, and the latter into the element.

(c) Group VII

(i) Homolytic reactions of halogens with organic compounds, initiated thermally or by irradiation, may lead to dehydrogenated products. Likewise, reactions of *N*-bromosuccinimide with organic compounds include oxidative effects [62]. More familiar for preparative purposes is the use of halogens as oxidants under mild conditions, in solvents, when they react heterolytically. Kinetics of bromine reactions have been comparatively studied, for example, with those of two-electron oxidants such as Hg(II) and Tl(III) [63].

Bromine or iodine finds frequent use for dehydrogenations, e.g., of hydroxylic compounds and hydrazine derivatives. Aqueous bromine remains a standard oxidant for converting an aldose into an aldonic acid [52]. Recent applications of iodine include the conversion of hydrazones into azines [64].

$$2RR'C{=}N{-}NH_2 \xrightarrow{(-4H)} RR'C{=}N{-}N{=}CRR' + N_2$$

and dehydrogenation of hydrazino-compounds, with accompanying insertion of iodine [65];

$$RNH{-}NH_2 + 2I_2 \rightarrow RI + 3HI + N_2$$

Mechanistic studies of oxidations by halogens have centred particularly on the alternatives of hydride-ion abstraction, and of reaction through an intermediate hypobromite, when bromine interacts with hydroxylic compounds. The literature has been summarized by R. Stewart [1]; investigations of the subject continue to be reported [66].

(ii) In varying degree, the following halogen-containing oxyanions find use as oxidants for organic compounds: hypohalites; chlorite; chlorate, bromate, iodate; periodate. Of these oxidants, periodate has attracted particular attention in recent years. This is due partly to continued application of its fission reactions [6] with glycols, carbohydrates, and amino-alcohols, and also to newer applications involving oxidations of phenols, and of nitrogenous and sulphur-containing organic compounds; the subject has lately been reviewed [67].

Reactions of periodate with monohydric and dihydric phenols have been studied in particular by E. Adler and his co-workers. Various kinds of products, resulting from dehydrogenation, fission and oxygenation processes, have been isolated; e.g., 2,4-dimethylphenol gives *o*- and *p*-hydroxycyclohexadienones, the former of which dimerizes by a Diels–Alder reaction.

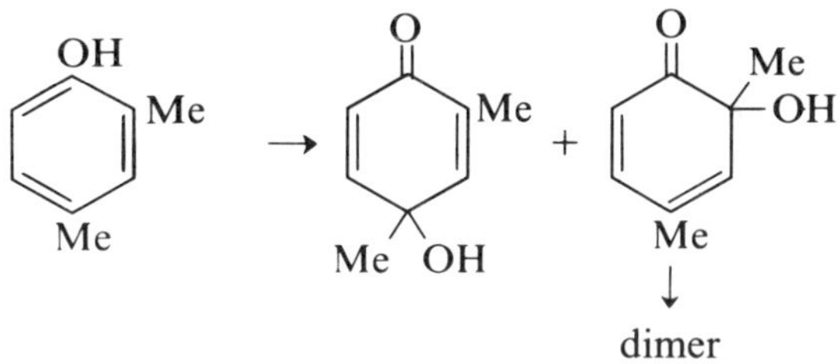

Periodate is useful for converting sulphides into sulphoxides.

Continued use is found for the modified form of periodate oxidation, introduced by R. U. Lemieux in 1955, whereby a catalytic amount of permanganate in the reaction mixture leads to fission of an olefinic bond via diol formation; some other hydroxylating oxidants may be used as catalysts in place of permanganate.

3.3 METAL SALTS, OXIDES, OXYANIONS, AND COMPLEXES

This is the largest general class of oxidants. It comprises numerous compounds which are capable of ready transformation to a lower oxidation level and which contain transition elements or certain metallic elements adjacent to them in the Periodic Table, i.e., metals of the first transition series, including copper; those of the second transition series, including silver; those of the third transition series, beginning with cerium among the lanthanides and leading to mercury, thallium, lead, and bismuth (at which point the discussion links with that of Group V non-metal oxidants given in Section 3.2). Comments are given below, in subsections (b)–(r), on compounds of seventeen of these metals; some of the reagents have longstanding importance, while others have been relatively recently employed as oxidants for organic compounds. Studies of reagents within this general category seem likely to provide the greatest scope for further developments in oxidation chemistry. This is a consequence of the numerous metals showing variable valency, the sequence of oxidation levels sometimes displayed by a particular metal, and the variety of atoms, groups, and molecules which may serve as a ligands at a particular level.

(a) General Considerations

(i) The variety of types of oxidant containing a particular metal may be illustrated by reference to manganese compounds [2]. The common oxidant is a solution of the oxyanion, permanganate, MnO_4^-, containing Mn(VII); the dioxide, MnO_2, containing Mn(IV), is a reagent for heterogeneous oxidations; when used as an oxidant in the presence of sulphuric acid the dioxide may dissolve, producing the unstable Mn(IV) salt, $Mn(SO_4)_2$; solutions of the pyrophosphate complex, $Mn(H_2P_2O_7)_3^{3-}$, have been employed to study oxidizing properties of Mn(III), a species which would disproportionate to Mn(II) and Mn(IV) in the absence of a complexing agent.

If, as a second example, iron is considered, the choice of oxidant is commonly between solutions of ferric chloride and of potassium ferricyanide. Both contain six-coordinate Fe(III), but in the former case the Fe^{3+} ions carry a mobile and varied population of ligands (H_2O, OH^-, Cl^-, etc.), whilst in the latter case the oxidant $Fe(CN)_6^{3-}$ contains six tightly coordinated CN^- ligands. Many other iron(III) complexes, with varying structural characteristics and stability constants, are available; e.g., the tris-1,10-phenanthroline complex, ferriin, has been used as an oxidant. There are other compounds, little explored as oxidants, which contain iron in higher oxidation states, e.g., iron(VI) in ferrate ions, FeO_4^{2-}.

(ii) The reactivity of organic species, whether in oxidations or in other types of transformation, may be profoundly modified if they become coordinated as ligands in metal complexes. This has become a topic of major importance in chemistry; recent research interests are exemplified by the contents of a symposium held in 1962 [68]. Oxidative and reductive aspects of transformations in ligands were documented in a contribution to another symposium, held in 1964 [69].

A common mode of oxidation may be summarized in the general form (L representing various ligands): organic substrate (A)+complexed metal ion (M^nL_x)→unstable complex ($M^nL_{x-1}A$)→oxidized product of A+complexed reduced metal ion ($M^{n-1}L_x$). Production of an intermediate metal complex, with displacement of an existing ligand, may therefore be an essential step in an oxidation process. The formulation of likely intermediates is prominent in discussions of mechanisms in recent publications, e.g., those of W. A. Waters and his co-workers [2] concerning oxidations by manganese(III) or vanadium(V). There are, however, alternative modes of oxidation involving metal complexes. One may be summarized in the general form: $M^nL_x + A \rightarrow M^nL_x \cdots A \rightarrow M^{n-1}L_x$ + oxidized product of A. In such cases, collision between the reactants is considered to be followed by an electron transfer, relayed to the metal through the enveloping ligands L_x, which do not undergo displacement. One function of these ligands could be the stabilization of an oxidation state of the metal which would otherwise be labile in the reaction medium. There are further types of organic oxidation (e.g., by oxygen; cf. p. 56), in which the function of a transition-metal ion is essentially that of a catalyst, though its oxidation state perhaps fluctuates in an oxidation–reduction cycle. Metal complexes are likely to be involved as intermediates in such cases, but mechanistic details are generally obscure.

Knowledge of mechanisms of redox reactions is most advanced in those cases in which both the oxidant and the reductant are complexed metal ions, e.g., reactions between Cr(II) and Co(III) complexes. Developments in this branch of inorganic chemistry, which have obvious relevance for organic oxidation processes, have been stimulated particularly by the studies of H. Taube; they have been most recently reviewed by A. G. Sykes [70]. Many of these inorganic redox reactions belong to the category termed by Taube 'inner-sphere'; i.e., the metal ions have a ligand in common, forming a bridge in the activated complex, within which electron transfer occurs; bridging ligands include carboxylate ions from a variety of organic acids.

(iii) Special features may exist in oxidations of organic compounds effected by metal oxides; the result may be dehydrogenation or oxygenation. Traditional reagents are the oxides CuO, Ag_2O, HgO, and PbO_2. The success and versatility of manganese dioxide [71], which was introduced in the 1950s for oxidation of the hydroxyl group in allylic alcohols, has directed attention to the potential value of other metal oxides; for example, silver(II) oxide and the so-called nickel peroxide have been recently examined as oxidants.

Whereas some oxides are employed as oxidants in solution (e.g., SeO_2, CrO_3, OsO_4, RuO_4), those referred to above are used as suspensions of finely-divided solids in liquid media. A summary of effects and problems encountered in the study of these heterogeneous systems has been given by W. F. Pickering [72]. Oxidation must begin on the oxide surface and may ultimately result in transformation of the entire solid lattice. Presumably these reactions are affected by factors which are of importance in gas–solid heterogeneous reactions, e.g., chemisorption, lattice imperfections, and variations in stoichiometry within the solid, but an additional complication is the presence of solvent at the interface. In a physico-chemical sense, the situation is similar in the case of other liquid–solid heterogeneous reactions involving insoluble metal-containing oxidants which are not oxides, e.g., silver(II) picolinate. A study by H. Taube [73] showed that reactions between various solid oxides (Mn_2O_3, MnO_2, Co_2O_3, CeO_2, Tl_2O_3, Pb_3O_4, PbO_2) and aquated chromium(II) ions differed in various respects from oxidations of this species effected by metal ions in homogeneous solution.

(iv) Biological oxidations do not fall within the scope of this review, but, as R. Stewart, W. A. Waters, and others have pointed out [1, 2], they show parallels in various respects with oxidations performed in the laboratory without the aid of enzymes. Transition metals are of common occurrence in enzymic systems in which oxidations occur. Accordingly, the study of biological transformations has stimulated research in which the chemist seeks to imitate a naturally occurring process under non-biological conditions.

A major development of this kind in recent years has been the recognition that the oxidative coupling of a phenol may be a key step in numerous biogenetic sequences producing alkaloids, pigments, lignins, tannins, and a variety of other products from plant or fungal sources. A fruitful period of synthetic work was inaugurated in 1956 by D. H. R. Barton's synthesis of usnic acid, a component of lichens, by a two-stage process involving oxidative coupling of the appropriate phenol with alkaline potassium ferricyanide; this is considered to be a homolytic process in which union of phenoxy radicals occurs. Subsequent syntheses of numerous natural products by several research groups, using similar methods, have been the subject of recent reviews [74].

The oxidants employed for these purposes are commonly of a type associated with one-electron transfer mechanisms, and their desired effect is hydrogen-abstraction and not oxygenation. Phenolic oxidations, however, are often very complicated [75] and may give a mixture of products, including polymers with a varying degree of complexity. Accordingly the yield of a desired dimeric product is often very low. Iron(III), in the form of potassium ferricyanide or ferric chloride, is the favoured type of oxidant for phenolic coupling, but compounds of several other metals have been successfully employed.

A brief consideration of applications of individual metal-containing oxidants now follows. They are presented in their Periodic-Table sequence.

(b) Titanium Compounds

Predominantly, the applications of titanium compounds in organic redox reactions are as reductants, in which case Ti(III) salts are used. Oxidation by Ti(IV) species requires a substrate with strong reducing properties, e.g., a carbanion. An example is to be found within the sequence of reactions which may occur between components of Ziegler–Natta polymerization catalysts, for instance when aluminium alkyls are used in conjunction with titanium tetrachloride.

$$TiCl_4 + AlEt_3 \rightarrow EtTiCl_3 + Et_2AlCl$$

$$EtTiCl_3 \rightarrow TiCl_3 + Et\cdot \ (\rightarrow \text{further products})$$

This type of reaction is applicable to the preparation of an organic dimer, as in the following case [76], in which the intermediate Ti(IV) aryl, prepared at $-70°$, decomposes to the biaryl and the pyrophoric Ti(II) aryl at $-10°$.

$$4LiPh + TiCl_4 \rightarrow TiPh_4 + 4LiCl$$

$$TiPh_4 \rightarrow TiPh_2 + Ph{-}Ph$$

(c) Vanadium Compounds

Until recently, the use of vanadium in organic oxidations was essentially restricted to its introduction into certain processes as a catalyst, generally in the form of the pentoxide, V_2O_5. The use of vanadium(V) as an oxidant for organic compounds in acidic aqueous solutions, containing the cation VO_2^+, or hydrated derivatives of this species, has been described mainly by W. A. Waters and J. S. Littler, in a series of publications dating from 1959. These authors [2] have lately reviewed oxidations by vanadium(V) in conjunction with those involving manganese(III) and cobalt(III). All three types of reagent need to be employed in fairly strongly acidic solution; they are oxidants by virtue of one-electron transfer processes; and their effects on organic substrates show a number of similarities.

Interest has been centred largely on mechanistic aspects of oxidation by vanadium(V), manganese(III), and cobalt(III) species, and as yet they have found only occasional employment for preparative purposes. Vanadium(V) oxidizes primary and secondary alcohols to carbonyl compounds; glycols undergo oxidative fission; α-hydroxyacids undergo oxidative decarboxylation; and reactions occur with aldehydes, ketones, phenols, amines, and with the more vulnerable carboxylic acids, such as oxalic and malonic. In general, oxidations are considered to be of hydrogen-abstracting type, as, for instance, in the following representation of the oxidation of a secondary alcohol.

$$VO_2^{+} + H_3O^{+} + R_2CH(OH) \rightleftharpoons [(HO)_3V \leftarrow (OH)CHR_2]^{2+} \longrightarrow$$

$$\dot{C}R_2{-}OH\ (\rightarrow \text{further products}) + H^{+} + V(OH)_3^{+}$$

(or alternative forms of vanadium(IV))

(d) Chromium Compounds

Chromic oxide and structurally related compounds are one of the longest known, most useful and versatile, and most extensively investigated types of oxidants for organic compounds. The present decade appears to have brought no major new developments, but the many established forms of the oxidant have continued to be used for new purposes, and there has been steady progress in the mechanistic studies of chromic oxidations which were initiated by F. H. Westheimer in the 1940s. K. B. Wiberg, a leading contributor to these investigations, has recently provided a comprehensive review of the subject [2]. J. Rocek, another contributor to current research, has surveyed the particular topic of aldehyde oxidation and compared chromium(VI) with other transition-metal oxidants for such reactions [77].

Typical oxidizing species containing Cr(VI) are oxyanions, $X{-}CrO_2{-}O^-$ (e.g., X = HO, AcO, and pyridinium cation, when chromic oxide is employed in water, acetic acid, or pyridine, respectively); dichromate ion, $\bar{O}{-}CrO_2{-}O{-}CrO_2{-}O^-$; and chromyl compounds, $X{-}CrO_2{-}X$ (e.g., X = Bu^{t}O in di-t-butyl chromate, Cl in chromyl chloride). Virtually all types of organic compounds are vulnerable under suitable conditions, but Cr(VI) oxidants have been used particularly for primary, secondary, and tertiary alcohols; diols; alkanes, alkenes, arylalkanes, and aromatic hydrocarbons; aldehydes and ketones. Oxidation steps, involving two-electron or one-electron reduction of the metal, proceeding through chromic esters or in other ways, and leading in various sequences to an overall change from Cr(VI) to Cr(III), have been abundantly discussed [1, 2].

(e) Manganese Compounds

Reference has been made, in subsections (a) and (c), to oxidations by permanganate, manganese dioxide, and manganic pyrophosphate. Because of its high oxidizing power and versatility, permanganate shares with chromic species a position of prime importance among oxidants for organic compounds. In various respects reactions of manganese-containing oxidants provide similar chemical problems to those of chromium-containing oxidants, and in both cases the discussion of oxidation mechanisms tends to be a dominant theme in recent publications. R. Stewart, a prominent investigator in the field, has recently surveyed oxidations by permanganate [2], and in an earlier review W. A. Waters considered oxidations by manganese species in conjunction with those by chromium species [78]. The two types of oxidant have also been reviewed by H. O. House [44].

Permanganate solutions are oxidants for the same wide range of organic substrates as chromic oxide and its derivatives. Recent results involving the dehydrogenation of benzylic amines [79] exemplify the applicability of permanganate in a relatively unexplored direction, in spite of its being a long-established oxidant.

Oxidation mechanisms are somewhat less well defined for manganese than for chromium species. Heterolytic reactions are commonly advocated, e.g., for the hydroxylation of olefins and for the conversion of alcohols into carbonyl compounds (a reaction for which Stewart has suggested the probability of hydride-ion transfer), but there is evidence of homolytic reactions in certain cases. In parallel with oxidations by Cr(VI), the overall reduction of Mn(VII) to Mn(IV) or Mn(II) occurs in a complicated sequence of steps involving a variety of oxidation levels. One reason for studies being undertaken with manganic pyrophosphate (or sulphate, or acetate) was to throw light on reactivity at the Mn(III) level in permanganate oxidations. Applications of manganese(III) have broad similarities to those of vanadate (see above), with the important exception that manganic pyrophosphate does not oxidize monohydric alcohols.

Heterogeneous oxidation by manganese dioxide continues to find new applications; for example, in the field of organonitrogen compounds it has been used to couple anthranilic acid to phenazine-1,6-dicarboxylic acid [80], to dehydrogenate thiazolines to thiazoles (a reaction which may be effected by running a solution down a column of the oxidant) [81], and to introduce carbonyl groups into amines [82]. Acetonitrile has been found to be an unsuitable solvent for oxidation because the manganese dioxide catalyses hydration of the nitrile to the amide [83]. It is evident that some oxidations may be achieved with relatively small amounts of manganese dioxide, whereas the use of a very large excess of the oxidant was a characteristic feature of processes described in earlier publications.

(f) Iron Compounds

As was indicated in subsection (a), the most important recent employment of iron(III) oxidants (ferricyanide or ferric chloride) has been for the coupling of phenols, particularly in imitation of biogenetic processes. The products of phenolic oxidations are not confined to dimers; for instance, trimers of novel structure are among derivatives obtained by F. R. Hewgill and his co-workers from phenols substituted by t-butyl and methoxy groups [84]. Applications of iron(III) oxidants to other classes of substrates may be exemplified by a final cyclization step reported for a porphyrin synthesis, involving intramolecular base-catalysed oxidative coupling at α-positions in pyrrole rings [85].

Reactions of ferric chloride with aromatic hydrocarbons, studied by P. Kovacic and his co-workers [86], result both in oxidative coupling and in chlorination; for instance, mesitylene gives chloromesitylene and bimesityl. The chlorination process

$$ArH + 2FeCl_3 \rightarrow ArCl + HCl + 2FeCl_2,$$

which incidentally provides a preparative method for anhydrous ferrous chloride, may be regarded as a type of oxidation; copper(II) halides behave similarly. Mechanisms of iron(III) oxidations are not well explored. In the cases of oxidations of cyclohexanone by ferriin and of thiols by ferricyanide, it is considered that electron transfer occurs without displacement of ligands from the metal by the substrate [87].

(g) Cobalt Compounds

Complexes of cobalt(III) are numerous, and they are prominent in the studies of inorganic redox reactions made by H. Taube [70]. The simple, highly reactive cobalt(III) salts, such as the sulphate and acetate, which are generally prepared by electrolytic oxidation, began to attract attention as oxidants in the 1950s. The principal contribution to their study as oxidants for organic compounds is a series of papers by W. A. Waters and his co-workers, dating from 1962, in which the experimental data are mainly of a kinetic character. This work is summarized in the review by Waters and Littler discussed in subsection (c). Cobaltic salts oxidize not only alcohols, aldehydes, and ketones, but also carboxylic acids and olefins.

(h) Nickel Compounds

The only species that needs to be considered is the so-called nickel peroxide, an oxidant for which several investigators have found applications since K. Nakagawa and his co-workers began publication on the subject in 1962 [88]. The following are typical oxidations which they have described.

$$RCH_2{-}OH \rightarrow RCHO$$

$$ArCHO + NH_3\ (\rightarrow ArCH(OH){-}NH_2) \rightarrow ArCONH_2 \text{ or } ArCN$$

$$2ArNH_2 \rightarrow ArN{=}NAr$$

$$Ar_2C{=}N{-}NH_2 \rightarrow Ar_2CN_2$$

$$2CHCl_3 \rightarrow CCl_3{-}CCl_3$$

The reagent, commonly prepared by treating alkaline nickel(II) sulphate solution with sodium hypochlorite, is used, like manganese dioxide, as a suspension in organic solvents. It is a black powder with a variable content of labile oxygen which needs to be assessed by iodometric titration. Treatment of alkaline nickel(II) sulphate with ozone is possibly a superior preparative method [89]. An example of the value of nickel peroxide in syntheses is the convection of 2-thiouracil into 1-methyl-2-thioorotic acid or 1-methylorotic acid, obtained in about 70 % yield by using two or six oxidizing equivalents, respectively, of the reagent [90]; more conventional oxidants degrade the pyrimidine ring.

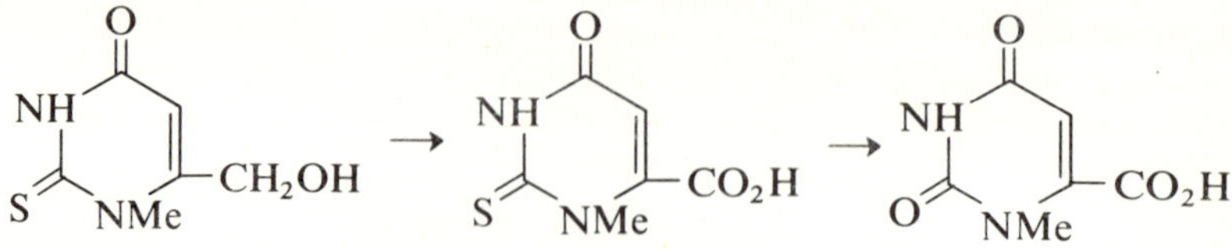

(i) Copper Compounds

Recent studies of oxidative effects of copper compounds in aromatic chemistry have been discussed in a review [13]. Dehydrogenative reaction is exemplified by the coupling of phenols with copper(II) salts in a melt at 200°, and oxygenation is exemplified by the production of phenols, or their esters, when basic or neutral copper(II) carboxylates are decomposed at 200–300°. Aqueous solutions of copper(II) salts, unlike those of iron(III), are relatively ineffective towards aromatic compounds. The review [13] also refers to halogenation by copper(II) halides (cf. ferric halides), to copper-catalysed oxidations by oxygen (cf. p. 56), and to the existence of copper(III) species, which may be labile intermediates in certain reactions, but appear to have limited oxidizing capacity when handled as complexes.

Copper(II) compounds find various applications in the oxidation of functional groups, e.g., $RNH{-}NH_2 \rightarrow RH$, and $-CO{-}CH(OH){-} \rightarrow -CO{-}CO$. A reaction of outstanding importance for organic syntheses, involving the use of copper compounds under various conditions, is the oxidative coupling of compounds containing acetylenic hydrogen, $RC{\equiv}CH \rightarrow RC{\equiv}C{-}C{\equiv}CR$. This subject has been reviewed by G. Eglinton [91], who in 1959 introduced the employment of copper(II) acetate in pyridine solution at ambient temperatures. This is the preferred reagent for converting diacetylenic compounds into cyclic polymers. The well-known investigations of F. Sondheimer and his co-workers, leading to syntheses of macrocyclic conjugated polyenes (annulenes), are of particular importance in this connection.

$CH{\equiv}C{-}(CH_2)_n{-}C{\equiv}CH \xrightarrow{Cu(II)}$ mixtures of cyclic polymers (dimers–heptamers) containing acetylenic bonds

↓ (rearrangement and partial reduction when $n = 2$)

$(CH{=}CH)_x$ (cyclic)

(j) Ruthenium Compounds

Ruthenium tetroxide, RuO_4, has found considerable use in organic syntheses since its introduction in 1953. It is a more vigorous oxidant than osmium tetroxide, is likewise poisonous, and is customarily used in carbon tetrachloride solution. A useful variation in oxidation procedure is to employ catalytic amounts of ruthenium trichloride with an excess of an oxidant, such as periodate, which is capable of maintaining the ruthenium species, the effective oxidant for the organic substrate, at the Ru(VIII) oxidation level. One of its recent applications is for conversions of the type $-CH(OH){-} \rightarrow -CO{-}$ in monosaccharides [92]; hydroxyl groups in which no change is desired must first be protected. The strong oxidizing action of ruthenium tetroxide is exemplified [93] by the fission of aromatic

rings, for example,

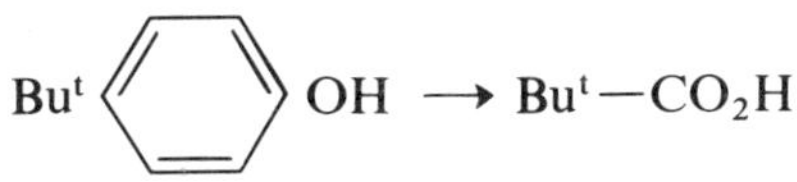

(k) Palladium Compounds

One of the most important developments during this decade has been the introduction of palladium(II) as an oxidant. This is mainly due to the development of an industrial process, with which J. Smidt is particularly associated, for the conversion of ethylene into acetaldehyde; it is complementary to the silver-catalysed gas-phase oxidation (p. 55) which produces ethylene oxide. Numerous investigations of olefin oxidation by palladium(II) solutions have been described since about 1960, and the subject has recently been covered by a review [94] which includes detailed discussion of possible mechanisms.

The industrial process is made economic by the use of copper(II) as a second oxidant, maintaining the supply of palladium(II), and by the use of air to replenish copper(II).

$$C_2H_4 + PdCl_2 + H_2O \longrightarrow CH_3CHO + Pd + 2HCl$$
$$Pd + 2CuCl_2 \longrightarrow PdCl_2 + 2CuCl$$
$$2CuCl + 2HCl + 0{\cdot}5O_2 \longrightarrow 2CuCl_2 + H_2O$$
$$(\text{Overall: } C_2H_4 + 0{\cdot}5O_2 \longrightarrow CH_3CHO)$$

If acetic acid is present instead of water, the product is vinyl acetate, whilst vinyl ethers and acetals result if alcohols are used.

In solutions containing chloride ion, four-coordinate Pd(II) is present as the ion $PdCl_4^{2-}$, which is in equilibrium with other complexes containing water, hydroxyl ion, and π-bonded ethylene as ligands. According to one view, the reaction may involve the switch of ethylene from a π-bonded to a σ-bonded state and an oxidative disruption of the resulting complex.

$$PdCl_4^{2-} + C_2H_4 \rightleftharpoons [PdCl_3(C_2H_4)]^- + Cl^-$$
$$[PdCl_3(C_2H_4)]^- + H_2O \rightleftharpoons [PdCl_2(H_2O)(C_2H_4)] + Cl^-$$
$$[PdCl_2(H_2O)(C_2H_4)] + H_2O \rightleftharpoons [PdCl_2(OH)(C_2H_4)]^- + H_3O^+$$
$$[PdCl_2(OH)(C_2H_4)]^- \rightleftharpoons HO{-}CH_2{-}CH_2{-}PdCl + Cl^-$$

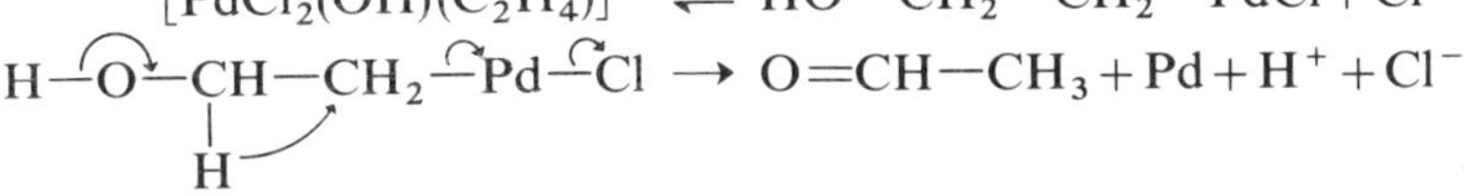

Other oxidative effects of palladium(II) include the coupling of benzene and its homologues [95]

$$2ArH + PdCl_2 + 2NaOAc \longrightarrow Ar{-}Ar + Pd + 2NaCl + 2AcOH$$

and the conversion of primary amines into isocyanates [96]

$$RNH_2 + CO + PdCl_2 \longrightarrow RNCO + Pd + 2HCl$$

(l) Silver Compounds

The traditional silver-containing oxidants [3] are monovalent species, e.g., Ag_2O, $AgNO_3$, and $Ag(NH_3)_2^+$; these continue to find application e.g., for oxidation of phenols and hydrazino compounds. In recent years considerable interest has been shown in compounds containing silver in higher oxidation states, e.g., argentic oxide, AgO, and organic complexes of silver(II) and silver(III) [97]. The oxide is precipitated if aqueous silver nitrate is oxidized, for instance by persulphate (p. 65), while four-coordinate silver(II) complexes are precipitated if, for example, heterocyclic bases are present during the oxidation. The oxide [72] does not appear to be a particularly useful oxidant, but it is reported to give good yields of carbonyl compounds from benzylic alcohols and arylalkanes when it is dissolved in strong acids [98]. A complex, silver(II) picolinate, $(C_5H_4N{\cdot}CO_2)_2Ag$, is an effective oxidant, e.g., for amines and phenols, at moderate temperatures, when used as a suspension in a polar solvent [99].

(m) Cerium Compounds

Cerium has a special position among the lanthanides because of its ability to provide stable tetravalent derivatives. Cerium(IV) salts are commonly used in acidic aqueous solutions, in which equilibrium exists between a variety of anionic and cationic cerium species. Such solutions are powerful oxidants, e.g., for alcohols and glycols, aldehydes and ketones, phenols, amines, and some aromatic hydrocarbons; studies of these reactions extend over forty years or more. The status of the oxidant is in some respects similar to that of the species vanadium(V), manganese(III), and cobalt(III), i.e., its oxidations involve single-electron transfer, and they have attracted much attention with respect to their kinetics and to the formulation of possible intermediate complexes; this work has lately been reviewed by W. H. Richardson [2]. Ceric salts have been recommended for various practical purposes, e.g., for the efficient oxidation of CH_3 and CH_2OH groups on aromatic rings [100].

(n) Osmium Compounds

Solutions of osmium tetroxide continue to be widely used for hydroxylation or cleavage of olefinic bonds or polycyclic aromatic nuclei [3, 28], sometimes in conjunction with other oxidants, such as periodate or hydrogen peroxide. Osmium chemistry has lately been reviewed [101].

(o) Mercury Compounds

Mercury(II) oxide has long been used for heterogeneous oxidations and is often the preferred reagent for dehydrogenative reactions with nitrogen-containing organic compounds, e.g., hydrazine derivatives. Mercuric acetate and other mercury(II) salts are widely used [3] for the additive mercuration of olefins and substitutive mercuration of aromatic compounds. An olefin sometimes undergoes oxidation by acetoxylation in the allylic position, whilst cyclohexanone gives the α-diketone [63]. The dehydrogenating power of mercuric acetate was demonstrated by N. J. Leonard's application of the reagent in the 1950s for oxidations of alkaloids and other amines. Dehydrogenation may be effected in cyclic compounds of other types, e.g., steroids and terpenes [102]. Two-electron transfer occurs [63] in oxidations by mercury(II):

$$Hg^{2+} + 2e \rightarrow Hg; \qquad Hg + Hg^{2+} \rightleftharpoons Hg_2{}^{2+}$$

(p) Thallium Compounds

Increasing attention is being given to oxidations by thallium(III) salts, particularly in comparison with oxidations by salts of mercury(II) and lead(IV), the neighbouring elements in the Periodic Table. Two-electron transfer, $Tl^{3+} + 2e \rightarrow Tl^{+}$, is similarly involved [63]. Mainly, oxidations of olefins have been studied [103]; ethylene gives the glycol and acetaldehyde, presumably via a Tl(III) complex; cyclohexene reacts additively with thallium triacetate to give *cis*- and *trans*-diacetates, substitutively to give the allylic acetate, and also undergoes ring contraction. The mechanism of ring contraction in cyclohexanone, giving cyclopentanecarboxylic acid, has been investigated [104].

(q) Lead Compounds

Lead dioxide, and in some cases the oxide Pb_3O_4, are useful reagents for heterogeneous oxidations [72], whilst lead tetra-acetate is a soluble oxidant of outstanding importance, and its applications are continually being extended. Oxidations by lead tetra-acetate have been frequently reviewed, and most recently [2] by R. Criegee, who pioneered the introduction of the reagent in the 1930s. In the same volume [2] as Criegee's review there is a general account by C. A. Bunton of the oxidative cleavage of glycols, effected by lead tetra-acetate, periodic acid, iodoso compounds, or sodium bismuthate. The acetoxylation of phenols by lead tetra-acetate, studied particularly by F. Wessely and his co-workers (giving acetoxycyclohexadienones, from which dihydric phenols may be obtainable) is a topic which is included in J. D. Loudon's survey of methods for the hydroxylation of phenols [105]. The numerous and varied applications of lead tetra-acetate have been well documented by L. F. and M. Fieser [3].

Two recent series of papers on the subject are from the research schools of R. O. C. Norman [106] and M. Lj. Mihailović [107]. These describe extensions of applications of the reagent, e.g., for oxidations of hydrazones and of amines, and they include discussion of the alternatives of one-electron or two-electron transformation of the oxidant ($Pb^{4+} + e \rightarrow Pb^{3+}$; $Pb^{4+} + 2e \rightarrow Pb^{2+}$), a subject which has long been a cause of argument. A recent variation in procedure [3] is use of lead tetra-acetate in conjunction with iodine, leading, for example, to reactions of the type $R{-}CO_2H \rightarrow RI$. Another variation [3] is use of the reagent in conjunction with boron trifluoride, whereby, e.g., α-acetoxylation of ketones occurs:

$$-CH_2-CO- \rightarrow -CH(OAc)-CO-$$

The known influence of pyridine on oxidations by lead tetra-acetate has lately been associated with formation of an isolable 1:1-complex [108].

(r) Bismuth Compounds

Bismuth is present in certain oxidation catalysts (p. 56). The oxidant sodium bismuthate, approximating in composition to $NaBiO_3$, is a sparingly soluble Bi(V) derivative which is prepared by oxidizing bismuth(III) hydroxide in alkali. It has found a limited amount of application since its introduction as a glycol-cleaving reagent by W. Rigby in 1950. A brief review of the literature by W. F. Pickering [72] demonstrates that it is also applicable to a variety of other types of oxidation.

(s) Catalytic Dehydrogenation by Transition Metals

Supported palladium catalysts were introduced in the 1930s for the dehydrogenation of alicyclic compounds [58]. A more recent development is a procedure for preparing heterocyclic 2,2′-biaryls by the catalysed dehydrogenative coupling of bases, such as pyridine, quinoline, and their homologues, generally carried out at the boiling point. The preferred catalyst is Raney nickel which has been degassed by heating *in vacuo*. Details are given in a series of papers by G. M. Badger, W. H. F. Sasse, and their co-workers, over a period of about ten years, beginning with their report of the preparation of 2,2′-bipyridyl in 1956 [109]. The yields are frequently good, but variable amounts of by-products, e.g., pyrroles from pyridines and indoles from quinolines, may be present.

A possible reaction mechanism involves transfer of an electron from the metal to an adsorbed pyridine molecule, giving an adsorbed radical-anion; coupling would produce adsorbed dihydrobipyridyl, from which hydrogen could be transferred to the metal.

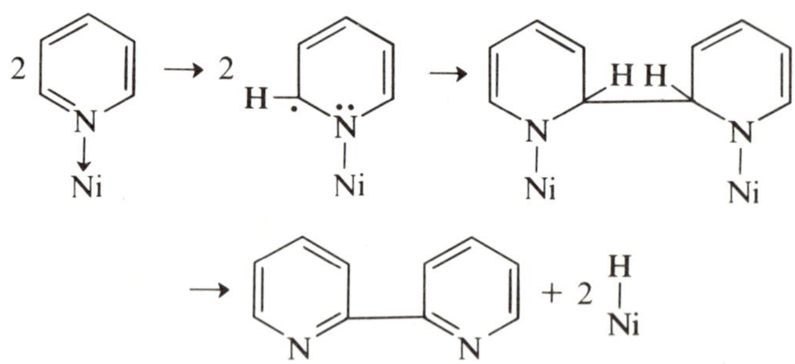

(t) Anodic Oxidations

Revival of interest in the electrolytic coupling reaction discovered by Kolbe (1849) is largely due to F. Fichter's studies in the 1930s and to later applications, by R. P. Linstead, B. C. L. Weedon, and their co-workers, for the synthesis of numerous long-chain hydrocarbons and acids, using mono-carboxylic acids, half-esters of dicarboxylic acids, or mixtures of these; Weedon has reviewed the subject [110].

More recent studies (some of which are summarized by R. Stewart [1]) deal with mechanistic aspects or with applications of the technique in new directions. Possible mechanisms may be summarized in the scheme

$$2R{-}CO_2^- \xrightarrow{-e} R{-}CO_2\cdot \nearrow CO_2 + R\cdot \rightarrow \text{coupled products}$$

$$R{-}CO_2\cdot \xrightarrow{-e} CO_2 + R^+ \rightarrow \text{other products}$$

Examples of substrates recently investigated are aromatic amines in aqueous acid [111], aromatic carboxylic acids [112], and β-diketones [113].

3.4 ORGANIC OXIDANTS

Many reactions between pairs of organic compounds are redox processes; enzymic reactions of this type have great biological importance. In the present review discussion is restricted to cases in which there may be practical advantages in using an organic compound as a consumable reagent for the oxidation of an organic substrate. Reagents may be broadly divided into two types: those which accept hydrogen and those which donate oxygen.

(a) Hydrogen-transfer Processes

(i) Metal-catalysed Processes

A typical procedure is that investigated by E. A. Braude, R. P. Linstead, and their co-workers, and reviewed by L. M. Jackman [114]; palladium, under mild conditions, catalyses hydrogen-transfer reactions in which the donor is cyclohexene (becoming benzene) and the acceptor may be, e.g., an olefinic, azo, or nitro group.

$$C_6H_{10}+2X \rightarrow C_6H_6+2XH_2$$

However, for practical purposes such reactions are to be regarded as reductions, since the desired product is the compound XH_2.

A related type of reaction, more recently studied [13], is hydrogen transfer from a variety of organic donors (phenols, amines, alkoxides, etc.) to aryl halides or α-halogenoketones, in suitable organic solvents, under the catalytic influence of copper species such as cuprous oxide.

$$\text{ArBr}+\text{XH}_2\ (\text{or } 2\text{YH}) \rightarrow \text{ArH}+\text{HBr}+\text{X} \quad (\text{or Y—Y})$$

Again, this is generally to be regarded as a reduction, giving ArH, though in some instances substantial yields of oxidized products (e.g., a dehydrogenated amine, X, or a coupled phenol, Y—Y) may be simultaneously available.

A process of different character, since it requires metal atoms to be combined in the reaction system, is the well-known transfer of hydrogen between aluminium alkoxides (or Grignard reagents) and a carbonyl group; this can be applied in the form of Oppenauer reactions to obtain oxidized products.

(ii) Photochemical Hydrogen Transfer

Hydrogen transfer, e.g., from alcohols or amines to ketones, is an effect frequently encountered in photochemistry (see Chapter 6). Phenols have recently been studied as donors in reactions of this type [115].

(iii) Carbonium Ions as Oxidants

It was recognized in the period 1956–1959, by P. D. Bartlett, R. Stewart and others [1], that certain oxidations may be achieved with the aid of carbonium ions (e.g., Ph_3C^+) which are neither too labile nor too stable to act as efficient hydride-ion acceptors. Alcohols, formic acid and cycloheptatriene (cf. p. 69) are examples of compounds thus oxidized. This method, involving, e.g., a triphenylmethyl salt or halide, continues to find use as a preparative method for dehydrogenated products [116].

(iv) Quinones, Triketones, Nitro Compounds, etc., as Oxidants

Without the aid of catalysts, organic compounds containing very readily reduced unsaturated groups may act as hydrogen-acceptors for certain other organic compounds [1]. Nitrobenzene is long established as a reagent for some aromatic side-chain oxidations and as a component in the Skraup reaction. Alloxan and ninhydrin are triketones which have found applications as oxidants; the chemistry of oxidations by the latter reagent has only fairly recently been clarified [117]. A compound containing an azo-linkage of high receptivity may also be effective; a recent example [118] is the use of 4-phenyl-1,2,4-triazoline-3,5-dione as an oxidant for alcohols.

Quinones, the main class of compounds to be considered, have acquired steadily increasing importance as oxidants since about 1930; their uses are discussed in Jackman's review [114]. The most suitable reagents are those with a high redox potential, e.g., 2,3-dichloro-5,6-dicyano-1,4-benzoquinone and tetrachloro-1,2-benzoquinone; diphenoquinones may also be used. Ultraviolet irradiation is occasionally employed to stimulate reactions. Quinones have found their chief applications as oxidants for cyclic molecules; for example, for aromatization of alicyclic compounds, dehydrogenation of dihydro derivatives of heterocyclic systems, and conversion of steroidal enones into dienones. They are finding applications in other directions, e.g., for the oxidative coupling of phenols and enolizable ketones [119].

(b) Oxygen-transfer Processes

In recent years one of the most prominent processes of this type has been the transfer of oxygen from a wide variety of organic compounds to phosphorus(III) derivatives, such as triphenylphosphine or trialkyl phosphites [120]. However, in the following typical example,

$$\underset{\text{(epoxide)}}{-\overset{|}{C}-\overset{|}{C}-} + P(OEt)_3 \rightarrow -\overset{|}{C}=\overset{|}{C}- + OP(OEt)_3$$

the practical consequence is the reduction of an epoxide, though the latter is incidentally serving as an organic oxidant for a phosphite.

In principle, potential organic oxidants with labile oxygen are best sought among species which yield stable molecules on the detachment of oxygen from a hetero-atom (typically N, S, or I). Many examples of such oxygen-transfer are to be found: in nitrogen chemistry, among the reactions of *N*-oxides, azoxy compounds, etc.; in iodine chemistry, among the reactions of iodoso compounds (well known for oxidations in the form of phenyliodoso diacetate, $PhI(OAc)_2$ [1]); and in sulphur chemistry, among the reactions of sulphoxides.

An outstanding example within these groups of compounds is dimethyl sulphoxide, introduced by N. Kornblum in 1957 for the conversion of alkyl tosylates and phenacyl halides into aldehydes.

$$Me_2SO + CH_2R-X \rightarrow Me_2\overset{+}{S}-O-CH_2R + X^-$$
$$\downarrow$$
$$Me_2S + RCHO + H^+$$

Oxidations of this type have lately been greatly extended in scope by procedural variations which make the method applicable to the oxidation of alcohols; further transformation of aldehyde into carboxylic acid, if the alcohol is primary, is avoided. The use of dimethyl sulphoxide as an oxidant depends upon intermediate formation of dimethylsulphoxonium salts. D. H. R. Barton facilitated this step by use of a chloroformate prepared from an alcohol [121].

$$Me_2SO + CH_2R{-}O{-}CO{-}Cl \rightarrow Me_2\overset{+}{S}{-}O{-}CO{-}O{-}CH_2R + Cl^- \downarrow Me_2S + CO_2 + RCHO + H^+$$

A superior method developed by J. G. Moffatt [122] involves preliminary reaction of dimethyl sulphoxide with dicyclohexylcarbodi-imide, in the presence of a good proton donor, e.g., phosphoric acid; the resulting adduct is efficiently transformed into the sulphoxonium salt by reaction with an alcohol.

$$NR'{=}C{=}NR' + Me_2SO + H^+ \rightarrow NHR'{-}C(=NR'){-}O{-}\overset{+}{S}Me_2 \xrightarrow{CH_2R-OH} NHR'{-}CO{-}NHR' + Me_2\overset{+}{S}{-}O{-}CH_2R$$

Other variations involve the use of dimethyl sulphoxide with acetic anhydride, when $Me_2\overset{+}{S}{-}OAc$ is a likely intermediate [123], or with sulphur trioxide [124].

REFERENCES

[1] WIBERG, K. B. In *Surveys of Progress in Chemistry* (ed. A. F. Scott), Vol. 1, pp. 211–248, Academic Press, New York (1963); WATERS, W. A. *Mechanisms of Oxidation of Organic Compounds*, Methuen, London (1964); STEWART, ROSS *Oxidation Mechanisms*, Benjamin, New York (1964); TURNEY, T. A. *Oxidation Mechanisms*, Butterworths, London (1965).

[2] Various authors in *Oxidation in Organic Chemistry* (ed. K. B. Wiberg), Part A, Academic Press, New York (1965).

[3] FIESER, L. F. and FIESER, M. *Reagents for Organic Synthesis*, Wiley, New York (1967).

[4] *Organic Reaction Mechanisms* (ed. B. Capon, M. J. Perkins and C. W. Rees), Interscience, New York (1965, 1966, 1967); Various authors, 'General methods', *A. Rep. chem. Soc.* (since 1950).

[5] RABJOHN, N. *Org. React.* **5**, 331–386 (1949).

[6] JACKSON, E. L. *Org. React.* **2**, 341–375 (1944).

[7] DAVIES, A. G. *Organic Peroxides*, Butterworths, London (1961).

[8] MARGOLIS, L. YA. *Adv. Catalysis* **14**, 429–501 (1963); VOGE, H. H. and ADAMS, C. R. *Adv. Catalysis* **17**, 151–221 (1966); SAMPSON, R. J. and SHOOTER, D. in *Oxidation and Combustion Reviews* (ed. C. F. H. Tipper), Vol. 1, pp. 223–302, Elsevier, Amsterdam (1965); CULLIS, C. F. and FISH, A. in *Chemistry of the Carbonyl Group* (ed. S. Patai), pp. 79–176, Interscience, New York (1966).

[9] In *Selective Oxidation Processes*, a symposium, Advances in Chemistry Series No. 51, American Chemical Society, Washington (1965).

[10] *Org. Synth.*, Coll. Vol. **4**, 493–496 (1963).

[11] HEYNS, K. and PAULSEN, H. *Adv. Carbohyd. Chem.* **17**, 169–221 (1962); *Newer Methods in Preparative Organic Chemistry* (ed. by W. A. Foerst), Vol. 2, pp. 303–335, Academic Press, New York (1963).

[12] BRACKMAN, W. *et al. Recl Trav. chim. Pays-Bas Belg.* **82**, 757–762 (1963); **84**, 1203–1229 (1965); **85**, 437–445 (1966); and related papers.

[13] BACON, R. G. R. and HILL, H. A. O. *Q. Rev. chem. Soc.* **19**, 95–125 (1965).

[14] BLUM, J., ROSENMAN, H. and BERGMANN, E. D. *Tetrahedron Lett.* 3665–3668 (1967).

[15] OTSUKA, S., NAKAMURA, A. and TATSUNO, Y. *Chem. Commun.* 836 (1967).

[16] HORNER, L. and DEHNERT, J. *Chem. Ber.* **96**, 786–797 (1963).

[17] WALLACE, T. J. *et al. J. org. Chem.* **29**, 888–891 (1964) and related papers.

[18] GOLLNICK, K. and SCHENCK, G. O. in *1,4-Cycloaddition Reactions* (ed. by J. Hamer) pp. 255–344, Academic Press, New York (1967).

[19] WILSON, T. *J. Am. chem. Soc.* **88**, 2898–2902 (1966) and references therein cited; MCKEOWN, E. and WATERS, W. A. *J. chem. Soc.* (*B*) 1040–1046 (1966).

[20] *Org. Synth.* **45**, 91–95 (1965); COLLINS, D. J. and HOBBS, J. J. *Aust. J. Chem.* **20**, 1905–1920 (1967).

[21] HUISGEN, R. *Proc. chem. Soc.* 357–369 (1961).

[22] BAILEY, P. S. *Chem. Rev.* **58**, 925–1010 (1958).

[23] STORY, P. R. *et al. J. Am. chem. Soc.* **87**, 737–741 (1965) and later papers.

[24] *Org. Synth.* **41**, 41–45 (1961); CALLIGHAN, R. H., TARKER, M. F. and WILT, M. H. *J. org. Chem.* **26**, 1379– 1382 (1961).

[25] In *Elucidation of Structures by Physical and Chemical Methods* (ed. K. W. Bentley) (*Technique of Organic Chemistry* ed. A. Weissberger, Vol. XI), Part 2, Interscience, New York (1963).

[26] LEE, J. B. and UFF, B. C. *Q. Rev. chem. Soc.* **21**, 429–457 (1967).

[27] PAYNE, G. B. *et al. J. org. Chem.* **26**, 651–659 (1961) and later papers.

[28] GUNSTONE, F. D. *Adv. org. Chem.* **1**, 103–147 (1960).

[29] ALDER, R. W. and WHITING, M. C. *J. chem. Soc.* 4707–4712 (1964).

[30] *Org. Synth.* **40**, 76–79 (1960).

[31] NORMAN, R. O. C. *et al. J. chem. Soc.* 2897–2905, 3119–3124 (1963) and later papers.

[32] *Org. Synth.* **40**, 90–92 (1960).

[33] KAHR, K. *Angew. Chem.* **72**, 135–137 (1960); *Chem. Ber.* **93**, 132–136 (1960).

[34] BACON, R. G. R. *et al. J. chem. Soc.* (*C*) 791–793 (1966) and later papers.

[35] CASPI, E., MALHOTRA, S. K., SHIMIZU, Y., MAHESHWARI K. and GASIC, M. J. *Tetrahedron* **22**, 595–602 (1966).

[36] CONNOR, J. A. and EBSWORTH, E. A. V. *Adv. inorg. Chem. Radiochem.* **6**, 279–381 (1964).

[37] SAUNDERS, B. C. *et al. Tetrahedron* **23**, 1867–1872 (1967) and earlier papers; SAUNDERS, B. C., HOLMES-SIEDLE, A. G. and STARK, P. B. *Peroxidase*, Butterworths, London (1964).

[38] MATSUURA, T. and OMURA, K. *Chem. Commun.* 127–128 (1966).

[39] BEHRMAN, E. J. *J. Am. chem. Soc.* **85**, 3478–3482 (1963).

[40] NEEDLES, H. L. and WHITFIELD, R. E. *J. org. Chem.* **31**, 341–342 (1966).

[41] BACON, R. G. R. *et al. J. chem. Soc.* (*C*) 1384–1387 (1966) and related papers.

[42] HOUSE, D. A. *Chem. Rev.* **62**, 185–203 (1962).

[43] *Org. Synth.* **40**, 18–20 (1960).

[44] HOUSE, H. O. *Modern Synthetic Reactions*, Benjamin, New York (1965).

[45] KWART, H. and HOFFMAN, D. M. *J. org. Chem.* **31**, 419–425 (1966).

[46] BOYD, D. R. and MCKERVEY, M. A. *Q. Rev. chem. Soc.* **22**, 95–122 (1968).

[47] ROBINSON, C. H., MILEWICH, L. and HOFER, P. *J. org. Chem.* **31**, 524–528 (1966); PEWS, R. G. *J. org. Chem.* **32**, 1628 (1967).

[48] SMITH, H., WEGFAHRT, P. and RAPOPORT, H. *J. Am. chem. Soc.* **90**, 1668–1669 (1968).

[49] Sagramora, L., Koch, P., Garbesi, A. and Fava, A. *Chem. Commun.* 985–986 (1967); Zwanenburg, B., Thijs, L. and Strating, J. *Recl Trav. chim. Pays-Bas Belg.* **86**, 577–588 (1967); *Tetrahedron Lett.* 3453–3456 (1967).

[50] Kochi, J. K. and Mains, H. E. *J. org. Chem.* **30**, 1862–1872 (1965) and related papers.

[51] *Org. Synth.* **44**, 12–15 (1964).

[52] Whistler, R. L. and Wolfrom, M. L. (Eds) *Methods in Carbohydrate Chemistry*, Vol. 2, Academic Press, New York (1963).

[53] Ogata, Y. *et al. Tetrahedron* **23**, 1007–1014 (1967) and related papers.

[54] Forrester, A. R. and Thomson, R. H. *J. chem. Soc.* (*C*), 1844–1846 (1966).

[55] Barton, D. H. R. *et al. Tetrahedron* **18**, 373–378 (1962) and related papers.

[56] *Org. Synth.* **43**, 101–104 (1963).

[57] Weedon, B. C. L. *et al. Tetrahedron* **8**, 221–260 (1960); Bonner, W. A. and Rewick, R. T. *J. Am. chem. Soc.* **84**, 2334–2337 (1962).

[58] Valenta, Z. in *Elucidation of Structures by Physical and Chemical Methods* (ed. K. W. Bentley) (*Technique of Organic Chemistry* ed. A. Weissberger, Vol. XI), Part 1, pp. 581–642, Interscience, New York (1963).

[59] Colebourne, N., Foster, R. G. and Robson, E. *J. chem. Soc.* (*C*) 685–688 (1967).

[60] Wegler, R., Kühle, E. and Schäfer, W. in *Newer Methods of Preparative Organic Chemistry* (ed. by W. Foerst), Vol. 3, pp. 1–51, Academic Press, New York (1964).

[61] Geluk, H. W. and Schlatmann, J. L. M. A. *Chem. Commun.* 426 (1967).

[62] Horner, L. and Winkelmann, E. H. in *Newer Methods of Preparative Organic Chemistry* (ed. by W. Foerst), Vol. 3, pp. 151–198, Academic Press, New York (1964).

[63] Littler, J. S. *J. chem. Soc.* 827–832 (1962).

[64] Barton, D. H. R., O'Brien, R. E. and Sternhell, S. *J. chem. Soc.* 470–476 (1962).

[65] Brown, D. M. and Jones, G. H. *J. chem. Soc.* (*C*) 252–256 (1967).

[66] Deno, N. C. and Potter, N. H. *J. Am. chem. Soc.* **89**, 3555–3556 (1967); Venkatasubramanian, N. and Thiagarajan, V. *Tetrahedron Lett.* 1711–1714 (1968).

[67] Sklarz, B. *Q. Rev. chem. Soc.* **21**, 3–28 (1967).

[68] *Reactions of Coordinated Ligands*, Advances in Chemistry Series No. 37, American Chemical Society, Washington (1963).

[69] Anbar, M. in *Mechanisms of Inorganic Reactions*, Advances in Chemistry Series No. 49, pp. 126–152, American Chemical Society, Washington (1965).

[70] Sykes, A. G. *Adv. inorg. Chem. Radiochem.* **10**, 153–245 (1967); cf. Taube, H. in Ref. [69], pp. 107–125.

[71] Evans, R. M. *Q. Rev. chem. Soc.* **13**, 61–70 (1959).

[72] Pickering, W. F. *Rev. pure appl. Chem.* **16**, 185–208 (1966).

[73] Zabin, B. A. and Taube, H. *Inorg. Chem.* **3**, 963–968 (1964).

[74] Scott, A. I. *Q. Rev. chem. Soc.* **19**, 1–35 (1965); Taylor, W. I. and Battersby, A. R. (Eds.), *Oxidative Coupling of Phenols*, Arnold, London (1967).

[75] Musso, H. *Angew. Chem. int. Edn* **2**, 723–735 (1963).

[76] Latjaeva, V. N., Razuvaev, G. A., Malisheva, A. V. and Kiljakova, G. A. *J. organometall. Chem.* **2**, 388–397 (1964).

[77] Rocek, J. in *Chemistry of the Carbonyl Group* (Ref. [8]), pp. 461–505.

[78] Waters, W. A. *Q. Rev. chem. Soc.* **12**, 277–300 (1958).

[79] Shechter, H. *et al. J. Am. chem. Soc.* **86**, 1701–1709 (1964).

[80] Morgan, L. R. and Aubert, C. C. *Proc. chem. Soc.* 73 (1962).

[81] Barton, M. A., Kenner, G. W. and Sheppard, R. C. *J. chem. Soc.* (*C*) 1061–1065 (1966).

[82] Henbest, H. B. *et al. J. chem. Soc.* (*C*) 995–996 (1966) and earlier papers.

[83] COOK, M. J., FORBES, E. J. and KAHN, G. M. *Chem. Commun.* 121–122 (1966).

[84] HEWGILL, F. R. *et al. J. chem. Soc.* 4987–4995 (1962) and later papers.

[85] DOLPHIN, D., JOHNSON, A. W., LENG, J. and VAN DEN BROEK, P. *J. chem. Soc.* (*C*) 880–884 (1966).

[86] KOVACIC, P. *et al. J. org. Chem.* **26**, 759–762 (1961) and related papers; *Inorg. Synth.* **6**, 172–173 (1960).

[87] LITTLER, J. S. and SAYCE, I. G. *J. chem. Soc.* 2545–2552 (1964); WIBERG, K. B., MALTZ, H. and OKANO, M. *Inorg. Chem.* **7**, 830–831 (1968).

[88] NAKAGAWA, K. *et al. J. org. Chem.* **27**, 1597–1601 (1962) and later papers; reviewed in Ref. [72].

[89] BELEW, J. S. and TEK-LING, C. *Chem. Commun.* 1100–1101 (1967).

[90] WARRENER, R. N. and CAIN, E. N. *Tetrahedron Lett.* 4953–4957 (1967).

[91] EGLINTON, G. and MCCRAE, W. *Adv. org. Chem.* **4**, 225–328 (1963); cf. *Org. Synth.* **45**, 39–42 (1965).

[92] OVEREND, W. G. *et al. J. chem. Soc.* (*C*) 1131–1136 (1966) and related papers; PARIKH, V. M. and JONES, J. K. N. *Can. J. Chem.* **43**, 3452–3453 (1965).

[93] CAPUTO, J. A. and FUCHS, R. *Tetrahedron Lett.* 4729–4731 (1967).

[94] AGUILÓ, A. *Adv. organometall. Chem.* **5**, 321–352 (1967).

[95] VAN HELDEN, R. and VERBERG, G. *Recl. Trav. chim. Pays-Bas Belg.* **84**, 1263–1273 (1965).

[96] STERN, E. W. and SPECTOR, M. L. *J. org. Chem.* **31**, 596–597 (1966).

[97] MCMILLAN, J. A. *Chem. Rev.* **62**, 65–80 (1962).

[98] SYPER, L. *Tetrahedron Lett.* 4193–4198 (1967).

[99] BACON, R. G. R. and HANNA, W. J. W. *J. chem. Soc.* 4962–4968 (1965); LEE, J. B. and CLARKE, T. G. *Tetrahedron Lett.* 415–422 (1967).

[100] TRAHANOVSKY, W. S. and YOUNG, L. B. *J. chem. Soc.* 5777–5778 (1965); *J. org. Chem.* **31**, 2033–2035 (1966).

[101] GRIFFITH, W. P. *Q. Rev. chem. Soc.* **19**, 254–273 (1965).

[102] ALLISON, J. M. *et al. J. chem. Soc.* 3353–3361, 5224–5230 (1961).

[103] ANDERSON, C. B. and WINSTEIN, S. *J. org. Chem.* **28**, 605–606 (1963); HENRY, P. M. *J. Am. chem. Soc.* **87**, 990–994 (1965) and later papers.

[104] WIBERG, K. B. and KOCH, W. *Tetrahedron Lett.* 1779–1782 (1966).

[105] LOUDON, J. D. in *Progress in Organic Chemistry* (eds J. W. Cook and W. Carruthers), Vol. 5, pp. 46–72, Butterworths, London (1961).

[106] NORMAN, R. O. C. *et al. J. chem. Soc.* (*C*) 735–739 (1967) and related papers.

[107] MIHAILOVIĆ, M. LJ. *Tetrahedron* **23**, 721–732 (1967) and related papers.

[108] PARTCH, R. and MONTHONY, J. *Tetrahedron Lett.* 4427–4431 (1967).

[109] BADGER, G. M. and SASSE, W. H. F. *Adv. heterocycl. Chem.* **2**, 179–202 (1963); SASSE, W. H. F. *et al. Aust. J. Chem.* **17**, 337–374 (1964) and related papers; *Org. Synth.* **46**, 5–10 (1966).

[110] WEEDON, B. C. L. in *Advances in Organic Chemistry* (eds R. A. Raphael, E. C. Taylor and H. Wynberg), Vol. 1, pp. 1–34, Interscience, New York (1960).

[111] ADAMS, R. N. *et al. J. Am. chem. Soc.* **84**, 2058–2068 (1962).

[112] BUNYAN, P. J. and HEY, D. H. *J. chem. Soc.* 3787–3790 (1960); 324–331 (1962).

[113] JOHNSTON, K. M. and STRIDE, J. D. *Chem. Commun.* 325 (1966).

[114] JACKMAN, L. M. *Adv. org. Chem.* **2**, 329–366 (1961).

[115] BECKER, H.-D. *J. org. Chem.* **32**, 2115–2144 (1967).

[116] DENO, N., SAINES, G. and SPANGLER, M. *J. Am. chem. Soc.* **84**, 3295–3297 (1962); HECK, R. F. *J. Am. chem. Soc.* **85**, 3387–3390 (1963); BLADON, P., PAUSON, P. L., PROCTOR, G. R. and RODGER, W. J. *J. chem. Soc.* (*C*) 926–929 (1966).

[117] MCCALDIN, D. J. *Chem. Rev.* **60**, 39–51 (1960).

[118] COOKSON, R. C., STEVENS, I. D. R. and WATTS, C. T. *Chem. Commun.* 744 (1966).

[119] Becker, H.-D. *J. org. Chem.* **30**, 982–994 (1965); Hay, A. S. *Tetrahedron Lett.* 4241–4243 (1965).
[120] Cadogan, J. I. G. *Q. Rev. chem. Soc.* **16**, 208–239 (1962).
[121] Barton, D. H. R., Garner, B. J. and Wightman, R. H. *J. chem. Soc.* 1855–1857 (1964).
[122] Moffatt, J. G. *et al. J. Am. chem. Soc.* **87**, 5561–5678 (1965); **88**, 1762–1765 (1966); *Org. Synth.* **47**, 25 (1967).
[123] Albright, J. D. and Goldman, L. *J. Am. chem. Soc.* **89**, 2416–2423 (1967).
[124] Parikh, J. R. and Doering, W. von E. *J. Am. chem. Soc.* **89**, 5505–5507 (1967).

Addendum

Recent reviews (relevant topics indicated in parentheses):

[125] Walker, D. and Herbert, J. D. *Chem. Rev.* **67**, 153–195 (1967) (dehydrogenation by 2,3-dichloro-5,6-dicyanobenzoquinone).
[126] Epstein, W. W. and Sweat, F. W. *Chem. Rev.* **67**, 247–260 (1967) (oxidation by dimethyl sulphoxide).
[127] *Oxidation of Organic Compounds*, Vols. I, II, and III, a symposium, Advances in Chemistry Series Nos. 75, 76, and 77, American Chemical Society, Washington (1968) (oxidation by oxygen).
[128] *Discuss. Faraday Soc.* **46**, 98–227 (1968) (homogeneous oxidations catalysed by metal ions).
[129] Sigel, H. *Angew. Chem. int. Edit.* **8**, 167–177 (1969) (reactions of hydrogen peroxide catalysed by copper(II) complexes).
[130] Tsuji, J. *Accounts Chem. Research* **2**, 144–152 (1969) (oxidations by palladium (II)).
[131] Brilkina, T. G. and Shushunov, V. A. *Reactions of Organometallic Compounds with Oxygen and Peroxides*, Iliffe, London (1969).

4

Synthetic Uses of Free Radicals

G. B. Gill, *Nottingham University*

The study of radicals and of their reactions has developed almost entirely over the last thirty-five years. The intermediacy of radicals in certain reactions was not generally accepted until the 1930s, and even then instances are known where important original research papers were rejected by some learned journals. One or two of these articles appeared years later after the presence of radicals had been conclusively established through e.s.r. studies. The subject has developed rapidly during the 1940s, '50s, and '60s (particularly over the last two decades), and the importance of radical reactions is now widely recognized. Indeed, at the present time about one-third of the output of the world's chemical industry relies on processes in which radical intermediates play an important part.

Since radical chemistry is now so large and diverse a part of organic chemistry, a number of important radical processes must be excluded from this article, and others mentioned only briefly. Photochemical, oxidation, and reduction processes are dealt with elsewhere in this book, and relatively brief mention will be made here of radical reactions in these categories. For the most part only those reactions in which the odd electron is localized on carbon, nitrogen, oxygen, sulphur and halogen atoms are discussed in any detail. Processes involving other types of radical, diradicals, and ground-state triplet species (i.e., carbenes and nitrenes) are largely excluded.

Reactions are discussed from the preparative point of view but experimental details are not given. Developments over the last ten years are the primary consideration, but reference to earlier work is often necessary in order to maintain a coherent picture, particularly since many of the important basic radical reactions were discovered early on in the development of radical chemistry. In order to avoid merely providing a catalogue of several (apparently unrelated) reactions, an attempt has been made to relate these syntheses to the main categories of radical reaction. Accordingly, this article is organized under the following headings.

(1) Main types of radical reaction.

(2) Radical-forming reactions (namely, thermal cleavage of covalent bonds, the radiation-induced cleavage of covalent bonds, and radical-forming redox reactions).

(3) Radical-destroying processes (namely, disproportionation, radical combination, and termination by electron transfer).

(4) Radical transfer processes (namely, intermolecular and intramolecular abstraction reactions, addition, aromatic substitution, displacement, elimination and fragmentation reactions, and radical rearrangements).

Under each heading the synthetic methods are discussed according to the transformations which are effected (e.g., chlorination, reductive decarboxylation). The reactions herein described refer almost entirely to reactions in solution, and mechanistic details are only occasionally given to illustrate the scope and limitations of particular synthetic methods. Since this book is intended for undergraduate chemists (as well as those who are more experienced in the science), certain sections are prefaced by a brief resumé of theoretical and practical considerations. Further, in order to convey as wide an appreciation of this general topic as possible, review articles are given prominence among the references.

4.1 MAIN TYPES OF RADICAL REACTION

Three main reaction categories may be distinguished.

(i) Reactions in which radicals are formed.

$$AB \longrightarrow A\cdot + B\cdot$$

(ii) Processes in which the site of the odd electron changes.

$$A\cdot + XY \longrightarrow AX + Y\cdot$$

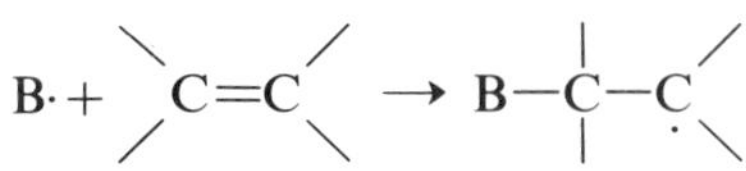

(iii) Reactions in which radicals are destroyed.

$$B\cdot + Y\cdot \longrightarrow BY$$

$$M^{n+} + B\cdot \longrightarrow \begin{cases} M^{(n-1)+} + B^{+} \\ M^{(n+1)+} + B{:}^{-} \end{cases}$$

When the radical transfer step (ii) is entirely absent from a particular reaction, the overall result is a non-chain process. When a few reactions of type (ii) intervene between radical formation (i) and radical destruction (iii), the chain length is short. Frequently, however, the rate of reaction (ii) is much greater than the rates of (i) or (iii), and accordingly the chain length of the process can become great. Since the initiation of chain reactions is often highly desirable and sometimes crucial in organic syntheses using radicals, a knowledge of the kinetics and energetics of such processes is extremely useful in the design of preparative procedures. The reader is referred to the few but excellent texts on radical chemistry [1–4].

Many radical reactions cannot be initiated directly by using thermal or photochemical activation. In such cases a radical initiator is employed. It is convenient to consider first the methods available for radical formation.

Table 4.1 Half-lives of some common initiators

Initiator	Half-life at T°C (hours)[a]						
	40	60	80	100	120	140	150
$Me_3C·O—OH$	—	—	—	24[b]	—	—	—
$Me_3C·O—O·CMe_3$	—	—	—	200	18	2·5	1·0
$Me_3C·O—O·CO·Me$	—	>8000					
$Me_3C·O—O·CO·Ph$	—	—	400	20	1·6	0·15	0·04
$Me_3C·O—O·CO·C(Me)_2Ph$	—	0·2	—	—	—	—	—
$Me·CO·O—O·CO·Me$	—	40	2	1[c]	—	—	—
$PhCH_2·CO·O—O·CO·CH_2Ph$	6[d]	—	—	—	—	—	—
$Ph·CO·O—O·CO·Ph$	—	90	6	0·5	0·06	0·005	—
$Me_2CHO·CO·O—O·CO·OCHMe_2$	—	3[e]	—	—	—	—	—
$^-O_3S·O—O·SO_3^-$	>1000	30	1·5	0·08	0·005	—	—
$Ph—N:N—CPh_3$	1·37[e]	—	0·065	—	—	—	—
$Me_2C(CN)—N:N—C(CN)Me_2$	750	30	1.8	0·1	0·01	—	—
$Me_3C·O—N:N—O·CMe_3$	—	0·5[f]	—	—	—	—	—

[a] The above data should only be used as a rough guide, since the rates of radical production may vary from solvent to solvent.
[b] Decomposes quantitatively in this time.
[c] 85°. [d] 0°. [e] 50°. [f] 65° [5].

4.2 RADICAL-FORMING REACTIONS

The method of radical formation employed is dependent upon the nature of the reactants, the type of reaction to be performed, and the proposed reaction conditions.

(a) The Thermal Cleavage of Covalent Bonds

Molecules containing weak bonds with dissociation energies of 40 kcal mol^{-1} or less dissociate at a convenient rate in solution at temperatures below 150°. The type of such initiators is inevitably somewhat restricted. The most frequently used compounds possess the peroxide (O—O) bond or the azo(N:N) linkage [1–4]. Some of the most popular initiators are listed in Table 4.1 together with details of the temperatures at which they produce radicals at a convenient rate. The elongated bonds are those which initially undergo homolytic cleavage.

In the practical use of such materials, the resulting radicals must be sufficiently reactive and must be produced at convenient and predictable rates. In designing experiments involving radical-chain processes, it is advisable to select conditions so that the half-life of the initiator is of the same order as the reaction time. In chain reactions the amount of initiator required can be very small, whereas in non-chain processes relatively large amounts are usually necessary. It is therefore expedient to carry out preliminary experiments on a small scale.

(b) The Radiation-Induced Cleavage of Covalent Bonds

Even strong covalent bonds may be cleaved if the organic molecule is able to absorb sufficient energy. This energy may be supplied by any form of electromagnetic radiation (ultraviolet light, visible light, or X-rays) or corpuscular radiation (α, β, or γ particles, neutrons, protons, or high-energy electrons). Provided the absorbed energy is not dissipated in other ways, homolytic dissociation may follow.

$$Me_3C{\cdot}O{-}Cl \xrightarrow[(300\text{–}320\text{ nm})]{h\nu} Me_3C{\cdot}O{\cdot} + Cl{\cdot} \tag{4.1}$$

$$CCl_4 \overset{\gamma}{\rightsquigarrow} Cl{\cdot} + {\cdot}CCl_3 \tag{4.2}$$

The most usual source of such radiation is ultraviolet or visible light, because of the cheapness and ease of handling of the necessary equipment. A necessary prerequisite is that the irradiated compound should have a convenient absorption band in the UV or visible spectrum corresponding to a dissociative electronic transition. One is thus largely interested in a small portion of the total area of photochemistry [6–8]. Quartz apparatus is essential for the region 230–300 nm of the UV spectrum, and from the point of view of light absorption the choice of solvent is also important.

The use of radiation in the chemical manufacturing industry is rapidly becoming more important, particularly since the cost of suitable isotopes (cobalt-60 and caesium-137) has decreased significantly in recent years.

The photochemical and radiolytic methods of radical production represent temperature-independent processes. As such they are complementary to the thermal method.

(c) Radical-Forming Redox Reactions

In redox reactions radicals can be produced by oxidation or by reduction. The rate of radical production can be controlled by varying the concentration of each member of the redox pair. These reactions are therefore not subject to the limitations experienced in the use of thermal initiators. The possibilities are briefly illustrated by the following equations.

OXIDATIONS

$$Fe^{3+} + HO{\cdot}OH \longrightarrow Fe^{2+} + HO{\cdot}O{\cdot} + H^{+} \qquad (4.3)$$

$$RCO_2^{-} \xrightarrow[(-e^{-})]{\text{anode}} RCO_2{\cdot} \longrightarrow R{\cdot} + CO_2 \qquad (4.4)$$

REDUCTIONS

$$Cu^{+} + RO{\cdot}OH \longrightarrow Cu^{2+} + RO{\cdot} + OH^{-} \qquad (4.5)$$

$$Ph{\cdot}CH{=}CH{\cdot}Ph \xrightarrow[(+H^{+},\,+e^{-})]{\text{cathode}} Ph{\cdot}\dot{C}H{\cdot}CH_2{\cdot}Ph \qquad (4.6)$$

Once radicals have been produced, two types of reaction are open to them: radical transfer reactions, and processes in which radicals are destroyed. Since the end product of a synthesis is not often a radical, it is convenient to consider the termination reactions first.

4.3 RADICAL-DESTROYING PROCESSES

Radical reactions are terminated by disproportionation, combination, and electron transfer reactions.

(a) Disproportionation

This mode of termination involves the transfer of a monovalent atom from one radical to another. The atom transferred, which is normally hydrogen, occupies a position β to the odd-electron centre in one of the two radicals. Syntheses are rarely designed in which the product molecule is formed by the disproportionation of two identical radicals, as in reaction (4.7), since the yield could never exceed 50%.

$$R{\cdot}CH_2{\cdot}CH_2{\cdot} + R{\cdot}CH_2{\cdot}CH_2{\cdot} \longrightarrow R{\cdot}CH{=}CH_2 + R{\cdot}CH_2CH_3 \qquad (4.7)$$

The synthetically important atom transfer processes are those in which the monovalent atom is transferred between two dissimilar radicals, as in (4.8). This type of disproportionation almost invariably follows on transfer processes, and so will be discussed in the next section. However, it should be

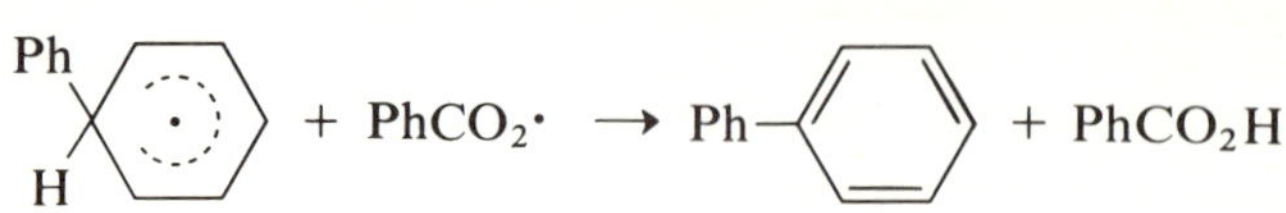

(4.8)

mentioned that disproportionation reactions frequently compete with radical combination in the termination process, and may therefore represent undesirable side reactions (4.9).

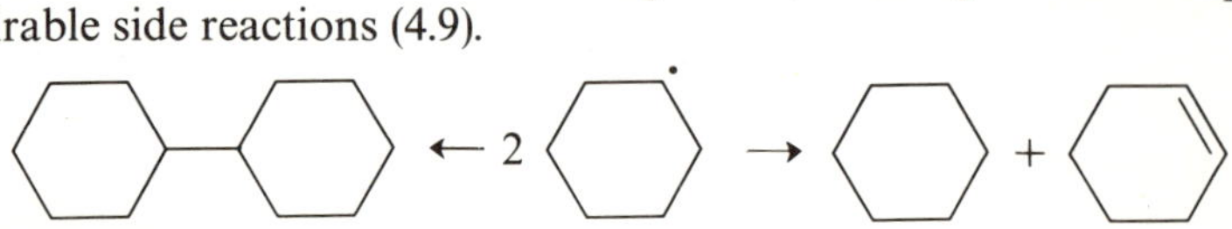

(4.9)

(b) Radical Combination

Termination by radical combination is not necessarily preceded by a chain reaction. In abstraction–combination reactions a pair of radicals must be generated for each molecule of dimer produced.

$$\begin{aligned} RH + A\cdot &\rightarrow R\cdot + AH \\ 2R\cdot &\rightarrow R{\cdot}R \end{aligned} \qquad (4.10)$$

The radical A· comes from the initiator, and therefore substantial amounts of initiator are required in such reactions. The choice of initiator depends upon the solvent employed and the dissociation energy of the R—H bond. In aqueous media Fenton's reagent ($FeSO_4 + H_2O_2$), a source of hydroxyl radicals, is commonly used. In organic solvents acetyl peroxide, di-t-butyl peroxide, and various peresters have found wide use. Since acetyl peroxide is a dangerous and unpredictable high explosive, it is best prepared and decomposed *in situ*.

In aqueous media, dimers of a variety of radicals may be prepared, including dimers of reactive radicals. The water molecule is relatively inert to attack by organic radicals because of the high dissociation energy (119 kcal mol^{-1}) of the HO—H bond. However, hydroxyl radicals are so reactive and unselective that in substrates with alternative abstraction sites hydrogen-removal from each is observed, and a mixture of dimeric compounds is formed. In contrast, only the combination of stabilized radicals can usually be achieved in organic media, since the solvent molecules are frequently susceptible to attack by reactive radicals such as methyl or phenyl. Such solvent-radical interactions are usually exothermic. Reactions (4.11) and (4.12) are examples, and the ΔH values are calculated from standard bond-dissociation energy data [1–4, 9, 10]. In these two cases

$$CH_3\cdot + CHCl_3 \rightarrow CH_4 + \cdot CCl_3; \quad \Delta H = -13{\cdot}8 \text{ kcal mol}^{-1} \qquad (4.11)$$

$$C_6H_5\cdot + \text{cyclo-}C_6H_{12} \rightarrow C_6H_6 + \text{cyclo-}C_6H_{11}\cdot; \quad \Delta H = -9 \text{ kcal mol}^{-1} \qquad (4.12)$$

the dimeric products obtained would be hexachloroethane and bicyclohexyl, respectively, and not ethane and biphenyl. The atom abstracted (normally hydrogen) by the initiator radical is usually adjacent to an unsaturated system (aryl, vinyl, carbonyl, etc.) or to polarizable hetero-atoms which possess lone pairs of electrons (—C̈l, >N̈—, —Ö—, etc.).

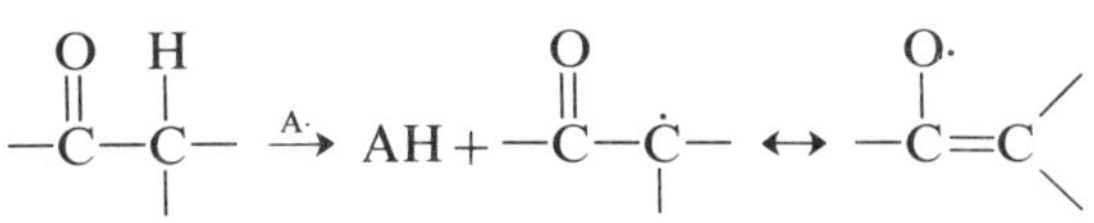

(4.13)

(4.14)

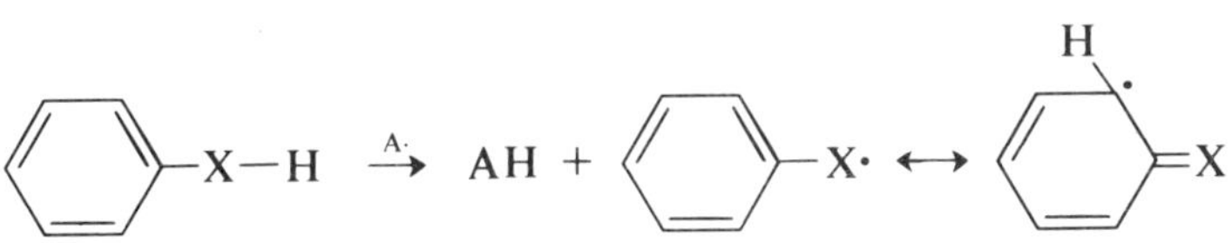

$$(X = -\overset{|}{\underset{|}{C}}-,\ -\ddot{O}-,\ -\ddot{S}-,\ \text{or}\ >\ddot{N}-)$$

$$RCH_2Cl \xrightarrow{A\cdot} AH + R\dot{C}H{-}\ddot{C}l \leftrightarrow RCH{=}\dot{C}l \qquad (4.15)$$

Interest in this area has been rather sporadic, although much definitive work remains to be done. Many of the most useful references are to be found in the literature prior to 1960 [11–20, 20a].

Phenols

The combination reactions of phenoxy radicals have been extensively studied on account of the great synthetic, biosynthetic [21–23], and commercial [24] importance of the reactions. The most versatile reagents for the oxidations are alkaline potassium ferricyanide and ferric chloride. Other reagents which have found use are manganese, nickel, and lead dioxides, silver oxide, lead tetra-acetate, potassium persulphate, Fenton's reagent, cerium(IV), vanadium(V), and copper(II) salts, horseradish peroxidase plus hydrogen peroxide, oxygen and ultraviolet light [25], and anodic oxidation. The primary process is the production of a relatively stable phenoxy radical by oxidative one-electron transfer (4.16). These radicals are not sufficiently reactive to attack the solvent (usually water or aqueous alcohol) or unchanged

$$PhOH + Fe^{3+} \longrightarrow PhO\cdot + H^+ + Fe^{2+}$$

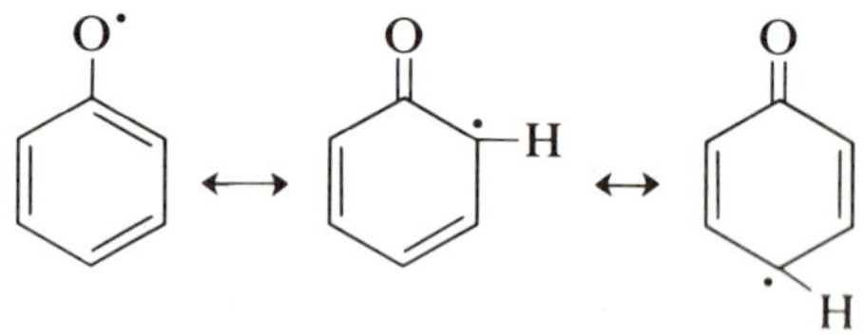

(4.16)

phenol, and dimerize by carbon–carbon coupling (yielding *o–o*, *o–p*, or *p–p* isomers), by carbon–oxygen coupling (at *o*- and *p*-positions), and only rarely by oxygen–oxygen coupling. Thus a simple phenol may give rise to a number of different dimeric products. The nature of the product is often influenced by the reaction conditions and the oxidizing agent. Certain phenol oxidations may go quite cleanly and in high yield. In many cases, however, a number of products are formed and the yield of individual components may be extremely low. This is in direct contrast to the reactions of the isoelectronic benzylic radicals, where nuclear coupling represents only a minor side reaction.

The oxidation of *p*-cresol with alkaline ferricyanide, ferric chloride, Fenton's reagent, copper(II) salts, or peroxidase–H_2O_2 (pH 6·5) yields the *ortho–ortho* coupled product, Pummerer's ketone (1), a trimer (2), and polymeric material [21, 22].

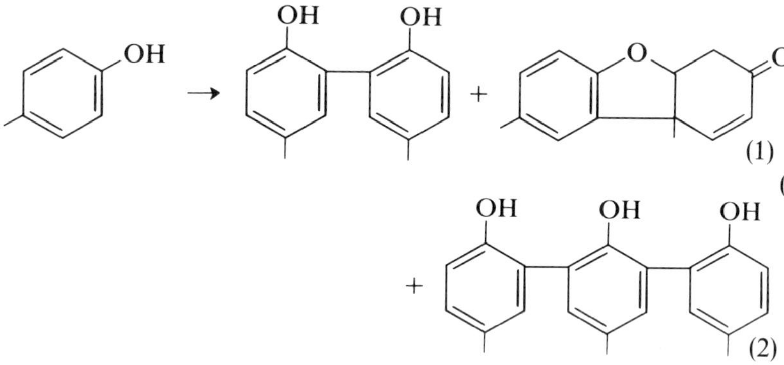

(1) (2) (4.17)

The ketone (1) is formed by *ortho–para* coupling of *p*-cresoxy radicals and is related to the lichen metabolite usnic acid (3), which has been synthesized in a similar manner [26].

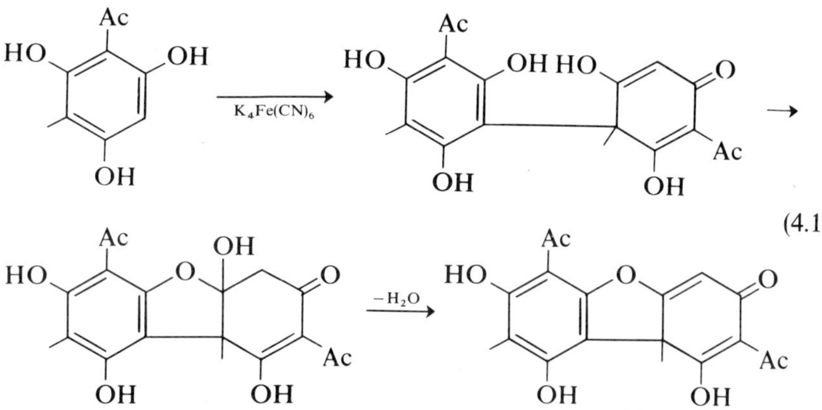

(3) (4.18)

When two of the three *ortho* and *para* positions of the phenol are occupied by substituents, *ortho–ortho* or *para–para* coupling sometimes proceeds in high yield [21].

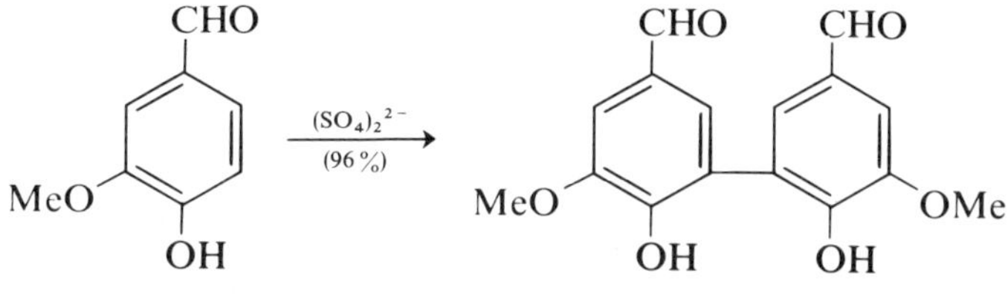

(4.19)

With 2,4,6-trisubstituted phenols possessing a *p*-methyl group a di-*p*-phenoxyethane is often formed. Sufficiently powerful oxidants cause further oxidation to stilbene quinones [27]. Diphenoquinones can arise from 2,6-disubstituted phenols in a similar sequence of reactions. Thus 2,6-dimethoxyphenol is oxidized by t-butyl perbenzoate to 3,3′,5,5′-tetramethoxydiphenoquinone [28].

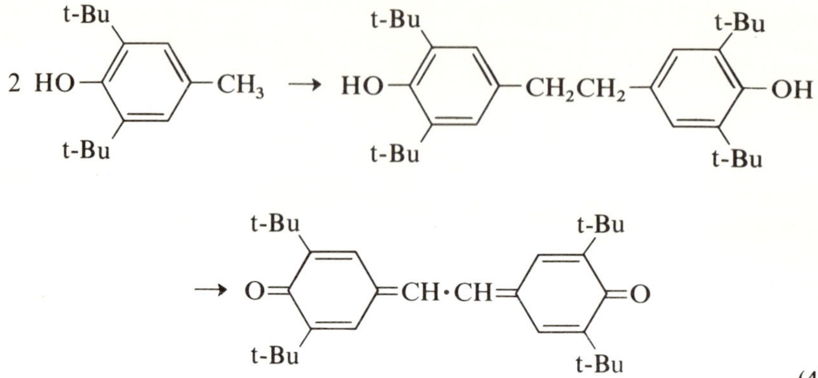

(4.20)

The assumption [29] that phenoxy radicals are also involved in the biosynthesis of many naturally occurring phenolic compounds has led to several *in vitro* syntheses of such compounds. These reactions are not restricted to monohydric phenols, but polyhydric phenols, amines, aminophenols, and carboxylic acids may react similarly [21]. Typical examples include the synthesis of thyroxine [30], (±)-griseofulvin [31], (±)-chimonanthine [32], umbelliferone [33], and erythroaphin-*fb* [34].

Electrochemical Coupling

Electrochemical oxidation and reduction methods are invaluable for the production of dimeric compounds by radical coupling at the electrode surfaces. Perhaps the best known oxidative reaction is the Kolbe synthesis (4.21), although other anions may be oxidized

$$\mathrm{RCO_2^-} \xrightarrow{\text{Pt anode}} [\mathrm{RCO_2\cdot}] \rightarrow \mathrm{R\cdot} \rightarrow \mathrm{R{\cdot}R} \qquad (4.21)$$

similarly, as in reaction (4.22), for example [35].

$$\begin{aligned} \mathrm{Me{\cdot}CO{\cdot}\bar{C}H{\cdot}CO_2Et} &\rightarrow \mathrm{Me{\cdot}CO{\cdot}\dot{C}H{\cdot}CO_2Et} \\ \mathrm{2Me{\cdot}CO{\cdot}\dot{C}H{\cdot}CO_2Et} &\rightarrow \begin{array}{c}\mathrm{Me{\cdot}CO{\cdot}CH{\cdot}CO_2Et} \\ | \\ \mathrm{Me{\cdot}CO{\cdot}CH{\cdot}CO_2Et}\end{array} \end{aligned} \qquad (4.22)$$

Reductive coupling at the cathode (reaction (4.23), for example [36]) is also a well-documented process.

$$\mathrm{ArCHO} \xrightarrow[(+e^-,\ +H^+)]{\text{Hg cathode}} \mathrm{Ar\dot{C}HOH} \rightarrow \mathrm{ArCH(OH){\cdot}CH(OH)Ar} \qquad (4.23)$$

These methods are discussed fully in Chapter 5.

Copper-catalysed Coupling

A number of important coupling reactions are promoted by copper metal and copper salts [23]. The development of these procedures has been largely empirical and their mechanisms are not well understood. In certain of these reactions free-radical mechanisms have been invoked, although such a distinction may be too rigid in view of the lack of information concerning electronic changes in the copper atom or ion. A further complication is the possible interconversion of organic ions and radicals by processes of the type:

$$R\cdot + Cu^{2+} \longrightarrow R^{+} + Cu^{+} \qquad (4.24)$$

Symmetrical alkyne dimers may be prepared by the Glaser coupling reaction [37], for example (4.25).

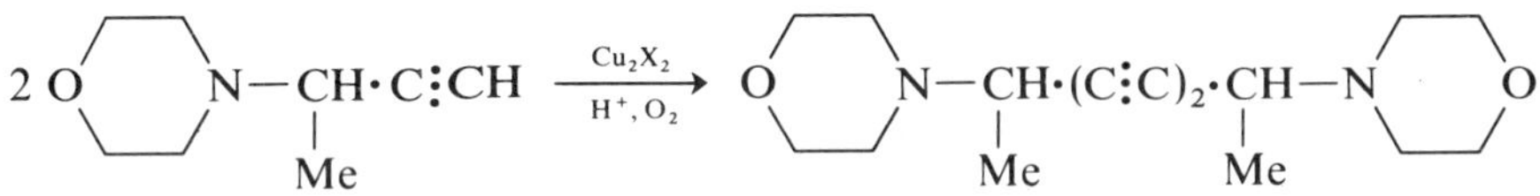

(4.25)

The Glaser coupling procedure is not well suited for unsymmetrical coupling, and in such instances the Cadiot–Chodkiewicz reaction [37], for example (4.26), or similar methods (e.g., (4.27)) may be successfully employed [38].

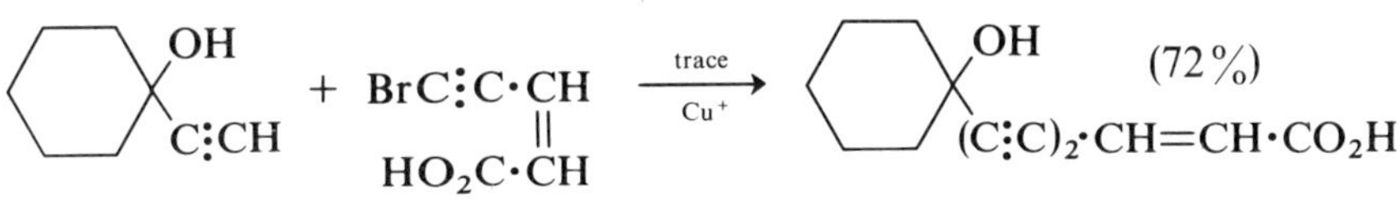

(4.26)

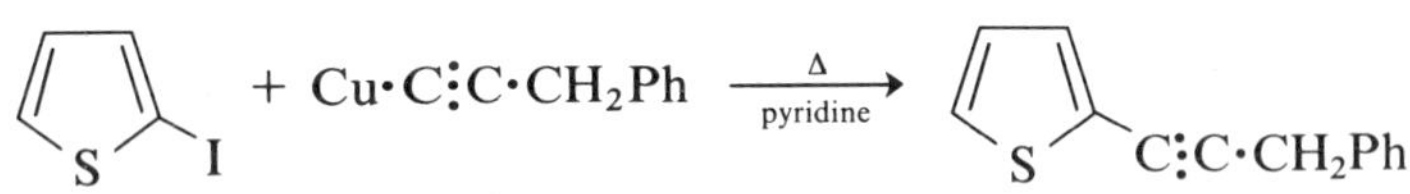

(4.27)

The use of a bis-terminal alkyne in the Glaser reaction leads to linear and to cyclic acetylenes by coupling reactions at both termini. Sondheimer and his collaborators [39] have utilized this cyclization of α,ω-diacetylenes as a

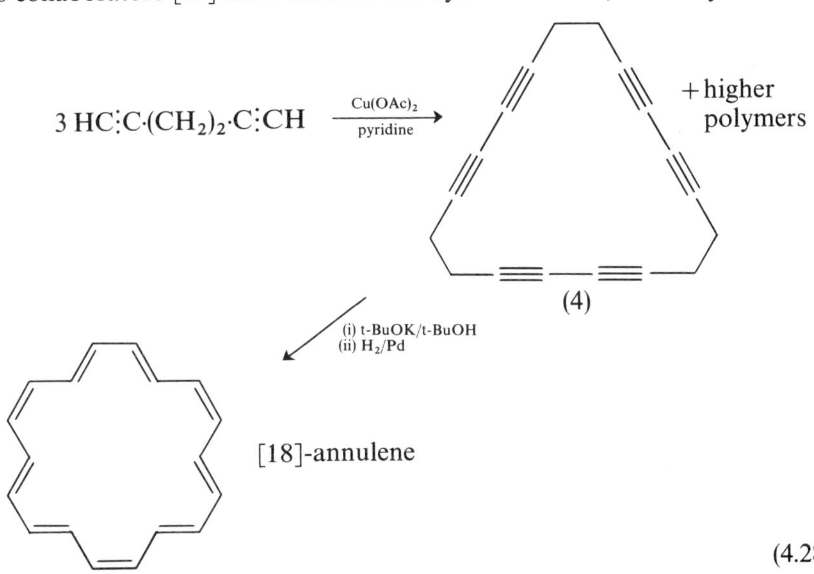

(4.28)

route to the synthesis of a number of annulenes. Thus [18]-annulene may be prepared from 1,5-hexadiyne. The cyclic trimer (4) on heating with potassium t-butoxide in t-butanol undergoes a prototropic rearrangement to a tri-dehydro-[18]-annulene, which on partial catalytic hydrogenation of the three acetylenic bonds yields [18]-annulene. Although the yield overall is only 1 %, an alternative synthesis from the C(18)-acyloin would require at least thirty-nine reaction steps.

The reductive coupling of aryl halides in the presence of copper metal (the Ullmann reaction) is a well-known procedure. The reaction is more useful for the preparation of symmetrical biaryls, but even in these cases the yield of dimer can be extremely low; compare, for example, reactions (4.29) and (4.30) [40].

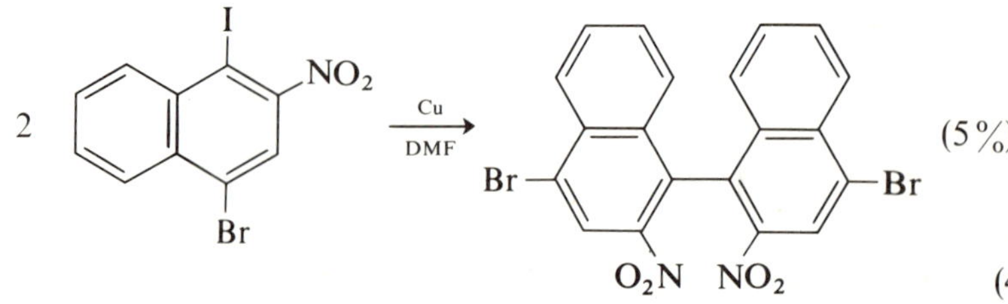

(4.29)

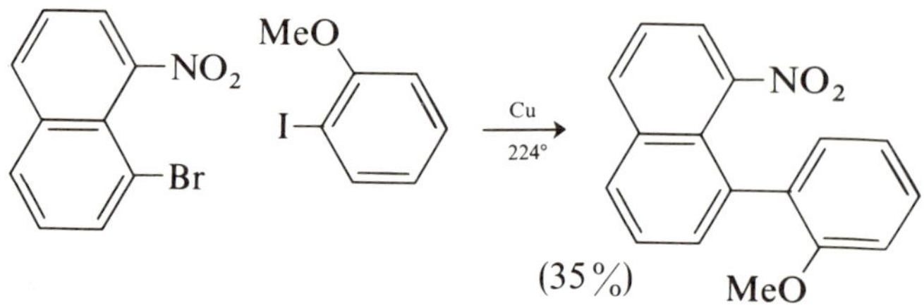

(4.30)

Recent work has indicated the intermediacy of organocopper compounds in these reactions. Thus, copper and silver aryls can be decomposed thermally in aprotic solvents to give symmetrical biaryls, (4.31) and (4.32). Unsymmetrical biaryls are obtained from mixed metal aryls, (4.33). Aryl radicals are implicated in these reactions, since the biaryl yield is depressed by the addition of a radical scavenger [40a].

$$2\mathrm{PhCu} \xrightarrow[\text{ether}]{35^\circ} \mathrm{Ph{\cdot}Ph} + \mathrm{PhH} \qquad (4.31)$$

$$(67\,\%) \quad (5\,\%)$$

$$2p\text{-}\mathrm{MeC_6H_4Ag} \xrightarrow[\text{ether}]{35^\circ} p\text{-}\mathrm{MeC_6H_4{\cdot}C_6H_4Me}\text{-}p' + \mathrm{PhMe} \qquad (4.32)$$

$$(70\,\%) \qquad (13\,\%)$$

$$\mathrm{PhCu} + p\text{-}\mathrm{MeC_6H_4Cu} \xrightarrow[\text{ether}]{35^\circ} \mathrm{PhPh} + p\text{-}\mathrm{MeC_6H_4Ph}$$

$$+\,p\text{-}\mathrm{MeC_6H_4{\cdot}C_6H_4Me}\text{-}p' \qquad (4.33)$$

The reactions between cuprous benzoate and aryl halides seem to be related processes (reaction (4.34), for example [41]).

$$p\text{-}\mathrm{MeC_6H_4Br} + \mathrm{PhCO_2Cu} \xrightarrow[\text{xylene}]{140\text{–}60^\circ} p\text{-}\mathrm{MeC_6H_4O{\cdot}CO{\cdot}Ph} \qquad (4.34)$$

$$(79\,\%)$$

Aryl Grignard reagents and aryl boronic acids may be oxidatively coupled by salts of copper(II) and of other transition-metal ions [23], (reaction (4.35), for example [41a]).

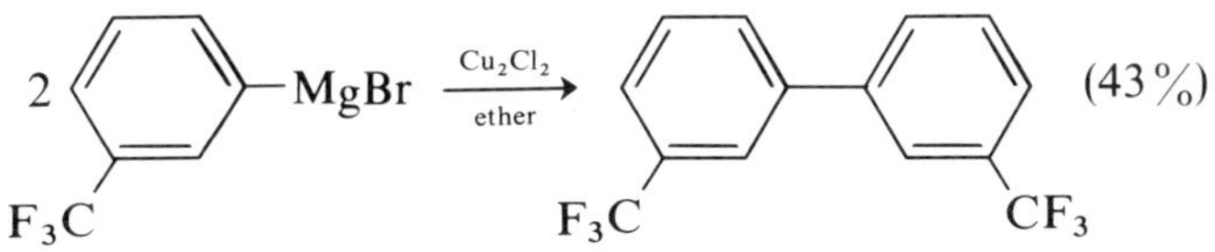

(4.35)

Very recently the replacement of the halogen atoms of aryl halides by methyl groups has been described [41b]. Methyl-lithium and cuprous iodide are the reagents, and the reactions are performed at 0° under nitrogen. Reaction (4.36) is typical [41b], and alkyl halides react similarly.

$$\text{PhI} \xrightarrow{(90\%)} \text{PhMe} \tag{4.36}$$

Coupling Induced by Alkali Metals

The acyloin condensation [42], an important sodium-induced reaction, appears on the basis of present evidence to involve radical intermediates, although this is not conclusive since definitive kinetic data and mechanistic studies are lacking [42, 43]. This interpretation may yet require review. The reaction is an important method for the construction of normal [C(5)–C(7)], medium [C(8)–C(11)], and large [>C(11)] carbocyclic rings, although many acyclic acyloins and a few small-ring acyloins have been prepared. The reaction involves the coupling of alkyl carboxylates under homogeneous or heterogeneous conditions [42]. Product yields are highly variable and somewhat dependent upon the rate of addition of the ester, reaction time, reaction conditions, and temperature. The reasons for this variability are not clear [42, 43]. However, many acyloin condensations proceed in excellent yield and represent an important method for circumventing the adverse entropy effects which attend and limit the general applicability of most cyclization procedures. Reactions (4.37) [44] and (4.38) [45] are illustrative of the general method.

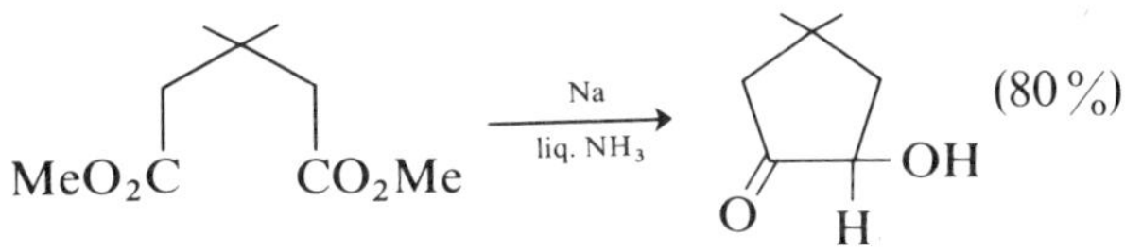

(4.37)

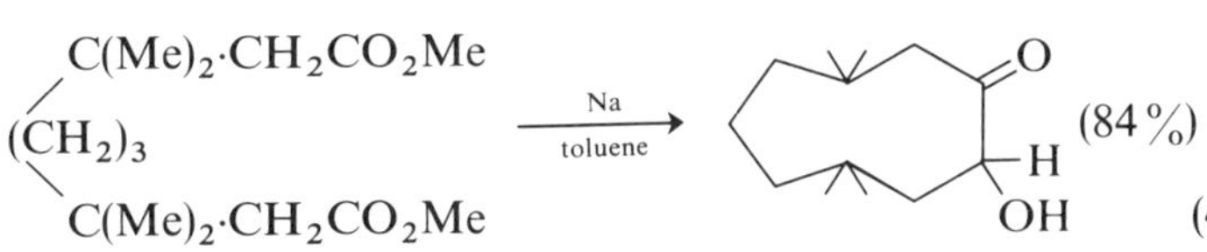

(4.38)

The acyloin condensation has been widely used for the construction of large rings, for example in the preparation of musks (4.39) [46].

The reductive coupling of organohalogen compounds in the presence of alkali metals (Wurtz reaction) is a process of limited synthetic utility. It is a reaction which is believed to occur through the formation of free radicals

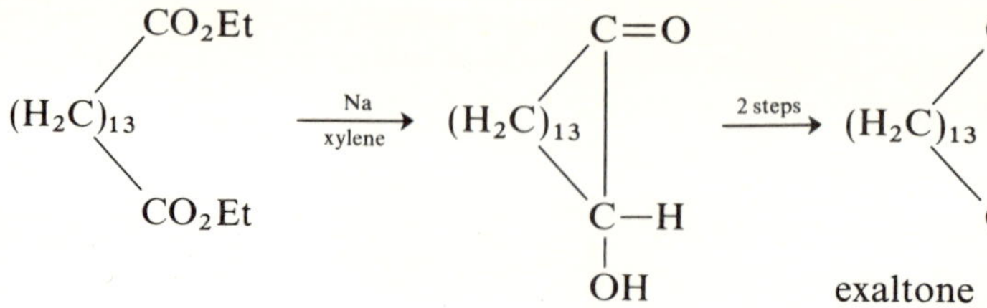

(4.39)

in several cases [47], but the evidence is not wholly conclusive for the reaction in its general form. The scope and limitations of the reaction and of the closely related Wurtz–Fittig synthesis are so widely known that further elaboration here is unnecessary. The similar reductive coupling of aldehydes and ketones (pinacol reaction) will also not be considered here.

THE COUPLING OF ORGANIC RADICALS WITH STABLE RADICALS

A number of organic radicals (e.g. galvinoxyl, (5)) are indefinitely stable in the solid state or in solution in the absence of other types of radical (e.g., oxygen). The stability of these radicals is in part due to the delocalization of the odd electron through the π-molecular orbitals of the molecule, and in part to steric effects. Galvinoxyl and other stable organic radicals react rapidly with less hindered radicals (e.g., $PhCO_2\cdot$ [48]), but the dimers are of little synthetic interest. These stable radicals are useful in mechanistic studies of radical-chain reactions [48], and certain types are of considerable technical importance, as in the prevention of autoxidation [24].

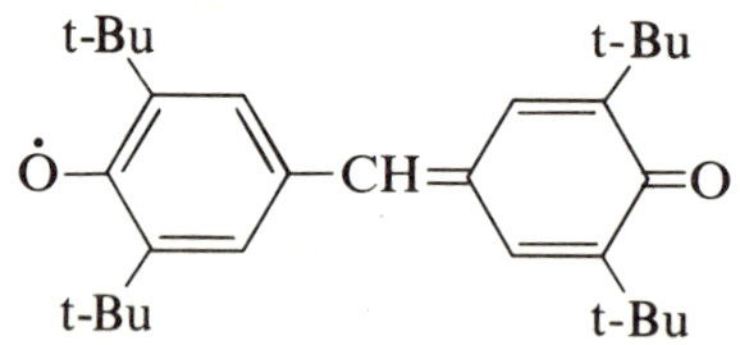

(5)

Among inorganic radicals, the best known are those which possess the three-electron bond N—O, as in NO, NO_2, and Frémy's salt. The oxidation of phenols to quinones by Frémy's salt (e.g., reaction (4.40) [48a]) is well known. The intermediate adduct (6) is not stable under the reaction conditions and decomposes to yield the quinone.

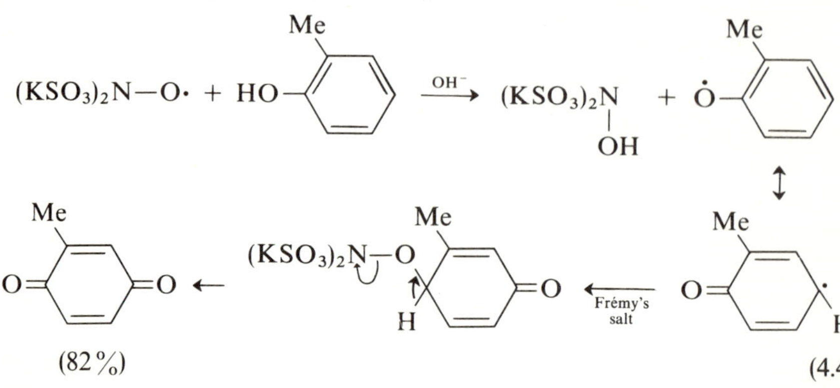

(4.40)

Nitric oxide is an efficient scavenger of organic radicals, but unfortunately the nitroso compounds produced may react further [48b]. The end-product is a trisubstituted hydroxylamine, reaction (4.41).

$$\mathrm{Me_2\dot{C}{\cdot}CN \xrightarrow{NO} Me_2C\begin{matrix}\diagup CN \\ \diagdown NO\end{matrix}}$$

$$\mathrm{\xrightarrow[2\,Me_2\dot{C}{\cdot}CN]{} (Me_2C{\cdot}CN)_2N{\cdot}O{\cdot}C(CN)Me_2} \quad (50\,\%) \tag{4.41}$$

Nitrogen dioxide is not so effective as a scavenger, but the nitro compounds so formed are more reluctant to react further with the organic radicals [48c]. It appears to be advantageous to add a little nitric oxide to

$$\mathrm{Me_2\dot{C}{\cdot}CN \xrightarrow{NO_2/NO} Me_2C(CN)NO_2} \quad (43\,\%) \tag{4.42}$$

the nitrogen dioxide. Presumably the organic radicals are trapped initially by the NO to give a nitroso compound, which is then rapidly oxidized by the excess NO_2 present (4.43).

$$\mathrm{R{\cdot} \xrightarrow{NO} RNO \xrightarrow{NO_2} RNO_2 + NO} \tag{4.43}$$

Nevertheless, even the reactive phenyl radical can be trapped by NO_2 [48d],

$$\mathrm{(\mathit{p}\text{-}O_2N{\cdot}C_6H_4CO{\cdot}O{-})_2 \longrightarrow \mathit{p}\text{-}O_2N{\cdot}C_6H_4{\cdot}}$$

$$\mathrm{\xrightarrow[NO_2/CCl_4]{} \mathit{p}\text{-}O_2N{\cdot}C_6H_4NO_2} \quad (11\,\%) \tag{4.44}$$

but the scope of this reaction has not been investigated in any detail.

In the case of the reactions with nitric oxide, if the initially formed nitroso compound can tautomerize to an oxime, then further reactions with the organic radicals may be averted. The reaction is then synthetically much more important, and examples are photochemical oximation (p. 124) and the Barton reaction (p. 127).

(c) Termination by Electron Transfer

Just as radicals are readily formed in redox systems, they are destroyed with equal facility. This limits the utility of some redox processes. However, there are many important examples of termination by redox transfer, and only one or two such cases can be considered here. The mechanisms of these processes are not always clear-cut and in several cases various mechanisms may operate, depending upon the nature of the substrate and the reaction conditions. Indeed, the copper-catalysed reactions considered above could equally well be considered under this heading.

The Perester Reaction

The metal-catalysed reactions of hydroperoxides [49], peroxides [49a], and peresters [50, 51] with organic substrates are useful methods for the selective oxidation of carbon atoms. The perester reaction is generally more selective, more versatile, and has a greater preparative utility, since the products (esters) are readily converted to other derivatives [51]. The hydroperoxide reaction supplements the perester reaction, however, since it can often still be used where the perester reaction fails [51].

The initial step of the reaction is the cleavage of the perester by the metal ion (usually Cu^+), which is only required in catalytic amounts, reaction (4.45).

$$\mathrm{R{\cdot}CO{\cdot}O{-}O{\cdot}CMe_3 + Cu^+ \rightarrow [R{\cdot}CO{\cdot}O{\cdot}Cu(II)]^+ + Me_3C{\cdot}O{\cdot}} \quad (4.45)$$

The t-butoxy radical so formed is a selective radical (unlike ·OH) and oxidizes the organic substrate by abstracting a labile hydrogen atom (4.46).

$$\mathrm{R'{-}\overset{|}{\underset{|}{C}}{-}H + Me_3C{\cdot}O{\cdot} \rightarrow R'{-}\overset{|}{\underset{|}{C}}{\cdot} + Me_3C{\cdot}OH} \quad (4.46)$$

The hydrogen atom removed is that which is attached to an atom which is α to an unsaturated centre or α to an hetero-atom possessing lone electron pairs (see p. 95). The reaction is then terminated by electron transfer (4.48, 4.49) or by ligand transfer (4.50) [51–53].

$$\mathrm{[R{\cdot}CO{\cdot}O{\cdot}Cu(II)]^+ \rightleftharpoons R{\cdot}CO{\cdot}O^- + Cu^{2+}} \quad (4.47)$$

$$\mathrm{R'{-}\overset{|}{\underset{|}{C}}{\cdot} + Cu^{2+} \xrightarrow{fast} R'{-}\overset{|}{\underset{|}{C}}{}^+ + Cu^+} \quad (4.48)$$

$$\mathrm{R'{-}\overset{|}{\underset{|}{C}}{}^+ + R{\cdot}CO{\cdot}O^- \xrightarrow{fast} R'{-}\overset{|}{\underset{|}{C}}{-}O{\cdot}CO{\cdot}R} \quad (4.49)$$

$$\mathrm{[R{\cdot}CO{\cdot}O{\cdot}Cu(II)]^+ + R'{-}\overset{|}{\underset{|}{C}}{\cdot} \xrightarrow{fast} R'{-}\overset{|}{\underset{|}{C}}{-}O{\cdot}CO{\cdot}R + Cu^+} \quad (4.50)$$

The peresters employed in the reaction are usually t-butylperacetate (R = Me) and t-butylperbenzoate (R = Ph), since these are commercially available. The scope of the reaction may be increased in some cases by the addition of an excess of an aliphatic acid to the reaction mixture; the ester of this acid is then obtained [50]. Some typical conversions are summarized in Table 4.2.

It will be noted that allylic rearrangements may occur (first example in Table 4.2). With cyclic and terminal olefins containing secondary hydrogens on the α-carbon atom relative to the double bond, products with the acyloxy group located at this position are formed predominantly or exclusively. However, with aliphatic olefins containing internal double bonds with no secondary hydrogens on the α-carbon atom (e.g., 2-butene), the predominant or exclusive product is an acyloxy derivative of the rearranged olefin [53]. t-Butyl ethers are occasionally formed in an interesting deviation of the reaction [51, 59]. The reaction fails for amines and for aliphatic aldehydes and esters, since mixtures of products are obtained. Benzylic esters react

Table 4.2 Reactions of active hydrogen-donors with peroxyesters in the presence of copper ions

Substrate	Perester[a]	Products	Yield (%)	Ref.
$Me(CH_2)_3CH{=}CH_2$	BPA	$Me(CH_2)_2CH(O{\cdot}CO{\cdot}Me){\cdot}CH{=}CH_2$	75	
		$Me(CH_2)_2CH{=}CHCH_2O{\cdot}CO{\cdot}Me$	14	[52]
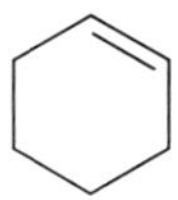	BPA	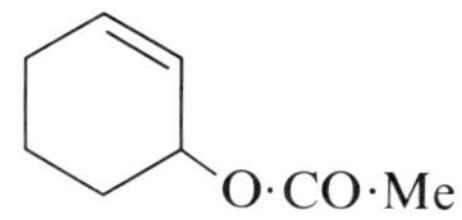	90	[52]
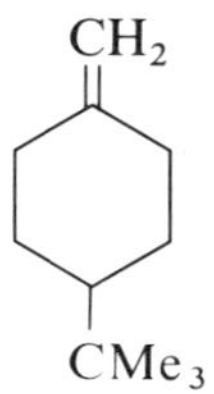	BPB	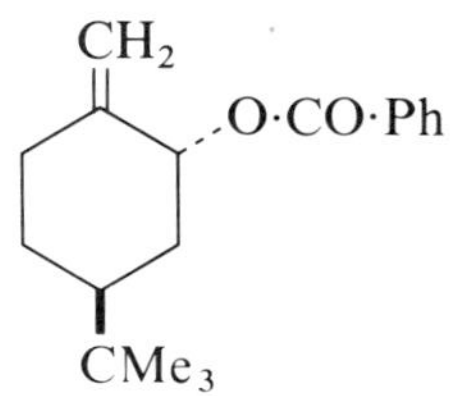 + trace *cis*-compound	>26	[54]
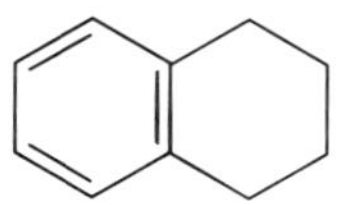	BPB	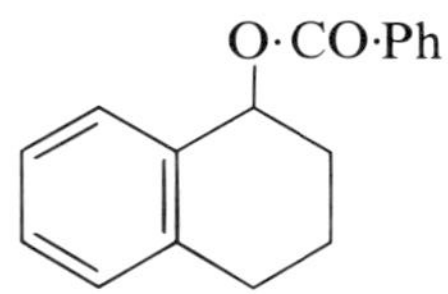	15	[55]
PhCHO	BPB	PhCO·O·COPh	70	[55]

Table 4.2 (continued)

Substrate	Perester[a]	Products	Yield (%)	Ref.
$Me(CH_2)_3S(CH_2)_3Me$	BPB	$Me(CH_2)_2CH(O{\cdot}CO{\cdot}Ph){\cdot}S(CH_2)_3Me$	66	[57]
$H_2C{=}CHCH_2O(CH_2)_3Me$	BPB	$H_2C{=}CH{\cdot}CH(O{\cdot}CO{\cdot}Ph){\cdot}O(CH_2)_3Me$	35	[57]
$PhOCH_3$	BPB	$PhOCH_2O{\cdot}CO{\cdot}Ph$	23	[57]
	BPB	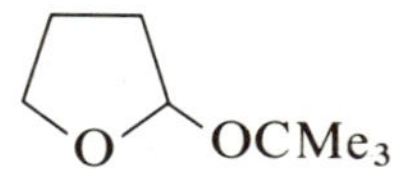	31–45	
		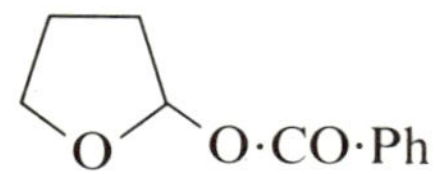	26	[56, 59]
	BPB + excess EtOH	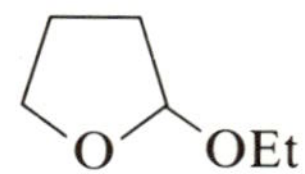	40	[59]
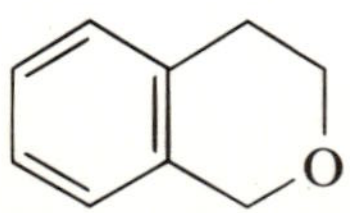	BPB	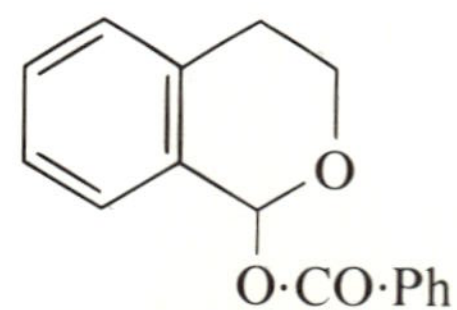	34	[60]
$HCO{\cdot}NMe_2$	BPA	$HCO{\cdot}N(Me)CH_2O{\cdot}CO{\cdot}Me$	25	[60]

smoothly to give acylals, reaction (4.51), for example [58, 60].

$$\mathrm{PhCH_2{\cdot}O{\cdot}CO{\cdot}Me + PhCO{\cdot}O{\cdot}CMe_3 \xrightarrow{Cu^+/Cu^{2+}} \underset{\underset{(35\%)}{O{\cdot}CO{\cdot}Me}}{PhCH}{\cdot}O{\cdot}CO{\cdot}Ph + Me_3C{\cdot}OH} \tag{4.51}$$

The α-acyloxy derivative obtained from the reaction of a perester with alkyl ethers [56, 57, 59, 61] may be converted to an acetal by treatment with excess of an alcohol (4.52) [61]. The reaction represents a method for the conversion of alkyl ethers to aldehydes.

$$\mathrm{R'{\cdot}CH_2{\cdot}O{\cdot}\underset{O{\cdot}CO{\cdot}R}{CHR'} + R''OH \rightarrow R'{\cdot}CH_2{\cdot}O{\cdot}\underset{OR''}{CHR'} \xrightarrow{R''OH} R'{\cdot}\underset{OR''}{CH}{\cdot}OR''} \tag{4.52}$$

(R′ = alkyl or aryl)

Ethers containing one or two tertiary hydrogens on the α-carbon atoms do not yield acyloxy derivatives. With excess alcohol present the corresponding ketals are obtained. Presumably the tertiary hydrogen atom is first substituted by an acyloxy group, and this is followed by the elimination of the carboxylic acid [61].

$$\mathrm{Me_2CH{\cdot}O{\cdot}CHMe_2 \xrightarrow[Cu^+/Cu^{2+}]{PhCO{\cdot}O{\cdot}O{\cdot}CMe_3} Me_2\underset{O{\cdot}CO{\cdot}Ph}{C}{\cdot}O{\cdot}CHMe_2 \xrightarrow{n\text{-}C_4H_9OH} Me_2\underset{\underset{(31\%)}{O{\cdot}C_4H_9}}{C}{\cdot}O{\cdot}C_4H_9 + PhCO_2H + Me_2CH{\cdot}OH} \tag{4.53}$$

Oxidation of Aralkanes

The oxidation of alkyl side-chains of aromatic compounds has received considerable attention in recent years because of the technical importance of the oxidation products. Several of the oxidative procedures are free-radical in nature, and a few examples have already been discussed. Further examples are considered under radical transfer processes. In this section the oxidations promoted by metal ions in a high valency state will be briefly mentioned.

The oxidation of aromatic side-chains by chromic acid, chromyl chloride, aqueous dichromate, chromyl acetate, and di-t-butyl chromate are well-known laboratory procedures [62]. The uses of these reagents have been thoroughly reviewed [62], as have the oxidations effected by permanganate [63] and by lead tetra-acetate [64]. These latter reagents also possibly oxidize aralkanes by radical pathways in some cases. The one-electron oxidations by vanadium(V), cobalt(III), and manganese(III) of a variety of organic compounds have been widely investigated, particularly from the kinetic viewpoint [65], but aralkanes have been largely neglected. In a recent study, Dewar has proposed that an initial electron transfer occurs in the oxidation of aralkanes by these reagents (i.e., the electron is removed from the highest occupied π-molecular orbital of the aromatic system).

$$\mathrm{ArR} + \mathrm{M}^{n+} \longrightarrow (\mathrm{ArR})^{+} + \mathrm{M}^{(n-1)+} \tag{4.54}$$

Reaction (4.55) is a typical example of the conversions which were effected [66].

$$p\text{-}\mathrm{MeO{\cdot}C_6H_4{\cdot}CH_3} + 2\mathrm{Mn(OAc)_3} \xrightarrow{\mathrm{AcOH}} p\text{-}\mathrm{MeO{\cdot}C_6H_4{\cdot}CH_2OAc} + \mathrm{AcOH} + 2\mathrm{Mn(OAc)_2} \tag{4.55}$$

Oxidations by cerium(IV) have also been widely studied, although again from the kinetic or analytical standpoint. Excess oxidant has often been employed and, on this account, formic acid is frequently the only product isolated [67]. The oxidation of the alkyl side-chains of aralkanes has been re-investigated more recently, and conditions suitable for the preparation of the α-carbonyl compounds (e.g., reactions (4.56) and (4.57)) have been discussed.

$$m\text{-}\mathrm{CH_3{\cdot}C_6H_4{\cdot}CH_3} \longrightarrow m\text{-}\mathrm{CH_3{\cdot}C_6H_4{\cdot}CHO} \quad (ca.\ 100\,\%) \tag{4.56}$$

$$\mathrm{PhCH_2CH_3} \longrightarrow \mathrm{PhCOCH_3} \quad (77\,\%) \tag{4.57}$$

In aqueous solvents the α-aryl carbinols are probably intermediates in the reaction, and these compounds are known to suffer ready oxidation to the α-aryl carbonyl compounds. Benzylic acetates can be isolated when glacial acetic acid is the solvent. The following reaction scheme (4.58) accounts for these observations [67, 68, 70], although a non-homolytic mechanism has also been proposed [69a].

$$\begin{array}{lrcl}
\text{(a)} & \mathrm{ArCH_3} + \mathrm{Ce^{IV}} & \longrightarrow & \mathrm{Ar\dot{C}H_2} + \mathrm{Ce^{III}} + \mathrm{H^+} \\
\text{(b)} & \mathrm{Ar\dot{C}H_2} + \mathrm{Ce^{IV}(OH)} & \longrightarrow & \mathrm{ArCH_2OH} + \mathrm{Ce^{III}} \\
\text{(c)} & \mathrm{ArCH_2OH} + \mathrm{Ce^{IV}} & \longrightarrow & \mathrm{Ar\dot{C}HOH} + \mathrm{Ce^{III}} + \mathrm{H^+} \\
\text{(d)} & \mathrm{Ar\dot{C}HOH} + \mathrm{Ce^{IV}(OH)} & \longrightarrow & \mathrm{ArCH(OH)_2} + \mathrm{Ce^{III}} \\
\text{(e)} & \mathrm{ArCH(OH)_2} & \longrightarrow & \mathrm{ArCHO} + \mathrm{H_2O}
\end{array} \tag{4.58}$$

Silver(III) ions react similarly [70], and both reagents are really only suitable for small-scale work.

Reduction of Organic Halides by Chromium(II) Salts

The reduction of organic halides (and of other organic compounds) by chromous salts has received considerable attention, especially for synthetic purposes in steroid and alkaloid systems [71]. More recent studies have been particularly concerned with the mechanistic aspects of such reactions, and also with a view to extending the scope of the reductions. The use of simple chromous salts for the reduction of halides is somewhat limited to the rather reactive allylic and benzylic halides, as well as to α-halocarbonyl and α-halocyano compounds [72]. Benzyl chloride, for example, can be reduced by chromium(II) to toluene and to bibenzyl under various conditions. The coupling reaction is favoured by low chromous-ion concentrations and by the use of benzyl iodide in place of the chloride; a mixture of benzyl chloride and sodium iodide is suitable for this purpose [71]. The low solubility of chromous salts in tetrahydrofuran (THF) makes this solvent

$$PhCH_2Cl \xrightarrow[\text{dioxan}]{Cr(ClO_4)_2} \underset{(77\%)}{PhCH_3} + \underset{(1{\cdot}8\%)}{PhCH_2CH_2Ph} \tag{4.59}$$

$$PhCH_2Cl \xrightarrow[\text{THF}]{CrCl_2} \underset{(0\%)}{PhCH_3} + \underset{(70\%)}{PhCH_2CH_2Ph} \tag{4.60}$$

invaluable for studying the coupling reaction, for example (4.60), [71]. The addition of the halide to a hot solution of chromous perchlorate in dioxan, on the other hand, gives predominantly toluene. The benzylchromium ion is undoubtedly formed in these reactions [72, 73]. Allyl halides react similarly, whereas alkyl halides are generally reduced to the alkanes with no observable coupling in somewhat slower reactions. Co-ordination of the chromium(II) ion with ligands such as ethylenediamine (en) greatly enhances its ability to reduce alkyl halides, particularly primary alkyl halides, to alkanes [72].

$$CH_3CH_2CH_2CH_2Br \xrightarrow[\text{DMF, 25°}]{Cr^{II}(en)} \underset{(ca.\ 100\%)}{CH_3CH_2CH_2CH_3} \tag{4.61}$$

Halides follow the usual reactivity sequence with this reagent; tertiary > secondary > primary, iodides > bromides > chlorides. Primary chlorides react slowly, and the chromium(II) ethanolamine complex (which reduces even water) is a more satisfactory reagent for these compounds [72].

This reduction of halides to hydrocarbons under relatively mild conditions has an obvious preparative importance since halides can be synthesized in a variety of ways. In a recent application of the reaction the readily available 9α-bromo-11β-hydroxy steroids have been converted to the medicinally important 11β-hydroxy steroids in good yield, reaction (4.62) for example

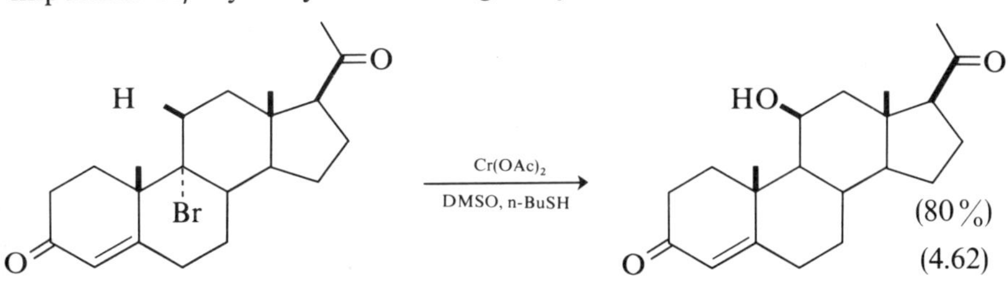

(4.62)

9α-bromo-11β-hydroxyprogesterone 11β-hydroxyprogesterone

[74]. Butanethiol, an efficient transfer donor of hydrogen, is added to minimize side reactions.

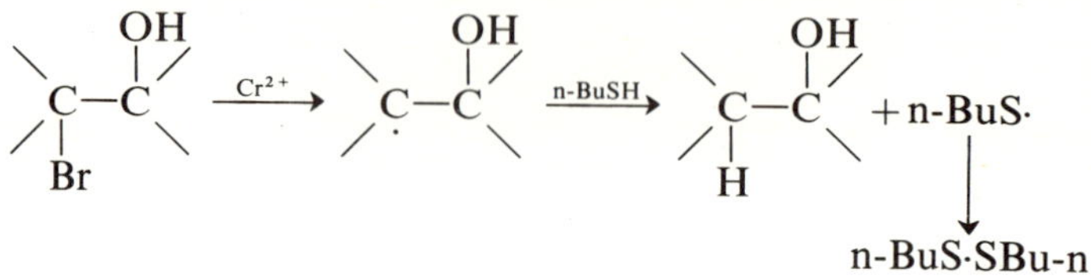

(4.63)

REDUCTION OF ORGANOHALOGEN COMPOUNDS BY ELECTROCHEMICAL METHODS

These processes are discussed in Chapter 5.

4.4 RADICAL TRANSFER PROCESSES

The main types of reaction in which the site of the odd electron changes from one atom or group to another atom or group are abstraction, addition, substitution, displacement, rearrangement, elimination, and fragmentation. Following upon these reactions are the termination processes which were discussed above.

(a) Abstraction

The monovalent peripheral atoms of a molecule are highly susceptible to abstraction by radicals, (4.64). The ease of abstraction of any particular

$$\mathrm{R\cdot + A{-}\underset{\displaystyle E}{\overset{\displaystyle B}{\underset{|}{\overset{|}{C}}}}{-}D \;\longrightarrow\; RA + \cdot\underset{\displaystyle E}{\overset{\displaystyle B}{\underset{|}{\overset{|}{C}}}}{-}D} \qquad (4.64)$$

atom is dependent upon two factors: the reactivity of the attacking radical (R·), and the stability of the newly formed radical (·CBDE). If the new radical is particularly stable then the dissociation energy of the C—A bond will be quite low. Abstraction reactions are usually exothermic, that is, D(R—A) is greater than D(C—A), and the feasibility of such processes can often be calculated by using bond dissociation energy data [1–4,9, 10]. The abstraction reactions may be intermolecular or intramolecular, and are very frequently chain reactions comprising initiation (4.65), propagation (4.66), and termination (4.67) steps. The desired product (RB) is normally

$$\mathrm{AB \longrightarrow A\cdot + B\cdot} \qquad (4.65)$$

(a) $$\mathrm{A\cdot + RH \longrightarrow AH + R\cdot}$$

(b) $$\mathrm{R\cdot + AB \longrightarrow RB + A\cdot} \qquad (4.66)$$

(a) $$\mathrm{2A\cdot \longrightarrow AA}$$

(b) $$\mathrm{A\cdot + R\cdot \longrightarrow AR}$$

(c) $$\mathrm{R\cdot + R\cdot \longrightarrow RR,\quad etc.} \qquad (4.67)$$

formed in the chain-propagation step. The termination reactions therefore lead to unwanted by-products and must be minimized, and this can be achieved only if the propagation steps have considerably greater rates than the termination reactions. Accordingly, the activation energies of reactions (4.66a, b) must be low and these steps must therefore be exothermic. This places a very real restriction on the types of abstraction reactions that can be performed.

The atom abstracted from the organic molecule (RH) is normally a hydrogen or halogen atom. Monovalent atoms located at allylic or benzylic positions are particularly susceptible to abstraction, even by comparatively unreactive radicals, as are aldehydic hydrogen atoms. Somewhat less reactive are atoms located α to a carbonyl group or α to polarizable hetero-atoms. The hydrogen atoms in saturated hydrocarbons are even less reactive, but, other things being equal, the reactivity falls in the sequence tertiary > secondary > primary. Vinylic hydrogen atoms are not usually abstracted by radicals, and the fluorine atoms in perfluoroalkanes are likewise extremely unreactive. The ease of abstraction therefore parallels the strengths of the bonds broken and the stabilities of the radicals so formed.

Most of the work in this area has been concerned with halogenation reactions, which are best considered first. Intramolecular halogenations, however, are considered together with other intramolecular abstraction processes (p. 126). Other types of syntheses are dealt with under the appropriate headings.

CHLORINATION

The reaction usually follows the basic pattern of equations (4.65–4.67) and external initiation is often not necessary. The chlorinating reagent (AB) has B equal to chlorine; A· can be a number of groups such as chlorine, hydroxy, alkoxy, amminium ($R-\overset{+\cdot}{N}H-R'$), sulphuryl ($\cdot SO_2Cl$), chloroiodophenyl ($Ph\dot{I}Cl$), trichloromethyl ($\cdot CCl_3$), amidyl ($RCO\cdot\dot{N}H$), sulphonamidyl ($RSO_2\dot{N}H$), and so on. The selectivity of the process depends entirely upon the nature of the radical A· (for a given organic substrate), since it is this radical which is involved in the abstraction process.

Substitution chlorinations are of tremendous practical importance, and most have been carried out with elemental chlorine. The reactions may be initiated by heat or light (wavelength less than 485 nm), and are always exothermic, so that heat must be dissipated by cooling or by dilution. The chlorine atom is notoriously unselective in its attack, and where abstraction from more than one site can occur mixtures of chlorinated products are inevitably formed. However, the selectivity is dependent upon the solvent and can be drastically altered by using a solvent with which the chlorine atom can form a complex [75]. The electrophilic chlorine atom is also sensitive to polar effects, and in the abstraction of hydrogen from a hydrocarbon chain, little attack tends to occur at sites α to powerful electron-withdrawing groups [76]. Chlorination by elemental chlorine, sulphuryl chloride, and *N*-chlorosuccinimide have been studied extensively, and more recent work has been largely concerned with mechanistic details of these processes. The synthetic utility of these reagents is exhaustively discussed elsewhere [76, 77]. One interesting example, which is worth noting here, illustrates nicely the sensitivity of the chlorine atom to the inductive withdrawal of electrons. In concentrated sulphuric acid, glycine does not undergo photochemical chlorination. Under similar conditions (±)-ornithine is chlorinated in the γ-position, alanine gives β-chloroalanine (78 %), and L-lysine gives γ-chloro-L-lysine (74 %), reaction (4.68) [78].

$$H_2N{\cdot}CH_2CH_2CH_2CH_2\underset{\displaystyle NH_2}{\underset{|}{C}}H{\cdot}CO_2H \xrightarrow[h\nu,\ Cl_2,\ 2\ \text{hours}]{H^+,\ 70^\circ} H_2N{\cdot}CH_2CH_2\underset{\displaystyle Cl}{\underset{|}{C}}H{\cdot}CH_2{\cdot}\underset{\displaystyle NH_2}{\underset{|}{C}}H{\cdot}CO_2H \qquad (4.68)$$

Reagents other than molecular chlorine are usually used in small-scale chlorination reactions only. Where only one type of abstractable hydrogen atom exists, as in cyclohexane, these more expensive reagents usually offer no advantage over chlorine. Their main advantage is that chlorination is often more selective. In addition, some reagents do not produce hydrogen chloride as a by-product (e.g., t-butyl hypochlorite or *N*-chlorosuccinimide) and the reaction may be carried out under acid-free conditions. However, systematic studies are lacking and most of the published work is largely concerned with the reactivity of the various sites in hydrocarbons towards attack by the radicals produced by these reagents [79–86].

Rearrangement reactions can accompany chlorinations (as in the chlorination of t-butyl bromide), equation (4.69) [85]. These rearrangements are considered in sub-section (f) below.

$$Me_3CBr \xrightarrow{\text{t-BuO}\cdot} Me_2\underset{\displaystyle Br}{\underset{|}{C}}{-}CH_2\cdot \rightarrow Me_2\dot{C}{-}CH_2Br \xrightarrow{\text{t-BuOCl}} Me_2\underset{\displaystyle Cl}{\underset{|}{C}}{\cdot}CH_2Br \qquad (4.69)$$

In addition to the above methods, chlorine can be introduced by halogen exchange [76]; reactions (4.70) and (4.71) are fairly typical [87, 88].

$$PhBr \xrightarrow{Cl_2,\ h\nu} PhCl \quad (68\,\%) \qquad (4.70)$$

$$CF_3I \xrightarrow{CCl_4,\ h\nu} CF_3Cl \quad (60\,\%) \qquad (4.71)$$

Bromination and Iodination

Most bromination reactions involving elemental bromine are endothermic processes, except for the reactions involving very weak carbon–hydrogen bonds. The reactions are frequently run at 0° to 25° using ultraviolet light, although occasionally an incandescent lamp is sufficient to initiate the radical reaction, particularly at higher temperatures. Peroxides accelerate the reactions, and oxygen also exhibits an accelerating influence on brominations, unlike many chlorinations. Because of its low reactivity, the bromine atom is a much more selective species than the chlorine atom.

Where several types of abstractable hydrogen atom are present in a substrate molecule, reaction occurs preferentially at the most weakly bonded atom. Therefore, unlike chlorination reactions using molecular chlorine, in brominations mixtures of monohalogenated and polyhalogenated compounds are less commonly formed. Allylic and benzylic sites are the most reactive, but allylic bromination is seldom achieved by using elemental bromine [76]. Although bromination reactions of the allylic position with bromine are known [76], the addition of bromine across the double bond is usually a much faster reaction. This addition can be limited by utilizing a very low concentration of bromine, for example by using *N*-bromosuccinimide (NBS) [76, 89, 90]. Other *N*-bromoamides can function similarly, and the practical use of these important reagents has been reviewed several times [76, 91]. NBS is invaluable for the allylic bromination of molecules which are sensitive to bromine, reaction (4.72) for example [92].

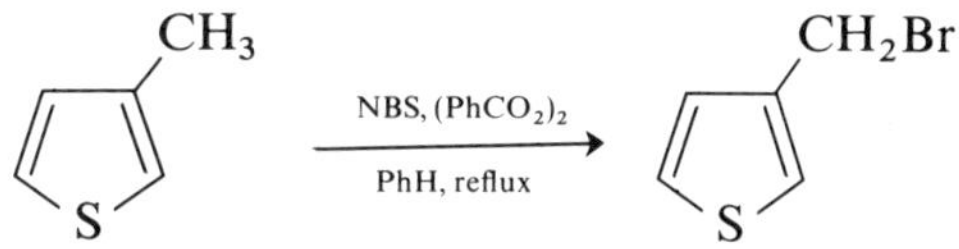

(4.72)

(70–80 %)

As in chlorination reactions, the radical A· need not necessarily be a bromine atom. Indeed, brominations have been achieved by using t-butylhypobromite [93]. In this case the site of bromine substitution is determined by the selectivity of the t-butoxy radical under the conditions of the reaction. The reactions therefore closely parallel those of the corresponding hypochlorite [93]. Since hypobromites are not so readily prepared, it appears that a mixture of the corresponding hypochlorite and bromotrichloromethane may be used as a brominating agent. Reaction (4.73c) must compete

$$
\begin{array}{lrcl}
\text{(a)} & \text{t-BuO}\cdot + \text{RH} & \longrightarrow & \text{t-BuOH} + \text{R}\cdot \\
\text{(b)} & \text{R}\cdot + \text{t-BuOCl} & \longrightarrow & \text{RCl} + \text{t-BuO}\cdot \\
\text{(c)} & \text{R}\cdot + \text{BrCCl}_3 & \longrightarrow & \text{RBr} + \cdot\text{CCl}_3 \\
\text{(d)} & \cdot\text{CCl}_3 + \text{t-BuOCl} & \longrightarrow & \text{CCl}_4 + \text{t-BuO}\cdot
\end{array} \qquad (4.73)
$$

successfully with reaction (4.73b), and a high CCl_3Br/t-BuOCl ratio must be maintained, since when the ratio is low the chloro compound is the predominant product (4.74) [93].

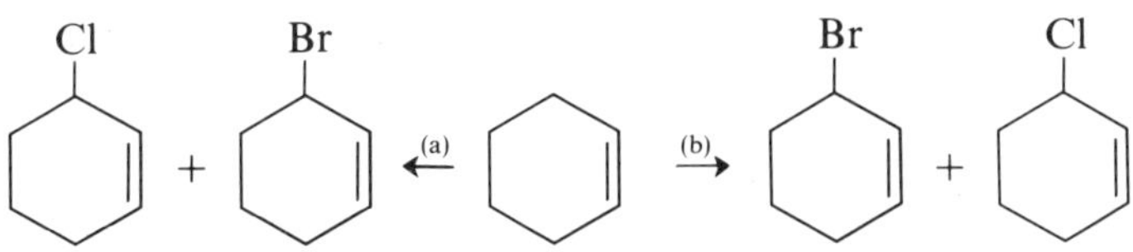

(4.74)

97:3 94:6

(a) Low ratio CCl_3Br/t-BuOCl
(b) High ratio CCl_3Br/t-BuOCl

The trichloromethyl radical may also be implicated in the hydrogen abstraction process, but the reaction is only exothermic for the removal of allylic and benzylic hydrogen atoms, since $D(Cl_3C—H)$ is quite low (90 kcal mol^{-1}). Nevertheless, trichloromethanesulphonyl bromide upon irradiation will brominate cyclohexane and cyclopentane to give the monobromo compounds in fair yield [94]. Unlike trichloromethanesulphonyl chloride chlorinations, in which the hydrogen-abstracting species is the $Cl_3C{\cdot}SO_2{\cdot}$ radical [82], in the brominations the active species is the $Cl_3C{\cdot}$ radical [94]. Accordingly, the specificities of the two processes are somewhat different.

Bromotrichloromethane, a liquid, is a useful source of $Cl_3C{\cdot}$ radicals, particularly as the reagent can also function as a solvent. It is cleaved by near ultraviolet radiation or by the usual thermal radical initiators such as azobisisobutyronitrile (AIBN). The trichloromethyl radical, rather than the

$$Cl_3CBr \xrightarrow[\text{or } \Delta/\text{AIBN}]{h\nu} Cl_3C{\cdot} + Br{\cdot} \tag{4.75}$$

bromine atom, is closely implicated in the hydrogen abstraction process, since the yield of chloroform is usually close to the yield of the final bromo compound [95]. It is an electrophilic and moderately bulky radical, and this is reflected in the nature and distribution of the brominated products. Toluenes are smoothly photobrominated to benzyl bromides, and very little polysubstitution occurs [95]. Norbornane gives a mixture of the 2-bromocompounds [96]. The ratio of *exo*:*endo* bromides (95:5) is much

$$p\text{-NC}{\cdot}C_6H_4CH_3 \xrightarrow[CCl_3Br]{h\nu} p\text{-NC}{\cdot}C_6H_4CH_2Br \quad (86\%) \tag{4.76}$$

higher than the ratio obtained by using elemental bromine (70:30). The increased stereospecificity of the CCl_3Br bromination may be ascribed to the relatively low accessibility of the intermediate 2-norbornyl radicals from the *endo* side for the bulkier reagent [96]. The synthetic utility of bromotrichloromethane has probably not been fully realized, but it suffers from one serious drawback which limits its general use—it rapidly adds to double bonds in radical-initiated reactions.

Hydrogen abstraction by iodine atoms is appreciably endothermic, and iodine atoms do not react with alkanes at moderate temperatures. Molecular iodine is an efficient scavenger of free radicals and the reaction can be used to prepare iodo compounds. The process is of very limited use, since the compounds from which the radicals are produced are more readily converted to iodo compounds by other methods.

$$R{\cdot} + I_2 \rightleftharpoons RI + I{\cdot} \tag{4.77}$$

The iodo compounds are therefore best prepared by other methods or by metathetical reactions on the chlorides or bromides (prepared as above) with potassium iodide in acetone, although even this reaction is not without its own shortcomings.

Fluorination

Elemental fluorine reacts with organic compounds with considerable violence, since both of the chain-propagation steps (reaction (4.66)) are highly exothermic [97]. Fluorination occurs in the dark even at $-78°$, and since simple homolysis of the F—F bond is too endothermic a reaction to be a significant initiation step at low temperatures (ΔH = 37 kcal mol^{-1}), a molecule-induced homolysis reaction is probably involved. Attempts to moderate the reaction by the suitable dilution of the fluorine with an inert

$$F_2 + RH \rightarrow HF + F\cdot + R\cdot; \qquad \Delta H \leqslant 5 \text{ kcal mol}^{-1} \tag{4.78}$$

gas or solvent, and by the use of various metal packings in specially designed reactors, were partially successful. However, the direct use of fluorine for the preparation of perfluoro compounds has now been largely discontinued in favour of three indirect methods. These are fluorination by halogen exchange [98] (by far the most widely used method), the Simons electrochemical process [99], and fluorinations with high-valency metallic fluorides [100]. The two latter methods may be free-radical in type and are usually used to prepare perfluoro-organic compounds. An important feature of the electrochemical method is that several functional groups retain their structural integrity through the reaction. The process is run at relatively low potential (< + 2·85 volts) so that free fluorine is not liberated at the anode.

$$CH_3COF \xrightarrow[\text{Ni anode}]{\text{HF}} CF_3COF \tag{4.79}$$

The importance of fluorochemicals can be judged from the buoyancy of the manufacturing industry [101], and from the current interest in their physiological action [102].

Reduction of Organic Halides with Organo-tin Hydrides

The n-butyl- and the phenyltin hydrides have been studied the most, and their effectiveness appears to follow the sequence: $Ph_2SnH_2 \cong$ n-$BuSnH_3 >$ $Ph_3SnH \cong$ n-$Bu_2SnH_2 >$ n-Bu_3SnH. The phenyltin hydrides are more reactive than the corresponding n-butyltin hydrides, and the di- and trihydrides are more active than the corresponding monohydrides [103]. The mechanism is outlined in equation (4.80), where I· is a radical derived from an initiator or from an adventitious impurity in the reagents. The nature of the termination reactions is not known. The reactions are generally

$$I\cdot + -\overset{|}{\underset{|}{Sn}}H \rightarrow -\overset{|}{\underset{|}{Sn}}\cdot + IH$$

$$-\overset{|}{\underset{|}{Sn}}\cdot + RX \rightarrow -\overset{|}{\underset{|}{Sn}}X + R\cdot$$

$$R\cdot + -\overset{|}{\underset{|}{Sn}}H \rightarrow RH + -\overset{|}{\underset{|}{Sn}}\cdot \tag{4.80}$$

remarkably free from complications, and low yields are usually due to incomplete reaction. The last step of reaction (4.80) must be extremely fast, since even the 2,2,2-triphenylethyl radical, which had not previously been observed as a kinetically free species since it is prone to undergo rearrangement, can be reduced by triphenyltin hydride.

$$Ph_3Sn\cdot + Ph_3CCH_2Cl \rightarrow Ph_3SnCl + Ph_3CCH_2\cdot$$

$$Ph_3CCH_2\cdot + Ph_3SnH \rightarrow Ph_3CCH_3 + Ph_3Sn\cdot \quad (4.81)$$

The yield of 2,2,2-triphenylethane varies from 27% to more than 90%, according to reaction conditions [104].

Aliphatic bromides react exothermically with tri-n-butyltin hydride or triphenyltin hydride in the absence of a solvent. Chlorides, even allylic and benzylic chlorides, are much less reactive and may require temperatures of 100° or above for reaction at a convenient rate. Geminal polyhalides can be reduced selectively in high yield to the monodehalogeno compounds. Although no cases of dehydrohalogenation of monohalides or geminal halides are known, vicinal dihalides are dehalogenated to olefins [105]. The tin hydride does not function like the more conventional reagent zinc metal in this respect since 1,3-dibromopropane gives no cyclopropane on treatment with tri-n-butyltin hydride. Halohydrins are reduced normally.

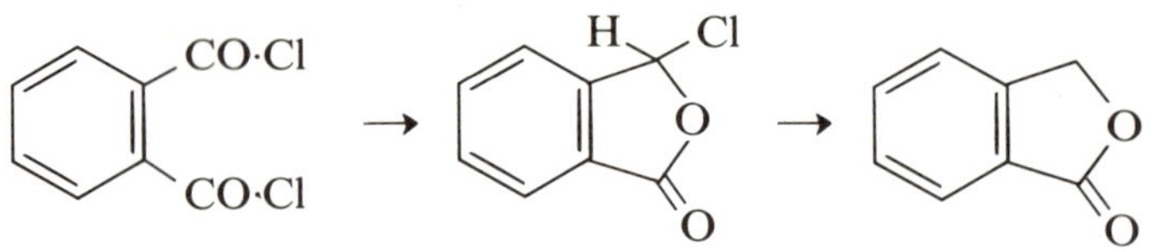

(4.82)

(55%)

Phthaloyl chloride is reduced to phthalide [105], whereas with succinoyl chloride the intermediate chlorolactone is not completely reduced [105]. Recent work on a variety of acyl halides has shown that in the presence or absence of a solvent the major reduction products are usually esters [106].

The combination of lithium aluminium hydride and an organo-tin halide is a convenient system for the generation of organo-tin hydrides [107]. The reagent may be formed *in situ* provided that the substrate has no functional groups which could be reduced by the lithium aluminium hydride. In favourable cases, one equivalent of $LiAlH_4$ and only catalytic quantities of organo-tin halide serve as the reductant since a low concentration of the tin hydride is maintained throughout the reaction. In this way, the necessity

$$R'_3SnH + RX \rightarrow RH + R'_3SnX$$

$$4R'_3SnX + LiAlH_4 \rightarrow LiAlX_4 + 4R'_3SnH \quad (4.83)$$

of preparing, handling, and storing organo-tin hydrides is avoided. In cases where groups are present which are susceptible to reduction by $LiAlH_4$, it is expedient to preform the tin hydride and then add the organic halide.

Some typical examples of the reaction are listed in Table 4.3. It can be seen that aryl chlorides are among the least reactive of halides. In such difficult cases it is advantageous, even necessary, to use a radical initiator, and sometimes in fairly substantial amounts [110]. It is also interesting to note that the reduction of a halide competes successfully with the hydrostannation of olefinic double bonds, the reduction of 1,1-dibromo-2-vinylcyclopropane being a case in point.

HALODECARBOXYLATION OF CARBOXYLIC ACIDS

The decarboxylation of the silver salts of carboxylic acids by molecular halogens under thermal activation (Hunsdiecker reaction [115]) has been known for some time to be a convenient route to organic halides. The reaction, which is normally carried out in an inert solvent such as carbon tetrachloride, involves the loss of one carbon atom from the substrate molecule.

$$\begin{array}{lrcl} \text{(a)} & RCO_2Ag+X_2 & \rightarrow & RCO_2X+AgX \\ \text{(b)} & RCO_2X & \rightarrow & RCO_2\cdot+X\cdot \\ \text{(c)} & RCO_2\cdot & \rightarrow & R\cdot+CO_2 \\ \text{(d)} & R\cdot+RCO_2X & \rightarrow & RX+RCO_2\cdot \\ \text{(e)} & R\cdot+X_2 & \rightarrow & RX+X\cdot \\ \text{(f)} & 2R\cdot & \rightarrow & RR, \quad \text{etc.} \end{array} \qquad (4.84)$$

When R is an alkyl group, reaction (4.84c) is normally a process of low or zero activation energy.

The Hunsdiecker reaction is of general application to straight-chain alkanoic acids. A substituent in any position other than the α-position does not interfere unless it is itself capable of reacting with the alkyl hypohalite. Thus, for example, the reaction is one of the best methods for the preparation of ω-haloesters. The reaction is likewise quite successful for phenyl or

$$RO_2C{\cdot}(CH_2)_n{\cdot}CH_2CO_2Ag \xrightarrow[X_2]{} RO_2C{\cdot}(CH_2)_nCH_2X \qquad (4.85)$$

deactivated phenyl alkanoic acids, but activated phenyl groups are halogenated in the ring. Secondary halides are usually obtained without undue difficulty from α-alkyl alkanoic acids, but α,α-dialkyl alkanoic acids usually give a variety of products and little of the t-halide. However, bridgehead carboxylic acids (e.g., adamantane-1-carboxylic acid) can often be decarboxylated smoothly, and α-keto acids and polyhalogeno acids usually react successfully. Aldehydes or ketones are obtained from α-hydroxy and α-amino alkanoic acids, depending upon the extent of the chain branching at the α-carbon atom (equations (4.86) and (4.87); R = alkyl, R′ = H or alkyl).

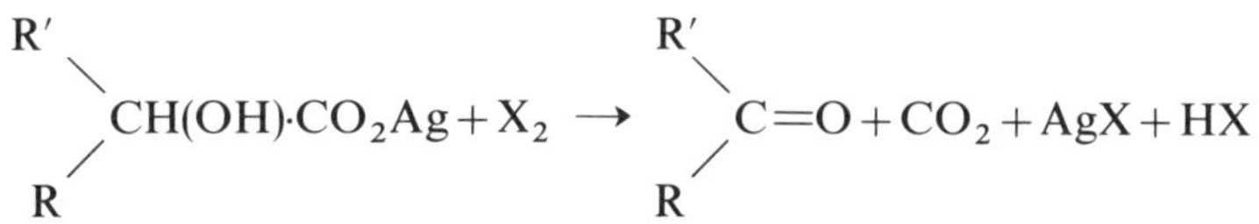

(4.86)

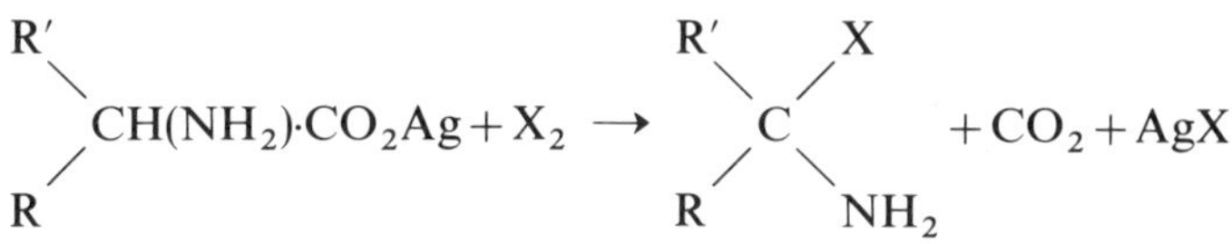

Table 4.3 Reduction of organic halides by organo-tin hydrides

Substrate	Hydride	Product	Yield (%)	Ref.
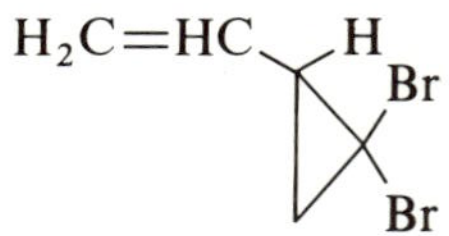	n-Bu_3SnH	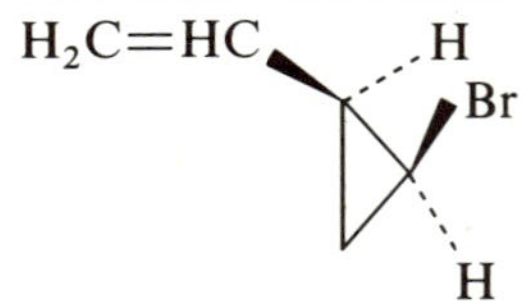	45	
		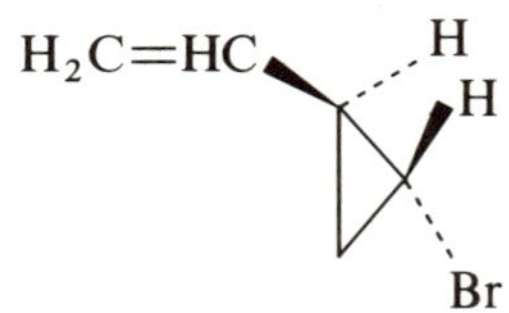	17	[108]
(+)-PhCHClMe	Ph_3SnD	(±)-PhCHDMe	76	[109, 110]
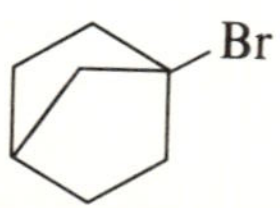	Ph_3SnH	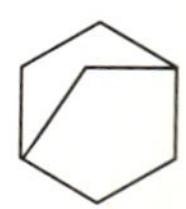	55	[112]

$PhCCl_3$	n-Bu_3SnH	$PhCHCl_2$	88	[107]
$PhCHCl_2$	n-Bu_3SnH	$PhCH_2Cl$	76	[107]
$PhCH_2Cl$	n-Bu_3SnH	$PhCH_3$	78	[107]
$MeCHBr{\cdot}CH_2Br$	n-Bu_3SnH	$MeCH{=}CH_2$	81	[107]
$CH_2Br{\cdot}CHOH{\cdot}CH_2Br$	n-Bu_3SnH	MeCHOHMe	85	[103]
$PhCOCH_2Cl$	Ph_3SnH	PhCOMe	86	[107]
PhCl	n-Bu_3SnH	PhH	1	[110]
PhCl	{ n-Bu_3SnH AIBN (1·5 mol %)	PhH	70	[110]
o-$Cl{\cdot}C_6H_4Br$	Ph_3SnH	PhCl	96	[113]
$MeCH_2COCl$	Ph_3SnH	$MeCO_2CH_2CH_2Me$	50	[106]
N-bromosuccinimide	Ph_3SnH	Succinimide	95	[114]
trans-9-chlorodecalin	n-Bu_3SnD	*cis*- + *trans*-9-D-decalin	—	[111]

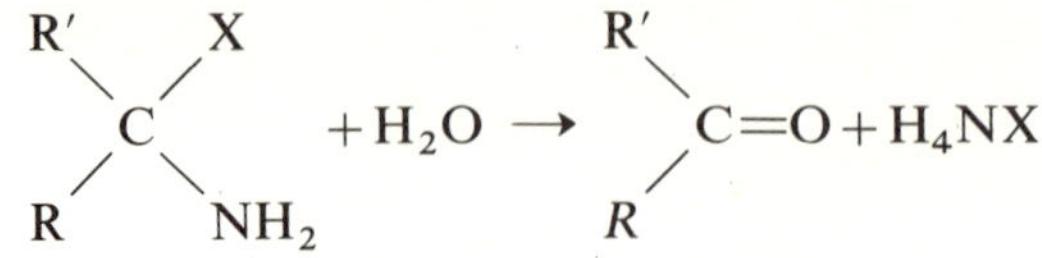

(4.87)

The decarboxylation of the silver salts of olefinic carboxylic acids is never successful because the intermediate hypohalite adds across the double bond (Prévost reaction [115]). With aromatic acids the yields of aryl halides are variable and greatly depend upon the reaction conditions.

Bromine has usually been the halogen of choice, although iodine has also found frequent use. It is important in using iodine to keep its concentration at least as high as that of the silver carboxylate. At low iodine:silver salt ratios (e.g., 1:2) esters are normally produced (Simonini reaction [115]).

$$2RCH_2CO_2Ag + I_2 \longrightarrow RCH_2CO_2CH_2R + 2AgI + CO_2 \qquad (4.88)$$

The reaction itself has no synthetic importance, but the intermediate complex of silver salt and hypoiodite ($RCO_2Ag{\cdot}RCO_2I$) formed from monocarboxylic acids not containing olefinic double bonds may be used in the more important Prévost reaction.

Yields in the Hunsdiecker reaction are frequently good, but the reaction is extremely sensitive to trace amounts of water, the presence of which leads to recovery of free acid. The preparation of dry silver carboxylates is not easy, since they are difficult to dry and are sensitive to heat. For these reasons, a number of modified procedures have been introduced which lower the cost and increase the reproducibility and scope of the halodecarboxylation, and often improve the yields of the halides.

The first of the modified procedures is the so-called Cristol–Firth method [116] which utilizes the free carboxylic acid and excess red mercuric oxide. The reactions are run in inert solvents (e.g., CCl_4, $C_2H_2Cl_4$) and usually at the reflux temperature. Relatively few halides have been synthesized by this method, but reactions (4.89)–(4.91) are typical [116, 117]. In each case bromine is the halogen employed, together with red mercuric oxide, although yellow mercuric oxide is almost as effective.

$$Me(CH_2)_{16}CO_2H \longrightarrow Me(CH_2)_{15}CH_2Br \quad (93\%) \qquad (4.89)$$

$$ClCH_2CO_2H \longrightarrow ClCH_2Br \quad (38\%) \qquad (4.90)$$

$$MeCH_2CH(Me)CO_2H \longrightarrow MeCH_2CHBr{\cdot}Me \quad (61\%) \qquad (4.91)$$

The halodecarboxylation may also be carried out by using lead tetra-acetate and iodine under irradiative conditions [118], or lead tetra-acetate and an inorganic halide (e.g., LiCl) [119]. Although both methods are applicable to primary, secondary and tertiary aliphatic carboxylic acids, the inorganic halide route seems to be preferable for tertiary acids. Iodides are of course formed when iodine is used, but the second method is best run using lithium chloride (1 mole/mole of $PbOAc_4$), when the alkyl chloride is obtained. Lithium bromide and iodide are less satisfactory reagents. In place of lead tetra-acetate and iodine, t-butyl hypoiodite may be employed [118]. With this reagent, primary, secondary and tertiary alkanoic acids, as well as the hitherto intractable glutaric and adipic acids, can be decarboxylated without difficulty.

Reductive Decarboxylation of Carboxylic Acids

There are conventional heterolytic methods available for reductive decarboxylation, but they often require forcing conditions or are not of general applicability. The decarboxylation can be carried out in two stages by using radical reactions. The first involves the conversion of the acid to a halide, using a modified Hunsdiecker reaction. In the second the halide can be reduced by the organo-tin hydride method. A further two- or three-stage method which has been developed recently involves the conversion of the acid to the t-butyl perester, which is then allowed to decompose at a moderately high temperature in the presence of a good hydrogen-donor (normally the solvent) [120].

$$\text{(a)} \quad RCO_2H \xrightarrow[\text{(ii) t-BuOOH}]{\text{(i) SOCl}_2} RCO_2{\cdot}OCMe_3$$

$$\text{(b)} \quad RCO_2{\cdot}OCMe_3 \xrightarrow[130°\text{, cymene}]{} RH + CO_2 + Me_3COH \qquad (4.92)$$

The mechanism of reaction (4.92b) is as follows:

$$\text{(a)} \quad RCO_2{\cdot}OCMe_3 \xrightarrow{\Delta} RCO_2{\cdot} + {\cdot}OCMe_3$$

$$\text{(b)} \quad RCO_2{\cdot} \rightarrow R{\cdot} + CO_2$$

$$\text{(c)} \quad R{\cdot} + SH \rightarrow RH + S{\cdot}$$

$$\text{(d)} \quad Me_3CO{\cdot} + SH \rightarrow Me_3COH + S{\cdot}$$

$$\text{(e)} \quad 2S{\cdot} \rightarrow S{\cdot}S \qquad (4.93)$$

where SH represents a solvent molecule. The crucial step is (4.93b) which must be fast so that reactions of type (4.94) can be avoided. When R is an alkyl group the activation energy of step (4.93b) is very low and the loss of

$$RCO_2{\cdot} + SH \rightarrow RCO_2H + S{\cdot} \qquad (4.94)$$

carbon dioxide very rapid. When R is phenyl, however, the loss of CO_2 is not as fast, and side reactions may become a complication. The reaction appears to be useful but so far has found only a few applications; (4.95) and (4.96) are examples [120, 121]. The elegant synthesis of the elusive compound cubane is an outstanding case [121]. Cubane has also been prepared by a shorter but similar route from cubane-1,4-dicarboxylic acid [122].

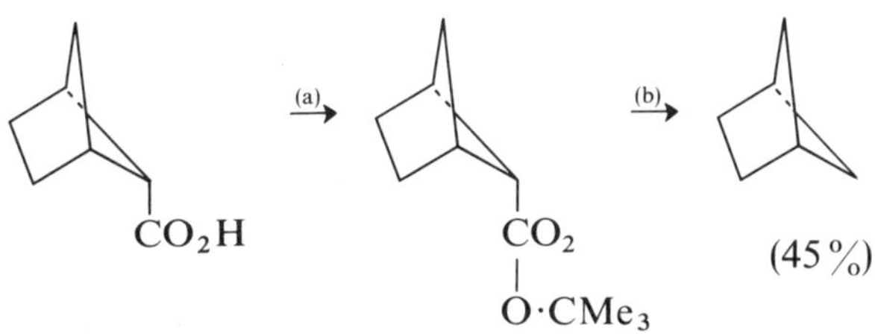

(4.95)

(a) (i) $SOCl_2$; (ii) t-BuOOH
(b) Cymene solution, 115–130°

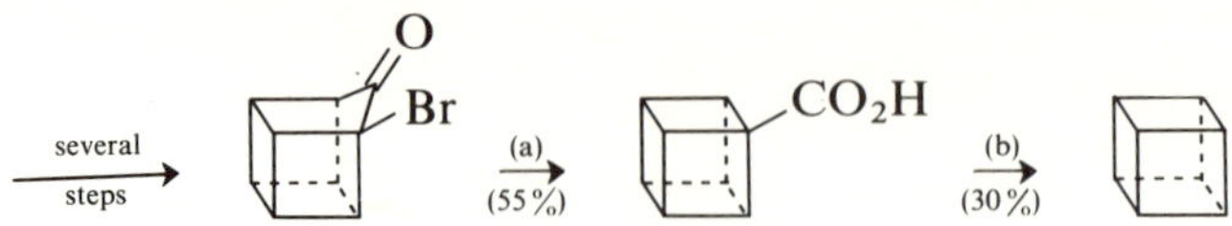

(4.96)

(a) 25% KOH (Favorskii reaction)
(b) (i) $SOCl_2$; (ii) t-BuOOH/pyridine; (iii) 150° in di-isopropylbenzene

AUTOXIDATION

The reaction of molecular oxygen with organic compounds under conditions which do not involve combustion is termed autoxidation [123]. The reaction normally proceeds at fairly moderate temperatures, and for the majority of cases is undoubtedly free-radical in character. It is probably the most common, perhaps even the most important, of all radical processes. Autoxidation is a key step in certain biological reactions and it has been developed technically into several important modern processes in chemical industry [124]. Direct oxidation by molecular oxygen is also responsible for the deterioration of a whole range of materials, and the inhibition of autoxidation has also, therefore, been the subject of intensive study [123, 125, 126].

A number of functional groups are susceptible to attack by oxygen [1, 2, 123, 124], but only reaction at C—H bonds will be considered here. The reactions are catalysed by light, radical initiators, transition-metal ions, and often by the hydroperoxide products themselves, which can lead to an autocatalytic enhancement of the reaction rates. Inhibition of the reaction by radical-scavengers points to a radical-chain mechanism [125, 126]. Although complex mixtures of products are frequently formed, the simple scheme below is fairly typical of many autoxidation reactions.

$$
\begin{aligned}
\mathrm{I\cdot + RH} &\rightarrow \mathrm{R\cdot + IH} \\
\mathrm{R\cdot + O_2} &\xrightarrow{\text{fast}} \mathrm{ROO\cdot} \\
\mathrm{ROO\cdot + RH} &\rightarrow \mathrm{ROOH + R\cdot} \\
\mathrm{ROO\cdot + R\cdot} &\rightarrow \mathrm{ROOR} \\
\mathrm{R\cdot + R\cdot} &\rightarrow \mathrm{R{\cdot}R}, \quad \text{etc.} \\
\mathrm{2ROO\cdot} &\rightarrow \mathrm{ROOR + O_2}, \quad \text{etc.}
\end{aligned}
\tag{4.97}
$$

The peroxides so formed can react further under suitable conditions [1, 2, 123, 124, 127].

The alkylperoxy radicals are weakly reactive and usually only labile C—H bonds are attacked. Polar effects have been demonstrated in their reactions with a number of substrates [128]. Transition state (7) explains the stabilization by electron-releasing substituents in R′·, whilst (7a) accounts for the preference for reaction at a site which leads to the production of the most stable radical R′· in the reaction. When both effects complement each other the reactivity of a C—H bond towards the peroxy radical can be very high, as in the case of aldehydes and of ethers. The hazards arising from the last reaction should be well noted [129]. In saturated alkanes, usually only the tertiary C—H bonds are attacked to give good yields of hydroperoxide, whereas allylic and benzylic sites are readily autoxidized [130–134].

$$\text{ROO}\cdot + \text{R}'\text{H} \longrightarrow \underset{(7)}{[\text{ROO}:^{-}\text{H}^{+}\cdot\text{R}'} \leftrightarrow \underset{(7a)}{\text{ROO}:\text{H}\cdot\text{R}']} \longrightarrow \text{ROOH} + \text{R}'\cdot \qquad (4.98)$$

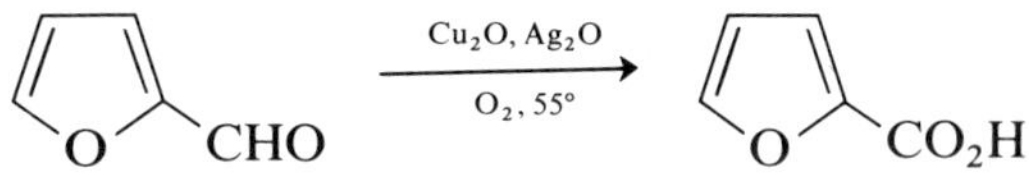

(4.99)

(86–90%)[130]

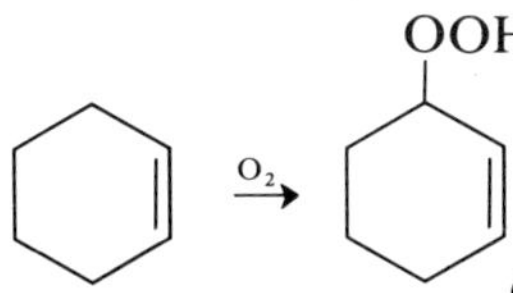

(4.100)

(96%)[132]

The reactions are often carried out somewhat below 100° so that thermal decomposition of the hydroperoxide is minimized whilst a useful rate of autoxidation is maintained. The hydroperoxide moiety can be converted to other functional groups by standard chemical procedures. An interesting example is the synthesis of 6-hydroxycyclodecanone from *trans*-4a-hydroperoxydecahydronaphthalene by treatment with sulphuric acid in t-butanol [134].

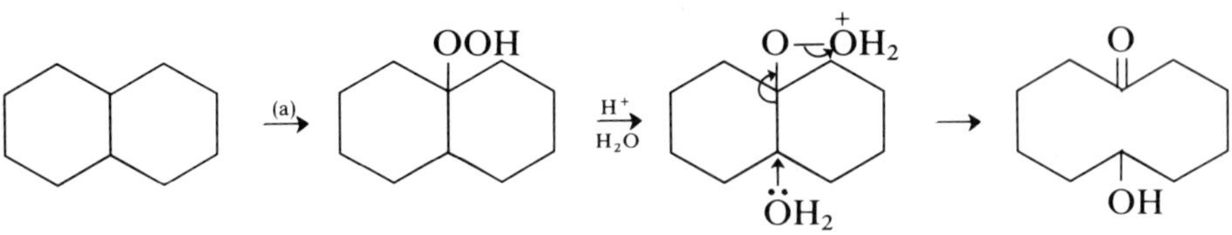

(26%) (25%)

(a) O_2, Co^{2+}, 125°, 8 hours (4.101)

Autoxidation reactions are not limited to the abstraction of hydrogen atoms by peroxy radicals, nor are all such oxidations radical-chain processes [1, 2]. Indeed, some autoxidations do not involve organic radical intermediates at all [135]. The additions of oxygen and peroxy radicals to olefinic double bonds are well-known phenomena, and the oxidation of carbanions and quinone anions have been studied [131, 136]. A recent application of the reaction of oxygen with carbanions has involved the use of dimethylsulphoxide (DMSO) as the solvent [136]. The DMSO probably

$$\text{Ph}_3\text{CH} \xrightarrow[\text{t-BuOK}]{\text{excess}} \begin{cases} \xrightarrow{\text{pyridine}} \text{Ph}_3\text{COOH} \quad (80\%) \\ \xrightarrow[\text{t-BuOH}]{\text{DMSO}} \text{Ph}_3\text{COH} \quad (95\%) \end{cases} \qquad (4.102)$$

reduces the hydroperoxide anion to alkoxide anion and thus facilitates the removal of the hydroperoxide before decomposition reactions occur.

$$ROO^- + Me_2SO \rightarrow RO^- + Me_2SO_2 \qquad (4.103)$$

Surprisingly, the use of DMSO as a solvent for radical reactions has attracted very little attention.

Other Intermolecular Abstraction Reactions

The nitration of saturated and unsaturated compounds, aralkanes and aromatic compounds by nitric acid and by the oxides of nitrogen has been extensively studied in both liquid and gaseous phases. Radical mechanisms have been proposed for many nitration processes [137], and several of the reactions are of commercial importance. Work in this complex field has been reviewed several times in recent years [76, 137], and nitration reactions are therefore not considered further here.

The photochemical reaction of nitrosyl chloride with hydrocarbons in the presence of hydrogen chloride at room temperature produces oximes [76, 138–140]. The reactions involve radical intermediates [140], and proceed in good yield. The following examples are fairly typical.

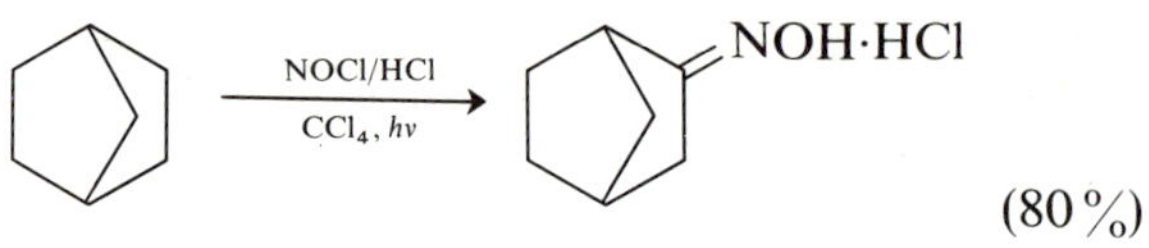

(4.104)

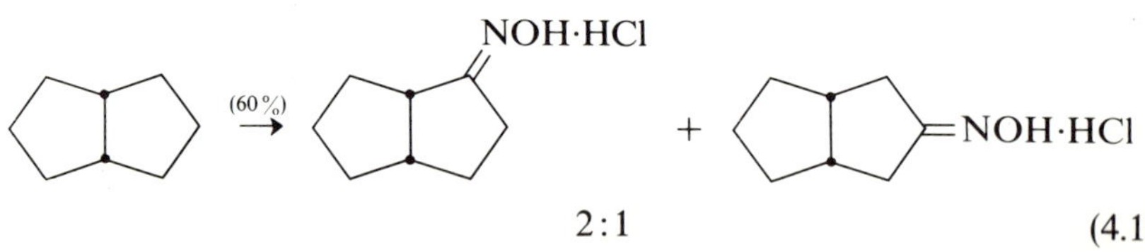

(4.105)

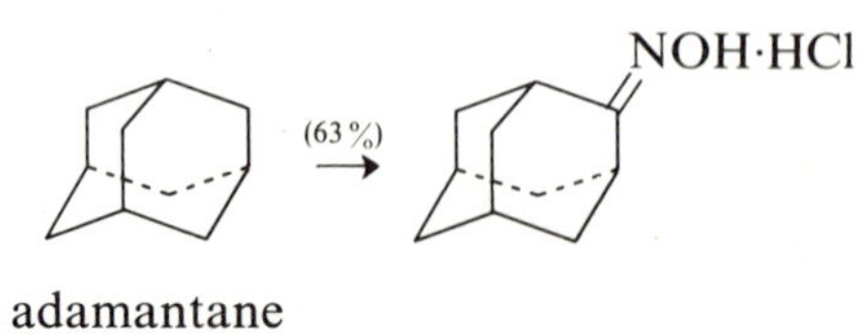

(4.106)

The photoximation of cyclohexane is the only photochemical reaction run on an industrial scale. The Japanese company, Toyo Rayon, first developed the process in 1962 and utilize a 20 kW mercury arc. The cyclohexane is irradiated for three hours at 5–30° in the presence of nitrosyl chloride. The cyclohexanone oxime so formed is converted to ω-caprolactam by reaction with sulphuric acid (Beckman rearrangement), which is then used in the manufacture of nylon-6.

The interesting and important feature of these reactions is that oximation occurs preferentially at secondary carbon atoms. This is contrary to the majority of the reactions discussed above, in which radical attack always occurs mainly at tertiary sites. Oximes are also produced in the corresponding reactions with nitrosyl sulphonic acid [76].

Nitriles can also be prepared by a somewhat similar process, namely the ultraviolet light-initiated reaction of cyanogen chloride with organic compounds. At the present time, nitriles have been prepared from saturated hydrocarbons and from ethers [141]. The yields are good but the actual conversions are only modest. The thiocyanide group can be introduced in a similar manner [76].

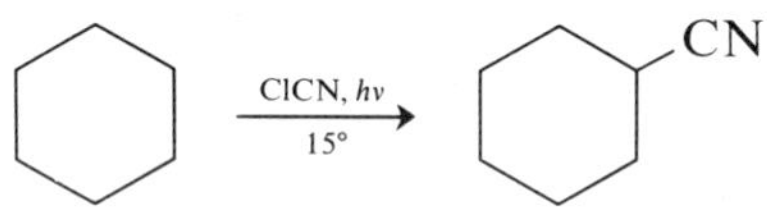

(96%) (4.107)

$$CH_3CH_2OCH_2CH_3 \rightarrow CH_3CH_2OCH(CN)CH_3 \quad (95\%) \quad (4.108)$$

$$CH_3OCH_2CH_2OCH_3 \rightarrow CH_3OCH_2CH(CN)OCH_3 \quad (45\%)$$
$$+ CH_3OCH_2CH_2OCH_2CN \quad (45\%) \quad (4.109)$$

There are several other types of intermolecular abstraction reaction which cannot be considered in detail here. Triplet carbenes and nitrenes, for example, are known to react with C—H and C—C bonds by insertion [142–147]. The reactions probably have a considerable radical character. Radicals, particularly those derived from peroxides, can perform dehydrogenation reactions, reactions (4.110)–(4.112), for example [148, 149].

$$RCH_2{-}\underset{\underset{OH}{|}}{N}{-}CH_2R + \text{t-BuOOH} \rightarrow RCH{=}\underset{\underset{O}{\downarrow}}{N}{-}CH_2R + \text{t-BuOH} \quad (4.110)$$

(R = Me or Ph)

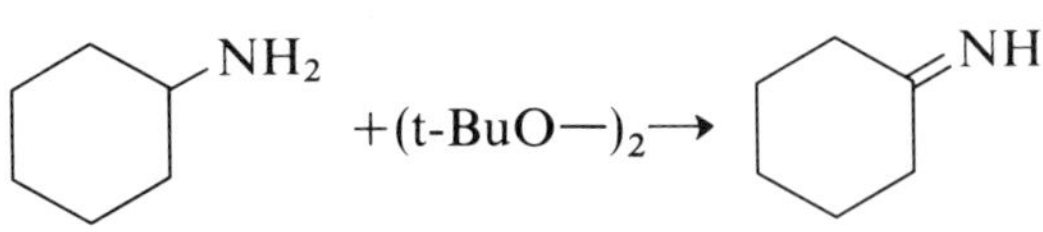

(4.111)

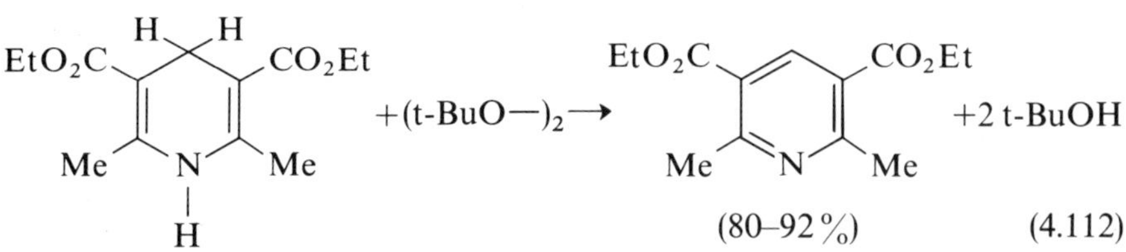

(80–92%) (4.112)

Peroxides are the most convenient sources of the radicals, since the substrate is oxidized in two distinct steps.

(a) $R_2CHNH_2 + \text{t-BuO}\cdot \rightarrow R_2\dot{C}NH_2 + \text{t-BuOH}$

(b) $R_2\dot{C}NH_2 + (\text{t-BuO}{-})_2 \rightarrow R_2C{=}NH + \text{t-BuOH} + \text{t-BuO}\cdot$ (4.113)

INTRAMOLECULAR RADICAL ABSTRACTION REACTIONS

In the last few years interest in these reactions has grown rapidly because they offer means of introducing functionality at saturated carbon atoms in a very specific manner. The reactions succeed where intermolecular abstraction reactions would fail or would be of very limited use. Many of the examples discussed below are of steroidal systems; most of the definitive work has been carried out using steroids as precursors.

The reactions are similar in many respects. A radical centre is generated in one part of a molecule and, if the geometry of the molecule allows it, internal hydrogen transfer occurs from the δ-carbon atom. A six-membered chair-like transition state is essential for efficient transfer of the hydrogen atom [150–152]. Steroidal systems are particularly well suited to reactions of this type. The ring system has a rigid and known stereochemistry and,

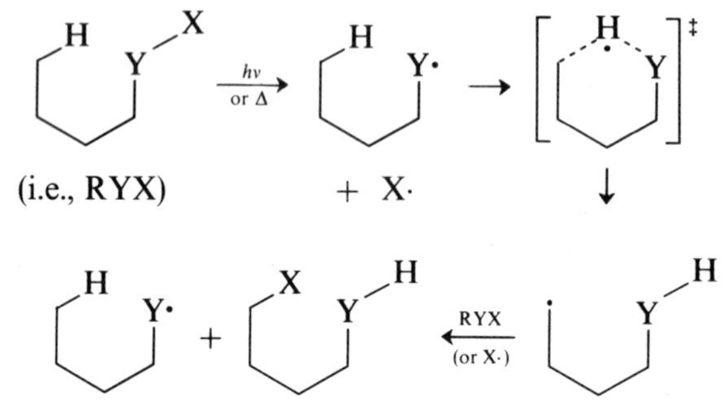

(4.114)

by the suitable siting of group YX, the radical Y· can be generated in the close vicinity of the hydrogen atom to be transferred. In principle C(18) can be attacked by radical $\dot{Y}$— attached to C(20), C(15), C(8), or C(11), and C(19) by the radical attached to C(11), C(6), C(4), or C(2). Typical examples of group YX are hypohalites (Y = O, X = halogen), nitrites (Y = O, X = NO), *N*-haloamides (Y = NH or O†, X = halogen), protonated

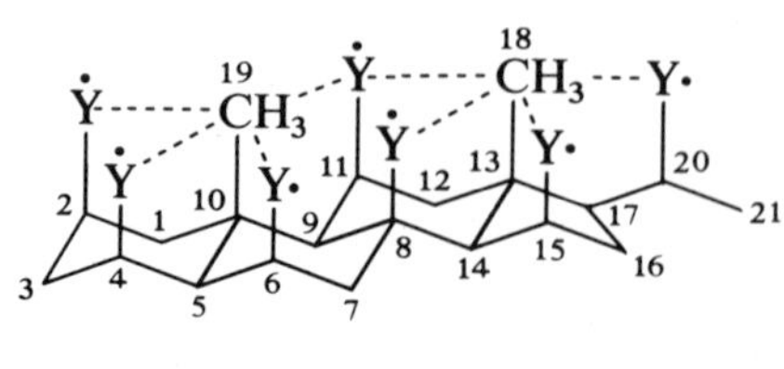

N-haloamines (Y = $\overset{+}{N}H_2$, X = halogen), *N*-nitrosoamides (Y = >N—, X = NO), and *N*-chlorosulphonamides (Y = NH, X = chlorine). In addition the group YX might also comprise a single atom which possesses two unpaired electrons, that is, YX could represent a triplet carbene or nitrene species (4.115) [2, 142–147, 153]. When X is halogen the processes appear to be radical-chain reactions; when X is not a halogen atom, the reactions

† The amide radical has two mesomeric forms: $-\overset{O}{\overset{\|}{C}}-\dot{N}H \leftrightarrow -\overset{O\cdot}{\overset{|}{C}}=NH$.

have little or no chain character. They can be initiated by heat or by light, but the photochemical procedures are very convenient and appear to be more reliable in the majority of cases.

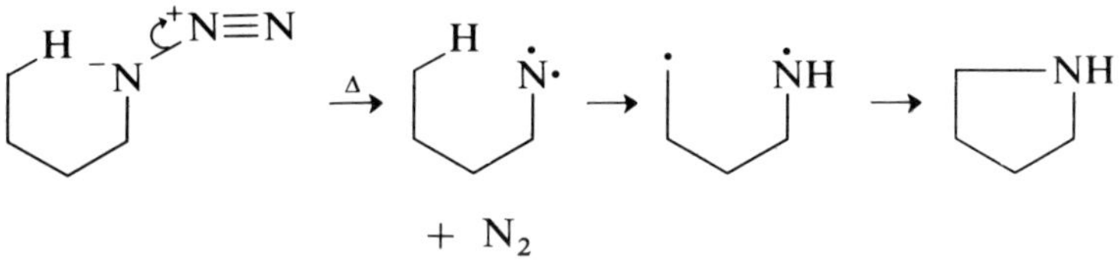

(4.115)

The photolytic reactions of nitrites (Barton reaction) are perhaps the best-known of these procedures, particularly in view of the number of spectacular syntheses which have been achieved with these compounds [150, 154–157]. The reaction is discussed in the chapter on photochemical methods.

Considerable interest has also been focused on the photolytic reactions of hypohalites. Generally speaking, the photochemical decompositions of aliphatic primary and secondary hypohalites have yielded mainly fragmentation or rearrangement products, and these reactions are considered later. Only tertiary hypohalites are considered here. The intramolecular reactions lead to δ-halohydrins which can be cyclized smoothly to derivatives of tetrahydrofuran [150].

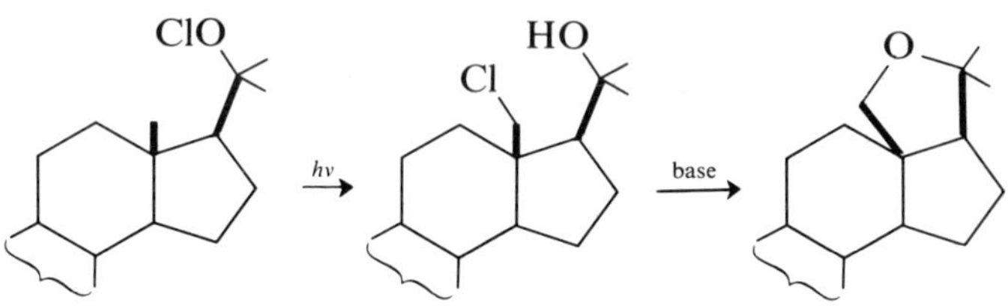

(4.116)

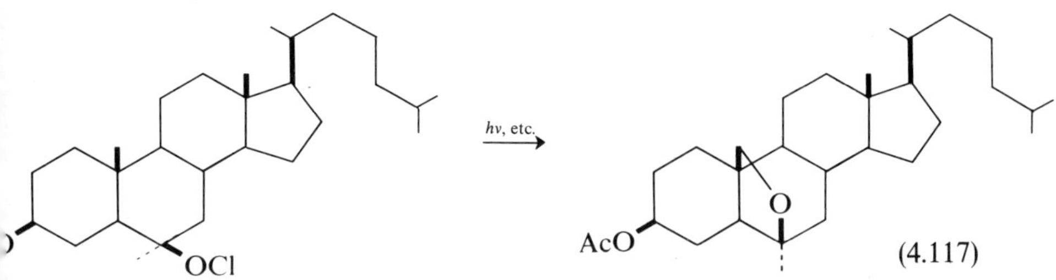

(4.117)

In the aliphatic series, if the δ-hydrogen atom is attached to a terminal carbon atom, cleavage reactions are relatively important, reaction (4.118) for example [158]. When the δ-hydrogen atom is not primary the hydrogen-transfer proceeds normally, (4.119) and (4.120) [159, 160]. In the steroid series, fragmentations of tertiary alkoxy radicals are less probable since

$$CH_3CH_2CH_2C(Me)_2OCl \xrightarrow{h\nu} Me_2CO + \underset{(59\%)}{CH_3CH_2CH_2Cl} \qquad (4.118)$$

$$CH_3CH_2CH_2CH_2C(Me)_2OCl \xrightarrow{h\nu} CH_3\underset{\displaystyle Cl}{\underset{|}{C}}H{\cdot}CH_2CH_2C(Me)_2OH \quad (55\%) \qquad (4.119)$$

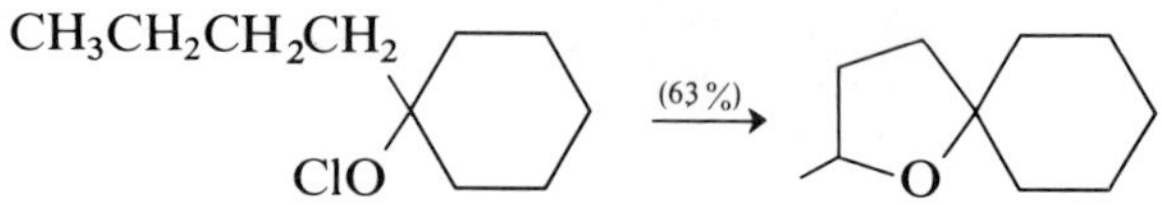

(4.120)

either a methyl radical would have to be eliminated, or one of the rings cleaved. In addition, the requisite transition-state conformation can be achieved with a low expenditure of energy and with a high probability.

Of hypobromites and hypoiodites, only the t-butyl compounds are known. Other tertiary hypobromites can be prepared *in situ* by an exchange reaction between the t-alcohol and t-butylhypobromite. Hypoiodites, however, are more conveniently generated by the reaction of the t-alcohol with iodine and either mercuric oxide or lead tetra-acetate under irradiative conditions [161]. The reactions of the hypoiodites are generally more complex that the corresponding reactions of hypochlorites.

Iodoamides are prepared in a similar fashion to tertiary hypoiodites, namely by the reaction of the amide (usually primary) with iodine and lead tetra-acetate under irradiation. An alternative and preferable method is to use t-butyl hypochlorite and iodine as the source of electrophilic iodine. The iodoamides can be isolated but are usually formed *in situ* and decomposed photolytically without isolation. The photolysis products are iminolactones (or complexes of these compounds) which, upon hydrolysis by base and subsequent re-acidification, yield γ-lactones [162]. Examples of the iodoamide reaction are summarized in equations (4.122)–(4.124). Reaction (4.124)

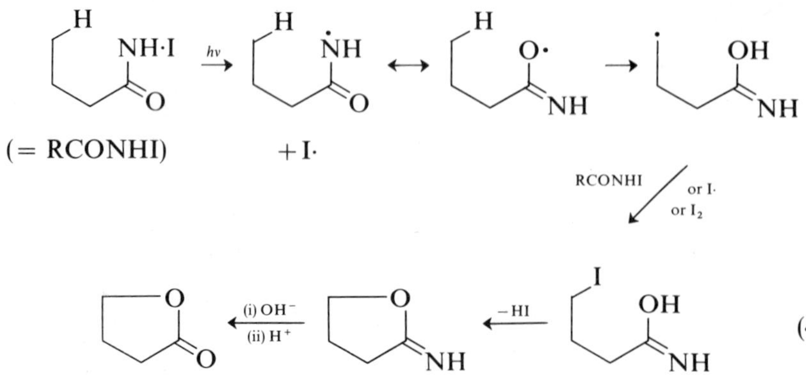

(4.121)

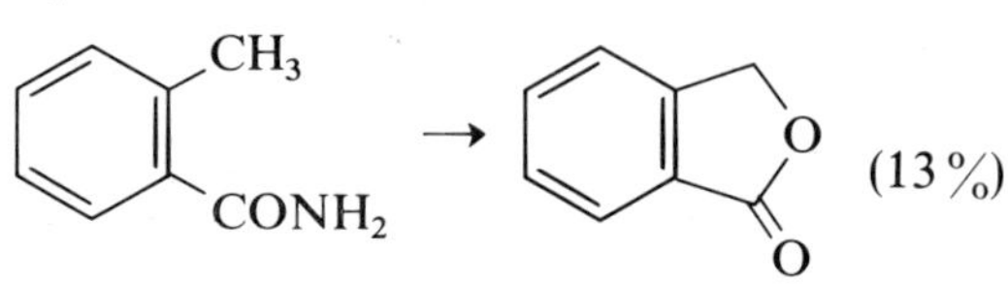

(4.122)

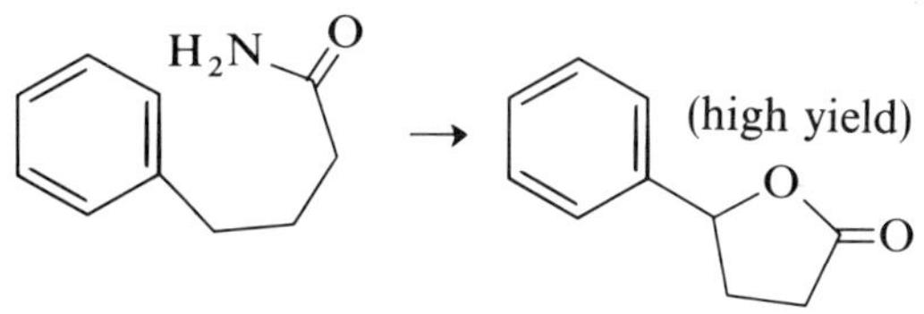

(4.123)

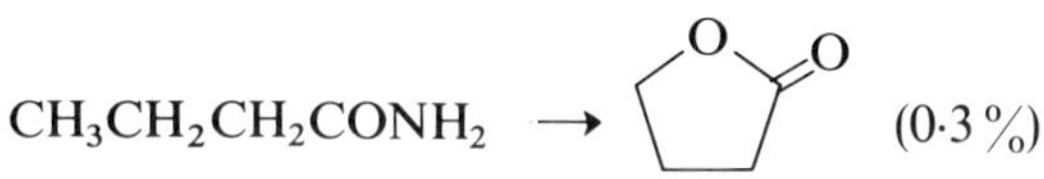

(4.124)

again illustrates the point that unactivated primary hydrogen atoms are of low reactivity, and the radical centre and the methyl hydrogens have to be retained in close proximity to each other (as in steroidal systems) for such reactions to occur in good yield. *N*-chloroamides also react similarly under photolytic conditions, but shorter-wavelength ultraviolet light is necessary to effect the cleavage, and quartz apparatus is essential. The chloroamide procedure is convenient and is comparable with the favourable cases of the iodoamide reaction. Interestingly, δ-lactones have also been isolated from these reactions, though in lower yield than the γ-lactones. Collinearity of the three reaction centres (C···H···O) is more easily realized in the seven-membered transition state, though the six-membered transition state is attended by a more favourable entropy factor. The reactions can also be initiated by trace quantities of cuprous chloride, which functions as a one-electron reducing agent [163].

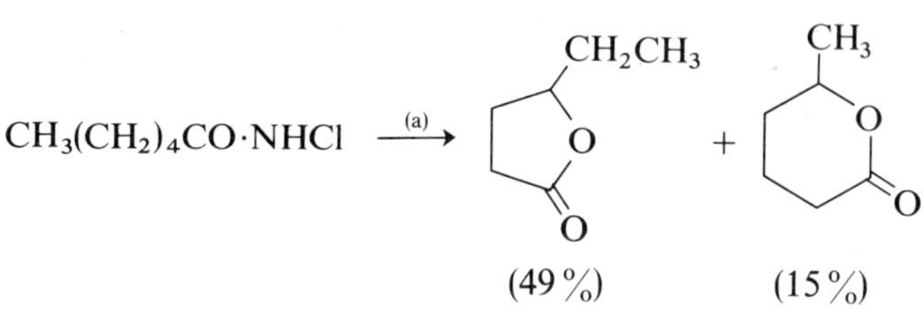

(4.125)

(a) (i) CH_2Cl_2 solution, 20°, *hν*; (ii) H^+

The substitution of the nitrogen atoms of iodoamides by alkyl or aryl groups is not advantageous, since very low yields of the iminolactone are produced [162]. In contrast, *N*-alkyl-*N*-bromoamides can be rearranged photolytically to *N*-alkyliminolactones in quite good yield. The products are insoluble in the solvents employed and are therefore readily isolated.

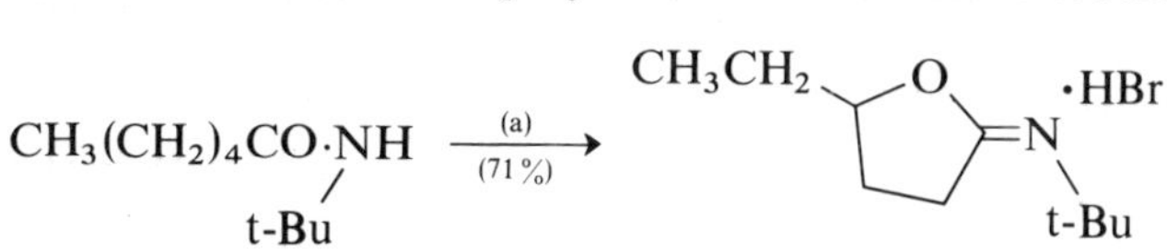

(4.126)

(a) (i) t-BuOBr, CCl_4; (ii) 10 min, *hν*; (iii) Δ [164]

An important advantage of this procedure over the iodoamide method is that the halogen acid does not interfere with the reaction. In the iodoamide method, because of reaction (4.127), the product yield is limited to 50% unless an excess of the oxidant is employed [162].

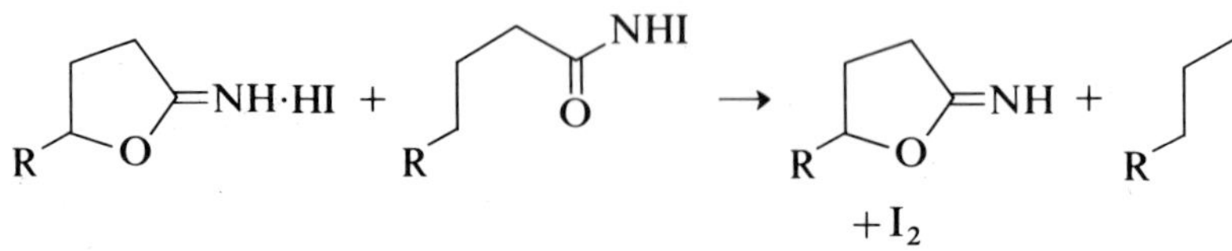

(4.127)

The photo-rearrangement of *N*-chlorosulphonamides appears to be a very similar process to the above, since both γ- and δ-chlorosulphonamides are formed [165]. It is a promising method for the synthesis of sultams, reaction (4.129) for example [165].

$$\mathrm{CH_3CH_2CH_2CH_2SO_2N(Cl){\cdot}Bu\text{-}t} \xrightarrow[\text{benzene}]{h\nu,\ 10\text{–}15^\circ} \mathrm{CH_3CH(Cl){\cdot}CH_2CH_2SO_2NHBu\text{-}t} \tag{4.128}$$

(4.129)

The photolytic rearrangement of *N*-chloroamines to γ- and δ-chloroamines in strongly acidic media (the Hofmann–Löffler–Freytag reaction) has been known for a considerable time [166]. The mechanism and scope of the reaction, however, have only been established in recent years. The γ- and δ-chloroamines can be isolated under special circumstances, but are usually cyclized directly to derivatives of pyrrolidine and piperidine, (4.130).

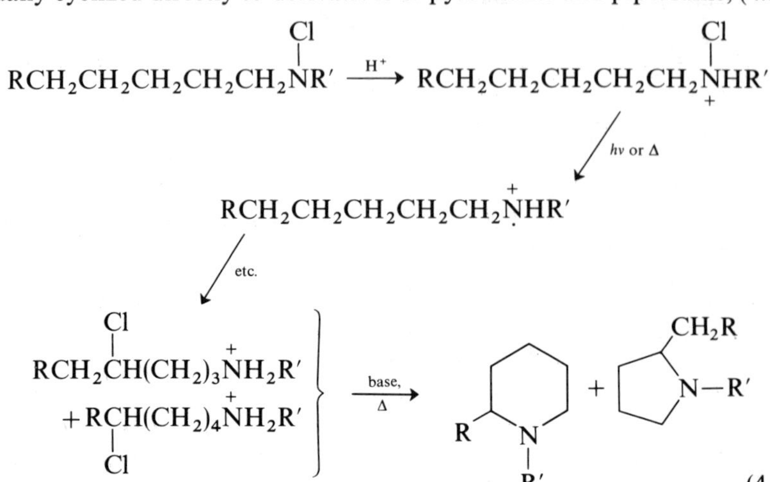

(4.130)

The reactions are initiated by heat, light, or by redox transfer. Actually, the above mechanism is an oversimplification [166, 167].

The resulting amines are usually isolated by steam distillation, and the required tertiary amines are frequently contaminated with the secondary amine arising from disproportionation reactions of the intermediate radicals. The secondary amines are removed by the Hinsberg procedure. The following examples illustrate the general method [168–172]. Where there is a choice of abstractable hydrogen, the most labile C—H bond is cleaved. By using amides, cyclic secondary amines can be prepared. It is normal in these reactions to use thermal initiation, and it is probable that the amides are first hydrolysed to amines by the hot acid present in the reaction mixture [166].

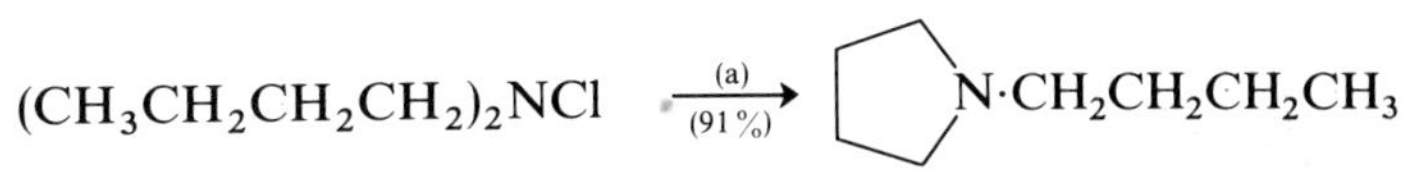

(4.131)

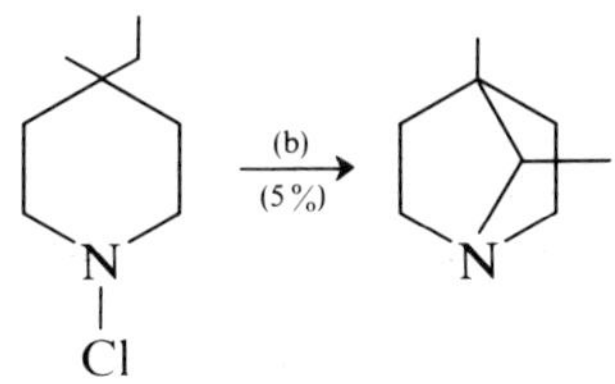

(4.132)

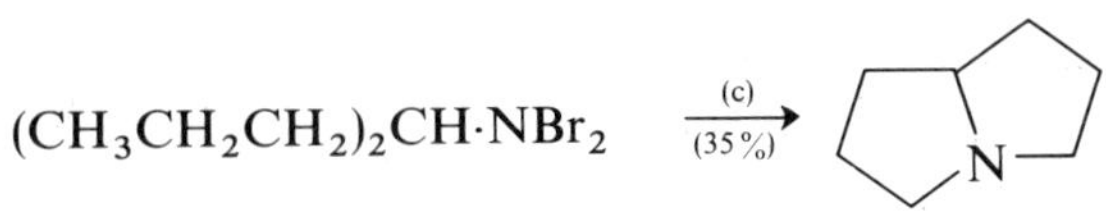

(4.133)

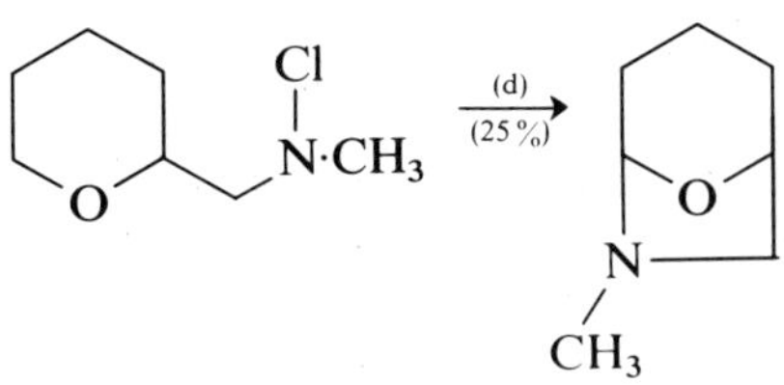

(4.134)

(a) Persulphate, Fe^{2+}, 85% H_2SO_4
(b) *hv*, 85% H_2SO_4, then Δ
(c) (i) *hv*, 85% H_2SO_4; (ii) NaOH
(d) *hv*, 2M H_2SO_4 in AcOH

(b) Addition Reactions

The addition of a radical A· to an olefinic double bond yields a new radical (8). The subsequent reactions of (8) depend upon the reaction conditions, the nature of the addendum (A—B), and on the nature of the olefin. If species AB is a very good transfer agent for B, that is, if B can be readily abstracted by (8), then the 1 : 1 adduct (9) will be formed predominantly, since step (b) will be very fast. On the other hand, if AB is a poor transfer agent,

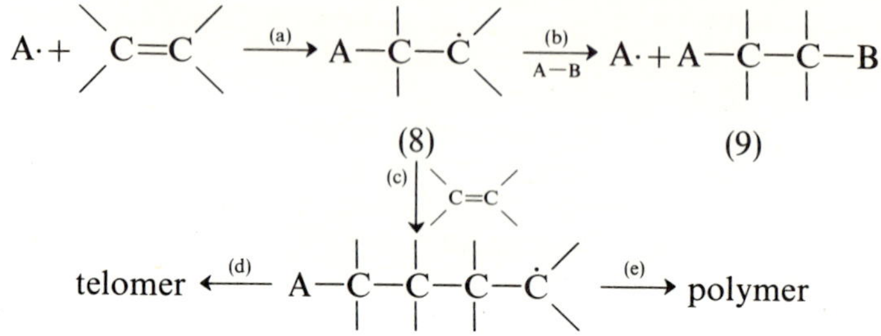

(4.135)

(10)

and if the olefin is susceptible to radical polymerization, then 2:1, 3:1, 4:1 (i.e., telomers) and higher adducts (polymers) will be formed, as reaction (c) and subsequent steps, (d) and (e), will be faster than reaction (b). In addition to these processes, particularly if the olefin possesses labile hydrogen atoms, abstraction reactions may also occur.

Polymerization is favoured by low concentrations of poor transfer agents [AB] and by substitution of the olefin by groups which are capable of stabilizing radicals (8) and (10), etc. The majority of polymerizable monomers are terminal olefins which bear substituents like —Cl, $—CO_2R$, —O·CO·R, $—C_6H_5$, and —CN attached to the β-carbon atoms [1–4, 173(a)]. Relatively few polysubstituted olefins show a tendency to polymerize. The polymerization of olefinic monomers, which is not limited to systems in which radicals are produced [173(b)], is of great importance and is still the subject of immense research effort. The field cannot even be superficially treated here, and the reader is referred to standard texts on the subject [1–4, 173]. In this section conditions suitable for the formation of the 1:1 adducts (9) only will be considered. A number of important reviews are available [76, 174–178], so that these reactions will be considered but briefly.

Conditions which favour the formation of 1:1 adducts (9) are high concentrations of reactive transfer agents (AB), and olefins with a low tendency to polymerize. The course of the addition reactions is thus controlled partly by the conditions and partly by the energetics of the individual reaction stages. In the latter connection, steps (4.135a and b) should both be exothermic. Step (a) is usually exothermic, since the decoupling of typical olefinic π-bonds requires about 63 kcal mol^{-1} and most C—A bonds (excepting C—I, for example) have dissociation energies which are greater than this [1–4, 9, 10]. Step (b), however, places a much more severe limitation on the types of addenda (AB) that can be successfully employed. For example, in the addition of HCl, H_2O, or NH_3 to ethylene, step (b) is endothermic in each case and the addition of these reagents to olefinic double bonds by a radical-chain mechanism is likely to be very difficult or impossible. In these circumstances the competing ionic addition reactions become much more important. Very few addenda can be successfully added in a 1:1 fashion to polymerizable olefins, but if a sufficiently large excess of addendum (AB) is used, then even poor transfer agents can be successfully added to normal (non-polymerizable) olefins.

The site of the initial attack by radical A· of AB on the olefin is determined by the relative stabilities of the two possible radicals (8); the more stable radical of the two is formed predominantly or exclusively. This results in the familiar anti-Markovnikov addition of unsymmetrical addenda to unsymmetrical olefins in radical addition reactions. When the A—B molecule is polyatomic there are several possible species of type A. The nature of the group A is again dependent upon stability differences among all the possibilities. Thus chloroform is a source of $Cl_3C\cdot$ radicals (not $\cdot CHCl_2$), and primary and secondary alcohols provide $R_2\dot{C}OH$ radicals (and not $R_2CHO\cdot$).

The use of an external initiator is frequently necessary. When the chain length is long, only trace quantities of initiator are used; when the chains are short, substantial amounts are necessary and a relatively large incorporation of initiator-derived radicals is to be expected in the final product.

A few recent examples of the literally hundreds of 1:1 additions that are known are listed in Table 4.4. The original papers and review articles should be consulted for a more complete indication of the scope of these processes [1–4, 76, 103, 174–188]. The examples also include additions to acetylenic triple bonds, intramolecular addition reactions, and addition across carbonyl groups.

In addition to the examples cited in Table 4.4, a few other reactions deserve specific mention. One of the best known of all radical additions is that of HBr to olefins. In recent years the Dow Chemical Co. has manufactured ethyl bromide by the radiation-induced addition of HBr to ethylene. The advantage over the more conventional synthesis using HBr and ethanol is that aqueous hydrobromic acid is not formed as a by-product, and the radical reaction can be run at higher efficiencies. The Dow synthesis was the first radiation-based heavy chemical process to be put into commercial practice.

Recently it has been demonstrated that various hydrocarbons can be added to carbon monoxide. Carbon tetrachloride is a most convenient solvent for the reactions, since it takes part in the radical-forming and radical transfer processes. The reactions are carried out under a high pressure of carbon monoxide in an autoclave, and can be initiated by peroxides or, more conveniently, by γ-radiation.

$$\begin{array}{lrcl} \text{(a)} & CCl_4 & \xrightarrow{\gamma} & Cl_3C\cdot + Cl\cdot \\ \text{(b)} & Cl_3C\cdot + RH & \rightarrow & Cl_3CH + R\cdot \\ \text{(c)} & R\cdot + CO & \rightarrow & R\dot{C}O \\ \text{(d)} & R\dot{C}O + CCl_4 & \rightarrow & RCO{\cdot}Cl + Cl_3C\cdot \\ \text{(e)} & R\cdot + CCl_4 & \rightarrow & RCl + Cl_3C\cdot, \quad \text{etc.} \end{array} \tag{4.136}$$

The reactions differ significantly from the addition reactions discussed above, since the π-bond is not destroyed. The step (4.136c) is most probably reversible, and a good transfer agent (CCl_4) is essential to the formation of the carbonylated product. The following examples illustrate the process [189]. The major by-products are alkyl chlorides derived from reaction (4.136e).

Table 4.4 Some typical radical addition reactions to multiple bonds

Unsaturated system	Addendum	Product	Yield (%)	Ref.
cis-MeCH=CHMe	DBr	*threo*-MeCHD·CHBrMe	95	[76]
$Me(CH_2)_3CH{=}CH_2$	*p*-MeC_6H_4Me	*p*-$M\dot{e}C_6H_4{\cdot}(CH_2)_6Me$	29	[76]
$CH_2{=}CH{\cdot}CN$	Et_3SnH	$Et_3SnCH_2CH_2CN$	100	[179]
$Me(CH_2)_5CH{=}CH_2$	$MeCH_2CHO$	$Me(CH_2)_6CH_2COCH_2Me$	64	[76]
$Me(CH_2)_5CH{=}CH_2$	$CH_2(CN)_2$	$Me(CH_2)_7CH(CN)_2$	31	[174]
$Me(CH_2)_3C{\equiv}CH$	$\{(n\text{-}Bu)_2NCl;\ H^+\}$	$Me(CH_2)_3CHCl{\cdot}CHO$ (by hydrolysis of intermediate adduct)	37	[182]
$(CH_2{=}CH{\cdot}CH_2)_2O$	CCl_3Br	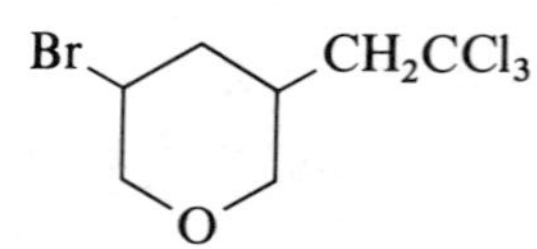	—	[175]

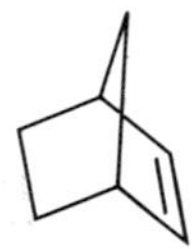	CH_3CH_2Cl	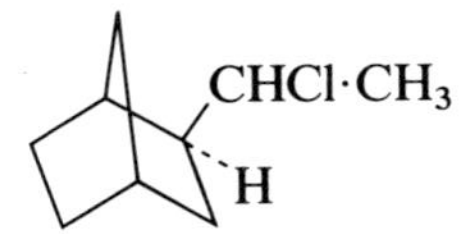	40	[180]
$CH_3CH{=}CH{\cdot}CO_2H$	Me_2CHOH	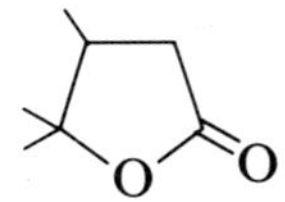	38	[76]
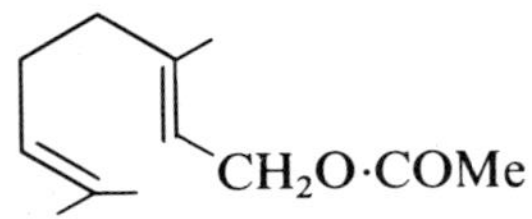	$(PhCO{\cdot}O{-})_2$ + Cu^+	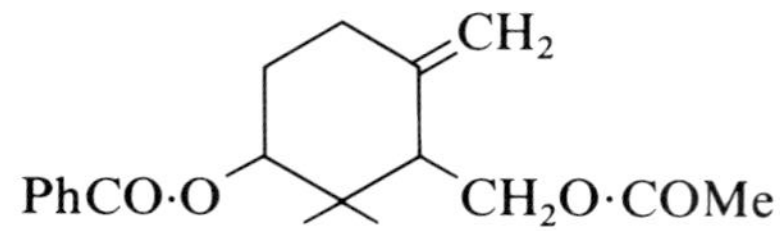	55–60	[181]
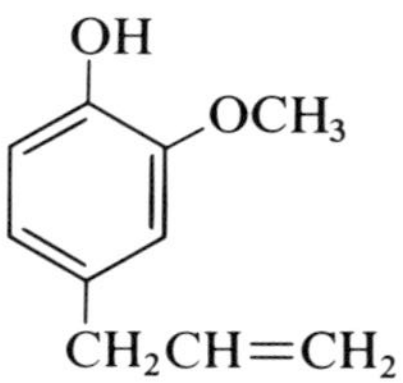	MeCOSH	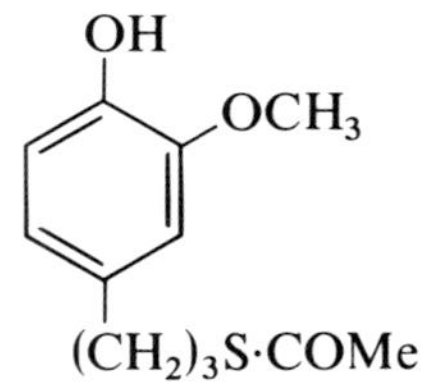	78	[176]
$HCO{\cdot}OCH_2CH{=}CH_2$	cyclo-C_6H_{12}	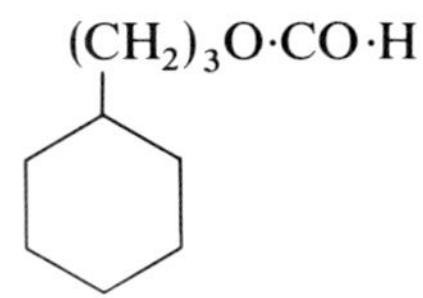	71	[183]

Table 4.4 (continued)

Unsaturated system	Addendum	Product	Yield (%)	Ref.
	$BrCH_2CO_2H$ H-donor solvent	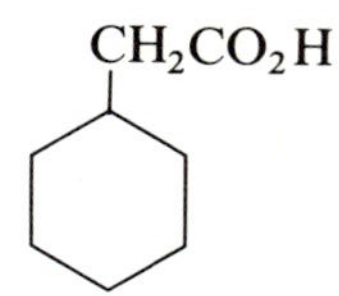	63	[184]
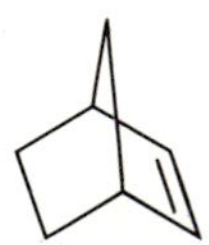	$Cl{\cdot}CO{\cdot}CO_2Et$ ($h\nu$)	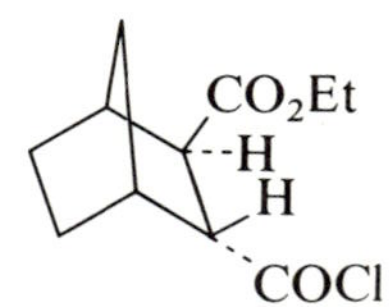	32	[185]
$HO_2C{\cdot}CH_2CH{=}CH_2$	n-$C_8H_{17}CO_2H$	$HO_2C(CH_2)_3CH(\text{n-}C_7H_{15}){\cdot}CO_2H$	37–52	[175]
$PhCH{=}CH{\cdot}CO_2H$	NO_2Cl	$PhCH(NO_2){\cdot}CH(Cl){\cdot}CO_2H$	43	[76]

$Et_3SiCH{=}CH_2$	$BrCH_2CO_2Et$	$Et_3SiCHBr{\cdot}(CH_2)_2CO_2Et$	45	[76]
$(CF_3)_2C{=}O$	PhCHO	$PhCO{\cdot}O{\cdot}CH(CF_3)_2$	50	[186]
	—	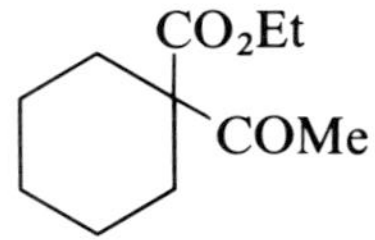	8	[174, 183]
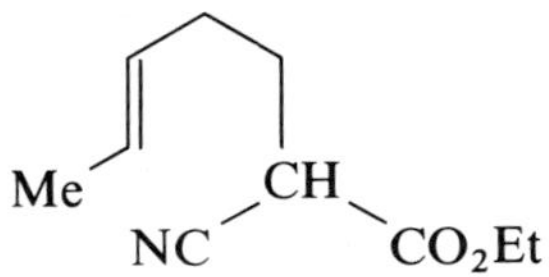	—		38	[187]
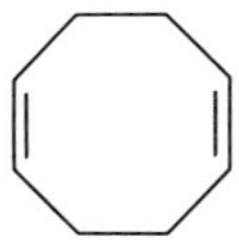	$HCONH_2$ (γ-radn.)	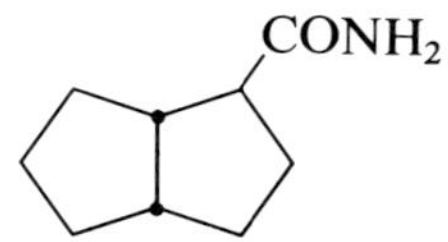	50	[186á]

$$\text{cyclo-}C_6H_{12} + CO \xrightarrow[CCl_4]{\gamma} \text{cyclo-}C_6H_{11}CO{\cdot}Cl \quad (47\%) \tag{4.137}$$

$$MeCH_2CH_2Me + CO \xrightarrow[CCl_4]{\gamma} \underset{\substack{| \\ CO{\cdot}Cl \\ (30\%)}}{MeCH}{\cdot}CH_2Me + Me(CH_2)_3CO{\cdot}Cl \quad (11\%) \tag{4.138}$$

A convenient method for the reductive acylation of ketones with acyl chlorides and triphenyltin hydride has been described [190]. The acyl chloride–triphenyltin hydride system is a source of acyl radicals which add to the oxygen atom of the ketonic carbonyl group. The product is formed by the efficient hydrogen-atom transfer from triphenyltin hydride to this intermediate radical (4.139). The reaction seems to be widely applicable and proceeds quantitatively for the systems so far investigated.

$$R^2CO{\cdot}Cl + Ph_3Sn{\cdot} \rightarrow R^2\dot{C}O + Ph_3SnCl$$

$$R^2\dot{C}O + R{\cdot}CO{\cdot}R^1 \rightarrow R{-}\underset{\substack{| \\ R^1}}{\dot{C}}{-}O{\cdot}CO{\cdot}R^2$$

$$R{-}\underset{\substack{| \\ R^1}}{\dot{C}}{-}O{\cdot}CO{\cdot}R^2 + Ph_3SnH \rightarrow R{-}\underset{\substack{| \\ R^1}}{CH}{\cdot}O{\cdot}CO{\cdot}R^2 \tag{4.139}$$

The addition of the elements of aryl halide (derived from the corresponding diazonium halide) to olefins has been widely investigated for a number of years. The reactions (Meerwein reaction) are conducted in the presence of copper salts and proceed most satisfactorily for olefins substituted by unsaturated electron-withdrawing groups (e.g., polymerizable monomers). The scope of the reaction has been extensively reviewed [191]. Reaction (4.140) is a recent example of the process [192], and illustrates the 1,4-addition to a 1,3-diene.

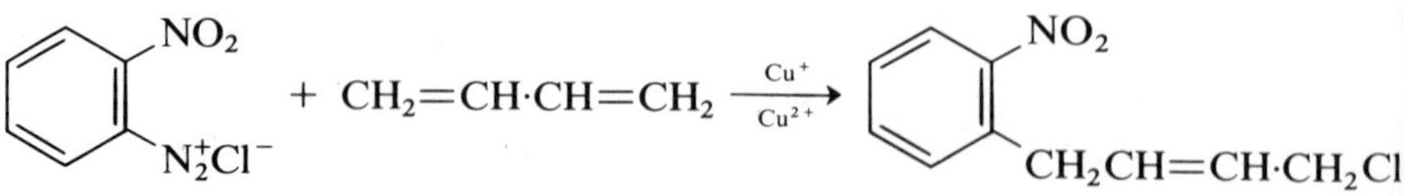

(72%) (4.140)

In the discussion on reaction (4.135) at the beginning of this section it was deduced that radical (8) could react to give a 1:1 adduct (step b) or telomeric and polymeric products (steps c, d, and e). One further reaction of (8) may be delineated, namely its combination with a similar or identical radical. This arises when (8) is unreactive and polymerization processes are minimized. Addition reactions to butadiene are of this type [193], and have been discussed elsewhere [2]. The scope of such processes has perhaps not been fully explored, even though they are of preparative interest.

The addition reactions of carbenes and nitrenes across π-bonds [2, 142–147] are of considerable importance, and in particular the reaction of carbenes with olefins is one of the best methods for preparing derivatives of cyclopropane. To the extent that triplet species are involved in certain of these reactions, they may reasonably be classified as radical processes.

(c) Aromatic Substitution

In principle, during homolytic aromatic substitution, a group (Y) attached to the aromatic nucleus is displaced by the attacking radical (R·). The

$$\mathrm{R\cdot + ArY \longrightarrow ArR + Y\cdot} \qquad (4.141)$$

group (Y) is most usually a hydrogen atom. The mechanism of most homolytic substitutions is complex and many points of detail are still controversial. The scheme set out in equation (4.142) is therefore an oversimplification, but it illustrates several points of interest regarding the synthetic applications of these reactions. In addition, reaction may also be possible at the substituent (A) itself (abstraction or addition). Similarly, if R· is an alkyl radical, then further attack at the benzylic hydrogen atoms of the aralkane (12) can occur, and on this account alkylation is often best achieved by other synthetic procedures.

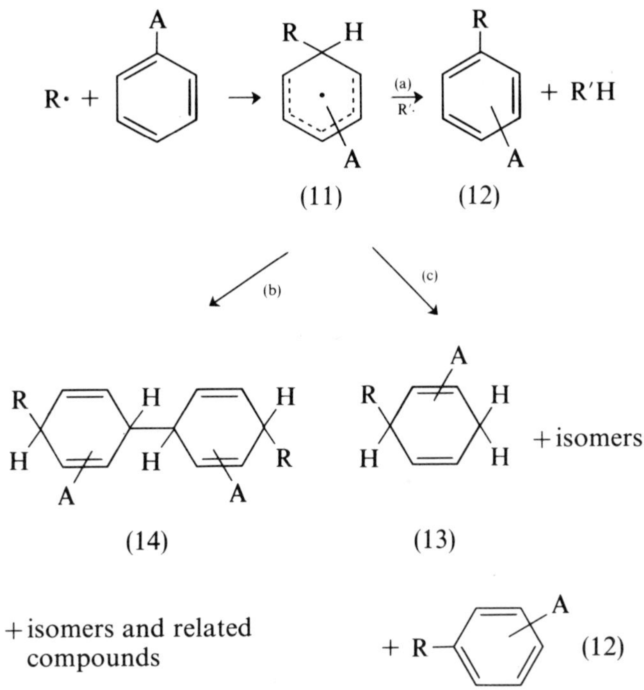

(4.142)

The steps (4.142b and c) are usually undesirable side reactions and are minimized by the transfer of the allylic hydrogen atom from (11) to a suitable radical, or to a molecular or ionic oxidizing agent (R′· in the above scheme; step a) to give the desired product (12). When the group (A) is not a hydrogen atom, three isomeric products (12) are obtained. The isomer ratio is dependent mainly upon the nature of (A), but the electronic nature of the attacking radical is also important [1–4, 194, 195]. In many cases, however, this ratio is in the region *o*-, 50–70%, *m*-, 5–30%, *p*-, 10–30%. When (A) is hydrogen, only one product of type (12) is, of course, possible. In the substitution of polysubstituted benzenes, the number of possible isomeric products is equal to the number of different nuclear sites within the benzenoid compound. This usually, but not always, excludes the sites occupied by substituent groups, since these groups are not usually displaced. Substituted benzenes and polycyclic aromatic hydrocarbons are more susceptible to radical attack than benzene itself. Where reactive *meso*-positions exist (as in the 9 and 10 positions in anthracene and phenanthrene), radical addition at these sites frequently occurs.

Much of the work in this area has been aimed at the synthesis of unsymmetrical biaryls, since the Ullmann reaction fails in many cases. Various additives may be used to advantage; they take on the role of (R′·) in (4.142a). For example, oxygen, various nitro, nitroso, and azo compounds, and metal salts have proven effective for the oxidation of radicals of type (11). The aryl radicals (R·) can be generated by a number of standard procedures, as in the thermal cleavage of diaroyl peroxides [(Ar·CO·O—)$_2$], arylazo compounds [ArN=NAr′], diaryloxy derivatives of trivalent iodine [ArI(O·CO·Ar′)$_2$], and lead(IV) derivatives of aryl carboxylic acids [Pb(O·CO·Ar)$_4$], and by the photolytic cleavage of aryl-iodine, aryl-mercury, and aryl-lead bonds [1–4, 194, 195].

The following examples are fairly typical of the various synthetic procedures available for the production of biaryls. The several by-products formed in each case (e.g., CO_2 and $ArCO_2H$ in the decomposition of diaroyl peroxides) are not included in the equations.

Diaroyl Peroxides [48, 196]

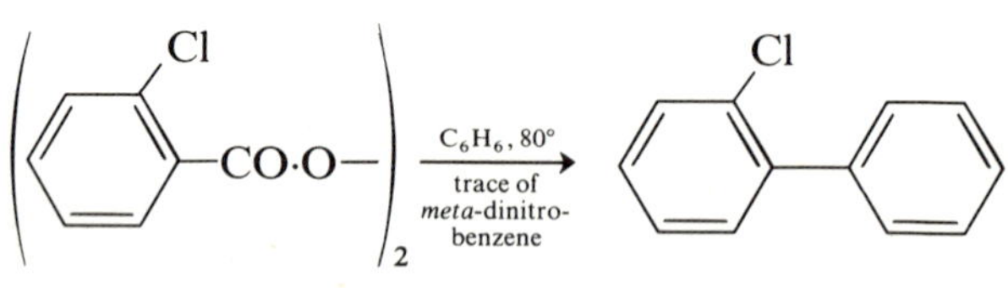

(51%)

(4.143)

Gomberg Reaction [197]

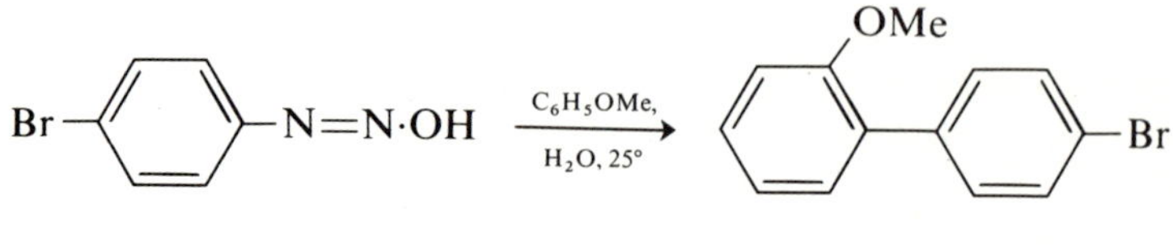

(27%)+isomers

(4.144)

PSCHORR REACTION [198]

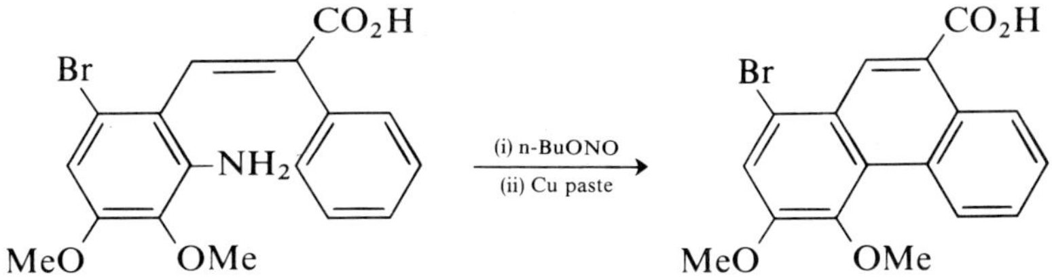

(72–82%)

(4.145)

GRIGNARD REAGENT–COBALTOUS CHLORIDE PROCEDURE [199]

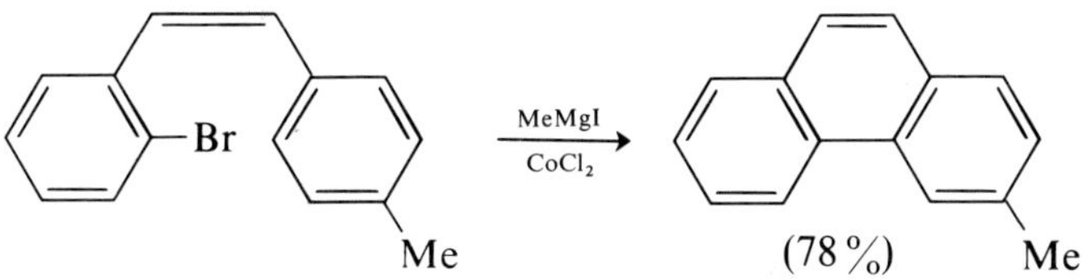

(4.146)

PHOTOLYSIS OF IODOAROMATIC COMPOUNDS [200, 201]

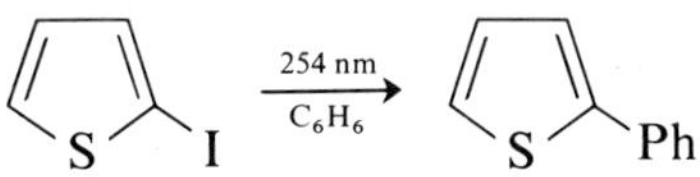

(70%) (4.147)

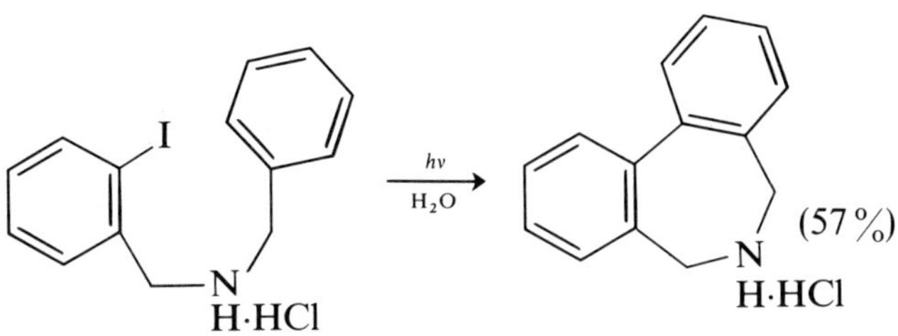

(57%)

(4.148)

Useful radical substitution reactions of aromatic compounds are not limited to the introduction of aryl substituents by any means. The photolysis of iodo compounds can be adapted for the introduction of various chemical groupings, reactions (4.149)–(4.151) for example [202], and the hydroxyl

$$PhC{\equiv}CI \xrightarrow[C_6H_6]{h\nu} PhC{\equiv}CPh \quad (50\%) \tag{4.149}$$

$$ICH_2CO_2H \xrightarrow[C_6H_6]{h\nu} PhCH_2CO_2H \quad (67\%) \tag{4.150}$$

$$ICN \xrightarrow[C_6H_6]{h\nu} PhCN \quad (50\%) \tag{4.151}$$

function can be introduced either directly [203] by utilizing hydroxy radicals [e.g., (4.152)], or indirectly [203a, 204] by first preparing a phenolic ester [e.g., (4.153), (4.154)] which can be subsequently hydrolysed. The last example

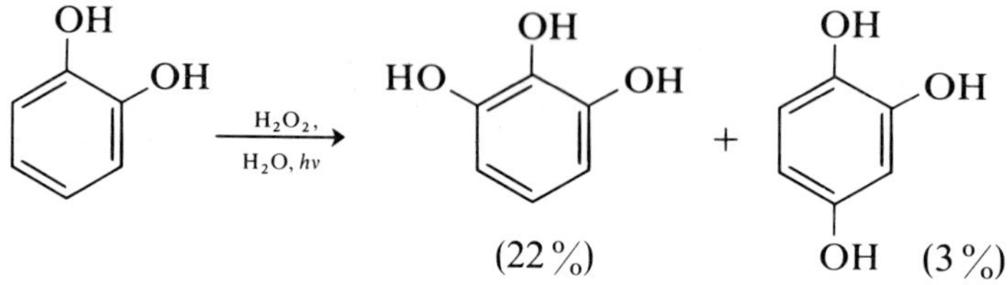

(4.152)

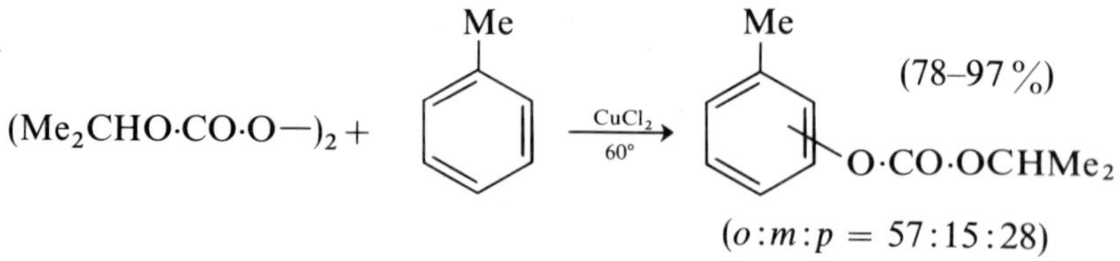

(4.153)

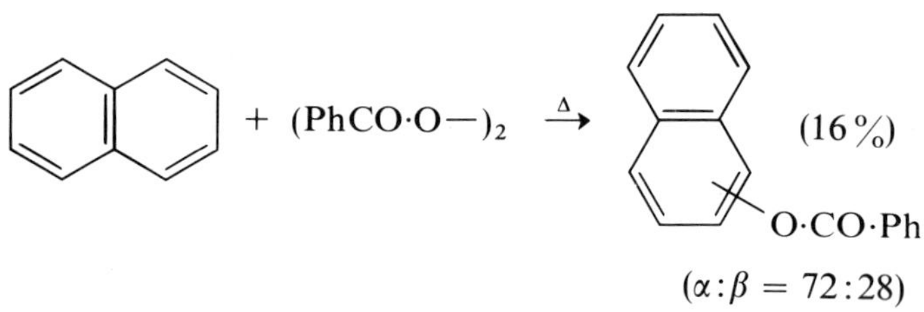

(4.154)

illustrates the intermediacy of aroyloxy radicals in the thermal decomposition of diaroyl peroxides. These radicals are relatively unreactive and decarboxylate fairly readily, particularly at temperatures much above 70°. With reactive substrates, however, esters are formed much more readily—as with naphthalene, for example. Even with unactivated nuclei esterification may occur to a reasonable extent if intramolecular substitution is possible [205, 206].

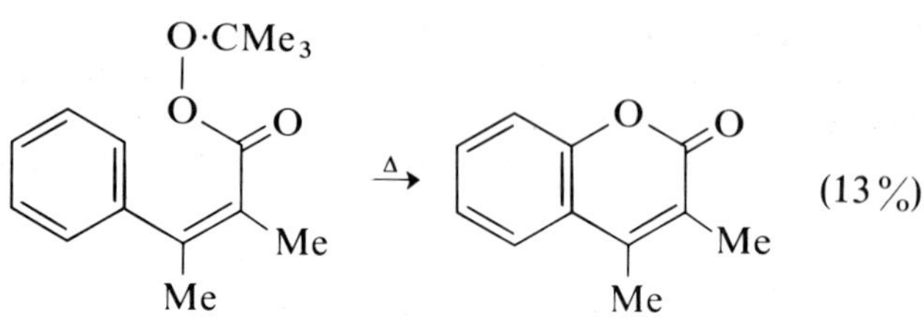

(4.155)

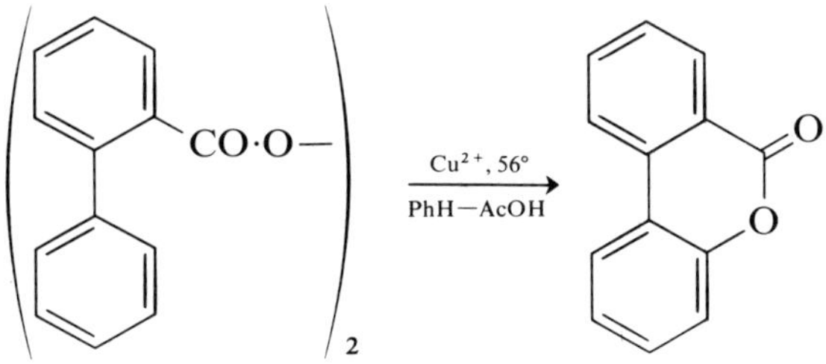

(41–46%)

(4.156)

Tertiary amines can be prepared by the ferrous-sulphate-catalysed decomposition of protonated secondary *N*-chloroamines [207], reaction (4.157) for example, and the carboxyamide grouping can be introduced by the acetone-photosensitized decomposition of formamide in the presence of appropriate aromatic substrates [208], reaction (4.158).

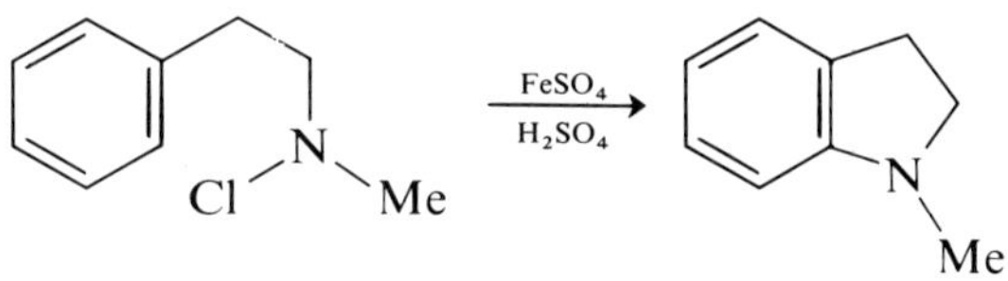

(27%)

(4.157)

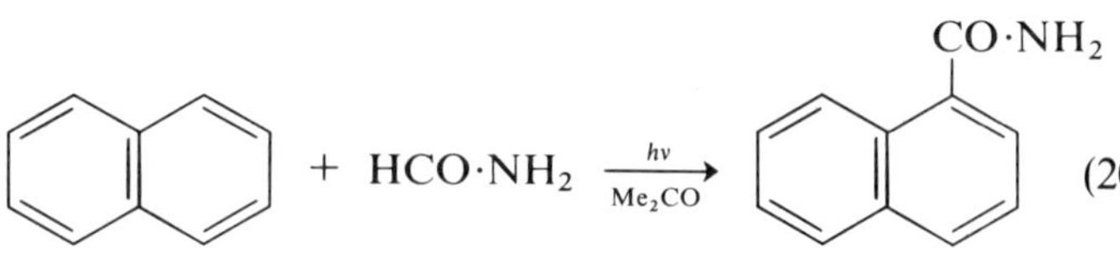

(20%)

(4.158)

Carbenes and nitrenes can also be usefully employed, and their insertion reactions into aromatic C—H bonds to give substitution products, or their addition across two carbon atoms of the aromatic nucleus to furnish substitution, addition, and ring-expanded products, are well documented [2, 142–147].

Substitution reactions of unsaturated compounds are not limited to those of aromatic and hetero-aromatic compounds. However, olefins normally yield addition products, since addition is usually energetically more favourable than substitution in these systems. On the other hand quinones, a group of compounds which are quite efficient as radical-scavengers, are readily substituted by a variety of radicals. Intense interest was shown in these substitutions, particularly in alkylation, during World War II in connection with the search for new anti-malarial drugs. More recently, the radical alkylation of a quinone was the crucial step in an ingenious synthesis of the naturally occurring tanshinones [208a].

(d) Displacement

Abstraction is essentially a radical displacement reaction on a monovalent atom. Radical displacement on carbon, however, is very rare [209–211], and is the radical counterpart of an S_N2 reaction (sometimes designated S_R

$$R\cdot + \gt C{-}Y \rightarrow R{-}C\lt + Y\cdot \tag{4.159}$$

or S_H2). Reaction (4.160) is a recent and interesting example of this process

$$I{\cdot}CH_2CH_2CH_2I + (PhCO{\cdot}O{-})_2 \xrightarrow[C_6H_6]{116^\circ} \text{cyclo-}C_3H_6 + 2PhI + 2CO_2 \tag{4.160}$$

[209]. Only a few examples have been recorded of radical displacements on other tetravalent atoms [212], and even for trivalent atoms the reaction is almost unknown [213]. More work in this area is clearly desirable before possible synthetic applications can be delineated.

Displacement at oxygen and sulphur, on the other hand, is fairly well known [1–4, 48, 214, 215], and is most common for systems possessing the O—O or S—S linkage. Thus, reaction (4.161) is responsible for the induced decomposition of peroxides, and reaction (4.162) finds use in the control of polymer molecular weights (R· = polymer radical) [1–4].

$$R\cdot + R'CO{\cdot}O{-}O{\cdot}COR' \rightarrow R{\cdot}O{\cdot}COR' + R'CO{\cdot}O\cdot \tag{4.161}$$

$$R\cdot + R'S{-}SR' \rightarrow RSR' + R'S\cdot \tag{4.162}$$

(e) Elimination and Fragmentation Reactions

Reactions of the type (4.163) are normally regarded as fragmentations. Reaction (4.164) is more properly an elimination reaction, but few examples are known. There is, however, e.s.r. evidence for this type of behaviour in

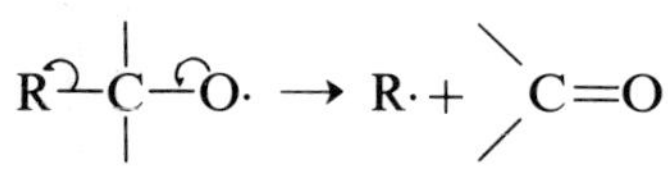

(4.163)

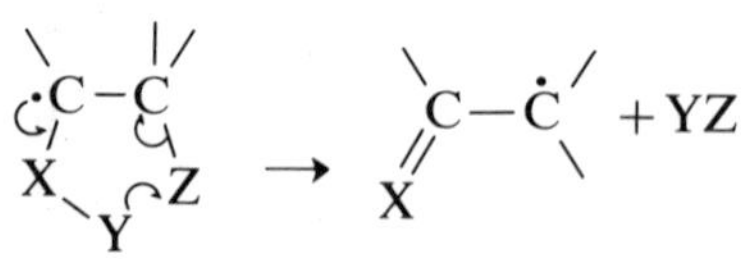

(4.164)

the oxidation of 1,2-diols (X = O, Y = H, Z = OH) and 1,2-chlorohydrins (X = O, Y = H, Z = Cl) [216].

Fragmentation reactions are much more common, although radicals show much less tendency to fragment than do more electron-deficient species. The fragmentations follow the basic pattern set out in (4.163) in that a group β to the odd-electron centre is ejected with the concomitant formation of an unsaturated molecule. The stability of the new radical is frequently the feature which determines which β bond is cleaved. Hydrogen atoms and aryl radicals are rarely ejected if there are other processes which can occur.

The cleavage reactions of alkoxy radicals, which are usually obtained from the photolytic or thermal decomposition of hypohalites, have been widely investigated [159], and are of preparative and degradative interest. Primary alkoxy radicals (from primary hypochlorites) fragment to formaldehyde and an alkyl radical which then abstracts a chlorine atom from a further molecule of the hypochlorite (4.165). This reaction could be useful

$$\begin{aligned} RCH_2O\cdot &\rightarrow R\cdot + CH_2O \\ R\cdot + RCH_2OCl &\rightarrow RCl + RCH_2O\cdot, \quad \text{etc.} \end{aligned} \qquad (4.165)$$

in an alternative route for the reductive decarboxylation of aliphatic acids (4.166) [217].

$$RCO_2H \rightarrow RCH_2OH \rightarrow RCH_2OCl \rightarrow RCl \rightarrow RH \qquad (4.166)$$

With t-alkoxy radicals, the fragmentation occurs according to a well-defined pattern, as illustrated by reactions (4.167) and (4.168) [159], in that the most stable alkyl radical is the one ejected. The cleavage of 1-substituted cycloalkoxy radicals (4.168) suggests a useful synthetic route to ω-chloroalkyl ketones.

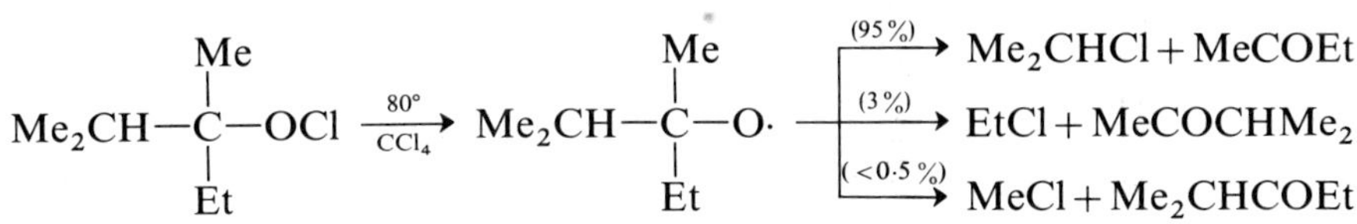

(4.167)

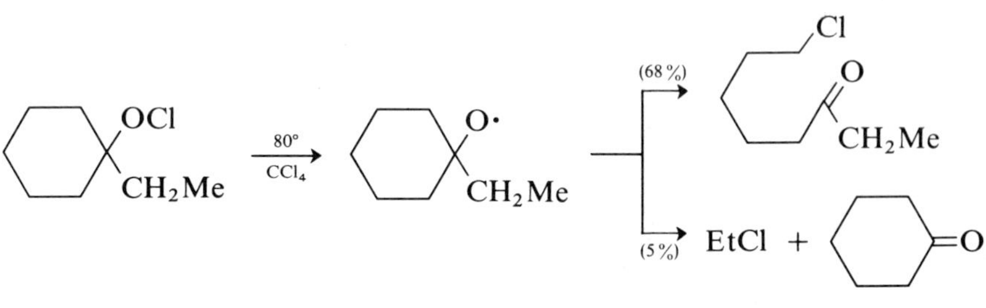

(4.168)

In the case of triarylcarbinyl radicals ($Ar_3CO\cdot$), rearrangement rather than fragmentation reactions occur.

The fragmentation of acyloxy radicals (4.169) was referred to earlier in connection with the thermal decomposition of diacyl peroxides, the decarboxylation of organic carboxylic acids, and with the Kolbe synthesis.

$$R{\cdot}CO{\cdot}O\cdot \rightarrow R\cdot + CO_2 \qquad (4.169)$$

An extension of this basic reaction has found use in an intriguing new cheap synthetic route to macrocyclic compounds by the thermal or photolytic cleavage of ketone peroxides. The thermal decomposition of cyclohexane diperoxide (obtained from cyclohexanone and hydrogen peroxide), for example, yields cyclodecane (44%), 10-decanolactone (23%), and cyclohexanone (21%). The following mechanism is thought to apply [217a].

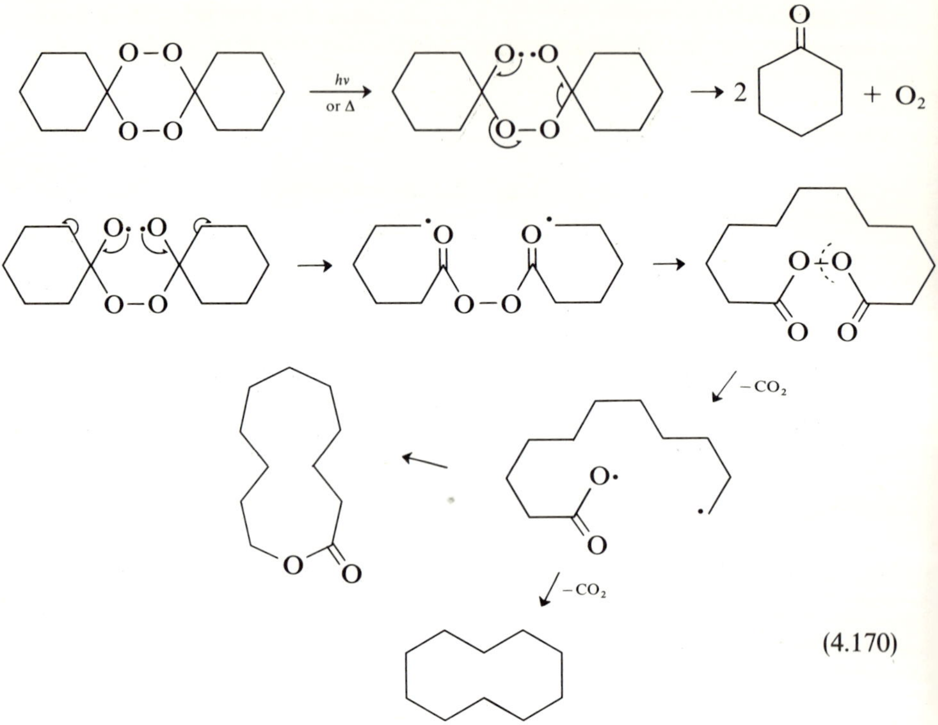

(4.170)

With the cyclic trimeric peroxide (cyclohexane triperoxide), the products are cyclopentadecane (16%), 16-hexadecanolactone (1%), and cyclohexanone (15%) in the thermally initiated reaction. A different product distribution is obtained in the photolytic processes. The method appears to be applicable to the preparation of oxygen-containing macrocycles also.

Acyl radicals decarbonylate fairly readily, particularly at temperatures in excess of 100°. The reaction is an example of α cleavage. Aldehydes are therefore a convenient high-temperature source of alkyl radicals and have been used extensively in studies of radical rearrangements [218].

$$R{-}\dot{C}{=}O \rightarrow R\cdot + CO \qquad (4.171)$$

The process of radical fragmentation is essentially the reverse of radical addition. Indeed, some radical additions, particularly those of weakly reactive radicals (e.g., bromine and iodine atoms and thiyl radicals), are reversible. This can lead to *cis-trans* isomerization when an olefin is involved in the reaction. The synthesis of stereo-regular naturally occurring polyenes (e.g., carotenoids) is but one use of radical-catalysed *cis-trans* isomerization reactions.

(f) Rearrangements

Radical rearrangements are less common than those of more electron-deficient species, but the 1,2-shift of a chlorine or bromine atom, the 1,2-

shift of an aromatic or vinyl group, or the 1,5-hydrogen transfer reaction (considered earlier) are known. The evidence for each of these types of rearrangement has been critically reviewed [218, 219].

The driving force for rearrangement processes is the production of a more stable radical, (4.172)–(4.174). When the new radical has an identical or

$$-\overset{|}{\underset{|}{C}}-\overset{\cdot}{\underset{|}{C}}-CCl_3 \longrightarrow -\overset{|}{\underset{|}{C}}-\overset{Cl}{\overset{|}{C}}-\dot{C}Cl_2 \qquad (4.172)$$

$$Ph_3C{-}O\cdot \longrightarrow Ph_2\dot{C}{-}O{-}Ph \qquad (4.173)$$

$$Ph_3C{\cdot}CH_2\cdot \xrightarrow{(96\%)} Ph_2\dot{C}{-}CH_2Ph \qquad (4.174)$$

But

$$Me_3C{\cdot}CH_2\cdot \not\longrightarrow Me_2\dot{C}{-}CH_2Me \qquad (4.175)$$

lower stability as compared with the original, rearrangement occurs only to a small extent (4.176), if at all. Halogen-bridged radicals have been

$$Ph{\cdot}{}^{14}CH_2CH_2\cdot \xrightarrow{(3{\cdot}3\%)} \cdot{}^{14}CH_2CH_2Ph \qquad (4.176)$$

invoked to explain the high stereospecificity of certain bromination reactions. For example, bromination of the optically active chloride (15) gives optically active 1-chloro-2-bromo-2-methylbutane [220]. This interpretation is still highly controversial.

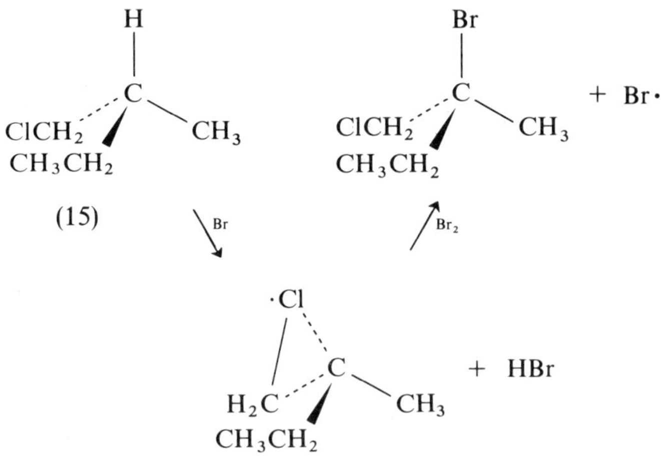

(4.177)

In summary, although 1,2-shifts are theoretically interesting, the preparative utility of the processes is not clear at the present time [218, 219].

The definition of a rearrangement which has been adopted in this section is that the change in the site of the odd-electron centre should also be accompanied by the *instantaneous* and *complete* transfer of an atom or group.

Accordingly, reactions (4.178)–(4.180) are not here classified as rearrangements but are regarded as intramolecular addition [(4.178), (4.179)] or fragmentation [(4.179), (4.180)] processes. The classification adopted varies

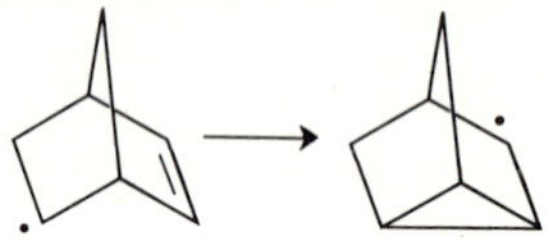

(4.178)

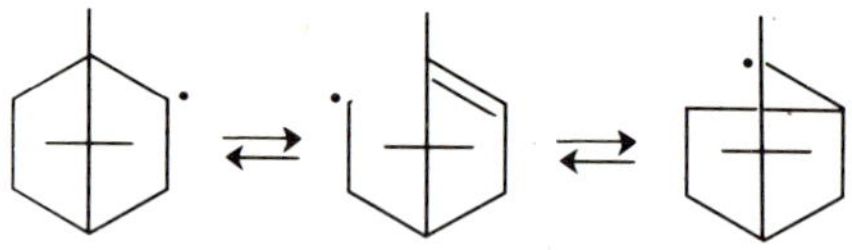

(4.179)

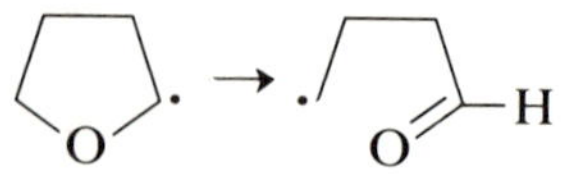

(4.180)

according to one's point of view [218, 219]. Similarly, the apparent 1,4- and 1,5-aryl shifts, (4.181) and (4.182) respectively [221, 222], are here considered as successive intramolecular substitution and fragmentation reactions. The

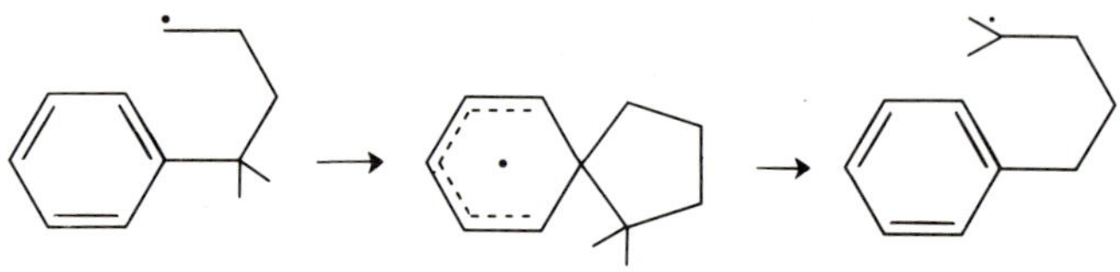

(4.181)

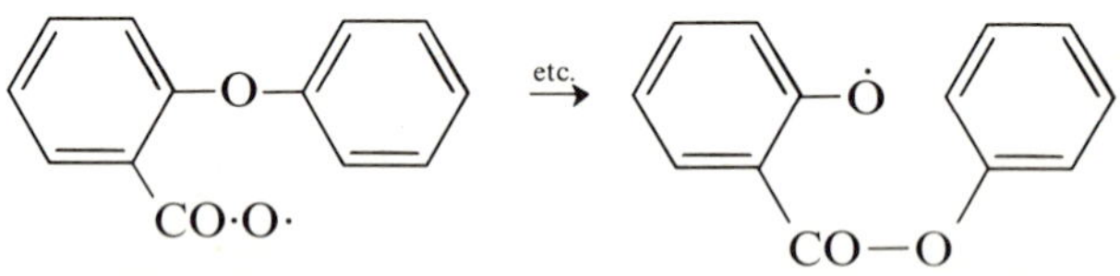

(4.182)

occurrence of reaction (4.182) militates against the synthesis of depsidones by intramolecular aroyloxylation reactions [21, 223].

REFERENCES

[1] Walling, C. *Free Radicals in Solution*, Wiley, New York (1957).
[2] Stirling, C. J. M. *Radicals in Organic Chemistry*, Oldbourne Press, London (1965).
[3] Pryor, W. A. *Introduction to Free Radical Chemistry*, Prentice-Hall, Englewood Cliffs, N.J. (1966).
[4] Pryor, W. A. *Radical Reactions*, McGraw-Hill, New York (1966).
[5] Kiefer, H. and Traylor, T. G. *Tetrahedron Lett.* 6163 (1966).
[6] Calvert, J. G. and Pitts, J. N., Jr. *Photochemistry*, Wiley, New York (1966).
[7] Kan, R. O. *Organic Photochemistry*, McGraw-Hill, New York (1966).
[8] Turro, N. J. *Molecular Photochemistry*, Benjamin, New York (1965).

[9] Weast, R. C., Selby, S. M. and Hodgman, C. D. (Ed.), *Handbook of Chemistry and Physics*, The Chemical Rubber Co., Cleveland, Ohio (1965).
[10] Vedeneyev, V. I., Gurvich, L. V., Kondrat'yev, V. N., Medvedev, V. A. and Frankevich, Ye. L. *Bond Energies Ionisation Potentials and Electron Affinities*, Arnold, London (1966).
[11] Kharasch, M. S. and Gladstone, M. T. *J. Am. chem. Soc.* **65**, 15 (1943).
[12] Coffman, D. D., Jenner, E. L. and Lipscomb, R. D. *J. Am. chem. Soc.* **80**, 2864 (1958).
[13] Kharasch, M. S., Jensen, E. U. and Urry, W. H. *J. org. Chem.* **10**, 386 (1945).
[14] McBay, H. C. and Tucker, O. *J. org. Chem.* **19**, 869 (1954).
[15] Kharasch, M. S., McBay, H. C. and Urry, W. H. *J. org. Chem.* **10**, 394 (1945).
[16] Farmer, E. H. and Moore, C. G. *J. chem. Soc.* 149 (1951).
[17] Johnston, K. M. and Williams, G. H. *J. chem. Soc.* 1168 (1960); see also Huang, H. M. and Lim, P. K. K. *J. chem. Soc.* (*C*) 2432 (1967).
[18] Schwetlick, K., Jentzsch, J., Karl, R. and Wolter, D. *J. prakt. Chem.* **25**, 95 (1964).
[19] Schwetlick, K., Geyer, W. and Hartmann, H. *Angew. Chem.* **72**, 779 (1960).
[20] Beckwith, A. L. J. *Aust. J. Chem.* **13**, 244 (1960).
[20a] Cookson, R. C., Hudec, J. and Mirza, N. A. *Chem. Commun.* 824 (1967).
[21] Scott, A. I. *Q. Rev. chem. Soc.* **19**, 1 (1965).
[22] Musso, M. *Angew. Chem. int. Edn.* **2**, 723 (1963).
[23] Bacon, R. G. R. and Hill, H. A. O. *Q. Rev. chem. Soc.* **19**, 95 (1965).
[24] Ingold, K. U. *Chem. Rev.* **61**, 563 (1961); Wexler, H. *Chem. Rev.* **64**, 591 (1964).
[25] Johnson, A. W. and Tam, S. W. *Chemy Ind.* 1425 (1964).
[26] Barton, D. H. R., DeFlorin, A. M. and Edwards, O. E. *J. chem. Soc.* 530 (1956).
[27] For example, see Bauer, R. H. and Coppinger, G. M. *Tetrahedron* **19**, 1201 (1963).
[28] Walling, C. T. and Hodgdon, R. B. *J. Am. chem. Soc.* **80**, 228 (1958).
[29] Barton, D. H. R. *Proc. chem. Soc.* 293 (1963).
[30] Yip, C. and Klebanoff, S. J. *Biochim. biophys. Acta* **74**, 747 (1963); Ljunggren, J.-G. *Biochim. biophys. Acta* **107**, 434 (1965).
[31] Day, A. C., Nabney, J. and Scott, A. I. *J. chem. Soc.* 4067 (1961).
[32] Scott, A. I., McCapra, F. and Hall, E. S. *J. Am. chem. Soc.* **86**, 302 (1964).
[33] Scott, A. I., McCapra, F., Dodson, P. A. and Meyers, M. B. *J. Am. chem. Soc.* **85**, 3702 (1963).
[34] Cameron, D. W., Chan, H. W.-S. and Hildyard, E. M. *J. chem. Soc.* (*C*) 1832 (1966).
[35] Binns, T. D., Brettle, R. and Parkin, J. G. *Chem. Commun.* 409 (1965).
[36] Grimshaw, J. and Ramsey, J. S. *J. chem. Soc.* (*C*) 653 (1966).
[37] Eglinton, G. and McCrae, W. *Adv. org. Chem.* **4**, 225 (1963).
[38] Atkinson, R. E., Curtis, R. F. and Taylor, J. A. *J. chem. Soc.* (*C*) 578 (1967); see also Stephens, R. D. and Castro, C. E. *J. org. Chem.* **28**, 3313 (1963).
[39] Sondheimer, F. *Pure appl. Chem.* **7**, 363 (1963).
[40] Fanta, P. E. *Chem. Rev.* **64**, 613 (1964).
[40a] Hashimoto, H. and Nakano, T. *J. org. Chem.* **31**, 891 (1966).
[41] Cohen, T. and Lewin, A. H. *J. Am. chem. Soc.* **88**, 4521 (1966).
[41a] Markarian, M. *J. Am. chem. Soc.* **74**, 1858 (1952).
[41b] Corey, E. J. and Posner, G. H. *J. Am. chem. Soc.* **89**, 3911 (1967).
[42] Finley, K. T. *Chem. Rev.* **64**, 573 (1964).
[43] Finley, K. T. and Sasaki, N. A. *J. Am. chem. Soc.* **88**, 4267 (1966).
[44] Kwart, H. and Ford, J. A. *J. org. Chem.* **24**, 2060 (1959).
[45] Blomquist, A. T. and Miller, G. A. *J. Am. chem. Soc.* **83**, 243 (1961).
[46] Mathur, H. H. and Bhattacharyya, S. C. *J. chem. Soc.* 114, 3505 (1963).

[47] Bryce-Smith, D. *J. chem. Soc.* 1603 (1956); Müller, E. and Kiedaisch, W. *Annln Chem.* (*Liebigs*) **632**, 28 (1960).
[48] Gill, G. B. and Williams, G. H. *J. chem. Soc.* 995, 7127 (1965); *J. chem. Soc.* (*B*), 880 (1966).
[48a] Teuber, H.-J. and Rau, W. *Chem. Ber.* **86**, 1036 (1953).
[48b] Gingras, B. A. and Waters, W. A. *J. chem. Soc.* 1920 (1954).
[48c] Tilney-Bassett, J. F. and Waters, W. A. *J. Chem. Soc.* 3129 (1957).
[48d] Gill, G. B. and Williams, G. H. *J. chem. Soc.* 5756 (1965).
[49] Kharasch, M. S. and Sosnovsky, G. *Tetrahedron* **3**, 105 (1958); Kharasch, M. S. and Fono, G. *J. org. Chem.* **23**, 324 (1958); **24**, 72 (1959).
[49a] Kochi, J. K. *Tetrahedron* **18**, 483 (1962).
[50] Kharasch, M. S., Sosnovsky, G. and Yang, N. C. *J. Am. chem. Soc.* **81**, 5819 (1959) and earlier papers.
[51] Sosnovsky, G. and Lawesson, S.-O. *Angew. Chem. Int. Edn.* **3**, 269 (1964).
[52] Walling, C. and Zavitsas, A. A. *J. Am. chem. Soc.* **85**, 2084 (1963).
[53] Kochi, J. K. *J. Am. chem. Soc.* **84**, 774, 3271 (1962); Denney, D. Z., Appelbaum, A. and Denney, D. B. *J. Am. chem. Soc.* **84**, 4969 (1962): Denney, D. B., Napier, R. and Cammarata, A. *J. org. Chem.* **30**, 3151 (1965).
[54] Cross, B. and Whitham, G. H. *J. chem. Soc.* 1650 (1961).
[55] Sosnovsky, G. and Yang, N. C. *J. org. Chem.* **25**, 899 (1960).
[56] Sosnovsky, G. *Tetrahedron* **13**, 241 (1961).
[57] Lawesson, S.-O., Berglund, C. and Grönwall, S. *Acta chem. scand.* **15**, 249 (1961).
[58] Lawesson, S.-O. and Berglund, C. *Ark. Kemi* **16**, 287 (1960).
[59] Lawesson, S.-O. and Berglund, C. *Acta chem. scand.* **14**, 1854 (1960); *Ark. Kemi* **17**, 475 (1961).
[60] Lawesson, S.-O. and Berglund, C. *Ark. Kemi* **20**, 225 (1963).
[61] Sosnovsky, G. *J. org. Chem.* **25**, 874 (1960); Lawesson, S.-O. and Berglund, C. *Ark. Kemi* **17**, 465 (1961); *Angew. Chem.* **73**, 65 (1961).
[62] Wiberg, K. B. (Ed.). *Oxidation in Organic Chemistry*, Chapter 2, Academic Press, New York (1965).
[63] Stewart, R. in *Oxidation in Organic Chemistry* (ed. K. B. Wiberg), Chapter 1, Academic Press, New York (1965).
[64] Criegee, R. in *Oxidation in Organic Chemistry* (ed. K. B. Wiberg), Chapter 5, Academic Press, New York (1965).
[65] Waters, W. A. and Littler, J. S. in *Oxidation in Organic Chemistry* (ed. K. B. Wiberg), Chapter 3, Academic Press, New York (1965).
[66] Dewar, M. J. S., Dietz, R. and Hunt, R. L. *J. Am. chem. Soc.* **88**, 5473 (1966).
[67] Richardson, W. H. in *Oxidation in Organic Chemistry* (ed. K. B. Wiberg), Chapter 4, Academic Press, New York (1965).
[68] Syper, L. *Tetrahedron Lett.* 4493 (1966).
[69] Trahanovsky, W. S. and Young, L. B. *J. chem. Soc.* 5777 (1965).
[69a] Trahanovsky, W. S. and Young, L. B. *J. org. Chem.* **31**, 2033 (1966).
[70] Syper, L. *Tetrahedron Lett.* 4193 (1967).
[71] Slaugh, L. H. and Raley, J. H. *Tetrahedron* **20**, 1005 (1964).
[72] Kochi, J. K. and Mocadlo, P. E. *J. Am. chem. Soc.* **88**, 4094 (1966).
[73] Kochi, J. K. and Davis, D. D. *J. Am. chem. Soc.* **86**, 5264 (1964); Kochi, J. K. and Buchanan, D. *J. Am. chem. Soc.* **87**, 853 (1965).
[74] Barton, D. H. R., Basu, N. K., Hesse, R. H., Morehouse, F. S. and Pechet, M. M. *J. Am. chem. Soc.* **88**, 3016 (1966).
[75] Russell, G. A. *J. Am. chem. Soc.* **80**, 4987, 4997, 5002 (1958); *Tetrahedron* **8**, 101 (1960); Walling, C. and Mayahi, M. F. *J. Am. chem. Soc.* **81**, 1485 (1959).
[76] Sosnovsky, G. *Free Radical Reactions in Preparative Organic Chemistry*, Macmillan, New York (1964).

[77] SOSNOVSKY, G. *Curr. Stud. free Radical React. preparative org. Chem.* **1**, 57 (1967).
[78] KOLLONITSCH, J., ROSEGAY, A. and DOLDOURAS, G. *J. Am. chem. Soc.* **86**, 1857 (1964); see also KOLLONITSCH, J., DOLDOURAS, G. A. and VERDI, V. F. *J. chem. Soc.* (*B*) 1093 (1967).
[79] WALLING, C. and THALER, W. *J. Am. chem. Soc.* **83**, 3877 (1961).
[80] WALLING, C. and JACKNOW, J. *J. Am. chem. Soc.* **82**: (a) 6108; (b) 6113 (1960).
[81] BANKS, D. F., HUYSER, E. S. and KLEINBERG, J. *J. org. Chem.* **29**, 3692 (1964).
[82] HUYSER, E. S. and GIDDINGS, B. *J. org. Chem.* **27**, 3391 (1962); see also HUYSER, E. S., SCHIMKE, H. and BURHAM, R. L. *J. org. Chem.* **28**, 2141 (1963).
[83] MINISCI, F., GALLI, R., GALLI, A. and BERNARDI, R. *Tetrahedron Lett.* 2207 (1967).
[84] TANNER, D. D. and VAN BOSTELEN, P. B. *J. org. Chem.* **32**, 1517 (1967).
[85] SKELL, P. S., ALLAN, R. G. and GILMOUR, N. D. *J. Am. chem. Soc.* **83**, 504 (1961).
[86] FULLER, A. E. and HICKINBOTTOM, W. J. *J. chem. Soc.* 3228, 3235 (1965).
[87] MILLER, B. and WALLING, C. *J. Am. chem. Soc.* **79**, 4187 (1957).
[88] BANUS, J., EMELÉUS, H. J. and HASZELDINE, R. N. *J. chem. Soc.* 3041 (1950).
[89] MCGRATH, B. P. and TEDDER, J. M. *Proc. chem. Soc.* 80 (1961).
[90] PEARSON, R. E. and MARTIN, J. C. *J. Am. chem. Soc.* **85**, 354, 3142 (1963); SKELL, P. S., TULEEN, D. L. and READIO, P. D. *J. Am. chem. Soc.* **85**, 2850 (1963); HEDAYA, E., HINMAN, R. L. and THEODOROPULOS, S. *J. Am. chem. Soc.* **85**, 3052 (1963); WALLING, C., RIEGER, A. L. and TANNER, D. D. *J. Am. chem. Soc.* **85**, 3129 (1963); RUSSELL, G. A. and DESMOND, K. M. *J. Am. chem. Soc.* **85**, 3139 (1963); HEDAYA, E., HINMAN, R. L., KIBLER, L. M. and THEODOROPULOS, S. *J. Am. chem. Soc.* **86**, 2727 (1964); KOENIG, T. and BREWER, W. *J. Am. chem. Soc.* **86**, 2728 (1964).
[91] HORNER, L. and WINKELMANN, E. H. in *Newer Methods of Preparative Organic Chemistry* (ed. W. Foerst), Vol. 3, p. 151, Academic Press, New York (1964); *Angew. Chem.* **71**, 349 (1959); see also FILLER, R. *Chem. Rev.* **63**, 21 (1963).
[92] CAMPAIGNE, E. and TULLAR, B. F. *Org. Synth.*, Coll. Vol. **4**, 921 (1963).
[93] WALLING, C. and PADWA, A. *J. org Chem.* **27**, 2977 (1962).
[94] PINNELL, R. P., HUYSER, E. S. and KLEINBERG, J. *J. org. Chem.* **30**, 38 (1965).
[95] HUYSER, E. S. *J. Am. chem. Soc.* **82**, 391 (1960).
[96] KOOYMAN, E. C. and VEGTER, G. C. *Tetrahedron* **4**, 382 (1958).
[97] TEDDER, J. M. *Adv. Fluor. Chem.* **2**, 104 (1961).
[98] BARBOUR, A. K., BELF, L. J. and BUXTON, M. W. *Adv. Fluor. Chem.* **3**, 181 (1963).
[99] BURDON, J. and TATLOW, J. C. *Adv. Fluor. Chem.* **1**, 129 (1960).
[100] STACEY, M. and TATLOW, J. C. *Adv. Fluor. Chem.* **1**, 166 (1960).
[101] HAMILTON, J. M. *Adv. Fluor. Chem.* **3**, 117 (1963).
[102] SAUNDERS, B. C. *Adv. Fluor. Chem.* **2**, 183 (1961).
[103] KUIVILA, H. G. *Adv. organometall. Chem.* **1**, 47 (1964).
[104] KAPLAN, J. *J. Am. chem. Soc.* **88**, 4531 (1966).
[105] KUIVILA, H. G. *J. org. Chem.* **25**, 284 (1960).
[106] KUPCHIK, E. J. and KIESEL, R. J. *J. org. Chem.* **31**, 456 (1966).
[107] KUIVILA, H. G. and MENAPACE, L. W. *J. org. Chem.* **28**, 2165 (1963).
[108] SEYFERTH, D., YAMAZAKI, H. and ALLESTON, D. L. *J. org. Chem.* **28**, 703 (1963).
[109] KUIVILA, H. G., MENAPACE, L. W. and WARNER, C. R. *J. Am. chem. Soc.* **84**, 3584 (1962).
[110] MENAPACE, L. W. and KUIVILA, H. G. *J. Am. chem. Soc.* **86**, 3047 (1964).
[111] GREENE, F. D. and LOWRY, N. N. *J. org. Chem.* **32**, 882 (1967).
[112] KUPCHIK, E. J. and KIESEL, R. J. *J. org. Chem.* **29**, 764 (1964).
[113] LORENZ, D. H., SHAPIRO, P., STERN, A. and BECKER, E. I. *J. org. Chem.* **28**, 2332 (1963); see also ROTHMAN, L. A. and BECKER, E. I. *J. org. Chem.* **25**, 2203 (1960).
[114] KUPCHIK, E. J. and LANIGAN, T. *J. org. Chem.* **27**, 3661 (1962).

[115] WILSON, C. V. *Org. React.* **9**, 332 (1957); JOHNSON, R. G. and INGHAM, R. K. *Chem. Rev.* **56**, 332 (1956).
[116] CRISTOL, S. J. and FIRTH, W. C. *J. org. Chem.* **26**, 280 (1961).
[117] DAVIS, J. A., HERYNK, J., CARROLL, S., BUNDS, J. and JOHNSON, D. *J. org. Chem.* **30**, 415 (1965).
[118] BARTON, D. H. R., FARO, H. P., SEREBRYAKOV, E. P. and WOOLSEY, N. F. *J. Chem. Soc.* 2438 (1965).
[119] KOCHI, J. K. *J. Am. chem. Soc.* **87**, 2500 (1965).
[120] WIBERG, K. B., LOWRY, B. R. and COLBY, T. H. *J. Am. chem. Soc.* **83**, 3998 (1961).
[121] EATON, P. E. and COLE, T. W. *J. Am. chem. Soc.* **86**, 3157 (1964).
[122] BARBORAK, J. C., WATTS, L. and PETIT, R. I. *J. Am. chem. Soc.* **88**, 1328 (1966).
[123] LUNDBERG, W. O. (Ed.), *Autoxidation and Antioxidants*, Vols. I and II, Wiley, New York (1961, 1962).
[124] TWIGG, G. H. *Chemy Ind.* 4 (1962); MANLY, T. D. *Chemy Ind.* 12 (1962); CULLIS, C. F. *Chemy Ind.* 23 (1962); JOHN J. A. and WEYMOUTH, F. J. *Chemy Ind.* 62 (1962); LANDAU, R. and SIMON, R. W. *Chemy Ind.* 70 (1962).
[125] INGOLD, K. U. *Chem. Rev.* **61**, 563 (1961).
[126] SCOTT, G. *Chemy Ind.* 271 (1963).
[127] CULLIS, C. F. and FISH, A. in *Chemistry of the Carbonyl Group* (ed. S. Patai), p. 95, Interscience, London (1966).
[128] RUSSELL, G. A. *J. chem. Educ.* **36**, 111 (1959).
[129] STEERE, N. V. in *Chemistry of the Ether Linkage* (ed. S. Patai), p. 691, Interscience, London (1967).
[130] HARRISSON, R. J. and MOYLE, M. *Org. Synth.* Coll. Vol. **4**, 493 (1963).
[131] HOUSE, H. O. *Org. Synth.,* Coll. Vol. **4**, 367 (1963).
[132] BATEMAN, L. *Q. Rev. chem. Soc.* **8**, 147 (1954).
[133] KNIGHT, H. B. and SWERN, D. *Org. Synth.*, Coll. Vol. **4**, 895 (1963).
[134] HOLMQUIST, H. E., ROTHROCK, H. S., THEOBALD, C. W. and ENGLUND, B. E. *J. Am. chem. Soc.* **78**, 5339 (1956).
[135] HOCK, H., KROPF, H. and ERNST, F. *Angew. Chem.* **71**, 541 (1959).
[136] RUSSELL, G. A. and BEMIS, A. G. *J. Am. chem. Soc.* **88**, 5491 (1966).
[137] TITOV, A. I. *Tetrahedron* **19**, 557 (1963).
[138] MÜLLER, E., PADEKEN, H. G., SALOMON, M. and FIEDLER, G. *Chem. Ber.* **98**, 1893 (1965).
[139] MÜLLER, E. and FIEDLER, G. *Chem. Ber.* **98**, 3493 (1965).
[140] MÜLLER, E. and TRENSE, U. *Tetrahedron Lett.* 2045 (1967).
[141] MÜLLER, E. and HUBER, H. *Chem. Ber.* **96**, 670, 2319 (1963).
[142] KIRMSE, W. *Carbene Chemistry*, Academic Press, London (1964).
[143] HINE, J. *Divalent Carbon*, Ronald Press, New York (1964).
[144] CHINOPORUS, E. *Chem. Rev.* **63**, 235 (1963).
[145] ROZANTSEV, G. G., FAINZIL'BERG, A. A. and NOVIKOV, S. S. *Russ. chem. Revs* **2**, 69 (1965).
[146] HORNER, L. and CHRISTMANN, A. *Angew. Chem. int. Edn.* **2**, 599 (1963).
[147] ABRAMOVITCH, R. A. and DAVIS, B. A. *Chem. Rev.* **64**, 149 (1964).
[148] DE LA MARE, H. E. and COPPINGER, G. M. *J. org. Chem.* **28**, 1068 (1963).
[149] HUYSER, E. S., BREDEWEG, C. J. and VAN SCOY, R. M. *J. Am. chem. Soc.* **86**, 4148 (1964).
[150] AKHTAR, M. *Adv. Photochem.* **2**, 263 (1964), and references therein.
[151] HEUSLER, K. and KALVODA, J. *Angew. Chem. int. Edn.* **3**, 525 (1964).
[152] ACOTT, B. and BECKWITH, A. L. J. *Aust. J. Chem.* **17**, 1342 (1964).
[153] SMOLINSKY, G. and FEUER, B. I. *J. org. Chem.* **30**, 3216 (1965).
[154] BARTON, D. H. R., BEATON, J. M., GELLER, L. E. and PECHET, M. M. *J. Am. chem. Soc.* **82**, 2640 (1960); **83**, 4076 (1961).

[155] BARTON, D. H. R. and BEATON, J. M. *J. Am. chem. Soc.* **82**, 2641 (1960).
[156] ROBINSON, C. H., GNOJ, O., MITCHELL, A., OLIVETO, E. P. and BARTON, D. H. R. *Tetrahedron* **21**, 743 (1965).
[157] KABASAKALIAN, P. and TOWNLEY, E. R. *J. Am. chem. Soc.* **84**, 2724 (1962).
[158] WALLING, C. and PADWA, A. *J. Am. chem. Soc.* **83**, 2207 (1961).
[159] GREENE, F. D., SAVITZ, M. L., OSTERHOLTZ, F. D., LAU, H. H., SMITH, W. N. and ZANET, P. M. *J. org. Chem.* **28**, 55 (1963).
[160] WALLING, C. and PADWA, A. *J. Am. chem. Soc.* **85**, 1597 (1963).
[161] MEYSTRE, CH., HEUSLER, K., KALVODA, J., WIELAND, P., ANNER, G. and WETTSTEIN, A. *Experientia* **17**, 475 (1961); *Helv. chim. Acta* **45**, 1317 (1962).
[162] BARTON, D. H. R., BECKWITH, A. L. J. and GOOSEN, A. *J. chem. Soc.* 181 (1965); see also KUHN, L. P., KLEINSPEHN, G. P. and DUCKWORTH, A. C. *J. Am. chem. Soc.* **89**, 3858 (1967).
[163] BECKWITH, A. L. J. and GOODRICH, J. E. *Aust. J. Chem.* **18**, 747 (1965).
[164] NEALE, R. S., MARCUS, N. L. and SCHEPERS, R. G. *J. Am. chem. Soc.* **88**, 3051 (1966).
[165] OKAHARA, M., OHASI, T. and KOMORI, S. *Tetrahedron Lett.* 1629 (1967).
[166] WOLFF, M. E. *Chem. Rev.* **63**, 55 (1963).
[167] NEALE, R. S. and WALSH, M. R. *J. Am. chem. Soc.* **87**, 1255 (1965).
[168] COREY, E. J. and HERTLER, W. R. *J. Am. chem. Soc.* **82**, 1657 (1960).
[169] LUKES, R. and FERLES, M. *Colln. Czech. chem. Commun.* **20**, 1227 (1955).
[170] SCHMITZ, E. and MURAWSKI, D. *Chem. Ber.* **93**, 754 (1960).
[171] PARTCH, R. *Tetrahedron Lett.* 1361 (1966).
[172] WOLFF, M. E., KERWIN, J. F., OWINGS, F. E., LEWIS, B. B., BLANK, B., MAGNANI, A. and GEORGIAN, V. *J. Am. chem. Soc.* **82**, 4117 (1960).
[173] (a) HAM, G. E. (Ed.), *Vinyl Polymerisation*, Vol. 1, Arnold and Dekker, London (1967).
(b) RAVVE, A. *Organic Chemistry of Macromolecules*, Arnold and Dekker, London (1967); BURNETT, G. M. *Mechanism of Polymer Reactions*, Interscience, New York (1954).
[174] CADOGAN, J. I. G. and PERKINS, M. J. in *The Chemistry of the Alkenes* (ed. S. Patai), Chapter 9, Interscience, London (1964).
[175] WALLING, C. and HUYSER, E. S. *Org. React.* **13**, 91 (1963).
[176] STACEY, F. W. and HARRIS, J. F. *Org. React.* **13**, 150 (1963).
[177] RONDESTVEDT, C. *Org. React.* **11**, 189 (1960).
[178] BOHM, B. B. and ABELL, P. I. *Chem. Rev.* **62**, 599 (1962).
[179] LEUSINK, A. and NOLTES, J. G. *Tetrahedron Lett.* 335 (1966).
[180] TRECKER, D. J. and HENRY, J. P. *Chem. Commun.* 258 (1966).
[181] BRESLOW, R., GROVES, J. T. and OLIN, S. S. *Tetrahedron Lett.* 4717 (1966); 1837 (1968).
[182] NEALE, R. S. *J. Am. chem. Soc.* **86**, 5340 (1964); *Tetrahedron Lett.* 483 (1966).
[183] CADOGAN, J. I. G., HEY, D. H. and ONG, S. H. *J. chem. Soc.* 1932, 1939 (1965); CADOGAN, J. I. G., HEY, D. H. and SHARP, J. T. *J. chem. Soc.* (*B*) 803 (1967).
[184] KHARASCH, N., LEWIS, P. and SHARMA, R. K. *Chem. Commun.* 435 (1967).
[185] PAC, C. and TSUTSUMI, S. *Tetrahedron Lett.* 2341 (1965).
[186] URRY, W. H., NISHIHARA, A. and NIU, J. H. Y. *J. org. Chem.* **32**, 347 (1967).
[186a] KRAUCH, C. H., ROKACH, J. and ELAD, D. *Tetrahedron Lett.* 5099 (1967).
[187] JULIA, M. and LE GOFFIC, F. *Bull. Soc. chim. Fr.* 1550 (1965); JULIA, M. *Pure appl. Chem.* **15**, 167 (1967).
[188] WALLING, C., COOLEY, J. H., PONARAS, A. A. and RACAH, E. J. *J. Am. chem. Soc.* **88**, 5361 (1966).
[189] THALER, W. A. *J. Am. chem. Soc.* **88**, 4278 (1966); **89**, 1902 (1967).
[190] KAPLAN, L. *J. Am. chem. Soc.* **88**, 1833, 4970 (1966).
[191] RONDESTVEDT, C. *Org. React.* **11**, 189 (1960).

[192] NOLAND, W. E. and SELLSTEDT, J. H. *J. org. Chem.* **31**, 345 (1966).
[193] LANGKAMMERER, C. M., JENNER, E. L., COFFMANN, D. D. and HOWK, B. W. *J. Am. chem. Soc.* **82**, 1395 (1960) and earlier papers.
[194] WILLIAMS, G. H. *Homolytic Aromatic Substitution*, Pergamon, London (1960).
[195] WILLIAMS, G. H. *Chemy Ind.* 1286 (1961).
[196] HEY, D. H., PERKINS, M. J. and WILLIAMS, G. H. *Chemy Ind.* 83 (1963); see also CHALFONT, G. R., HEY, D. H., LIANG, K. S. Y. and PERKINS, M. J. *Chem. Commun.* 367 (1967).
[197] BACHMANN, W. E. and HOFFMANN, R. A., *Org. React.* **2**, 224 (1944).
[198] DETAR, D. F. *Org. React.* **9**, 409 (1957).
[199] TIECCO, M. *Chem. Commun.* 555 (1965).
[200] WOLF, W. and KHARASCH, N. *J. org. Chem.* **30**, 2493 (1965); see also KHARASCH, N. and SHARMA, R. K. *Chem. Commun.* 106 (1966).
[201] JEFFS, P. W. and HANSEN, J. F. *J. Am. chem. Soc.* **89**, 2798 (1967); see also YONEMITSU, O., TOKUYAMA, T., CHAYKOVSKY, M. and WITKOP, B. *J. Am. chem. Soc.* **90**, 776 (1968).
[202] KHARASCH, N. and GÖTHLICH, L. *Angew. Chem. int. Edn.* **1**, 459 (1962).
[203] MATSUURA, T. and OMURA, K. *Chem. Commun.* 127 (1966).
[203a] KOVACIC, P. and KURZ, M. E. *J. Am. chem. Soc.* **88**, 2068 (1966); **89**, 4960 (1967); *Chem. Commun.* 321 (1966).
[204] DAVIES, D. I., HEY, D. H. and WILLIAMS, G. H. *J. chem. Soc.* 1878 (1958); see also NAKATA, T., TOKUMARU, K. and SIMAMURA, O. *Tetrahedron Lett.* 3303 (1967).
[205] KAMPMEIER, J. A. and FANTAZIER, R. M. *J. Am. chem. Soc.* **88**, 1959 (1966).
[206] KOCHI, J. K. and GILLIOM, R. D. *J. Am. chem. Soc.* **86**, 5251 (1964); see also BROWN, P. M., RUSSELL, J., THOMSON, R. H. and WYLIE, A. G. *J. chem. Soc.* (*C*) 842 (1968).
[207] MINISCI, F. and GALLI, R. *Tetrahedron Lett.* 433 (1965); 2531 (1966).
[208] ELAD, D. *Tetrahedron Lett.* 77 (1963).
[208a] BAILLIE, A. C. and THOMSON, R. H. *J. chem. Soc.* (*C*) 48 (1968).
[209] KAPLAN, L. *J. Am. chem. Soc.* **89**, 1753 (1967).
[210] FRISWELL, N. J. and GOWENLOCK, B. G. *Chem. Commun.* 277 (1965).
[211] LAMBERT, R. M., CHRISTIE, M. I. and LINNETT, J. W. *Chem. Commun.* 388 (1967).
[212] JACKSON, R. A. *Chem. Commun.* 827 (1966).
[213] DAVIES, A. G. and ROBERTS, B. P. *J. chem. Soc.* (*B*) 17 (1967).
[214] KAMPMEIER, J. A. and EVANS, T. R. *J. Am. chem. Soc.* **88**, 4096 (1966).
[215] PRYOR, W. A. and GUARD, H. *J. Am. chem. Soc.* **86**, 1150 (1964).
[216] BULEY, A. L., NORMAN, R. O. C. and PRITCHETT, R. J. *J. chem. Soc.* (*B*) 849 (1966); see also KAMPMEIER, J. A., GEER, R. P., MESKIN, A. J. and D'SILVA, R. M. *J. Am. chem. Soc.* **88**, 1257 (1966).
[217] GREENE, F. D. and LOWRY, N. N. *J. org. Chem.* **32**, 882 (1967).
[217a] STORY, P. R., DENSON, D. D., BISHOP, C. E., CLARK, B. C. and FARINE, J.-C. *J. Am. chem. Soc.* **90**, 817 (1968).
[218] WALLING, C. in *Molecular Rearrangements* (ed. P. de Mayo), Vol. 1, p. 407, Interscience, London (1963).
[219] FREIDLINA, R. KH. *Adv. Free Radical Chem.* **1**, 211 (1965).
[220] SKELL, P. S. *Organic Reaction Mechanisms*, p. 131, Chemical Society (London) Special Publication No. 19 (1965) and references cited therein.
[221] WINSTEIN, S., HECK, R., LAPPORTE, S. and BAIRD, R. *Experientia* **12**, 138 (1956).
[222] THOMSON, R. H. and WYLIE, A. G. *Proc. chem. Soc.* 65 (1963).
[223] LEWIS, J. R. *J. chem. Soc.* 2533 (1962).

5

Electrochemical Methods

R. BRETTLE, *Sheffield University*

The passage of an electric current through an electrolytic cell by the agency of an applied potential provides, at the electrodes, a means for the transfer of electrons. At the positively charged anode the discharge of a chemical species with the loss of one or more electrons leads to an oxidation, and conversely the discharge of a species at the negatively charged cathode with the gain of one or more electrons results in a reduction. It is not surprising that many useful preparative oxidations and reductions of organic compounds can be accomplished by electrochemical methods. Since in any particular electrolysis the organic chemical interest is normally restricted to only one half of the coupled oxidation–reduction process which is taking place, it is possible to discuss electrochemical oxidations and reductions separately.

In many systems the potential at which a species exchanges a certain number of electrons with an electrode can be determined by polarography, and coulometric analysis allows the number of electrons being exchanged per unit species to be measured for the process at that particular potential. This kind of experiment, which frequently forms the basis for a micro-electroanalytical method assists in the choice of optimum conditions for macro-scale electrochemical preparations, and provides important information about the mechanisms of the oxidation or reduction steps involved.

ELECTROCHEMICAL OXIDATIONS

The most widely investigated type of electrochemical oxidation involves the discharge of a carboxylate anion. In its simplest form the electrolysis gives a product which arises by the coupling of two fragments formed by the oxidative decarboxylation of such anions.

$$2R{\cdot}CO_2^- \xrightarrow{-2e} R_2 + 2CO_2$$

This is the basis of the Kolbe electrosynthesis, one of the most useful and versatile coupling reactions known [1]. The coupling process normally only occurs at a smooth platinum anode; virtually all other materials are unsatisfactory. The nature of the cathode is much less critical; platinum or mercury cathodes are commonly used. Much of the earlier work on the Kolbe electrosynthesis was conducted under what would not now be regarded as the optimum conditions. For example, many of the electrolyses were performed in aqueous solution, whereas methanol, which will also dissolve sodium carboxylates, has proved to be a better general solvent. It is customary to use the sodium salt of the acid in these electrolyses, but electrolyses have also been performed with ammonium salts, and the use of triethylammonium salts may be advantageous, particularly in non-methanolic solvent systems [2].

In order to understand which acids give preparatively useful yields of the coupled product it is necessary to consider a simple rationalization of the processes involved in the anodic oxidation of a carboxylate anion. The primary electrode process, which occurs at potentials above a critical value of +2·3 volts, may be represented as

$$R{\cdot}CO_2^- \xrightarrow{-e} R^{\cdot} + CO_2$$

without specifying in more detail the exact nature of the decarboxylation step. The coupled product only arises by the dimerization of the radicals $R^{\cdot}$ at the surface of the anode.

$$2R^{\cdot} \rightarrow R_2$$

By far the most important process which may intervene and so compete with the coupling process is the further oxidation of the intermediate radical to the corresponding cation, though this has only been recognized during the last decade.

$$R^{\cdot} \xrightarrow{-e} R^+$$

In general this further oxidation will occur when the cation so formed is a relatively stable carbonium ion of the type typically produced in an S_N1 reaction by the rapid solvolytic ionization of a halide or arenesulphonate, for example a tertiary cation such as the t-butyl cation, or one stabilized by electron-donating substituents [3]. To the extent that this process occurs, radical coupling will be suppressed and typical carbonium-ion-derived products will be formed.

Other competing processes, which are generally much less significant, are those in which the intermediate radical abstracts hydrogen from the solvent, and in which the intermediate radical disproportionates, that is, abstracts hydrogen from a like radical. The former process leads, for example, to the formation of a little methane, as well as ethane, during the electrolysis of acetates [4].

$$CH_3CO_2^- \xrightarrow[-CO_2]{-e} CH_3^{\cdot} \qquad CH_3^{\cdot} + H{-}R \rightarrow CH_4 + R^{\cdot}$$

$$2CH_3^{\cdot} \rightarrow C_2H_6$$

The latter explains the formation of methyl nonanoate and methyl nonenoate during electrolyses with methyl hydrogen sebacate [5].

$$CH_3O_2C[CH_2]_8CO_2^- \xrightarrow{-e,\ -CO_2} CH_3O_2C[CH_2]_8^{\cdot}$$

$$\longrightarrow CH_3O_2C[CH_2]_6CH{=}CH_2 + CH_3O_2C[CH_2]_6CH_2CH_3$$

It will thus be apparent that acids of the general type RCH_2CO_2H give good (in practice $> 50\%$) yields of the coupled product in the Kolbe electrosynthesis. A very large number of functionally substituted acids has been electrolysed, and high yields of the coupled products have been obtained. For example, the cheapness and availability of sebacic acid makes the anodic oxidation of methyl hydrogen sebacate, followed by a hydrolysis, an excellent route to octadecanedioic acid [6].

$$CH_3O_2C[CH_2]_8CO_2H \rightarrow CH_3O_2C[CH_2]_{16}CO_2CH_3$$
$$\rightarrow HO_2C[CH_2]_{16}CO_2H$$

Very little detailed work has been done on the electrolysis of racemic acids, where the coupled products can exist in diastereoisomeric forms, but ($\pm$)-methyl hydrogen β-methylglutarate, for example, on electrolysis gives a mixture of the *erythro*- and *threo*-forms of dimethyl 3,6-dimethylsuberate [7] and it is clear that other coupled products from racemic starting materials will similarly be mixtures, though in many cases this has not been demonstrated.

$$CH_3O_2C{\cdot}CH_2{\cdot}CH(CH_3){\cdot}CH_2{\cdot}CO_2H$$
($\pm$)-form
$$\rightarrow CH_3O_2C{\cdot}CH_2{\cdot}CH(CH_3){\cdot}[CH_2]_2{\cdot}CH(CH_3){\cdot}CH_2{\cdot}CO_2CH_3$$
threo- and *erythro*-forms

Racemization occurs only when the chiral centre is in the α-position to the carboxyl group [8], but in such cases, of course, carbonium-ion-derived products predominate.

The Kolbe electrosynthesis becomes a more versatile synthetic method when mixtures of two acids are used, and the unsymmetrical coupled product can be isolated.

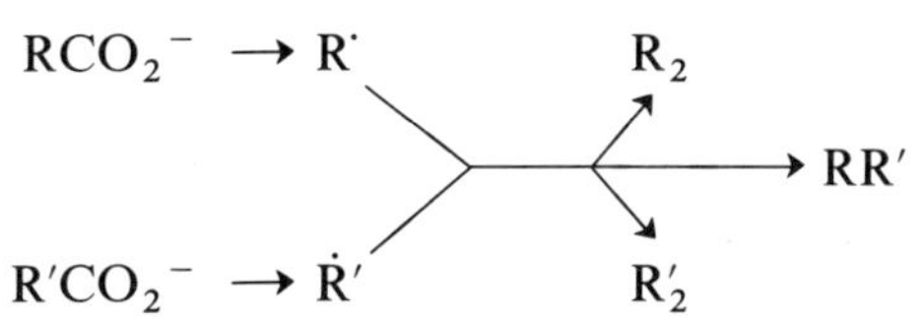

No systematic study has been made of the factors controlling the formation of the unsymmetrical product in such cases, but it has often been found that the use of a large excess of one acid (for example, if only one of the acids, RCO_2H, is readily available) facilitates the conversion of the second acid, $R'CO_2H$, into the unsymmetrical product RR' and reduces the amount of the unwanted symmetrical product R'_2 which is formed.

In the following sections a number of typical examples of Kolbe crossed coupling reactions, which illustrate the scope of the method, are discussed.

The crossed product from the electrolysis of a mixture of hexanoic acid and (−)-[methyl hydrogen β-acetoxyglutarate], in which there is no α-substituent and in which the chiral centre is in the β-position with respect to the carboxyl group, on hydrolysis gave (+)-3-hydroxynonanoic acid. Acetylation of this, and further electrolysis with methyl hydrogen undecanedioate, gave a second crossed product, which on hydrolysis gave (+)-12-hydroxyoctadecanoic acid, the antipode of the hydrogenation product of naturally occurring ricinoleic acid. Since the absolute configuration of the starting (−)-[methyl hydrogen β-acetoxyglutarate] had previously been established, this sequence served to establish the absolute configuration at the chiral centre in the natural product [9].

$$CH_3O_2C{\cdot}CH_2{\cdot}\overset{*}{C}H(OCOCH_3)CH_2{\cdot}CO_2H \quad + \quad CH_3[CH_2]_4CO_2H$$

$$\rightarrow CH_3[CH_2]_5\overset{*}{C}H(OCOCH_3)CH_2{\cdot}CO_2CH_3$$

$$CH_3[CH_2]_5\overset{*}{C}H(OCOCH_3)CH_2{\cdot}CO_2H \quad + \quad CH_3O_2C[CH_2]_9CO_2H$$

$$\rightarrow CH_3[CH_2]_5\overset{*}{C}H(OCOCH_3){\cdot}[CH_2]_{10}{\cdot}CO_2CH_3$$

$$\downarrow$$

$$CH_3[CH_2]_5{\cdot}\overset{*}{C}H(OH){\cdot}[CH_2]_{10}{\cdot}CO_2H$$

The Kolbe electrosynthesis has been particularly useful when applied to the preparation of unsaturated compounds. With alkenoic acids in methanol it has been observed that mixtures of positional isomers may result when a platinum cathode is used and the electrolyte becomes alkaline at the end of the reaction because of the formation of methoxide ion. This trouble does not arise when a mercury cathode is used, and the solution never becomes strongly alkaline [10]. Chaulmoogric acid has been synthesized from (+)-cyclopent-2-en-1-ylacetic acid as shown, thereby establishing a correlation between the chiral centres in the starting material and the product [11].

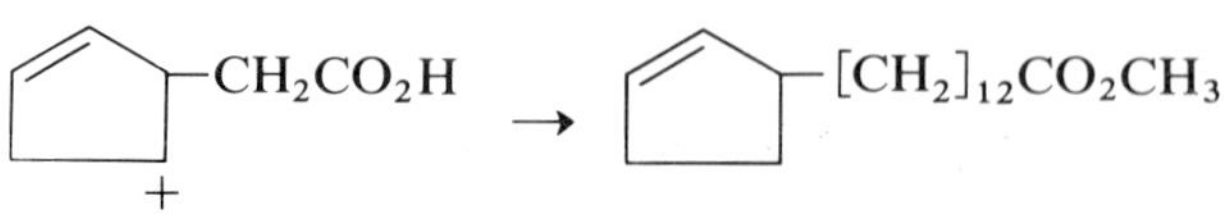

$$CH_3O_2C[CH_2]_{11}{\cdot}CO_2H$$

The stereochemistry of a double bond is retained in the product, making the Kolbe electrosynthesis a very convenient chain extension method for the preparation of alkenoic acids of known geometry, e.g. [12],

$$CH_3[CH_2]_7CH\overset{cis}{=}CH[CH_2]_7CO_2H \quad + \quad HO_2C[CH_2]_2CO_2CH_3$$

$$\rightarrow CH_3[CH_2]_7CH\overset{cis}{=}CH[CH_2]_9CO_2CH_3$$

The anodic oxidation of the simplest βγ-unsaturated acid, but-3-enoic acid, leads to the normal Kolbe coupling, despite the possibility that the intermediate allyl radical might be further oxidized to the allyl cation [10]. With other $\beta\gamma$-unsaturated acids the intermediate radical produced by the anodic decarboxylation of the anion can dimerize in three possible ways, so that even in the self-coupling reaction three isomeric products are formed.

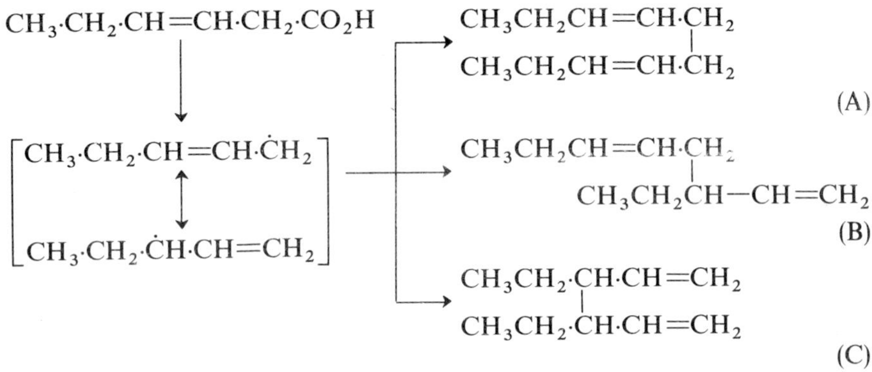

The configuration of the disubstituted double bonds is largely retained in products (A) and (B); for example, *trans*-hex-3-enoic acid gives *trans*, *trans*-deca-3,7-diene (A) [10]. However, when an allylic radical is generated by the electrolysis of a precursor containing a monosubstituted double bond, the disubstituted double bonds formed in the coupled products are of mixed geometry.

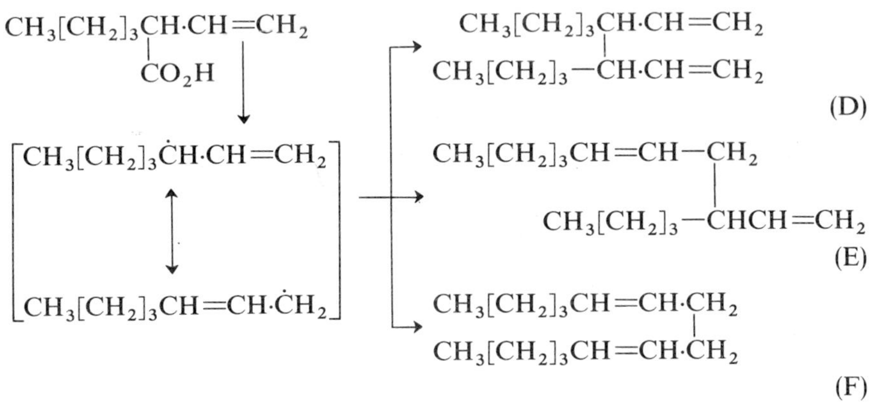

For example, (E) is a mixture of *cis*- and *trans*-forms [10]. When an appropriately substituted double bond is present, cyclization of the intermediate radical has been observed, e.g.,

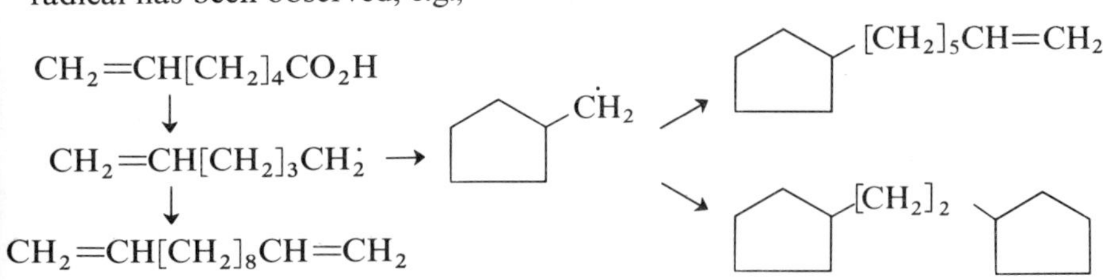

This cyclization is irreversible, since cyclopentylacetic acid undergoes Kolbe coupling without detectable ring-opening [10].

A final example illustrates one of the very few uses of a heterocyclic acid in the Kolbe electrosynthesis.

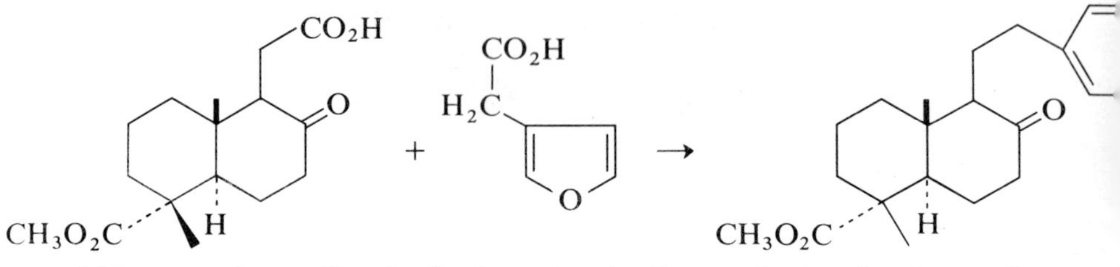

This crossed coupling is the key step in the synthesis of a furanoditerpenoid [2(b)].

The examples so far cited have illustrated Kolbe coupling reactions in which the starting acids have carried alkoxycarbonyl, acyloxy, or keto groups, or contained olefinic unsaturation. Many other acids of the general type RCH_2CO_2H have been used successfully in this reaction, including acids bearing acetal, ether, nitrile, acylamino or *vic*-diol groups, or containing acetylenic unsaturation [1]. The conditions for the successful use of ω-halogenoacids in the reaction need to be carefully controlled. For example, 11-bromoundecanoic acid gives the coupled product in methanol at 50°, but at 65° the major product is methyl 11-bromoundecanoate [13].

$$Br[CH_2]_{10}CO_2H \xrightarrow{50^\circ} Br[CH_2]_{20}Br$$
$$Br[CH_2]_{10}CO_2H \xrightarrow{65^\circ} Br[CH_2]_{10}CO_2CH_3$$

Certain β- and γ-hydroxyacids do not undergo Kolbe coupling. For example, the γ-hydroxyacid shown gives an unsaturated ketone, although the corresponding acetoxyacid couples normally.

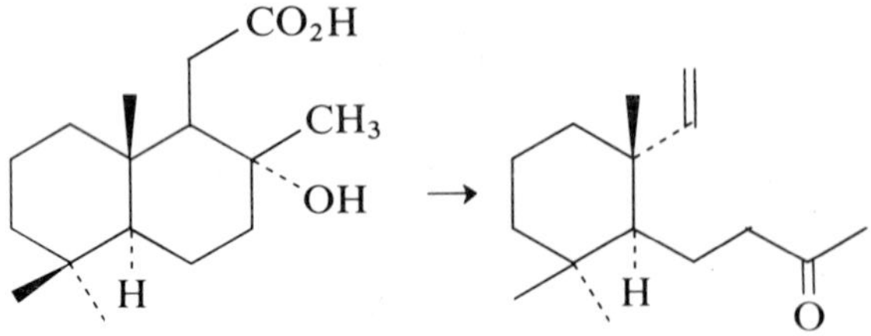

This fragmentation probably involves the production of a cation, though a radical mechanism has not been ruled out [14].

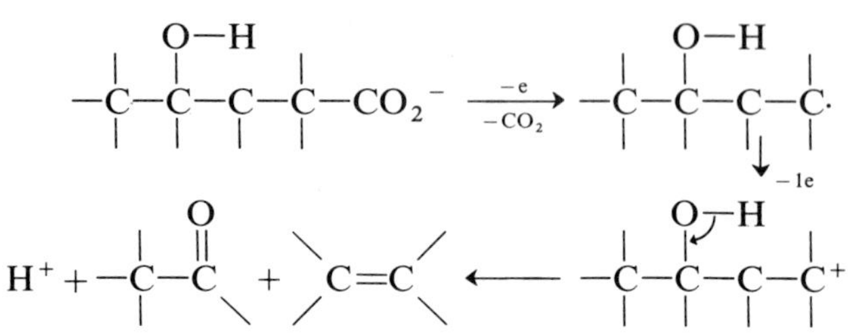

A related cationic process is observed with certain β-hydroxyacids.

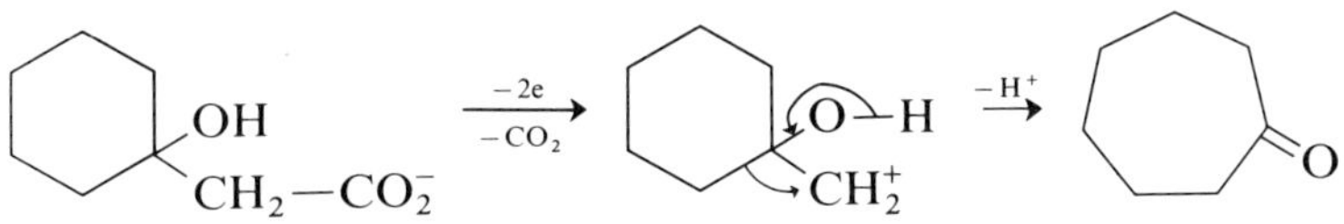

This provides a useful method for the ring expansion of cyclic ketones, since the initial hydroxyacids can be prepared by a Reformatski reaction on the lower homologue of the product ketone [3].

Not all acids with an α-substituent fail to undergo the Kolbe coupling reaction. For example phenylacetic [15], phenoxyacetic [16] and trifluoracetic acids, and other acids of the type $R{\cdot}CF_2{\cdot}CO_2H$ [17] give useful yields of self-coupled products. Half-esters of malonic acid can also be employed. In this case disproportionation of the intermediate radical is a major side-reaction, but when this is not possible as, for example, with ethyl hydrogen t-butylmalonate, an extremely high yield of coupled product is obtained [18].

$$C_2H_5O_2C{\cdot}\underset{}{\overset{C(CH_3)_3}{\overset{|}{C}}}H{\cdot}CO_2^- \xrightarrow[-CO_2]{-e} C_2H_5O_2C{\cdot}\overset{C(CH_3)_3}{\overset{|}{C}}H{\cdot}$$

$$\longrightarrow C_2H_5O_2C{\cdot}\overset{(CH_3)_3C}{\overset{|}{C}}H{\cdot}\overset{C(CH_3)_3}{\overset{|}{C}}H{\cdot}CO_2C_2H_5$$

meso- and ($\pm$)-forms

The corresponding cyano-acetic acid on electrolysis undergoes the usual oxidative decarboxylation, but the normal C $\rightarrow$C coupling occurs only to a minor extent. The major product arises by C $\rightarrow$N coupling of two of the intermediate radicals to give a ketene-imine which is transformed by non-electrolytic processes into an *N*-acylglycinonitrile [19].

$$NC{\cdot}\overset{C(CH_3)_3}{\overset{|}{C}}H{\cdot}CO_2^- \xrightarrow[-CO_2]{-e} [(CH_3)_3C{\cdot}\dot{C}H{\cdot}C{\equiv}N \leftrightarrow (CH_3)_3C{\cdot}CH{=}C{=}\dot{N}]$$

$$\downarrow$$

$$(CH_3)_3C{\cdot}CH_2{\cdot}CO{\cdot}NH{\cdot}\overset{C(CH_3)_3}{\overset{|}{C}}H{\cdot}CN \leftarrow (CH_3)_3CCH{=}C{=}N{\cdot}\overset{C(CH_3)_3}{\overset{|}{C}}H{\cdot}CN$$

Low yields of the Kolbe coupled product are also obtained with certain secondary acids, e.g. [20],

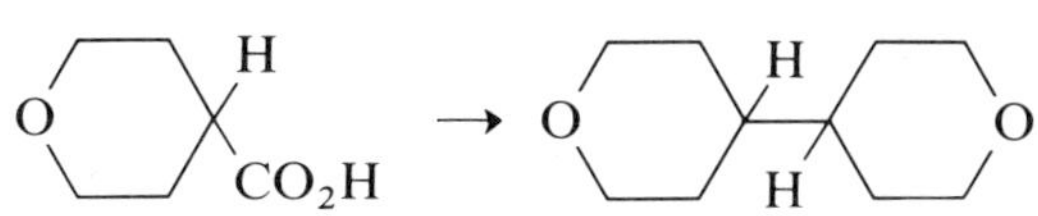

Aromatic acids in which the carboxyl group is directly attached to an aromatic ring give none of the Kolbe coupled product, but this is not surprising, as the intermediate aryl radicals would be expected to abstract

hydrogen from a solvent like methanol.

$$C_6H_5CO_2^- \xrightarrow[-CO_2]{-e} C_6\dot{H}_5 \xrightarrow{CH_3OH} C_6H_6$$

If pyridine is present, it is attacked by the aryl radicals and a mixture of the arylpyridines is formed [21]. Other radicals produced by the electrochemical oxidative decarboxylation of carboxylic acids will likewise attack aromatic and other π-electron systems. For example, a minor product from the electrolysis of sodium acetate in the presence of an olefin results from the addition of a methyl group and an acetate group across the double bond [22].

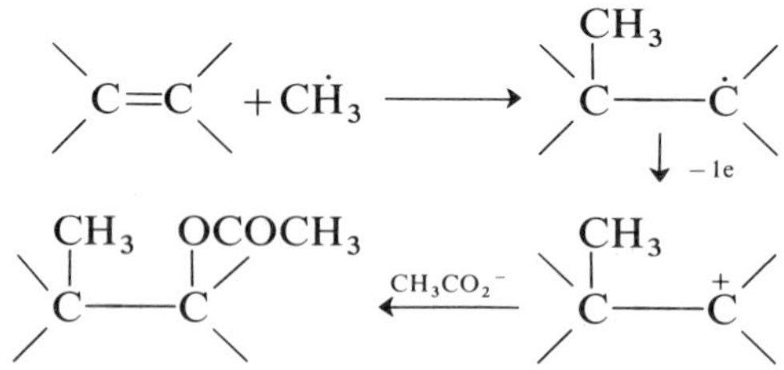

The radical produced by the initial addition of an electrochemically generated radical to buta-1,3-diene may either dimerize or combine with a second electrochemically generated radical [23].

$$R{\cdot}CO_2^- \xrightarrow[-CO_2]{-e} R^{\cdot}$$

$$R^{\cdot} + CH_2{=}CH{\cdot}CH{=}CH_2 \longrightarrow R{\cdot}CH_2{\cdot}CH{=}CH{\cdot}\dot{C}H_2$$

$$R{\cdot}CH_2{\cdot}CH{=}CH[CH_2]_2CH{=}CH{\cdot}CH_2{\cdot}R \qquad R{\cdot}CH_2{\cdot}CH{=}CH{\cdot}CH_2{\cdot}R$$

In many early experiments on the electrochemical oxidation of secondary or tertiary acids and α-substituted acids it was merely reported that the Kolbe coupled product was formed only in very low yield, or not at all, and the actual nature of the products was not established. In the majority of these cases, and in many more recent studies in which the products have been fully investigated, a carbonium ion is formed as an intermediate. This has been rationalized above in terms of a further one-electron oxidation of the radical formed by the one-electron oxidative decarboxylation of the carboxylate anion, although in fact it is not known for certain when or how the carbon dioxide is eliminated during the formation of the carbonium ion from the carboxylate anion. In cases where a carbonium-ion pathway is followed it is not necessary to use a platinum anode; a carbon one may be substituted [24].

The subsequent fate of such electrochemically generated carbonium ions follows the usual pattern. In methanol, the commonest solvent in recent work, solvolysis results in the formation of methyl ethers, which are usually accompanied by some olefin formed in a competing elimination reaction. This type of process has been used as a method for decarboxylating secondary and tertiary acids, where the carbonium-ion pathway is followed. The case of isostevic acid is illustrated [25].

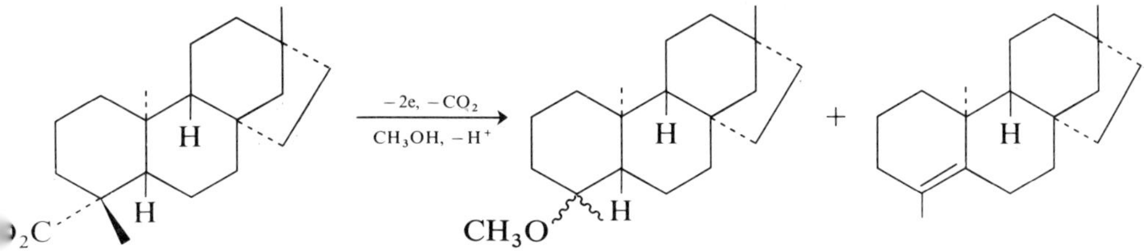

Treatment of the mixture with hydrogen bromide converts the ether into a mixture of olefins; hydrogenation then gives a single saturated product. In aqueous solution reaction of a carbonium ion with the solvent leads to an alcohol; in the case of secondary acids the further electrochemical oxidation of the product alcohol to a ketone has been observed. A carbonium ion can also react with any other nucleophile present in the system. One such species always present is the starting carboxylate anion. This accounts for the frequent occurrence of esters amongst the products.

$$R{\cdot}CO_2^- \xrightarrow[-CO_2]{-2e} R^+ \xrightarrow{R{\cdot}CO_2^-} R{\cdot}CO_2R$$

One useful synthetic method involving carbonium ions is the Ritter reaction, in which a nitrile acts as a nucleophile, and it has been shown that this reaction can be conducted with electrochemically generated carbonium ions. For example, the electrolysis of 2,2-dimethylpropionic acid in acetonitrile containing a little water leads through the t-butyl cation to *N*-acetyl-t-butylamine [26].

$$(CH_3)_3C{\cdot}CO_2^- \longrightarrow (CH_3)_3\overset{+}{C} \xrightarrow{CH_3CN} (CH_3)_3C{\cdot}N{=}\overset{+}{C}{\cdot}CH_3 \xrightarrow{H_2O} (CH_3)_3C{\cdot}NH{\cdot}CO{\cdot}CH_3$$

In this case the carbonium ion could easily be prepared in other ways, for example by the protonation of 2-methylpropene in strong acid, but the electrochemical method can also be used to generate less readily available carbonium ions, e.g. [26],

$$CH_3O_2C{\cdot}C(CH_3)_2{\cdot}C(CH_3)_2{\cdot}CO_2^-$$
$$\xrightarrow[CH_3CN,\ H_2O]{-2e,\ -CO_2} CH_3O_2C{\cdot}C(CH_3)_2{\cdot}C(CH_3)_2{\cdot}NH{\cdot}CO{\cdot}CH_3$$

Some other less accessible carbonium ions can also be prepared electrochemically. Somewhat surprisingly, for example, bridgehead carbonium ions can be obtained in this way [27].

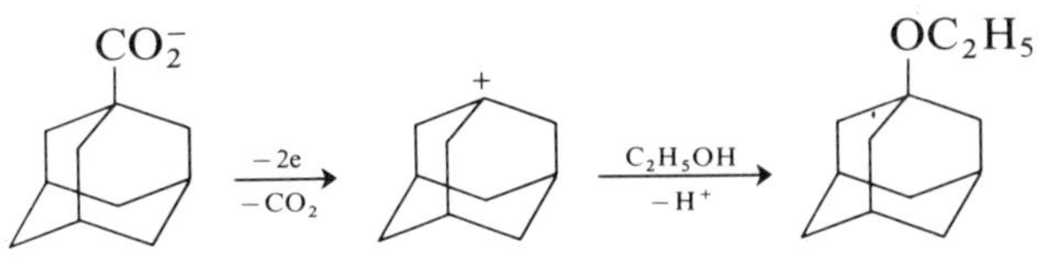

By the use of graphite anodes even primary carboxylate anions can be oxidatively decarboxylated to give carbonium ions as intermediates, and the radical Kolbe coupling is almost completely suppressed. For example, the electrolysis of aqueous sodium acetate at a carbon anode gives a very high yield of methyl acetate [24].

$$CH_3CO_2^- \xrightarrow[-CO_2]{-2e} CH_3^+ \xrightarrow{CH_3CO_2^-} CH_3CO_2CH_3$$

Similarly, the electrolysis of potassium dodecanoate at a carbon anode in methanol containing potassium chloride as a supporting electrolyte gives methyl undecanyl ether as the major product [28].

Typical carbonium-ion rearrangements are observed. For example, a major product from the electrochemical oxidation of cyclopropanecarboxylic acid is allyl cyclopropanecarboxylate, in which the allyl cation is formed by rearrangement of the cyclopropyl cation [29].

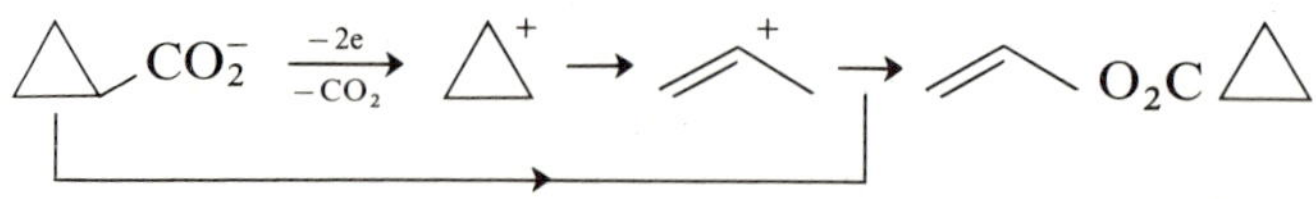

One of the problems in the field of carbonium-ion chemistry concerns the fact that although there is general agreement amongst the results of experiments in which, on a simple view, the same carbonium ion is being generated in various chemically different ways, nevertheless there can be substantial differences both in the ratios of the various products and in their stereochemistry. For example, carbonium ions produced by the solvolysis of halides or arenesulphonates tend to be of lower energy and hence to be more selective, leading to product mixtures containing fewer rearranged products than carbonium ions produced by the nitrous-acid deamination of amines. The establishment of the electrochemical route to carbonium ions provides a further method for producing carbonium ions, and the rather few results available so far suggest that such carbonium ions most resemble those formed in deamination reactions, though if anything they are of even higher energy. Two typical cases are considered below.

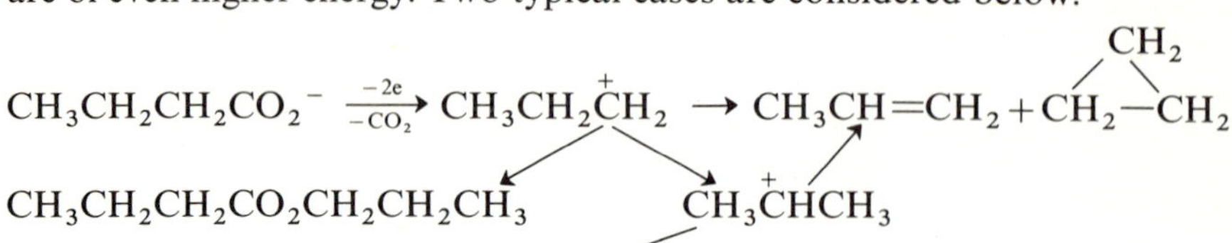

In this case the ratio of cyclopropane to propene is greater than in the deamination of 1-aminopropane [24].

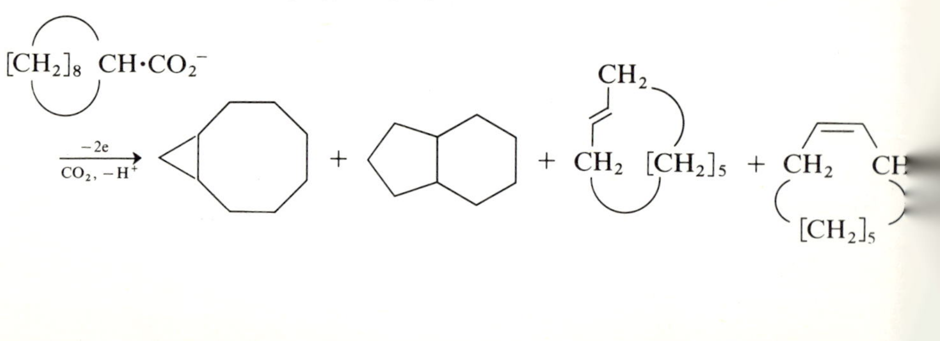

In the second case only the hydrocarbon products are shown. The results resemble those in the analogous deamination reaction, features of note being the 1,3- and 1,5- transannular eliminations, and the formation of substantial amounts of the *trans*-cyclic olefin [30].

Another major problem in the field of carbonium-ion chemistry concerns the structure of such cations as the norbornyl cation, for which both classical and non-classical formulations have been proposed. Cations of this type can be generated electrochemically, and analysis of the products derived from them provides evidence on the detailed nature of the carbonium-ion processes involved. For example, electrochemical oxidative decarboxylation of either *exo*- or *endo*-norbornane-2-carboxylic acid in methanol leads exclusively to *exo*-2-methoxynorbornane [3]; the same stereochemical result is observed in arenesulphonate solvolyses.

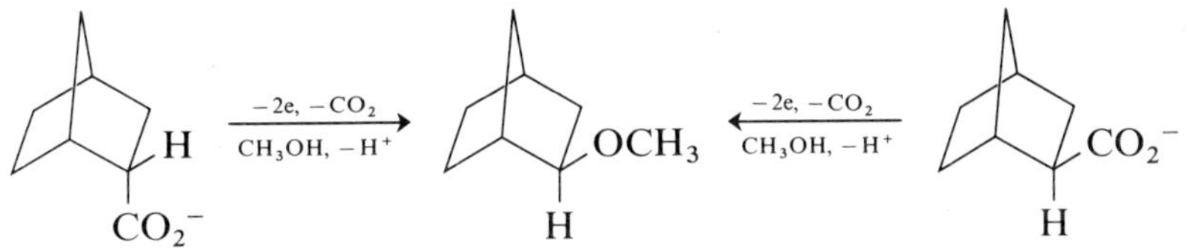

A second example is provided by work on the electrolysis of the bicyclo-[3,1,0]hexane-3-carboxylic acids in water [31].

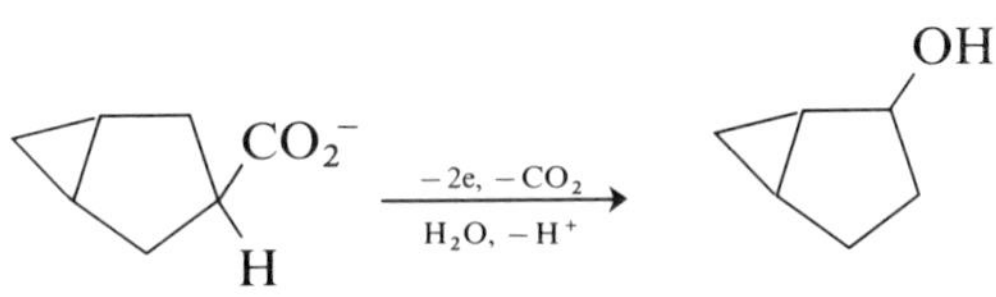

Only the major product is shown. In these experiments the product distributions differ from that observed, on the one hand, in experiments in which the classical cation is regarded as the intermediate and, on the other hand, in experiments in which the non-classical trishomocyclopropenyl cation has been implicated. It is clear that the study of the electrochemical initiation of carbonium-ion processes will increase understanding in this area.

Although phenylacetic acid gives the Kolbe coupled product on electrolysis, the presence of a *p*-methoxy group alters the course of the oxidation, and *p*-methoxybenzyl methyl ether is formed via the carbonium ion [32].

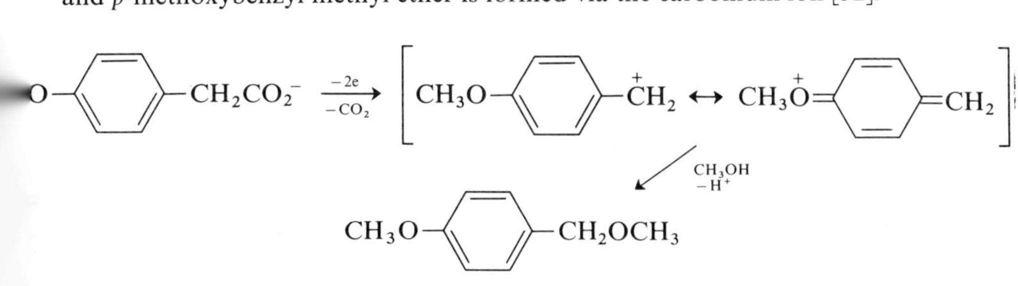

An α-methoxy-substituent also stabilizes the carbonium ion, so that electrolysis of α-methoxyphenylacetic acid in methanol leads to benzaldehyde dimethyl acetal. This process can be adapted to give mixed acetals, which are difficult to prepare except in this way [33].

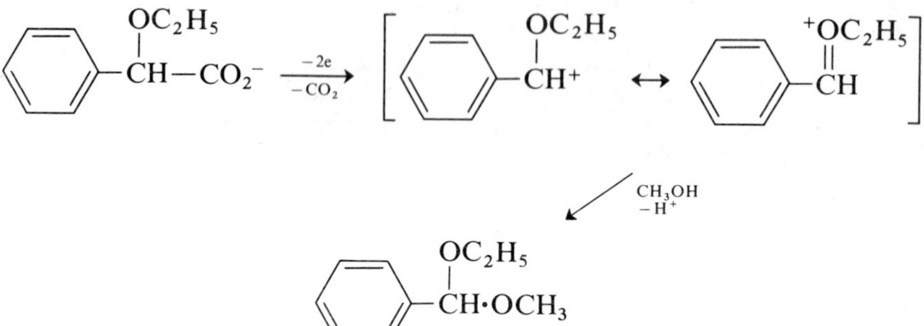

α-Hydroxyacids give carbonyl compounds on electrolysis, since a two-electron oxidative decarboxylation leads directly to the protonated form of the product. This process has been used to degrade D-glucuronic acid to D-arabinose [34].

$$HO{\cdot}CH_2[CHOH]_3CH(OH)CO_2^- \xrightarrow{-2e,\ -CO_2,\ -H^+} HO{\cdot}CH_2[CHOH]_3CHO$$

The α-amino-acid quinuclidine-2-carboxylic acid, on electrolysis in methanol, gives an ether via a carbonium ion [35].

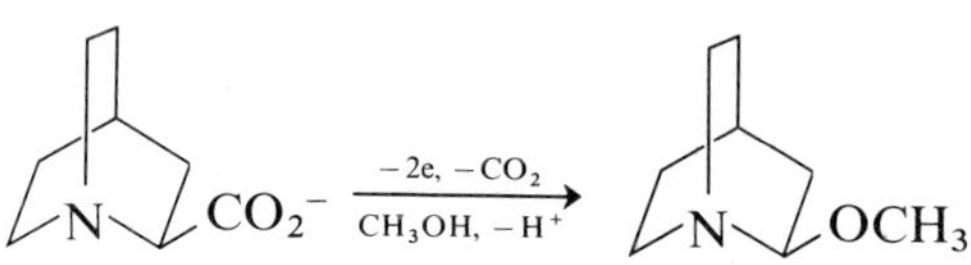

Other α-amino acids do not undergo this reaction, but α-*N*-acylamino-acids give α-*N*-acylamino-ethers, a class of compounds hard to prepare by alternative methods [36].

$$R{\cdot}CO{\cdot}NH{\cdot}CHR'{\cdot}CO_2^- \xrightarrow[CH_3OH,\ -H^+]{-2e,\ -CO_2} R{\cdot}CO{\cdot}NH{\cdot}CHR'{\cdot}OCH_3$$

In aqueous solution the corresponding α-acylamino-alcohol is formed, but this breaks down to give an aldehyde, R′CHO [37].

The electrochemical oxidative bisdecarboxylation of 1,2-dicarboxylic acids also involves the formation of cationic intermediates. 1,2-dihydrophthalic acid gives benzene in 70% yield, but this is a favourable case in which an aromatic system is formed [38]; cyclobutene is only formed in 5% yield from *cis*-cyclobutane-1,2-dicarboxylic acid [39]. The electrolysis of either *meso*- or (±)-1,2- diphenylsuccinic acid gives *trans*-stilbene, and this stereochemical result is in agreement with a non-concerted bisdecarboxylation involving a carbonium-ion intermediate [2(c)].

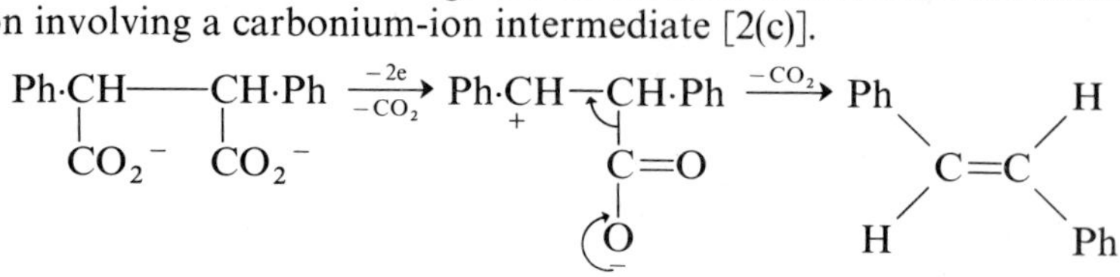

The formation of lactones from certain dicarboxylic acids can also be rationalized in terms of carbonium-ion intermediates [40].

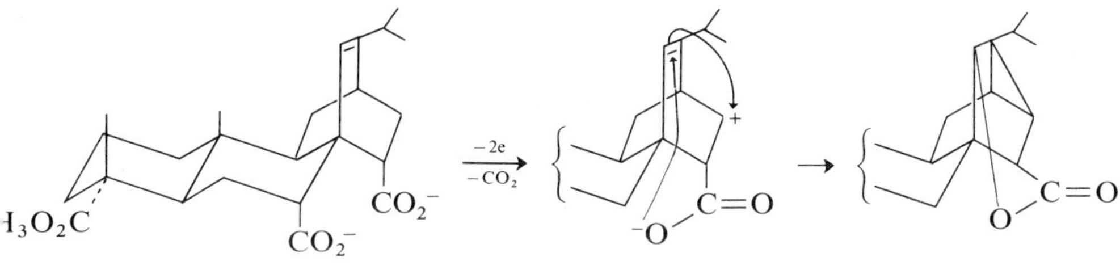

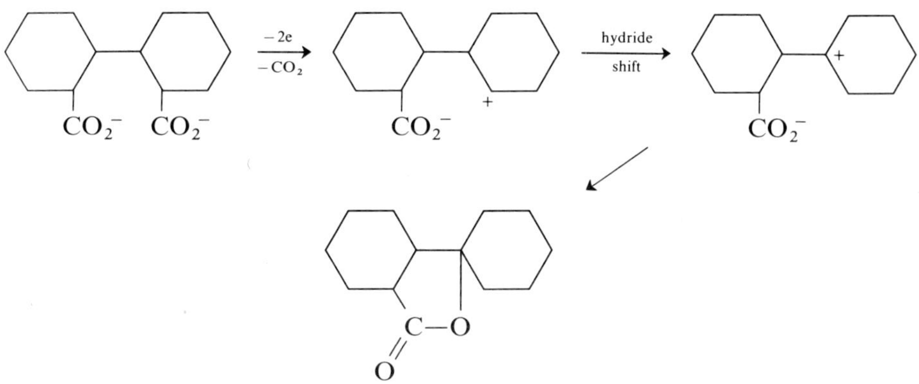

The only case in which the two α-carbon atoms of a dicarboxylic acid become directly linked with the formation of a cyclic system as a result of electrochemical oxidative bisdecarboxylation is that of *trans*, *trans*, *trans*-1,3-dicarboxy-2,4-dicarbomethoxycyclobutane, and this too involves a carbonium-ion intermediate [41].

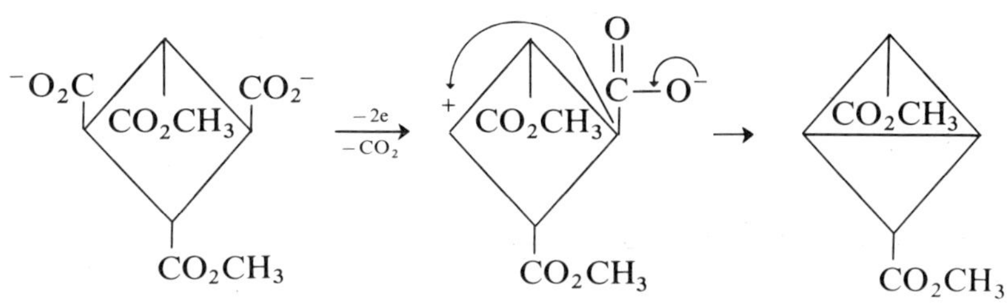

It should be emphasized that no cyclopropane is formed when the related 1,3-dicarboxylic acid, glutaric acid, is electrolysed [42].

Other anions apart from carboxylate anions can be oxidized electrochemically. For example, a phenolate ion will undergo a reversible one-electron loss to give a phenolate radical, and at a platinum anode C →C coupling can occur [43].

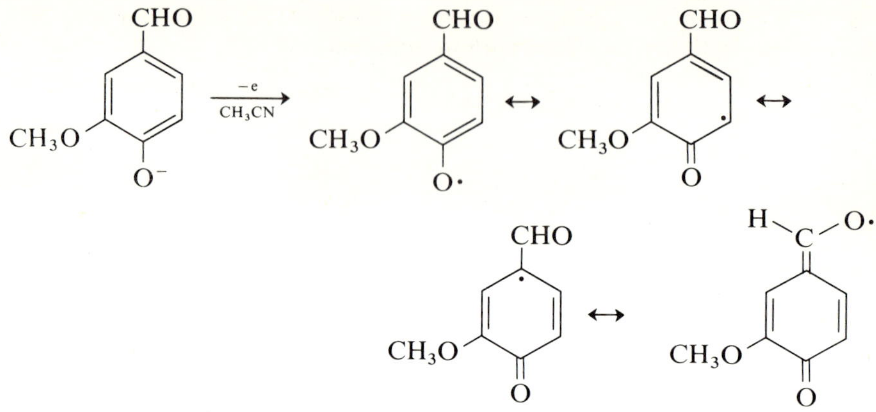

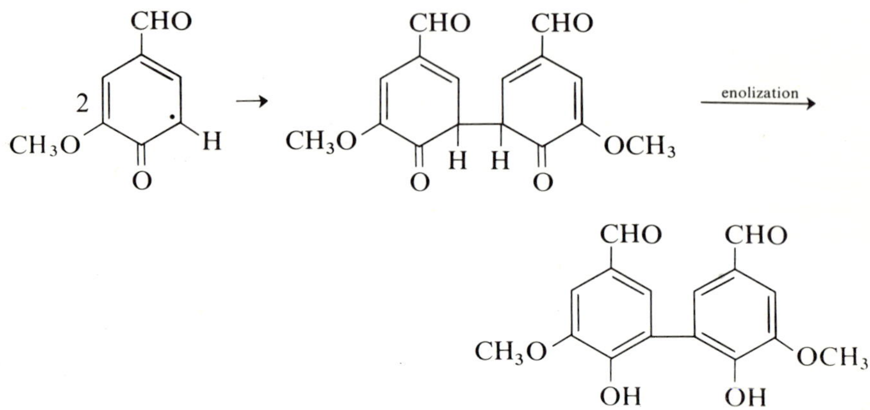

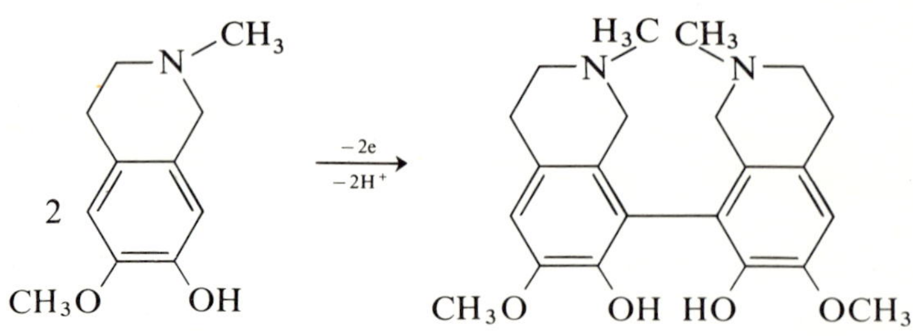

Other resonance-stabilized carbanions will undergo a similar electrochemical oxidative coupling, e.g. [44],

$$[C_2H_5O_2C]_3C^- \xrightarrow{-e} [C_2H_5O_2C]_3\dot{C} \rightarrow [C_2H_5O_2C]_3C{\cdot}C[CO_2C_2H_5]_3$$

$$C_2H_5{\cdot}\overset{-}{\underset{\substack{|\\ NO_2}}{C}}{\cdot}CH_3 \xrightarrow{-e} C_2H_5{\cdot}\underset{\substack{|\\ NO_2}}{\dot{C}}{\cdot}CH_3 \rightarrow C_2H_5{\cdot}\overset{\substack{CH_3\\ |}}{\underset{\substack{|\\ NO_2}}{C}}\text{——}\overset{\substack{CH_3\\ |}}{\underset{\substack{|\\ NO_2}}{C}}{\cdot}C_2H_5$$

In the second example [45], if the oxidation is carried out in the presence of a large excess of nitrite ions, the major product is 2,2-dinitrobutane, probably formed by the crossed coupling of the nitroalkyl radical with an electrochemically generated nitro radical.

Alcohols are unsuitable solvents for the preparation of the oxidative dimers of many carbanions because oxidation to the carbonyl compounds occurs, and these condense with the carbanions [46].

$$[CH_3O_2C]_2\bar{C}H \xrightarrow{-e} [CH_3O_2C]_2\dot{C}H \longrightarrow [CH_3O_2C]_2CH{\cdot}CH[CO_2CH_3]_2$$

$$[CH_3O_2C]_2\dot{C}H + CH_3OH \longrightarrow [CH_3O_2C]_2CH_2 + \dot{C}H_2OH$$

$$2\dot{C}H_2OH \longrightarrow CH_3OH + CH_2{=}O$$

$$CH_2{=}O + 2[CH_3O_2C]_2\bar{C}H \xrightarrow[-H_2O]{+2H^+} [CH_3O_2C]_2CH{\cdot}CH_2{\cdot}CH[CO_2CH_3]_2$$

The radicals formed by the electrolysis of Grignard reagents, which contain carbanionoid carbon, also attack the solvents used, e.g., ethers, to give complex mixtures of products, although the use of a lead anode for the electrolysis of an alkylmagnesium halide leads to the formation in high yield of the tetra-alkyl lead by reaction with the anode [47].

Many neutral molecules can also be oxidized electrochemically. Four reactions in which aromatic systems have been shown to undergo two-electron oxidations are illustrated here [48].

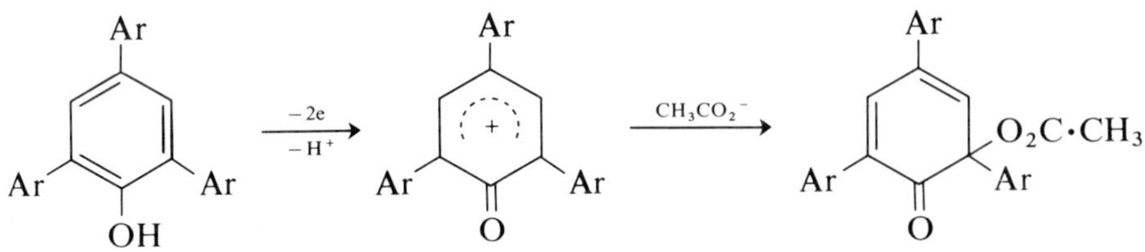

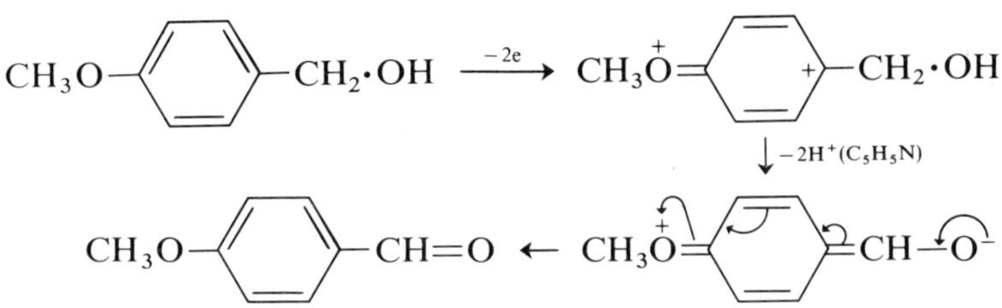

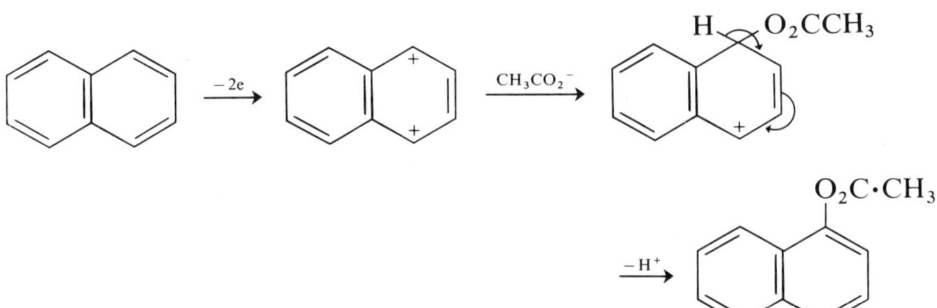

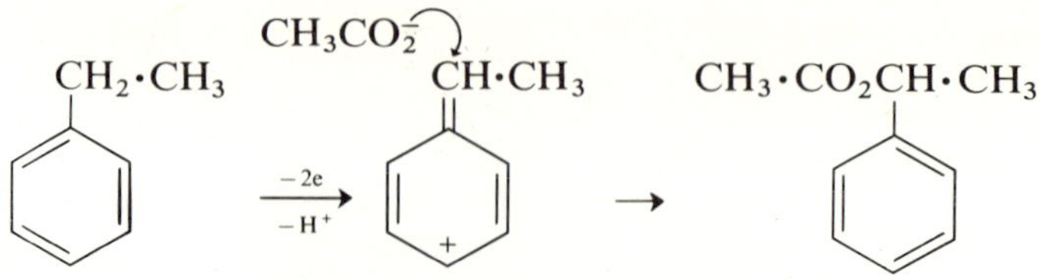

The timing of the various steps may not be accurately represented by these rationalizations, for example in respect of the loss of benzylic hydrogen in the formation of a monocation. Many other electrochemical reactions of aromatic systems can be rationalized in similar terms. Such electrochemically generated aromatic cations behave normally, and their solvolyses, or their reactions with other nucleophiles, e.g., cyanide ion [49] or alkyl cyanides in the Ritter reaction [50], have been observed.

The acetoxylation reactions illustrated above can be conducted at a potential below that necessary for the discharge of the acetate ion, showing that it is indeed the aromatic system which is being oxidized. These results suggest that certain reactions of aromatic carboxylic acids should be rationalized as the intramolecular analogues of reactions illustrated above, e.g. [51],

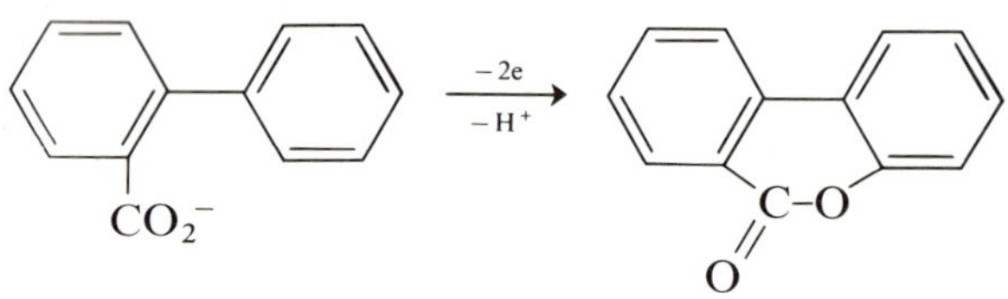

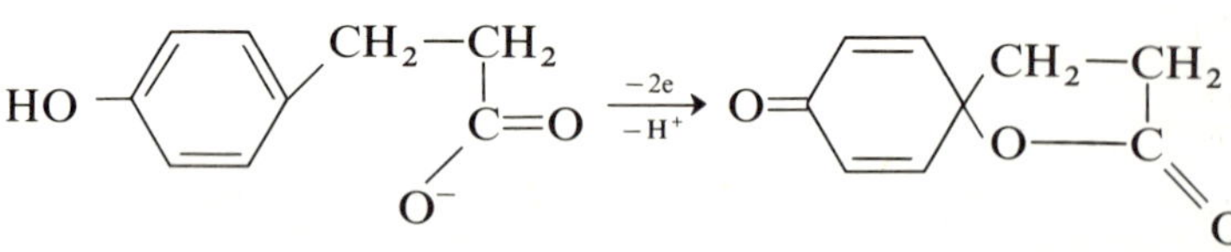

Certain carbon–oxygen phenyl migrations observed during the electrolysis of aromatic carboxylic acids can be similarly rationalized, e.g. [52],

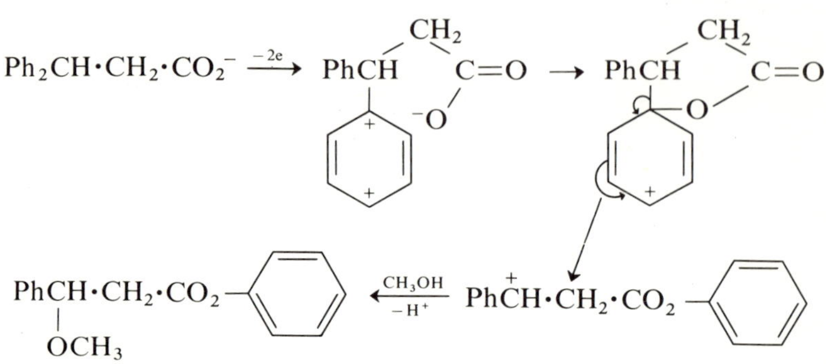

Other π-electron systems are similarly oxidized [53].

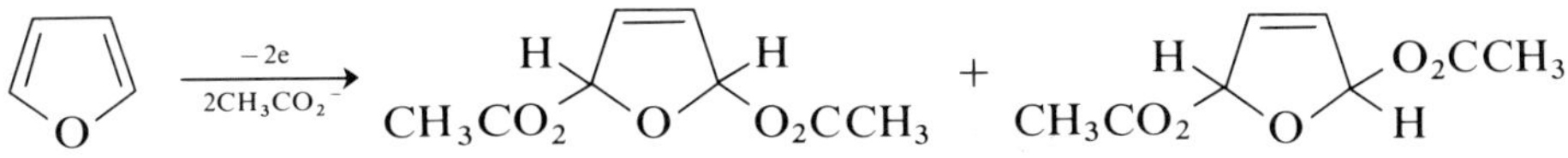

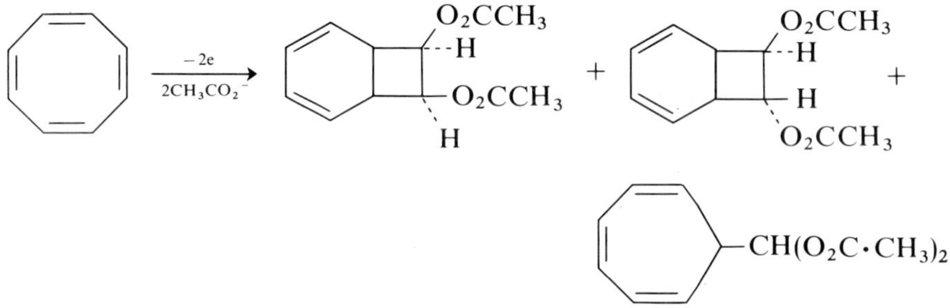

The addition of two acetate groups to stilbene is stereospecific and involves the electrochemical production of an acetoxonium ion [54].

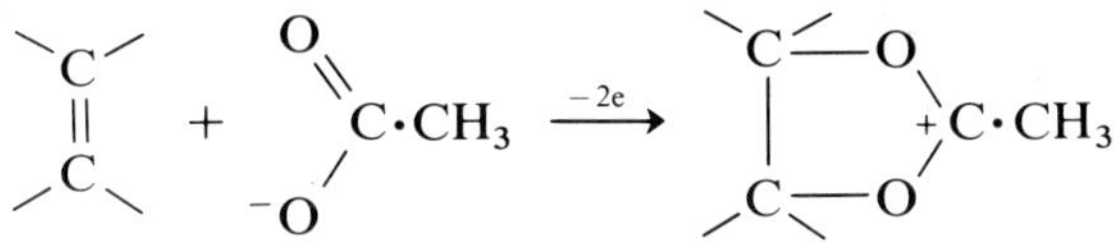

The observed stereospecificity is the same as for other solvolyses of the same acetoxonium ion, namely that, for example, *trans*-stilbene under anhydrous conditions gives the *meso*-diacetate, and under moist conditions the main product is the *threo*-glycol monoacetate. The yields with simple olefins are very low.

N, *N*-Dialkylamides also undergo anodic acetoxylation [55].

$$\mathrm{HCO{\cdot}N(CH_3)_2} \xrightarrow[-\mathrm{H}^+]{-2\mathrm{e}} \mathrm{HCO{\cdot}N(CH_3)\overset{+}{C}H_2}$$

$$\xrightarrow{\mathrm{CH_3CO_2^-}} \mathrm{HCO{\cdot}N(CH_3){\cdot}CH_2O_2CCH_3}$$

In many systems of the type discussed above, methoxylation has been observed when the oxidation is conducted in methanolic sodium methoxide. There are, however, cases where the results in methanol differ from those in acetic acid, e.g. [53(a)],

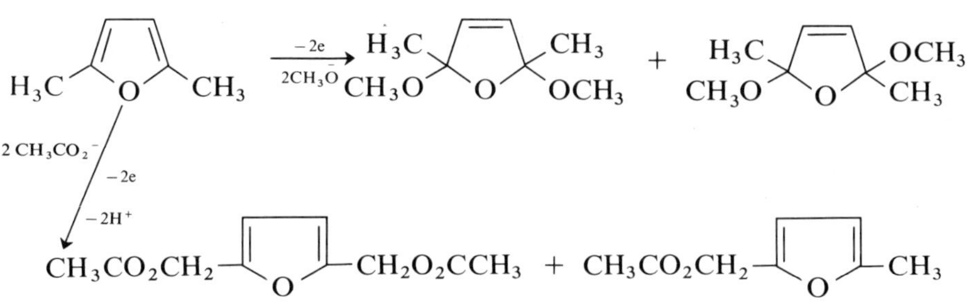

In this case it is possible that in methanol the methoxide ion, and not the aromatic system, is being oxidized, and that the products arise by an initial attack of a methoxy radical on the 1,4-diene system. The formation of a "dimeric" product from styrene by electrolysis in methanolic sodium methoxide [56] strongly suggests a radical intermediate, and the formation of the products can be rationalized as follows.

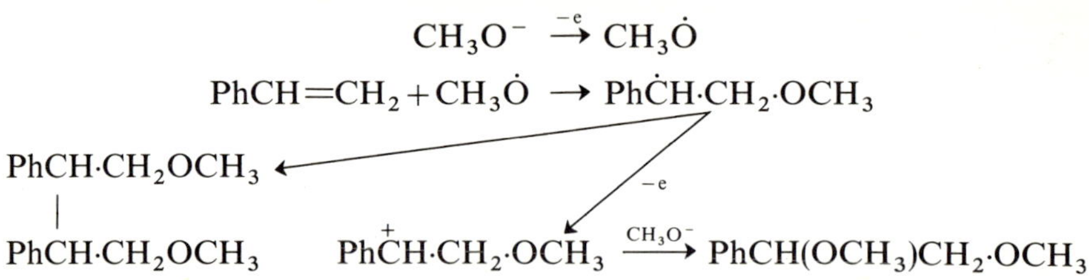

However, in many cases it seems probable that the methoxylations, like the acetoxylations, involve the initial two-electron oxidation of the substrate.

The 2,5-dimethoxy-2,5-dihydrofurans are useful products, since on acid hydrolysis they give 1,4-dicarbonyl compounds; a variety of furans has been electrochemically transformed into these derivatives, including some with electronegative substituents which cannot be methoxylated by the alternative chemical method [57]. Electrochemical methoxylation also provides a route to the otherwise rather inaccessible quinone ketals [58].

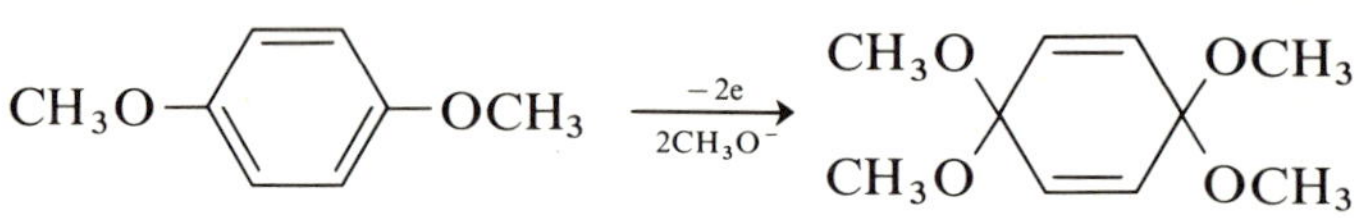

An interesting electrochemical oxidative bond cleavage has also been reported [58].

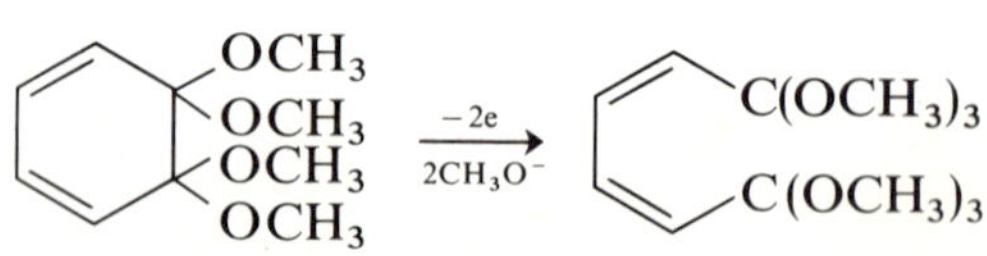

Finally, some other electrochemical oxidations should be mentioned. Primary and secondary alkanols can be oxidized in aqueous solution, e.g. [59],

$$(CH_3)_2CH{\cdot}CH_2{\cdot}OH \xrightarrow[PbO_2\ anode]{2N/H_2SO_4} (CH_3)_2CH{\cdot}CO_2H$$

The complete replacement of all the hydrogen atoms in an organic compound by fluorine atoms can generally be brought about by electrolysing a solution of the compound in anhydrous hydrogen fluoride at 0°. A steel cell is used, with nickel electrodes. The mechanism of the process is unknown, but at the potential used molecular fluorine is not evolved. Cleavage of carbon–carbon bonds is relatively common, and although the yields are often quite low they are frequently very useful preparatively, in view of the lack of alternative procedures [60].

ELECTROCHEMICAL REDUCTIONS [61]

A rather wider variety of compounds has been reduced electrochemically than has been electrochemically oxidized. The majority of such reductions can also be achieved, in much the same yields, by non-electrochemical methods. The critical factor in an electrochemical reduction is the cathode potential; the use of a potentiostat to control the potential is necessary much more often than in oxidations. The hydrogen overvoltage of the cathode is a limiting factor when the potential is not controlled; reducible organic substrates act as depolarizers, so that hydrogen is not preferentially discharged. A more negative potential, and hence greater reducing power, can be obtained with a cathode of high overvoltage, such as lead or mercury, than with a cathode of low voltage, such as platinum. Many electrochemical reductions involve a reversible transfer of electrons from the cathode to the substrate, so that a diaphragm to divide the cell into anode and cathode compartments is necessary. Electrochemical reductions are commonly conducted in aqueous solution, though it is often necessary to add co-solvents to increase the solubility of the substrates. The pH of the electrolyte often has a large effect on the course of an electrochemical reduction.

The nitro group is typical of many functional groups which can be reduced electrochemically. At cathodes of low overvoltage or at controlled weakly negative potentials the nitro group is reduced to the hydroxylamine, but at more negative potentials, or with cathodes of higher overvoltage, further reduction to the amine occurs, e.g. [62],

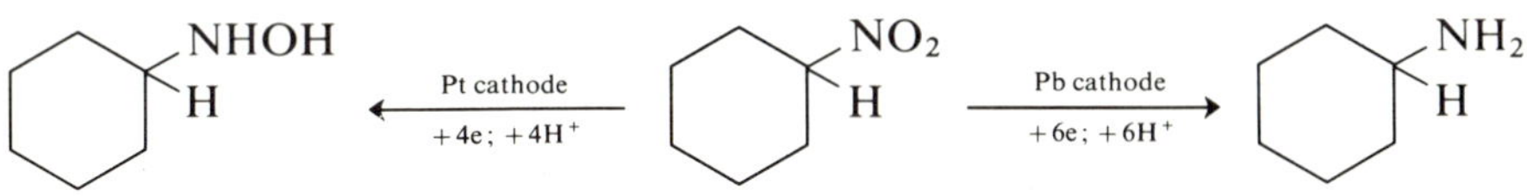

Such reductions are performed in acid solution. In the aromatic series, by using controlled potentials at a mercury cathode, it is possible to isolate either the hydroxylamine or the amine when electron-accepting substituents are present, but with electron-withdrawing substituents it is not possible to isolate the hydroxylamine, only the amine [63]. The *N*-arylhydroxylamines may undergo further reactions. For instance in hot, strongly acid solution, rearrangement to *p*-aminophenols occurs. This forms the basis of a useful quinone synthesis [64].

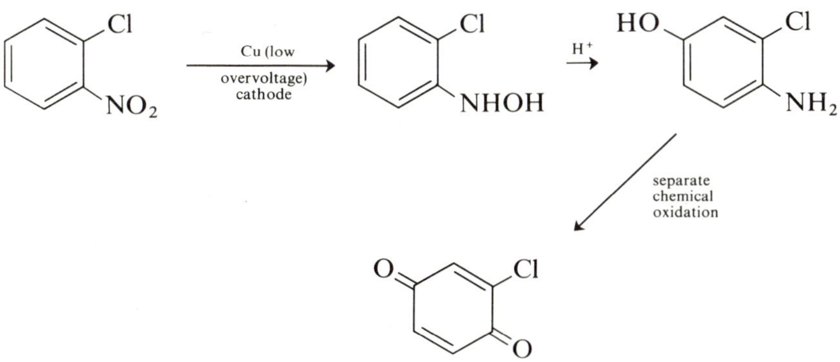

In alkaline solution the hydroxylamine gives rise to the azoxy compound, which frequently precipitates, and thus avoids further reduction. In conditions where the azoxy compound is kept in solution, further reduction to the azo or hydrazo compound can occur, and conditions for the preparation of all these types of compounds by the electrochemical reduction of aromatic nitro compounds have been devised [61].

The reduction of the nitramine, nitrourea, at a mercury cathode in aqueous sulphuric acid forms the basis for a manufacturing route to semicarbazide [65].

$$NH_2CONHNO_2 \xrightarrow{H^+} NH_2CONH\overset{+}{N}H_3$$

Nitroso compounds are also reduced in acid solution, and this provides a method of converting *N*-nitrosamines into hydrazines [66].

$$(CH_3)_2N{\cdot}NO \xrightarrow[\text{Pb cathode}]{H^+} (CH_3)_2N{\cdot}\overset{+}{N}H_3$$

Considerable attention has been given to the electrochemical reduction of carbon–oxygen double bonds. With ketones, at cathodes of high overvoltage, reduction may lead either to the corresponding secondary alcohol, or to the hydrodimer, the pinacol, e.g.,

$$Ph_2C(OH)C(OH)Ph_2 \leftarrow Ph_2C{=}O \rightarrow Ph_2CHOH$$

The product depends very much on the pH. In general, reduction in strongly acid solution leads to the pinacol, whereas reduction in neutral or slightly alkaline solution gives the alcohol. Pinacols are also formed in strongly alkaline solution. These results can be rationalized in the following way [67]. The pinacol formed in acid solution arises by the coupling of a radical formed by the transfer of an electron and a proton to a molecule of the ketone at the cathode.

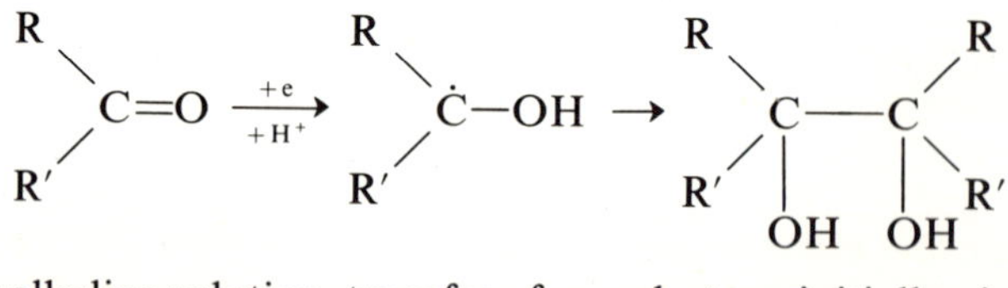

In strongly alkaline solution, transfer of one electron initially gives a radical anion, but this requires a more negative potential than is necessary when the transition state for the electron transfer also involves a proton, as in acid solution. Coupling of the anion radical gives the pinacol dianion, which is finally protonated.

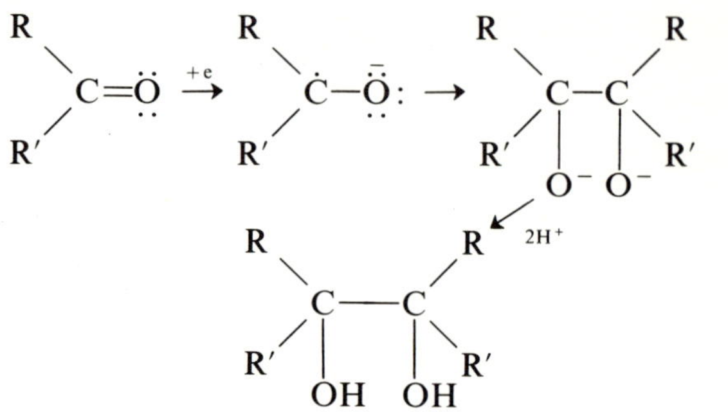

In neutral solution the radical anion is also produced, but becomes protonated whilst still absorbed by (or possibly bonded to) the cathode. At the potential required to effect the reduction to the radical anion, the radical formed on its protonation is further reduced to an anion, which is subsequently protonated to give the secondary alcohol.

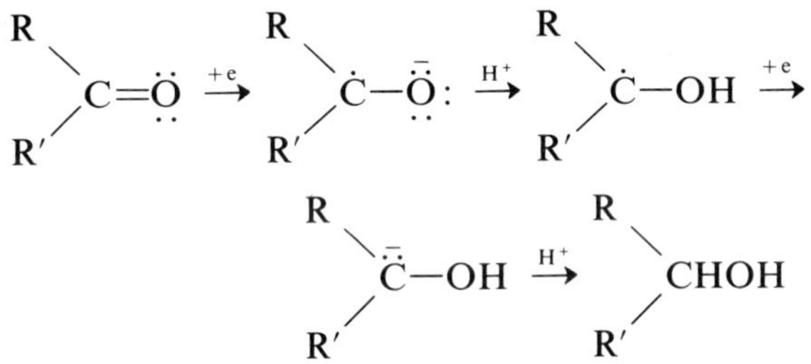

In the case of unsymmetrical ketones and aldehydes, mixtures of the *meso*- and ($\pm$)-forms of the pinacols are usually produced, although hydroxy-substituted benzaldehydes in alkaline solution give only the *meso*-form, a result attributable to the electrostatic repulsion between the phenoxide ions present in the dimerizing radicals [68].

Reduction of a mixture of acetone and acrylonitrile in aqueous acid, under conditions where acrylonitrile is not reduced, leads to a crossed product which must arise by the addition of an anion formed from the acetone to the acrylonitrile in a Michael type of reaction [69].

$$(CH_3)_2C{=}O \xrightarrow[+H^+]{+e} (CH_3)_2\dot{C}{-}OH \xrightarrow{+e} (CH_3)_2\bar{\ddot{C}}{-}OH \xrightarrow[H^+]{CH_2{=}CH{\cdot}CN} (CH_3)_2C(OH){\cdot}CH_2{\cdot}CH_2{\cdot}CN$$

Reduction of a keto group to a secondary alcohol has been observed during an attempted Kolbe electrolysis with *o*-benzoylbenzoic acid, when a high yield of 3-phenylphthalide was produced by lactonization of the initial product [70].

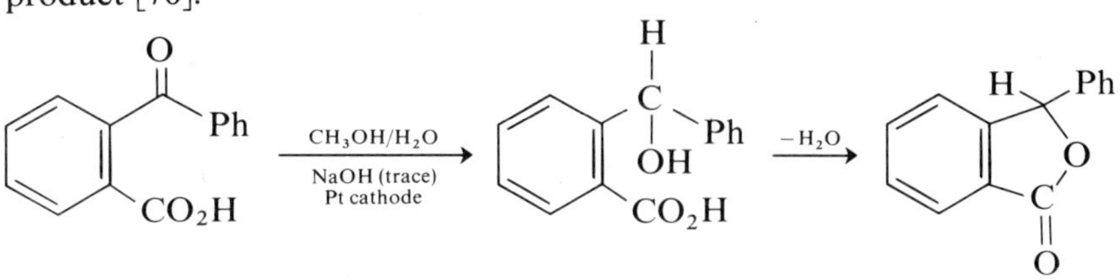

Sorbitol can be prepared by the electrochemical reduction of the masked aldehyde group in glucose, using a nickel–aluminium cathode and very concentrated aqueous solutions [71].

In general acids and acid derivatives are reduced to the primary alcohol, e.g. [72],

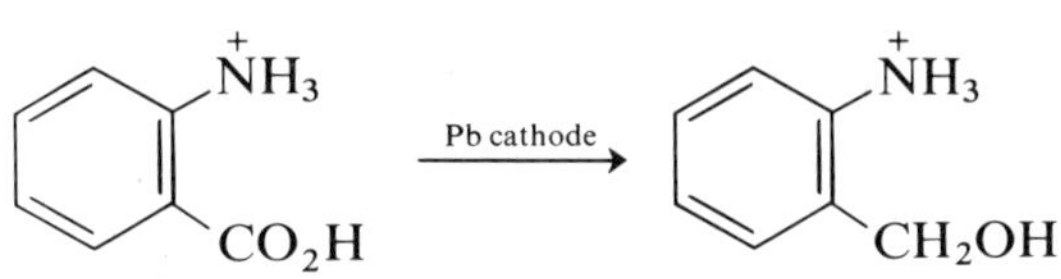

However, for certain acids and acid derivatives, and particularly for certain unsubstituted acid amides, the reduction can be stopped at the aldehyde stage, although in general aldehydes are more readily reduced electrochemically than acids. In these cases the reduction can be stopped at the aldehyde stage because the aldehyde forms a relatively stable hydrate, though prolonged electrolysis will result in reduction to the alcohol or to the pinacol, depending on the pH.

A typical example is provided by isonicotinamide [73].

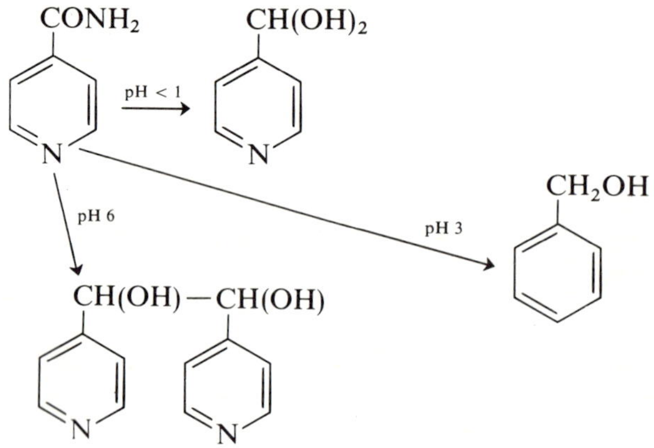

At a more alkaline pH, reduction of the pyridine ring occurs in preference to reduction of the amide group, with the formation of 1,4-dihydroisonicotinamide.

This type of behaviour is shown by acids and their derivatives in which the carbonyl group is influenced by a strong electron-withdrawing group, but in which the site of reduction is still the carbonyl group. It is possible, for example, to find conditions for the electrochemical reduction of oxalic acid to glyoxylic acid [74], although under more vigorous conditions reduction to glycollic acid [75] is observed.

$$\mathrm{HO_2C{\cdot}CH_2OH} \xleftarrow[\text{Pb cathode}]{40°,\ \mathrm{H_2SO_4}} \mathrm{HO_2C{\cdot}CO_2H} \xrightarrow[\text{Pb cathode}]{0°,\ \mathrm{H_2SO_4}} \mathrm{HO_2C{\cdot}CH(OH)_2}$$

Most of the known examples are provided by derivatives of π-electron-deficient nitrogen heterocyclic compounds, e.g. [76],

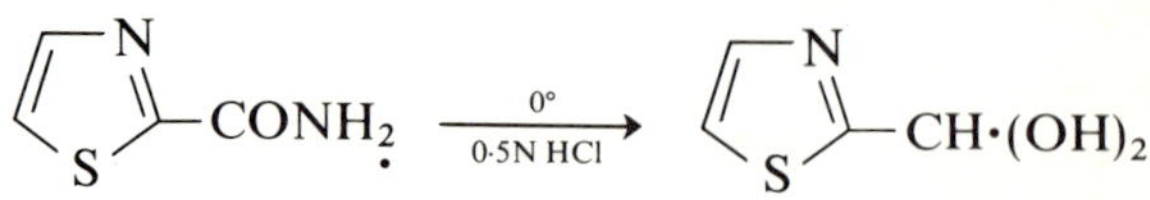

Under certain conditions an acid group may be reduced to a methyl group, e.g. [77],

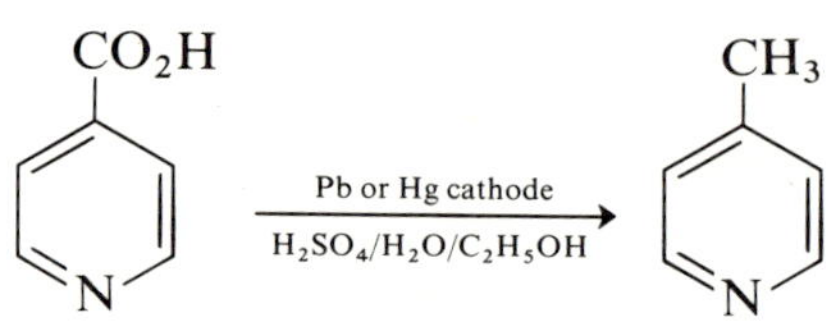

Such a reduction no doubt proceeds via the alcohol and involves the reductive cleavage of the carbon–oxygen bond.

The reduction of substituted amides can also lead to deoxygenated products [78].

Compounds containing carbon–nitrogen multiple bonds can be reduced electrochemically, for example enamines prepared *in situ* [79].

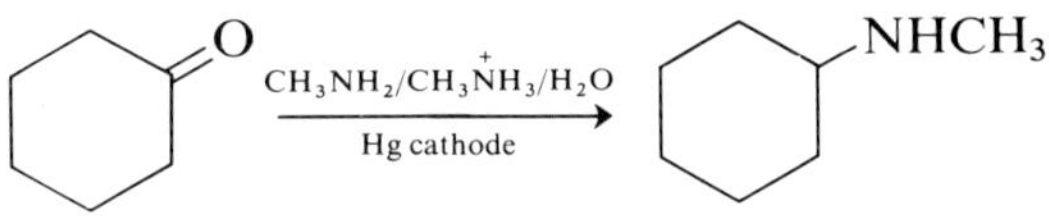

In general, compounds of the type RR′C=N[Y]R″ are reduced to the saturated compound when Y is carbon, but when Y is nitrogen or oxygen the most common type of product is the one formed by saturation of the double bond and reductive cleavage of the nitrogen–nitrogen or nitrogen–oxygen bond as well [79].

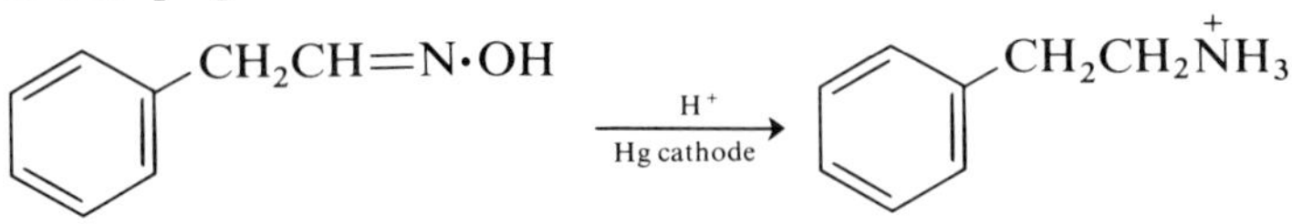

The electrochemical reduction of nitric acid to hydroxylamine can be conducted simultaneously with the reduction of oxalic acid to glyoxylic acid in such a way that the further reduction of the glyoxylic acid oxime so formed occurs, and glycine is obtained [80].

The nitrile group can be reduced electrochemically, e.g. [81],

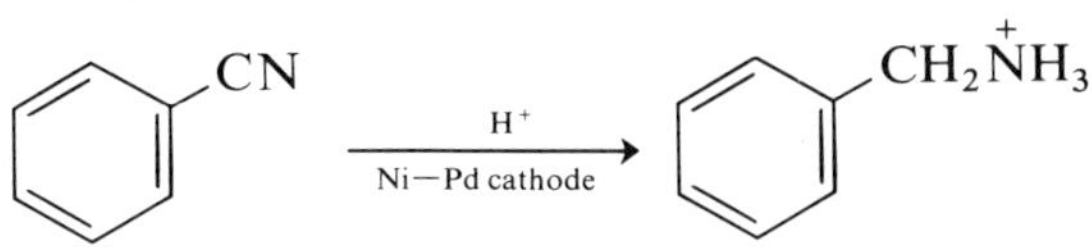

$$CH_2{=}CH{\cdot}CN \xrightarrow[\text{Zn, C, Pb, or Pt cathode}]{H^+,\ 45\text{–}60^\circ} CH_2{=}CH{\cdot}CH_2NH_2$$

Several aromatic systems can be reduced at cathodes of high overvoltage. An isolated benzene ring is not reduced, but polycyclic aromatic hydrocarbons and their derivatives are reduced, e.g. [82],

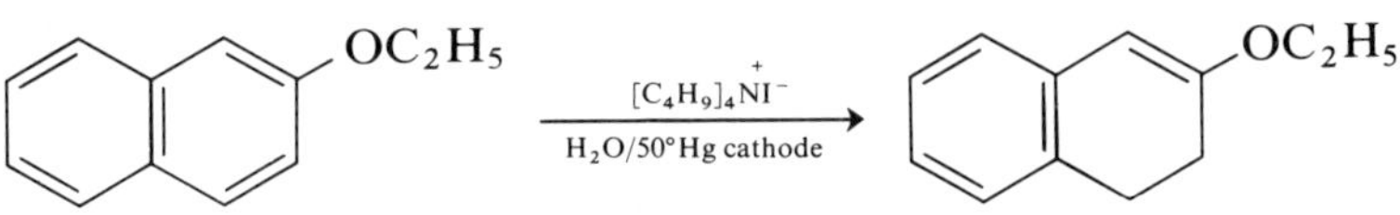

The above example illustrates the best available route to tetralone enol ethers, and hence tetralones. Another example is [83]:

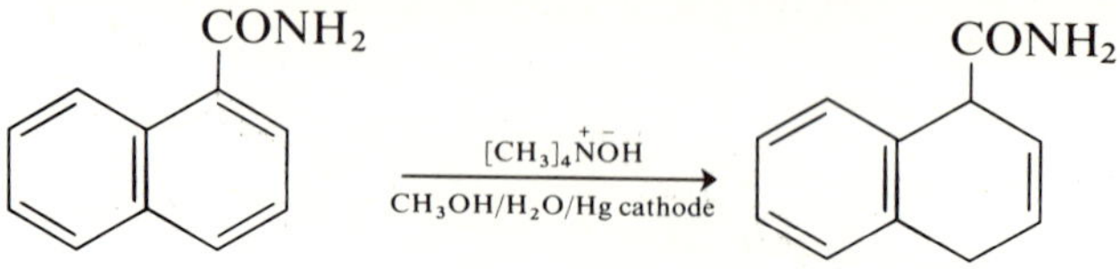

The formation of radical anions in the reduction of polycyclic aromatic systems is demonstrated by, for example, the isolation of a hydrodimer of phenanthrene [84].

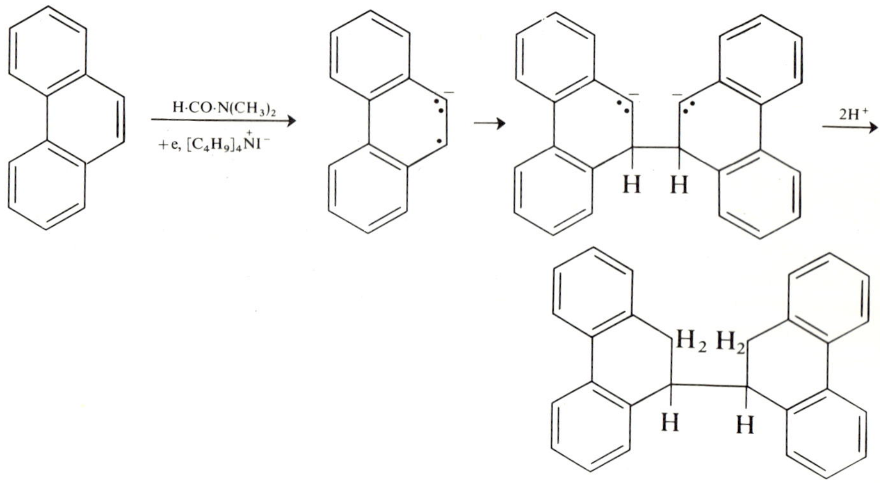

Intermediate anions can be trapped by conducting the reductions in the presence of carbon dioxide [84], e.g.,

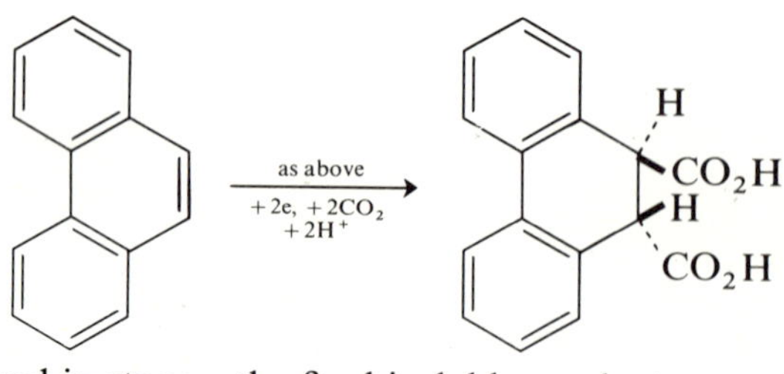

Pyridine is reduced in stages; the final isolable products are piperidine and 1,2,3,6-tetrahydropyridine [85].

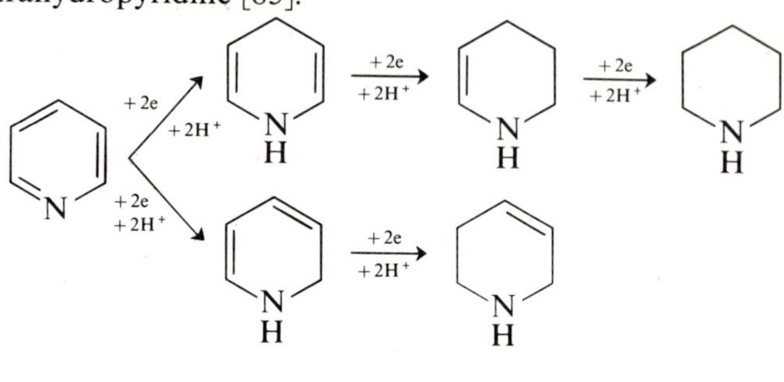

Pyridinium salts can also be reduced, e.g. [86],

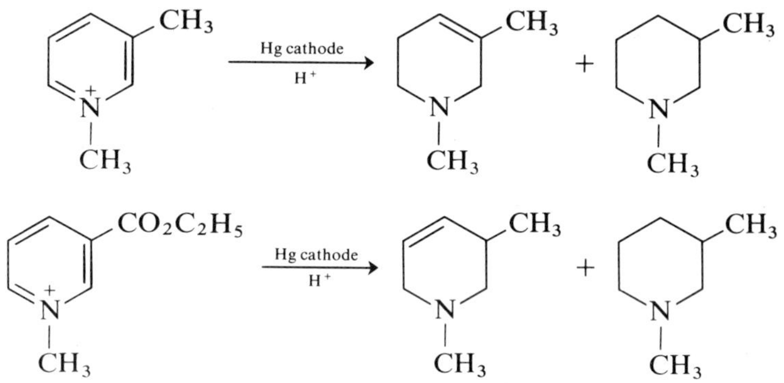

In the second example the simultaneous reduction of an ester is observed. The two reactions lead to isomeric partial reduction products.

The reduction of the porphin system is of interest as the products, which are oxidized back to porphins in air, are a class of dihydroporphins, the phlorins, in which the cyclic 18-π-electron aromatic system of the porphin nucleus has been broken [87].

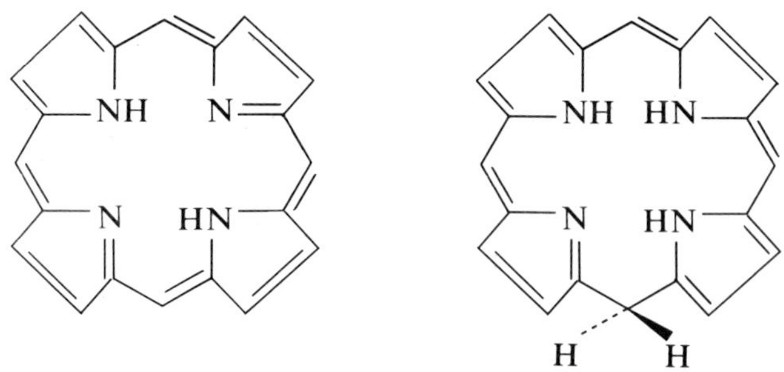

The porphin nucleus The phlorin nucleus

Aromatic systems can also be reduced electrochemically by other processes. With a platinum cathode, in acid solution, hydrogen ions are discharged, with the production of hydrogen atoms on the catalytic surface, and the results closely resemble those observed on catalytic hydrogenation with molecular hydrogen [88]. An electrochemical version of the Birch dissolving-metal reduction can be achieved by the electrolysis of alkali metal halides in solvents such as methylamine or hexamethylphosphoramide containing the aromatic substrate [89].

Unconjugated carbon–carbon double and triple bonds are not reducible at high-overvoltage cathodes, but a number of terminal acetylene derivatives undergo semi-hydrogenation at copper–silver cathode in dilute alkaline solution [90]. The process may involve organometallic compound-formation at the cathode.

$$(CH_3)_2C(OH){\cdot}C{\equiv}CH \rightarrow (CH_3)_2C(OH){\cdot}CH{=}CH_2$$

The reduction of diphenylacetylene at a mercury cathode does not stop at the olefin stage, as conjugated olefins are more easily reduced than conjugated acetylenes [84].

$$PhC{\equiv}CPh \rightarrow PhCH_2CH_2Ph + \begin{matrix} PhCH{\cdot}CH_2Ph \\ | \\ PhCH{\cdot}CH_2Ph \end{matrix}$$

When a double bond is conjugated with an electron-withdrawing group, several reactions are possible, namely reduction of the double bond to give the saturated monomer, reductive dimerization (EHD or electrochemical hydrodimerization), or polymerization initiated by species formed by the reduction of the monomer.

The simple reduction of several $\alpha\beta$-unsaturated acids has been reported, in both acid and alkaline solution [91].

$$PhCH{=}CH{\cdot}CO_2^- \xrightarrow[\text{Pb or Hg cathode}]{H_2O/NaOH} PhCH_2CH_2CO_2^-$$

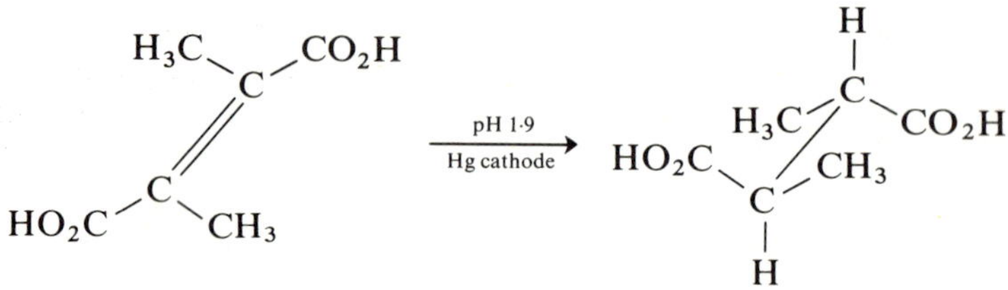

The reduction of dimethylfumaric acid is stereospecifically *trans*, as shown, and gives *meso*-$\alpha\alpha'$-dimethylsuccinic acid [91(b)].

Some $\alpha\beta$-olefinic nitriles have been reduced at the double bond without reduction of the nitrile group, e.g. [92],

$$CH_3CH{=}CHCN \xrightarrow[\text{Zn cathode}]{NaOH/H_2O} CH_3CH_2CH_2CN$$

The mechanism of such reductions is believed to involve initially the transfer of an electron and a proton to the double bond to give a radical, e.g.,

$$CH_2{=}CHX \xrightarrow[+H^+]{+e} CH_3\dot{C}HX$$

Further reduction of this intermediate radical gives an anion, protonation of which then gives the saturated product.

$$CH_3\dot{C}HX \xrightarrow[+H^+]{+e} CH_3CH_2X$$

At high olefin concentrations and when proton availability is low, the electron transfer to the olefin is probably assisted by a second molecule of olefin, so that the incipient radical anion adds, in a Michael type of reaction, to the second molecule to give a dimeric radical anion. This is certainly the case for electrochemical hydrocyclization (EHC), the intramolecular form of EHD [93].

$$CH_2{=}CHX \xrightarrow[CH_2=CHX]{+e} X\dot{C}H{\cdot}CH_2CH_2\bar{C}HX$$

The new radical anion may then be reduced to the dianion by a further one-electron transfer, and protonation then gives the hydrodimer. In the absence of sufficient protons, the dianion may initiate anionic polymerization of the olefin.

The experimental conditions for the successful electrochemical hydrodimerization of activated olefins have been worked out largely as a result of the work of Baizer and his group [94]. The success of the method depends on the use of hydrotropic salts such as tetraethylammonium toluene-*p*-sulphonate to obtain very concentrated water-containing solutions of the olefins, in which the cation is not discharged at the potential necessary for electron transfer to the olefin.

$$CH_2{=}CH{\cdot}CO_2C_2H_5 \xrightarrow[\text{Hg cathode } 40°]{[CH_3]_3\overset{+}{N}C_2H_5\ \text{Tos}^-/DMF/H_2O} C_2H_5O_2C[CH_2]_4CO_2C_2H_5$$

In all cases almost exclusive $\beta\beta$-coupling is observed, though some 10% of $\alpha\beta$-coupling is observed with methacrylonitrile [95].

$$CH_2{=}C(CH_3)CN \xrightarrow{\text{EHD}} NC{\cdot}CH(CH_3)CH_2CH_2CH(CH_3)CN + NC{\cdot}CH(CH_3)_2CH_2CH(CH_3)CN$$

In the case of activated 1,3-dienes the major coupling is usually in the δ-position [94], e.g.,

$$CH_2{=}CH{\cdot}CH{=}CH{\cdot}CN \xrightarrow[\text{(ii) catalytic hydrogenation}]{\text{(i) EHD}} NC[CH_2]_8CN$$

Electrochemical crossed hydrodimerization can also be accomplished [96]. When the two olefins have the same or nearly the same reduction potential, all three expected products are obtained, but when one is reduced more easily than the other, then only two products are obtained, e.g.,

$$C_2H_5O_2CCH\overset{cis}{=}CHCO_2C_2H_5\ (-1{\cdot}4V) \qquad CH_2{=}CH{\cdot}CN\ (-1{\cdot}9V)$$

$$\downarrow \text{EHD}$$

$$C_2H_5O_2C{\cdot}CH_2\underset{|}{C}H(CO_2C_2H_5){-}CH(CO_2C_2H_5){\cdot}CH_2CO_2C_2H_5 + C_2H_5O_2C{\cdot}CH_2CH(CO_2C_2H_5){\cdot}CH_2CH_2CN$$

In this case, if the potential is controlled at about $-1{\cdot}4$V, a Michael donor is only produced by reduction of diethyl maleate; the donor can then add to either of the conjugated olefins present.

Crossed coupling with a conjugated diene system leads predominantly to a straight-chain product by attack at the δ-position, e.g.,

$$\begin{matrix} NC{\cdot}CH{=}CH{\cdot}CH{=}CH_2 \\ + \\ CH_2{=}CH{\cdot}CO_2C_2H_5 \end{matrix} \xrightarrow[\text{(ii) catalytic hydrogenation}]{\text{(i) EHD}} NC[CH_2]_6CO_2C_2H_5$$

Another typical crossed coupling is this [97]:

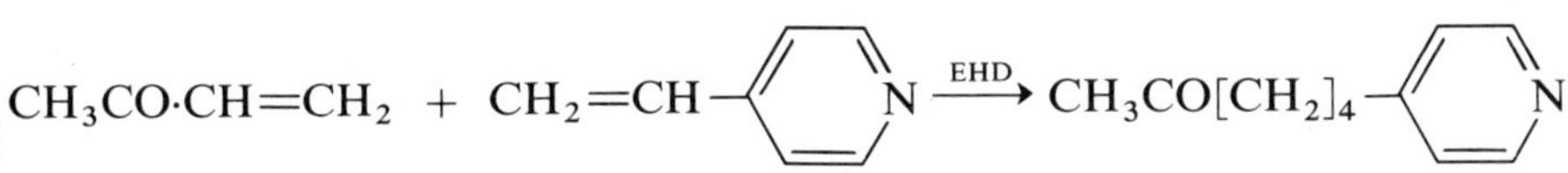

The two expected symmetrical products are also both formed.

Electrochemical hydrodimerization is also possible with the allylic tautomers of activated olefins [98]. For example, ethyl but-3-enoate gives the same product as is obtained from ethyl but-2-enoate. Such results would not be expected on the basis of the polarographic behaviour of the allylic tautomers, and the basis of this electrochemical tautomerization is not clear.

Bis-activated diolefins undergo electrochemical hydrocyclization (EHC) when the process leads to the formation of three-, four-, five- or six-membered rings, but larger rings are not produced at all [99]. The conditions employed are similar to those used for electrochemical hydrodimerizations. The concerted ring closure in which the one-electron reduction of one of the double bonds is assisted by the presence of the second double bond is particularly likely when the two double bonds are capable of parallel orientation, as in the formation of three- and five-membered, but not four-membered, rings. Some typical examples are as follows.

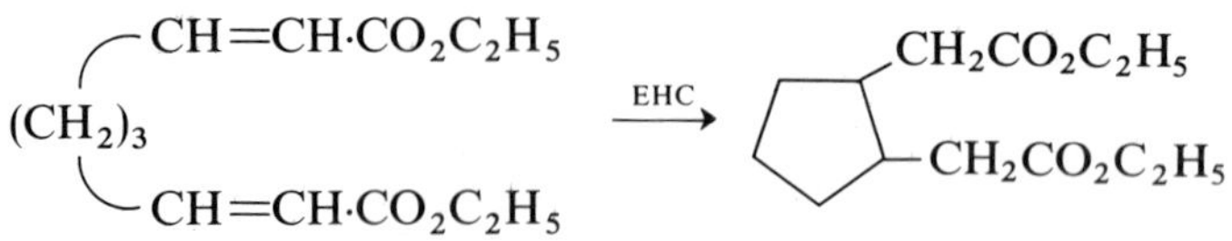

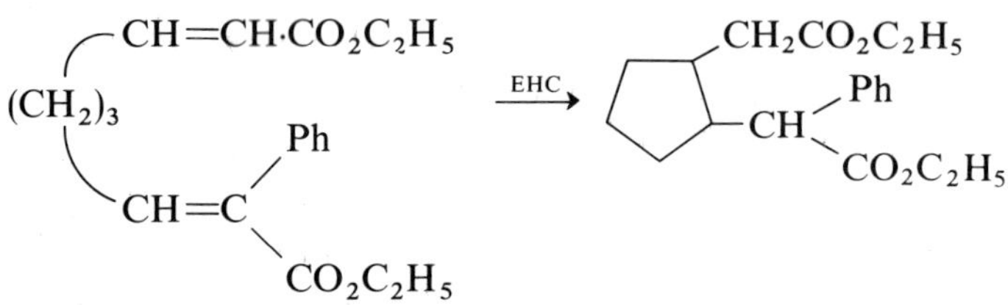

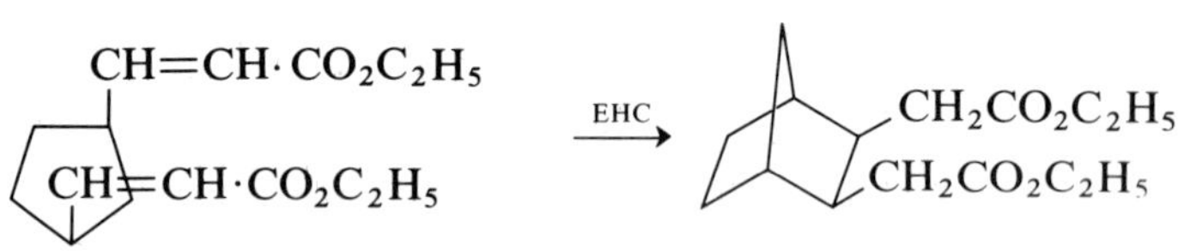

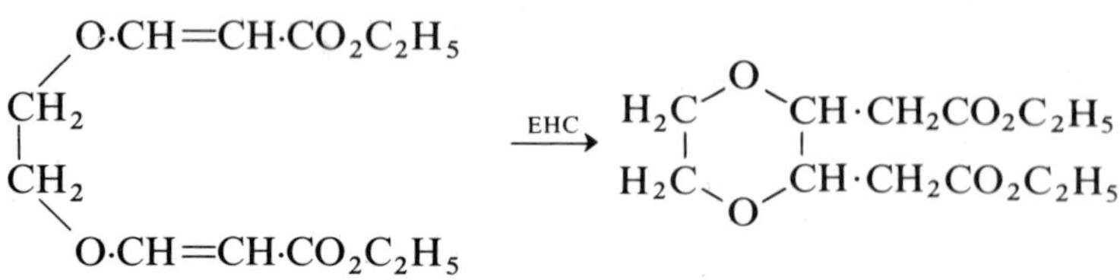

These examples illustrate the formation of carbocyclic and heterocyclic ring systems and a bridged ring system, and also include an example of an intramolecular mixed coupling reaction. The major product in those cases where EHC does not proceed in good yield is the product formed by electrochemical reduction of the two double bonds.

A related type of cyclization is one in which a carbanion is generated by the electrochemical reductive fission of a carbon–mercury bond [100].

$$\mathrm{RHgX} \xrightarrow{+2e} \ddot{\mathrm{R}}^- + \mathrm{Hg} + \ddot{\mathrm{X}}^-$$

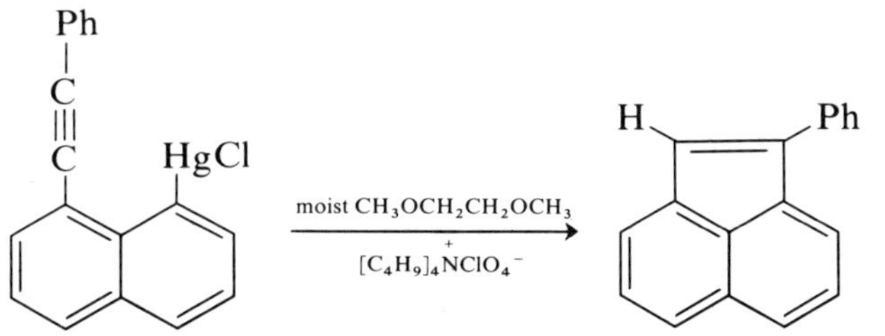

In this case intramolecular addition to the acetylenic bond competes successfully with protonation of the carbanion.

Another type of electrochemical cyclization is represented by the following example [99]. The yield here is low.

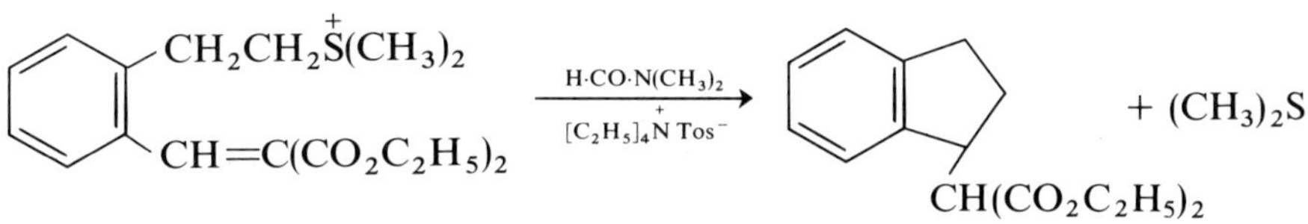

Although the electrochemical reductive cleavage of a bond from carbon to a good leaving group such as $-\overset{+}{S}R_2$ or $-\overset{+}{N}R_3$ is known, at the potential used in this reaction it is the double bond which is reduced. The resultant radical anion or anion then displaces the dimethyl sulphide with the formation of the indane ring system.

The electrochemical reductive cleavage of single bonds can occur by either one- or two-electron transfer processes. A two-electron process is exemplified above for the cleavage of a carbon–mercury bond. Certain quaternary ammonium salts, on the other hand, can be cleaved by a one-electron transfer [101].

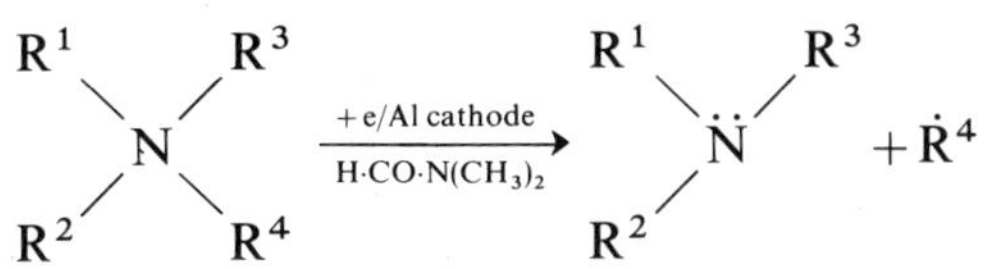

$$2\dot{R}^4 \longrightarrow R^4{\cdot}R^4$$

This reaction takes place when R^4 is a radical stabilized by electron delocalization, e.g., benzyl, and the product of radical coupling is formed.

Some other examples of the electrochemical reductive fission of single bonds have already been given, and further examples are shown below [76, 102].

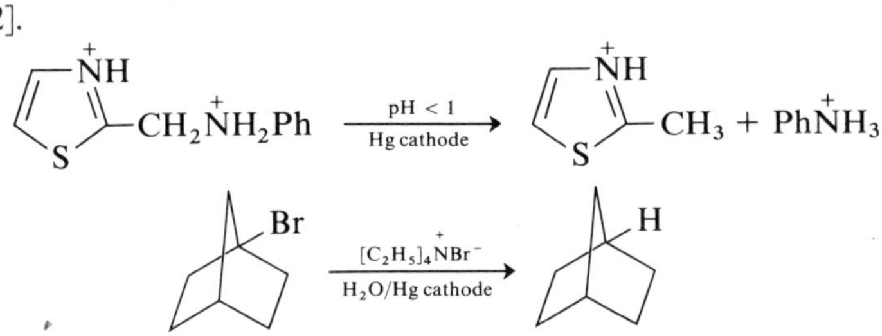

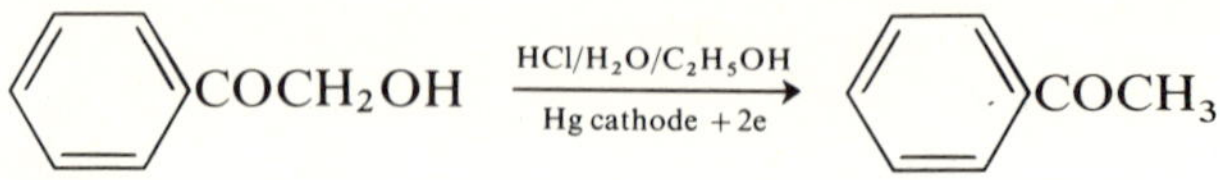

The last example [103] shows the selective reduction fission of a carbon–oxygen single bond in the presence of a carbon–oxygen double bond. The selective reductive cleavage of a carbon–chlorine bond forming part of a trichloromethyl group, in the presence of other reducible groups, can also be achieved [104], e.g.,

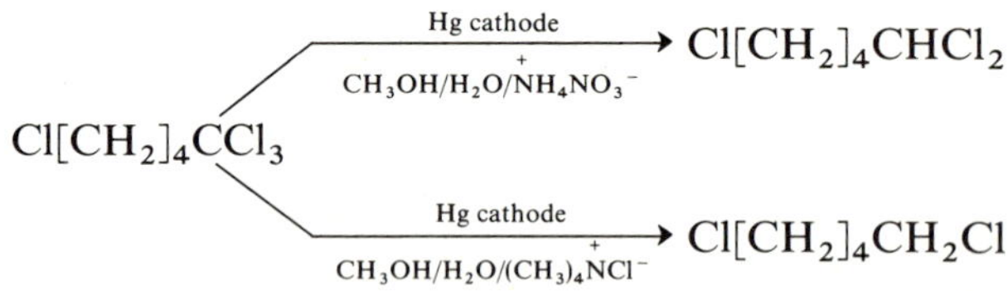

Both processes occur in very high yield; the extent of the dechlorination apparently depends solely on the nature of the supporting electrolyte.

REFERENCES

[1] SVADKOVSKAYA, G. E. and VOITKEVICH, S. A. *Russ. chem. Revs* **29**, 161 (1960); WEEDON, B. C. L. *Adv. org. Chem.* **1**, 1 (1960).

[2] (a) FINKELSTEIN, M. and PETERSEN, R. C. *J. org. Chem.* **25**, 136 (1960).
(b) PELLETIER, S. W., HAWLEY, L. B. and GOPINATH, K. W. *Chem. Commun.* 96 (1967).
(c) COREY, E. J. and CASANOVA, J. *J. Am. chem. Soc.* **85**, 165 (1963).

[3] COREY, E. J., BAULD, N. L., LA LONDE, R. T., CASANOVA, J. and KAISER, E. T. *J. Am. chem. Soc.* **82**, 2645 (1960).

[4] CLUSIUS, K. and SCHANZER, W., *Z. phys. Chem. Frankf.* **192A**, 273 (1943).

[5] BRETTLE, R. and HOLLAND, F. S., unpublished.

[6] SWANN, S. and GARRISON, W. E., *Org. Synth.* **41**, 33 (1961).

[7] LINSTEAD, R. P., LUNT, J. C. and WEEDON, B. C. L. *J. chem. Soc.* 3333 (1950).

[8] *Cf.* WALLIS, E. S. and ADAMS, F. H. *J. Am. chem. Soc.* **55**, 3838 (1933).

[9] SERCK-HANSSEN, K., *Chemy Ind.* 1554 (1958).

[10] GARWOOD, R. F., SCOTT, C. J. and WEEDON, B. C. L. *Chem. Commun.* 14 (1965).

[11] MISLOW, K. and STEINBERG, I. V. *J. Am. chem. Soc.* **77**, 3807 (1955).

[12] BAKER, B. W., HOUGHTON, R. P. and WEEDON, B. C. L., personal communication.

[13] WOOLFORD, R. G. *Can. J. Chem.* **40**, 1846 (1962).

[14] COREY, E. J. and SAUERS, R. R. *J. Am. chem. Soc.* **81**, 1743 (1959).

[15] LINSTEAD, R. P., SHEPHARD, B. R. and WEEDON, B. C. L. *J. chem. Soc.* 3624 (1952).

[16] FICHTER, F. and STENZL, H. *Helv. chim. Acta* **22**, 970 (1939).

[17] LEVIN, A. I., CHECHINA, O. N. and SOKOLOV, S. V. *J. gen. Chem. U.S.S.R.* **35**, 1776 (1965).

[18] EBERSON, L. *Acta chem. scand.* **13**, 40 (1959).

[19] EBERSON, L. *J. org. Chem.* **27**, 2329 (1962).

[20] BLOOD, C. T., CARTWRIGHT, N. J. and LINSTEAD, R. P. *J. chem. Soc.* 2268 (1952).

[21] HEY, D. H. and BUNYAN, P. J. *J. chem. Soc.* 3787 (1960).

[22] GOLDSCHMIDT, S. and STOCKEL, E. *Chem. Ber.* **85**, 630 (1952).

[23] LINDSEY, R. V. and PETERSON, M. L. *J. Am. chem. Soc.* **81**, 2073 (1959).

[24] KOEHL, W. J. *J. Am. chem. Soc.* **86**, 4686 (1964).

[25] WATERS, J. A., BECKER, E. D. and MOSETTIG, E. *J. org. Chem.* **27**, 4689 (1962).
[26] EBERSON, L. and NYBERG, K. *Acta chem. scand.* **18**, 1567 (1964).
[27] STEPANOV, F. N., BAKLAN, V. F. and GUTS, S. S. *Sintez prirod. Soed. Analogov Fragmentov, Akad. Nauk. SSSR. Otd. obshch. tekh. Khim.* 95 (1965) (*Chem. Abstr.* **65**, 627 (1966)).
[28] KRONENTHAL, R. L. U.S. Patent 2,760,926 (*Chem. Abstr.* **51**, 3329 (1957)).
[29] FICHTER, F. and REEB, H. *Helv. chim. Acta* **6**, 450 (1923); DEMYANOV, N. YA. *Zh. obshch. Khim.* **61**, 1861 (1929).
[30] TRAYNHAM, J. G. and DEHN, J. S. *J. Am. chem. Soc.* **89**, 2139, 2799 (1967).
[31] GASSMAN, P. G. and ZALAR, F. V. *J. Am. chem. Soc.* **88**, 2252 (1966).
[32] WLADISLAW, B. and VIERTLER, H. *Chemy Ind.* 39 (1965).
[33] WLADISLAW, B. and AYRES, A. M. J. *J. org. Chem.* **27**, 287 (1962).
[34] NEUBERG, C. *Biochem. Z.* **7**, 527 (1908).
[35] GASSMAN, P. G. and FOX, B. L. *J. org. Chem.* **32**, 487 (1967).
[36] LINSTEAD, R. P., SHEPHARD, B. R. and WEEDON, B. C. L. *J. chem. Soc.* 2854 (1951).
[37] FICHTER, F. and SCHMID, M. *Helv. chim. Acta* **3**, 704 (1920).
[38] PASQUINELLI, E. A. *An. Asoc. quim. argent.* **31**, 181 (1943) (*Chem. Abstr.* **38**, 5734 (1944)).
[39] PAUDLER, D. W., HERBENER, R. E. and ZEILER, A. G. *Chemy Ind.* 1909 (1965).
[40] ZALKOW, L. H. and BRANNON, D. R. *J. chem. Soc.* 5497 (1964); OVERBERGER, C. G. and KABASAKALIAN, P. *J. Am. chem. Soc.* **79**, 3182 (1957).
[41] VELLTURO, A. F. and GRIFFIN, G. W. *J. Am. chem. Soc.* **87**, 3021 (1965).
[42] VENZETTI, B. L. *Gazz. chim. ital.* **34**, 511 (1904).
[43] (a) VERMILLION, F. J. and PEARL, I. A. *J. electrochem. Soc.* **111**, 1392 (1964).
(b) BOBBITT, J. M., STOCK, J. T., MARCHAND, A. and WEISGRABER, K. H. *Chemy Ind.* 2127 (1966).
[44] ADICKES, F., BRUNNERT, W. and LÜCKER, O. *J. prakt. Chem.* [2] **130**, 163 (1931).
[45] BAHNER, C. T. *Ind. Engng. Chem.* **44**, 317 (1952).
[46] BRETTLE, R. and PARKIN, J. G., *J. chem. Soc.* (*C*) 1352 (1967).
[47] *Chem. Engng News*, p. 52, 7th Dec. (1964).
[48] (a) STEUBER, F. W. and DIMROTH, K. *Chem. Ber.* **99**, 258 (1966).
(b) LUND, H. *Acta chem. scand.* **11**, 491 (1957).
(c) EBERSON, L. and NYBERG, K. *J. Am. chem. Soc.* **88**, 1686 (1966).
(d) ROSS, S. D., FINKELSTEIN, M. and PETERSEN, R. C. *J. Am. chem. Soc.* **86**, 4139 (1964).
[49] KOYAMA, K., SUSUKI, T. and TSUTSUMI, S. *Tetrahedron Lett.* 627 (1965); PARKER, V. D. and BURGERT, B. E. *Tetrahedron Lett.* 4065 (1965).
[50] EBERSON, L. and NYBERG, K. *Tetrahedron Lett.* 2389 (1966).
[51] (a) KENNER, G. W., MURRAY, M. A. and TYLOR, C. M. B. *Tetrahedron* **1**, 259 (1957).
(b) SCOTT, A. I., DODSON, P. A., MCCAPRA, F. and MEYERS, M. B. *J. Am. chem. Soc.* **85**, 3702 (1963).
[52] BONNER, W. A. and MANGO, F. D. *J. org. Chem.* **29**, 430 (1964).
[53] (a) BAGGALEY, A. J. and BRETTLE, R. *Chem. Commun.* 108 (1966).
(b) EBERSON, L., NYBERG, K., FINKELSTEIN, M., PETERSEN, R. C., ROSS, S. D. and UEBEL, J. J. *J. org. Chem.* **32**, 16 (1967).
[54] MANGO, F. D. and BONNER, W. A. *J. org. Chem.* **29**, 1367 (1964).
[55] ROSS, S. D., FINKELSTEIN, M. and PETERSEN, R. C. *J. org. Chem.* **31**, 128 (1966).
[56] INOUE, T. and TSUTSUMI, S. *Bull. chem. Soc. Japan* **38**, 661 (1965).
[57] ELMING, N. *Adv. org. Chem.* **2**, 67 (1960).
[58] BELLEAU, B. and WEINBERG, N. L. *J. Am. chem. Soc.* **85**, 2525 (1963).
[59] FIOSHIN, M. YA. and DZHAFAROV, E. A. *Dokl. Akad. Nauk azerb. SSR* **21**, 25 (1965) (*Chem. Abstr.* **63**, 16207 (1965)).

[60] BURDON, J. and TATLOW, J. C. *Adv. Fluor. Chem.* **1**, 129 (1960).
[61] POPP, F. D. and SCHULTZ, H. P. *Chem. Rev.* **62**, 19, 1962.
[62] KHOMYAKOV, V. G., FIOSHIN, M. YA. and AVRUTSKAYA, I. *Zh. prikl. Khim., Leningr.* **38**, 1889 (1965) (*Chem. Abstr.* **64**, 1643 (1965)).
[63] LE GUYADER, M. *Bull. Soc. chim. Fr.* 1848 (1966).
[64] HARMAN, R. E. *Org. Synth.* Coll. Vol. **4**, p. 148, Wiley, New York (1963).
[65] INGERSOLL, A. W., BIRCHER, L. J. and BRUBAKER, M. M. *Org. Synth.*, Coll. Vol. **1**, 2nd Edn., p. 485, Wiley, New York (1941).
[66] HORWITZ, D. and CERWONKA, E. U.S. Patent 2,916,426 (1959) (*Chem. Abstr.* **54**, 6370 (1960)).
[67] ELVING, P. J. and LEONE, J. T. *J. Am. chem. Soc.* **80**, 1021 (1958).
[68] GRIMSHAW, J. and RAMSEY, J. S. *J. chem. Soc.* (*C*) 653 (1966).
[69] SUGINO, K. and NONAKA, T. *J. electrochem. Soc.* **112**, 1241 (1965).
[70] BUNYAN, P. J. and HEY, D. H. *J. chem. Soc.* 324 (1962).
[71] BELEN'KAYA, N. G. and BELOGERSKII, N. A. *Zh. obshch. Khim.* **19**, 1664 (1949).
[72] COLEMAN, G. H. and JOHNSON, H. L. *Org. Synth.* Coll. Vol. **3**, p. 60, Wiley, New York (1955).
[73] LUND, H. *Acta chem. scand.* **17**, 2325 (1963).
[74] TAFEL, J. and FRIEDRICHS, G. *Ber. dt. chem. Ges.* **37**, 3187 (1904).
[75] ERSHOV, B. P. and PYATNITSKAYA, G. N. *J. chem. Ind. U.S.S.R.* **18**, 13 (1941) (*Chem. Abstr.* **38**, 3912 (1944)).
[76] IVERSEN, P. E. and LUND, H. *Acta chem. scand.* **21**, 389 (1967).
[77] WIBAUT, J. P. and BOER, H. *Recl. Trav. chim. Pays-Bas Belg.* **68**, 72 (1949).
[78] YAMAZAKI, T. and NAGATA, M. *Yakugaku Zazzhi* **79**, 1222 (1959) (*Chem. Abstr.* **54**, 4596 (1960)).
[79] LUND, H. *Acta chem. scand.* **13**, 249 (1959).
[80] RAPSON, H. D. C. and BIRD, A. E. *J. appl. Chem. Lond.* **13**, 233 (1963).
[81] (a) KAWAMURA, F. and SUZULI, S. *J. chem. Soc. Japan* (*ind. Chem. Sect.*) **55**, 476 (1952) (*Chem. Abstr.* **48**, 3167 (1954)).
(b) VARSHAVSKII, S. L., TOMILOV, A. P., KAABAK, L. V. and SHANDRINOV, N. YA. USSR Patent 181,656 (1966) (*Chem. Abstr.* **66**, 2679, entry 28382n (1967)).
[82] DIAMOND, G. B. and SOFFER, M. D. *J. Am. chem. Soc.* **74**, 4126 (1952).
[83] LUND, H. *Acta chem. scand.* **16**, 525 (1962).
[84] WAWZONEK, S. and WEARING, D. *J. Am. chem. Soc.* **87**, 2067 (1959).
[85] FERLES, M. *Chemické Listy* **52**, 668 (1958) (*Chem. Abstr.* **52**, 13724 (1958)).
[86] FERLES, M. *Chemické Listy* **52**, 674 (1958) (*Chem. Abstr.* **52**, 13724 (1958)).
[87] INHOFFEN, H. H. and JÄGER, P. *Tetrahedron Lett.* 1317 (1964).
[88] LANGER, S. H. and LANDI, H. P. *J. Am. chem. Soc.* **86**, 4694 (1964).
[89] BENKESER, R. A., KAISER, E. M. and LAMBERT, R. F. *J. Am. chem. Soc.* **86**, 5272 (1964).
[90] TOMILOV, A. P. *Russ. chem. Revs* **31**, 569 (1962).
[91] (a) INGERSOLL, A. W. *Org. Synth.* Coll. Vol. **1**, 2nd Edn, p. 311, Wiley, New York (1941).
(b) ROSENTHAL, I., HAYES, J. R., MARTIN, A. J. and ELVING, P. J. *J. Am. chem. Soc.* **80**, 3051 (1958).
[92] TOMILOV, A. P., KAABAK, L. V. and VARSHAVSKII, S. L. *J. gen. Chem. U.S.S.R.* **33**, 2737 (1963).
[93] PETROVICH, J. P., ANDERSON, J. D. and BAIZER, M. M. *J. org. Chem.* **31**, 3897 (1966).
[94] BAIZER, M. M. and ANDERSON, J. D. *J. electrochem. Soc.* **111**, 223 and 226 (1964).
[95] JONES, G. C. and LEDFORD, T. H. *Tetrahedron Lett.* 615 (1967).
[96] BAIZER, M. M. *J. org. Chem.* **29**, 1670 (1964).
[97] ANDERSON, J. D., BAIZER, M. M. and PRILL, E. J. *J. org. Chem.* **30**, 1645 (1965).
[98] ORT, M. R. and BAIZER, M. M. *J. org. Chem.* **31**, 1646 (1966).

[99] Anderson, J. D., Baizer, M. M. and Petrovich, J. P. *J. org. Chem.* **31**, 3890 (1966).
[100] Kandil, S. A. and Dessy, R. E. *J. Am. chem. Soc.* **88**, 3027 (1966).
[101] Finkelstein, M., Petersen, R. C. and Ross, S. D. *J. Am. chem. Soc.* **81**, 2361 (1959).
[102] Sease, J. W., Chang, P. and Groth, J. L. *J. Am. chem. Soc.* **86**, 3154 (1964).
[103] Lund, H. *Acta chem. scand.* **14**, 1927 (1960).
[104] Nagao, M., Sato, N., Akashi, T. and Yoshida, T. *J. Am. chem. Soc.* **88**, 3447 (1966).

6

Photochemical Methods

E. V. BLACKBURN and C. J. TIMMONS, *Nottingham University*

Photochemical reactions have now made a wide range of compounds readily available to the synthetic chemist. Although most of these reactions are not commercially viable processes on the large scale, they may be used to prepare quantities of the order of 0·5 to 10 g which are adequate for many purposes.

An attempt has been made in this chapter to extract and summarize those reactions which are of synthetic interest from the wealth of photochemical literature available. Experimental conditions have been omitted, and no attempt has been made to catalogue all photochemical reactions. However, some attempt has been made to classify the reactions.

1. The photolysis of nitrite and hypohalite esters.
2. Addition reactions
3. Photocyclization of organic molecules.
4. Intramolecular electrocyclic transformations.
5. Photochemical rearrangements.
   (a) Sigmatropic rearrangements
   (b) Rearrangement of aromatic compounds
   (c) Rearrangement of *N*-Oxides
   (d) *Cis-trans* isomerizations
6. Aromatic substitutions.
7. Photochemical production of carbenes and nitrenes.

Photochemical oxidations and reductions are reviewed in Chapters 2 and 3 and will not be reviewed here. Examples have been taken from the literature up to May, 1968.

6.1 THE PHOTOLYSIS OF NITRITE AND HYPOHALITE ESTERS

The considerable difficulty in gaining access to the 18- and 19-substituted steroids, led D. H. R. Barton to explore the potentials of the photolysis of nitrite and hypohalite esters. An excellent synthetic route to these steroids has been evolved as a result of these investigations, and this reaction has become known as the Barton reaction [1–4].

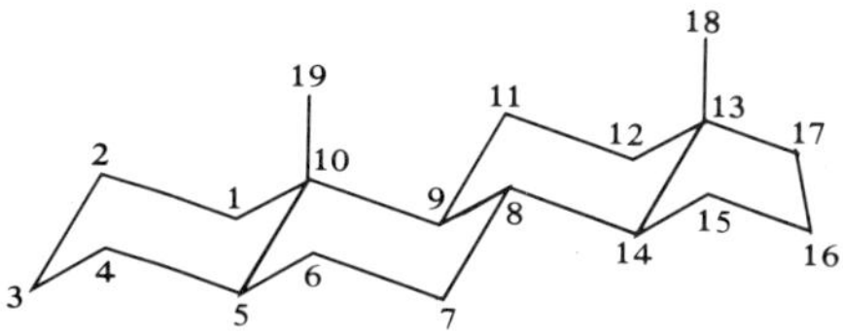

The primary reaction is the homolytic fission of an N—O or X—O bond.

$$\text{RONO} \longrightarrow \text{RO}\cdot + \text{NO}$$

$$\text{ROX} \longrightarrow \text{RO}\cdot + \text{X}\cdot$$

There are many reaction pathways open to the reactive alkoxy radical, and other reactions of synthetic interest as well as the Barton reaction have also been observed. However, as the Barton reaction is of such great importance, it will be discussed first.

In the Barton reaction, intramolecular hydrogen abstraction takes place to generate a free radical elsewhere in the molecule. This radical combines with nitric oxide to form nitroso dimers or oximes. The Barton reaction is summarized in Scheme 6.1.

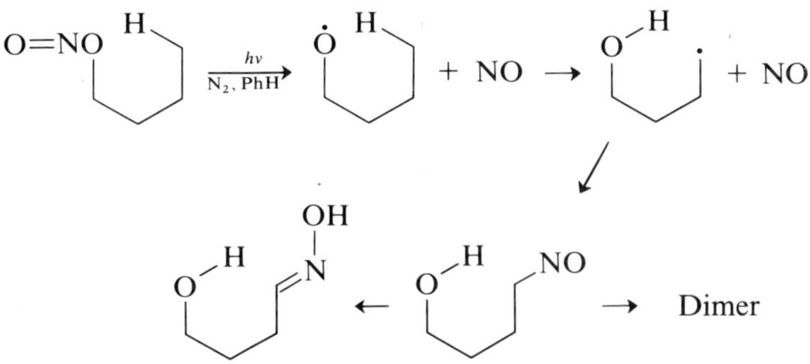

Scheme 6.1

Evidence has been presented for this "non-cage" free-radical mechanism [5], and it has been shown that a six-membered transition state is a requirement for the hydrogen abstraction step [6]. Radical exchange, which would be expected for this mechanism, has also been observed on photolysis in the presence of iodine or bromotrichloromethane [7].

The first photolysis of a nitrite reported by Barton was that of the 20β-nitrite of 3β-acetoxy-5α-pregnan-20β-ol (1) to the 18-oximino-5α-pregnan-3β-,20β-diol-3-acetate (2) in 34% yield [8], and since then many natural products have been prepared by this method, including β-amyrin [4(a)] and cycloartenol [4(b)].

The intramolecular rearrangement of the nitric oxide is facilitated by placing the nitrite group and C—H bond to be attacked in a 1,3-diaxial relationship. The nitrite of 20β-hydroxy-4-pregnene-3-one gave only 15% of 20β-hydroxy-18-oximino-4-pregnen-3-one on photolysis, whereas the 20α-nitrite gave the corresponding oxime in 60% yield [9].

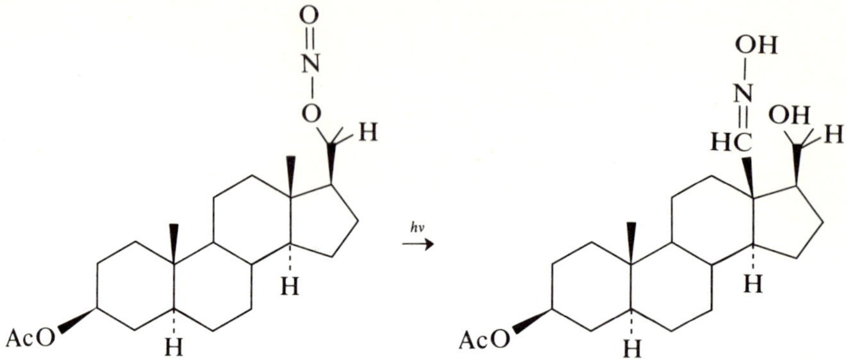

(1) (2)

In the elegant three-step synthesis of aldosterone acetate from corticosterone acetate, the hydrogen at C-18 is abstracted by the alkoxy radical at C-11β, and the postulated intermediate nitroso monomer (3), resulting from the recombination of the C-18 radical with nitric oxide, does not dimerize, presumably because of steric factors [10].

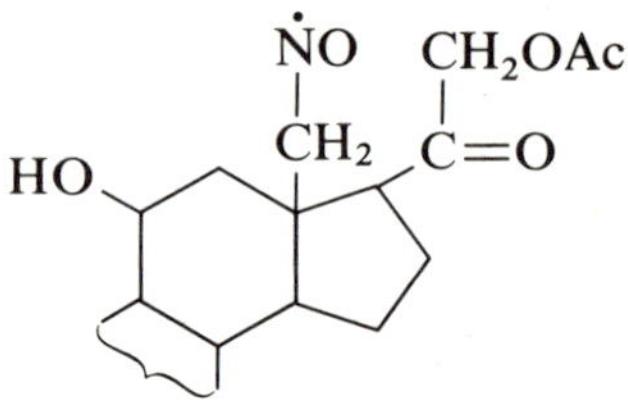

(3)

In the absence of steric factors unfavourable to dimerization, the intermediate mono nitroso precursors do dimerize and the oxime is usually generated by refluxing the dimer in isopropanol.

Irradiation of steroidal 11β-nitrites does not necessarily result in exclusive attack at C-18. Irradiation of corticosterone-3,20-bisethylene ketal-21-acetate-11β-nitrite gave both the 18- and the 19-oximes [11].

From the requirement of a six-membered transition state it can be seen that the C-18 position may be attacked by photolysis of nitrite esters at C-20, C-15, C-8, and C-11, while the C-19 position can be approached via nitrite esters at C-2, C-4, C-6, C-8, and C-11. Most of these routes have been utilized.

Functionalization of C-19 from C-6 in high yield has found useful application in the synthesis of the medically important 19-norsteroids from the readily available $\Delta^{5,6}$-steroids [12].

Intramolecular radical additions or rearrangement reactions may occur in preference to nitric oxide capture as a result of some structural feature of the molecule or a strategically located substituent, and may well be of synthetic interest. The alkyl radical intermediate may rearrange with C—C bond fission resulting in expansion of ring D. An example of this is the irradiation of the nitrite of 4-androstene-11β-ol-3,17-dione (4) in toluene to give

18-nor-D-homo-4,13(17a)-androstadiene-11β-ol-3,17-dione (5) [13]. Other 11β-nitrite-17-ones reacted in a similar manner. The proposed mechanism is given in Scheme 6.2.

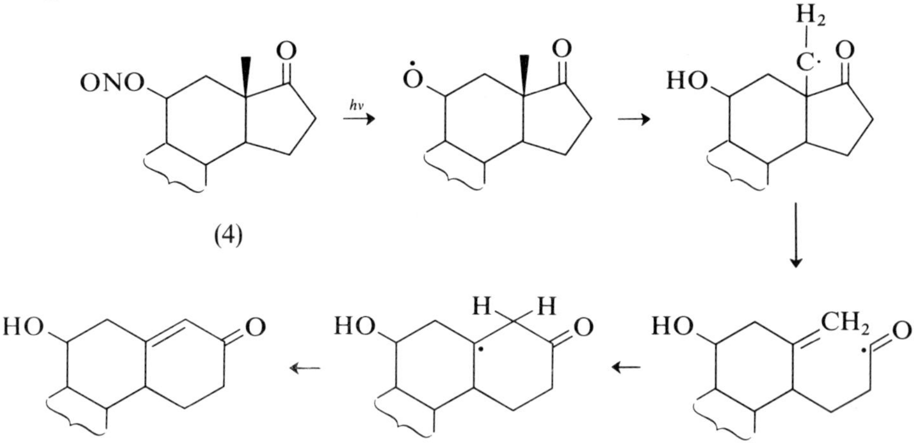

(4) (5)

Scheme 6.2

The alkyl radical intermediate may rearrange by addition to an αβ-unsaturated ketone. This has been observed in nitrites possessing a conjugated carbonyl group in ring A. The *syn* and *anti* oximino-ketones, (6) and (7), are formed during the irradiation of the nitrite of corticosterone acetate [10]. It seems probable that these compounds result from intramolecular attack at the C-19 methyl group to give (8), which cyclizes to the radical (9), and nitric oxide capture follows.

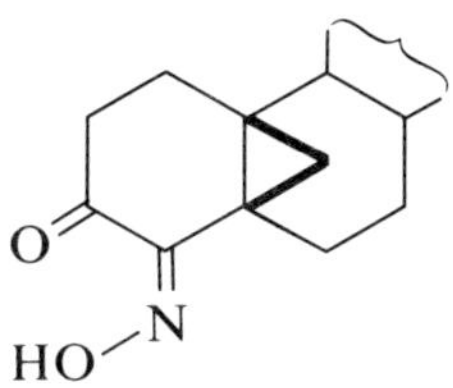

(6)

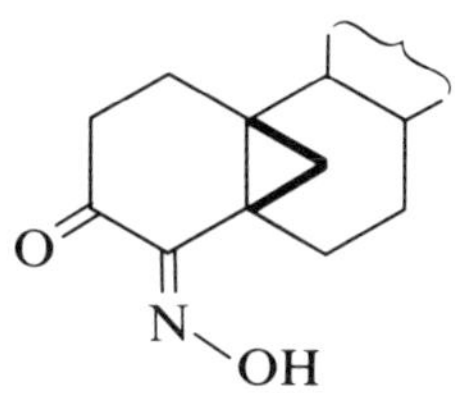

(7)

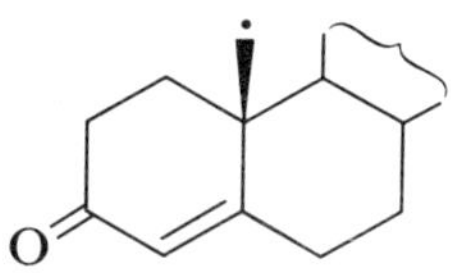

(8)

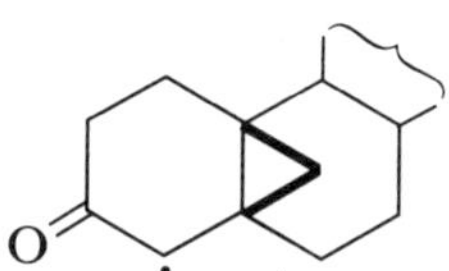

(9)

The corresponding alcohol is often encountered in the liquid-phase photolysis of a nitrite resulting from intermolecular hydrogen abstraction from other molecules. The extent to which this occurs depends upon the efficiency of the solvent as a hydrogen-donor.

The formation of aldehydes and ketones by the decomposition of primary and secondary nitrites has been widely observed and has been explained by alkoxy radical decomposition by loss of an α hydrogen or by a disproportionation process [14]. The nitrite of 3*O*,*N*-diacetyl-22,27-imino-17,23-oxidojervan-5-ene-3β,11β-diol (10) gave the compound (11), the 3β,11α-diol (12), and the 3β-ol-11-one (13) [15].

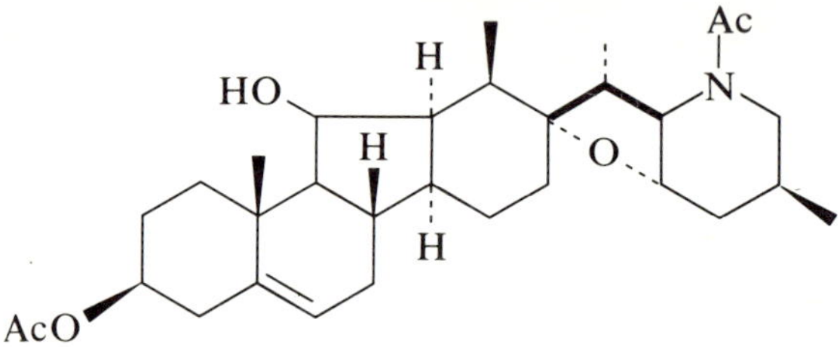

(10)

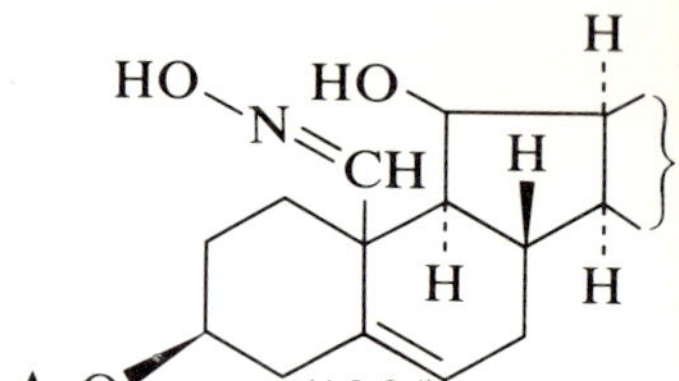

(13%)

(11)

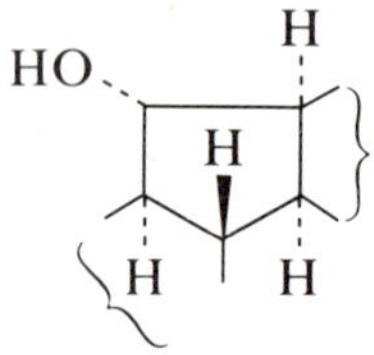

(26%)

(12)

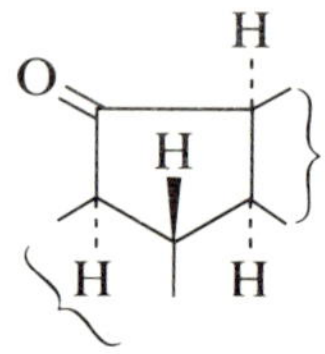

(31%)

(13)

The alkoxy radical derived from the photolysis of 2-phenyl-1-ethyl nitrite has been shown to undergo C—C bond fission and hence form a dimer [6].

$$PhCH_2CH_2ONO \rightarrow PhCH_2CH_2O\cdot + NO$$

$$PhCH_2\text{—}CH_2\text{—}O\cdot \rightarrow PhCH_2\cdot + CH_2O$$

$$PhCH_2\cdot + NO \rightarrow PhCH_2NO \rightarrow \text{Dimer}$$

In the case of 2-propyl nitrite, where methyl radicals are the only possible result of C—C bond fission, no nitroso dimers are observed. However, 3-pentyl nitrite, which can decompose to give ethyl radicals, forms dimeric nitrosoethane [16]. It is found that C—C bond fission reactions of secondary nitrites lead mainly to the formation of the larger of the two radicals.

Tertiary butyl nitrite failed to give nitrosomethane dimers, but the formation of acetone indicated the expected C—C bond cleavage and formation of methyl radicals. 2-Methyl-2-butyl and 2-methyl-2-pentyl nitrites give nitrosoethane and 1-nitrosopropane dimers respectively [16].

Hydroxamic acid derivatives may be formed by C—C bond fission of alicyclic alkoxy radicals. 5α-Androstane-3α,17β-diol-3α-acetate-17-nitrite (14) gave the hydroxamic acid (15) on irradiation [17]. The proposed mechanism is given in Scheme 6.3.

In cases where the nitrite to be photolysed is substituted by a certain type of oxygen-bearing moiety, an oxidative fission can occur. The neighbouring group may be a ketone, acetate, or alcohol. Thus photolysis of the nitrite of 3,3,21,21-bis-(ethylenedioxy)-5-pregnen-20β-ol (16) gives the etio-aldehyde

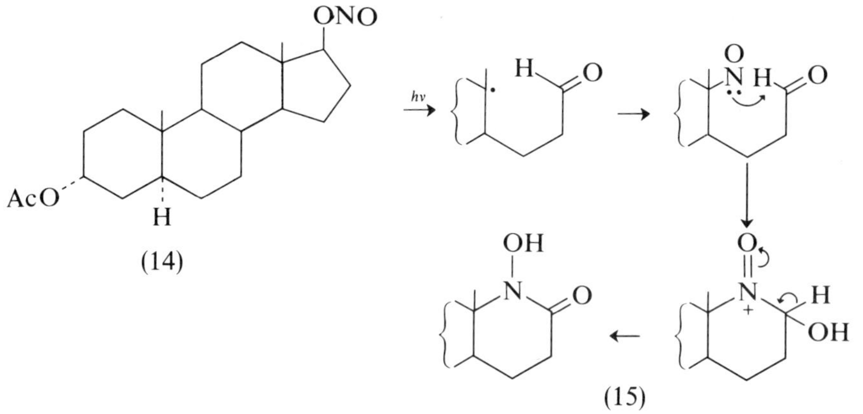

(14) (15)

Scheme 6.3

(17). This C—C bond fission is probably facilitated by the relative stability of the expelled fragment (18) [18]. Similarly, the steroidal glycol 17α,20β-dihydroxy-Δ^4-pregnen-3-one nitrite ester yielded Δ^4-androstene-3,17-dione on photolysis [2].

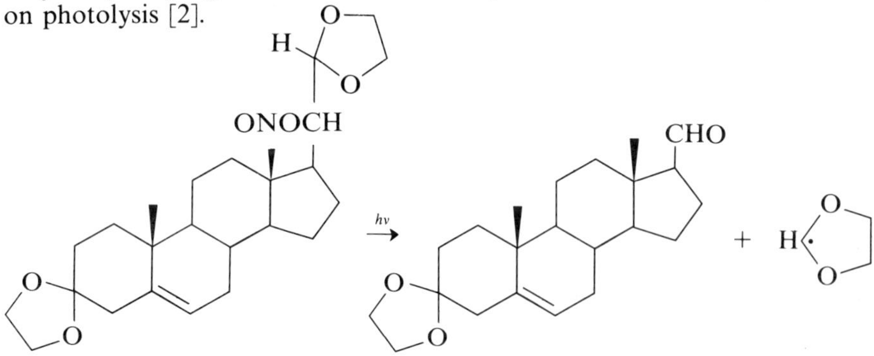

(16) (17) (18)

Certain steroidal alkaloid nitrite esters (e.g., (19)), substituted vicinally by a nitrogen moiety, undergo C—C bond fission on irradiation to give an aldehyde. This may provide a useful synthetic route to certain aldehydes which are inaccessible by other procedures [19].

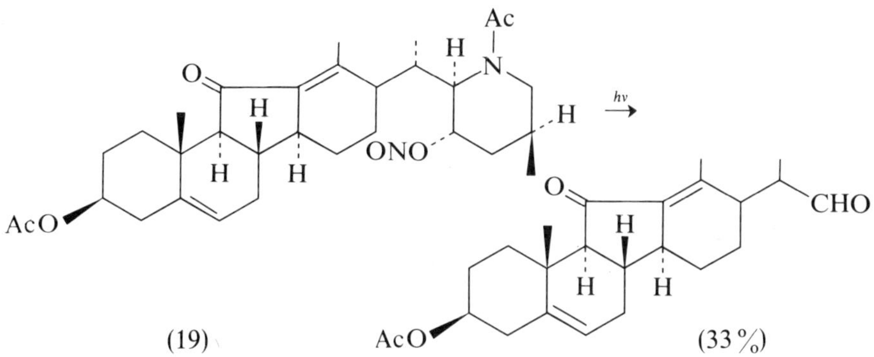

(19) (33%)

The alkoxy radical may associate with other radicals. Thus 1-octyl nitrite forms the nitrate on irradiation in the presence of oxygen [14].

The alkoxy radical may add to a double bond to form an epoxide. Thus, photolysis of 20α-hydroxy-$\Delta^{4,16}$-pregnadien-3-one nitrite gave (20) [20].

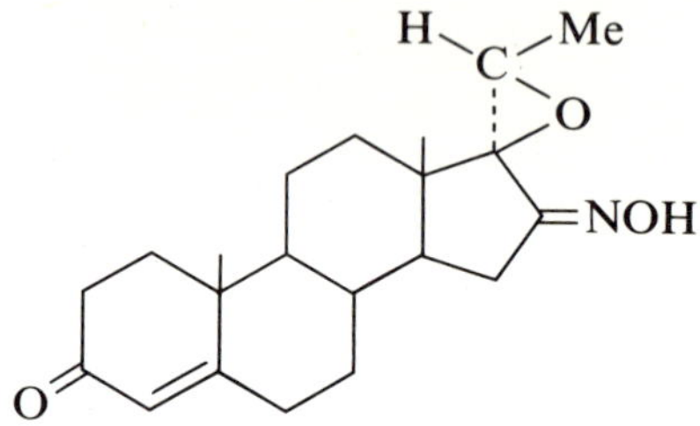

(20)

Another reaction of interest is the attack of the oximino group on neighbouring carbonyl groups [10].

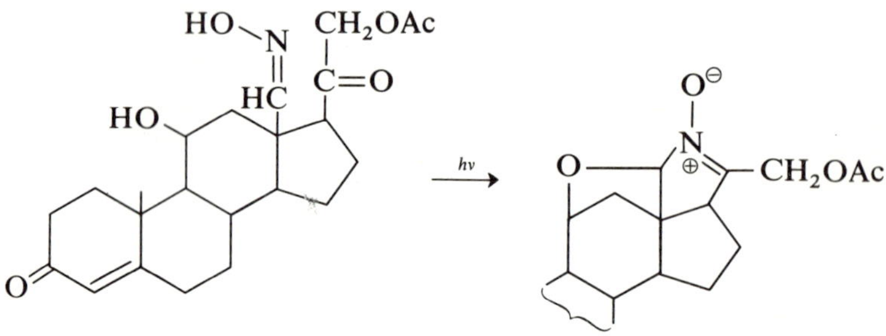

The photolysis of hypohalites proceeds in a similar fashion to that of the nitrite esters. Hypoiodites are usually made *in situ* with lead tetra-acetate and iodine. The decomposition reactions of the alkoxy radicals parallel those of the radicals obtained by the photolysis of nitrites. However, better yields are obtained by photolysis of nitrites.

Hypoiodites of carboxylic acids decarboxylate on photolysis to yield the corresponding alkyl iodides, and this provides a good method for decarboxylation [21].

$$(CH_2)_4(CO_2H)_2 \xrightarrow[I_2 \cdot Pb(OAc)_4]{h\nu} 2CO_2 + (CH_2)_2(CH_2I)_2$$

6.2 PHOTOADDITION REACTIONS

Photoaddition reactions have assumed an important position in organic synthesis and have made many ring systems readily available to the organic chemist [22, 23]. The simplicity of these reactions appeals to a synthetic chemist and is the method to be preferred to previous multi-step syntheses.

The first addition reaction to receive a detailed study was that of carbonyl compounds to olefins, which provided an excellent route to oxetans. Paterno

[24] reported the reaction, which was subsequently examined mechanistically by Büchi [25]; as a result, this reaction is often referred to as the Paterno–Büchi reaction.

Aromatic and aliphatic ketones, aldehydes, and esters readily add to olefins, allenes, and acetylenes to form oxetan analogues [26–29]. The yield of oxetan formed varies considerably and several side reactions can occur. Carbinols and pinacols are formed in the presence of protonic solvents [30]. A second side reaction is the possible dimerization of the olefin resulting from triplet-energy transfer from the carbonyl compound. Irradiation of norbornene gave varying amounts of dimer, depending on the triplet energy of the carbonyl sensitizer [31].

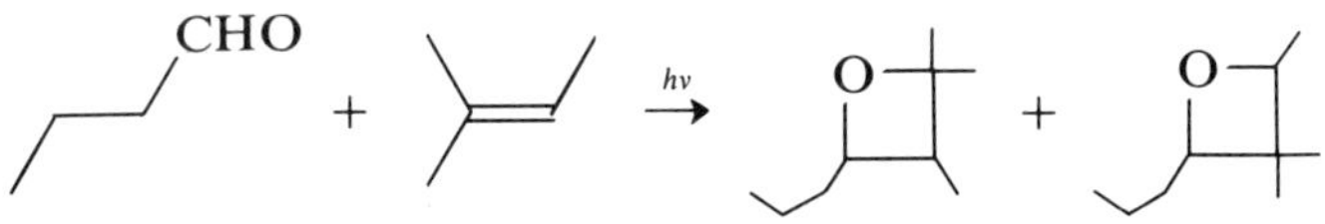

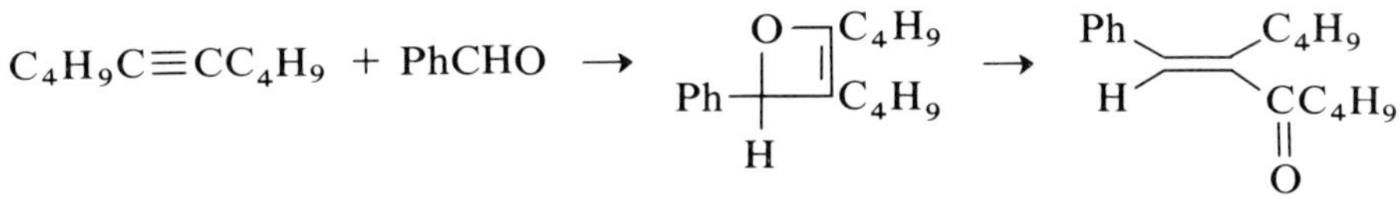

The major product of the cycloaddition reaction may be predicted on the basis of the most stable diradical intermediate [23].

Benzophenone (21) and furan (22) form a 1:1 adduct on irradiation [32]. Furan is unique in this way, as other heterocycles such as thiophen, pyrrole, oxazole, and isoxazole do not form analogous oxetans [33]. 2:1 adducts have been isolated from the irradiation of benzophenone and furan [34, 35].

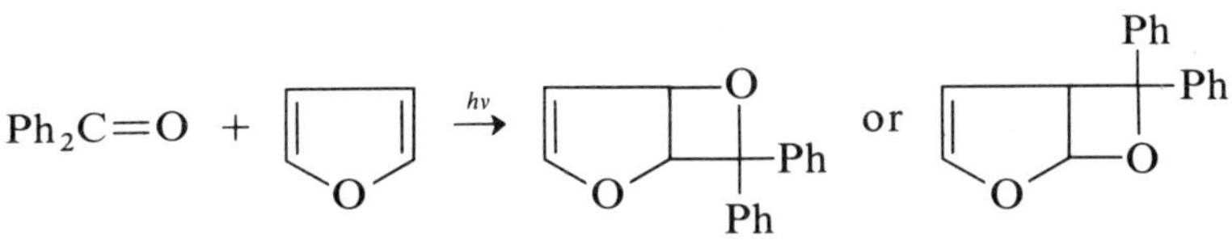

(21) (22)

The Paterno–Büchi reaction can be used to prepare oxetans having important functional groups. The photocycloaddition of acetyl cyanide (23) to an olefin gave a new oxetan (24) having a cyano group directly bonded to a carbon atom of a trimethylene oxide ring [36].

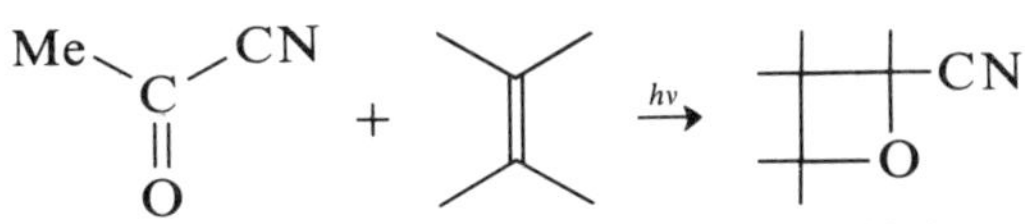

(23) (24)

1,4-Benzoquinones form oxetans with olefins and acetylenes. Irradiation of cyclo-octene and excess chloranil in benzene solution gave the expected oxetan (25) [37, 38]. The photoadduct (26) of *p*-benzoquinone and diphenylacetylene is probably formed via an oxetan intermediate [39].

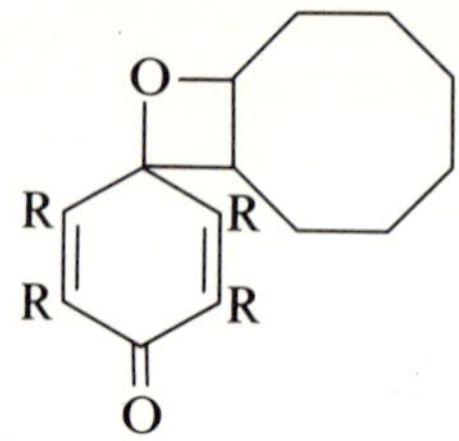

(25)

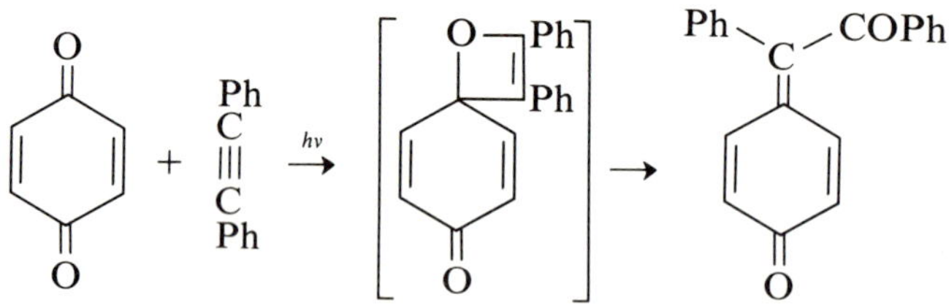

(26)

Thioketones form new olefins and thioaldehydes via the formation of cyclic adducts [40]. Thus, thiobenzophenone forms several interesting adducts (27–30) with α-phellandrene. This formation of a six-membered heterocyclic ring may be of synthetic interest [41].

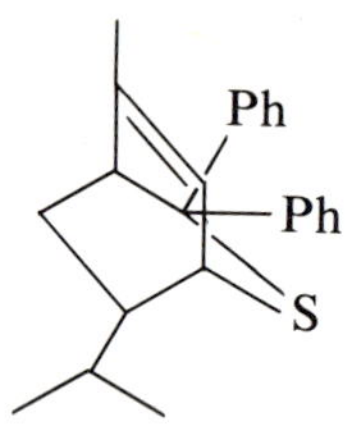

(27)

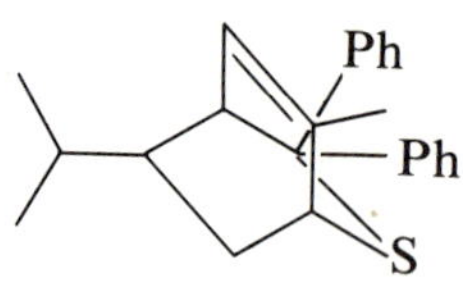

(28)

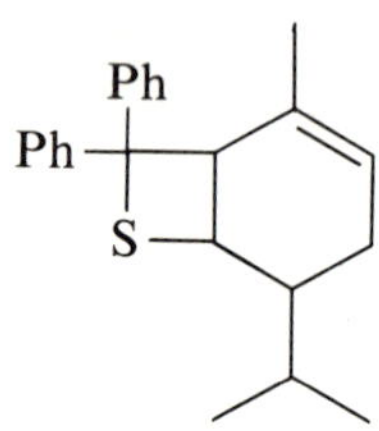

(29)

dibenzhydryl disulphide

(30)

Photocycloaddition reactions are of major importance in the formation of cyclobutane ring systems, and numerous cyclobutane derivatives have been prepared by this reaction [23]. Additions of olefins, ketones, quinones, and aromatic compounds are known.

αβ-Unsaturated ketones readily add to a wide variety of olefinic and acetylenic compounds [42–44]. This reaction is of synthetic use and has been utilized with success in the synthesis of many natural products. de Mayo and his co-workers have employed such a reaction in the synthesis of (±)-β-himachalene from (31) and (32) [45].

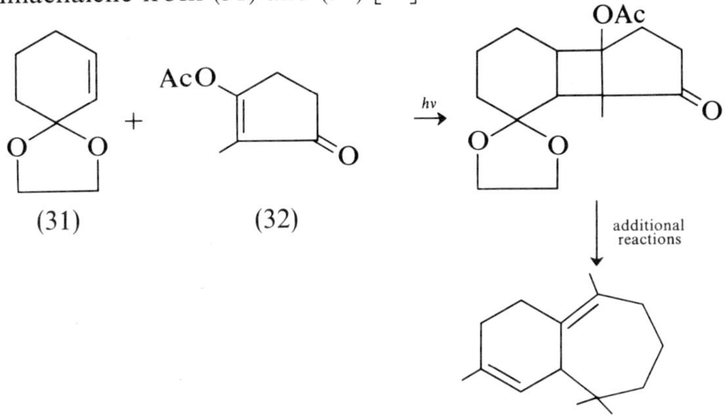

(31) (32)

An annotinine derivative (33) has been prepared in a similar way [46].

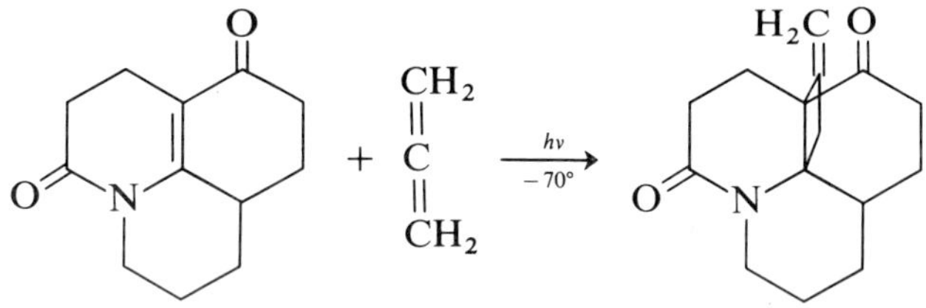

(33)

Ethylene itself undergoes photoaddition to cyclohexanones to give bicyclo[4,2,0]octanones without substituents on the four-membered rings. These compounds are useful as synthetic intermediates [47].

Photochemical cycloaddition of an enolic β-diketone to an olefin and subsequent thermal retroaldolization of the intermediate 2-acylcyclobutane to give a δ-diketone has become known as the de Mayo reaction and may well be of considerable synthetic interest in the future [48].

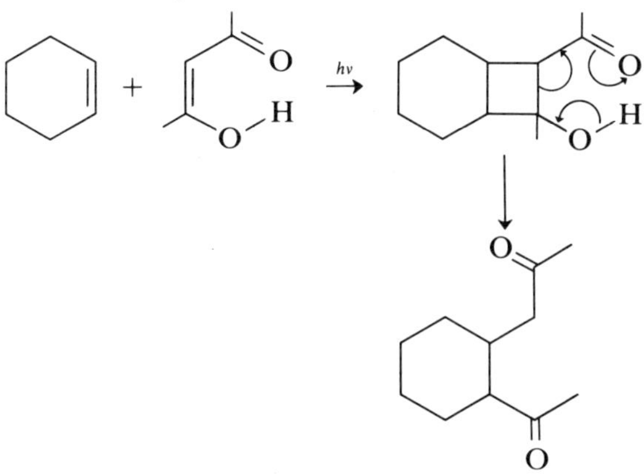

Acetylacetone can also be added to cyclic unconjugated dienes such as 1,5-cyclo-octadiene, but not to conjugated dienes. Additions of 2-acetylcycloalkanones to cyclic olefins, to give δ-ketones which can be cyclized by dehydration condensation to $\alpha\beta$-unsaturated ketones having three or more condensed rings, present considerable synthetic possibilities [49].

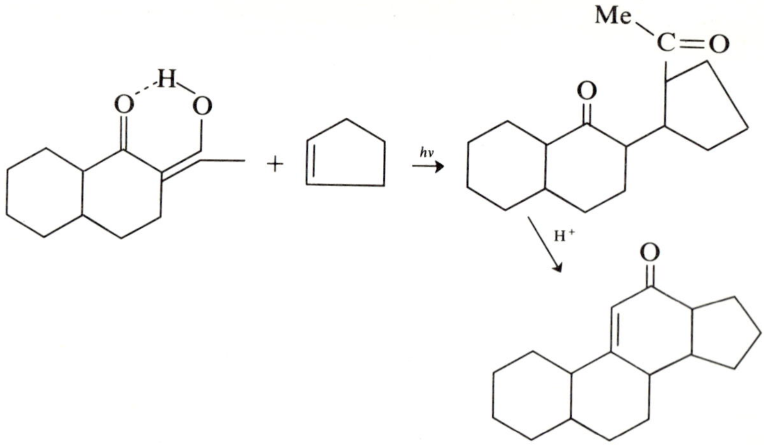

Corey has rigorously studied the photochemical reactions of 2-cyclohexenones with substituted olefins to determine the scope, orientation, and stereochemistry of the addition. Orientation of the substituent is such that the α carbon of 2-cyclohexenone becomes attached to the more nucleophilic carbon of the double bond. *Trans*-ring-fused products are often the major products formed. 1,1-Dimethoxyethylene, methoxyethylene, cyclopentene, isobutylene, allene, and acrylonitrile were increasingly less reactive to attack by excited 2-cyclohexenone [50].

Seven-membered ring ketones do not give photoadducts. However, *cis*-2-cycloheptenone undergoes *cis-trans* isomerization, and the *trans* isomer undergoes dimerization and mixed addition reactions in the dark [51].

The addition of β-nitrostyrene to olefinic compounds is stereospecific. The phenyl and nitro group are always *trans* to each other [52].

Internal cycloaddition reactions are important [53], and this route has been used in the synthesis of cubane [54], homocubyl derivatives, bullvalene, and several natural products [55].

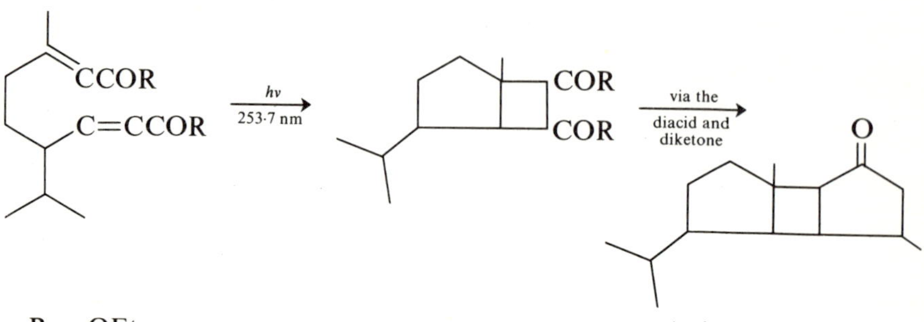

R = OEt

(±)-α-bourbonene

The photosensitized additions of olefins and acetylenes to maleic anhydrides and maleimides have been extensively studied [56]. The stereochemistry of the ring junction is *cis*, but that of the substituents may be *trans*. Maleic anhydride will also form adducts with cyclic 1,3-dienes [57, 58].

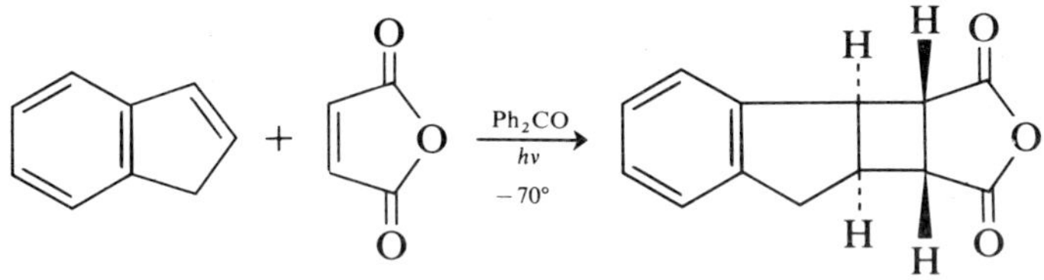

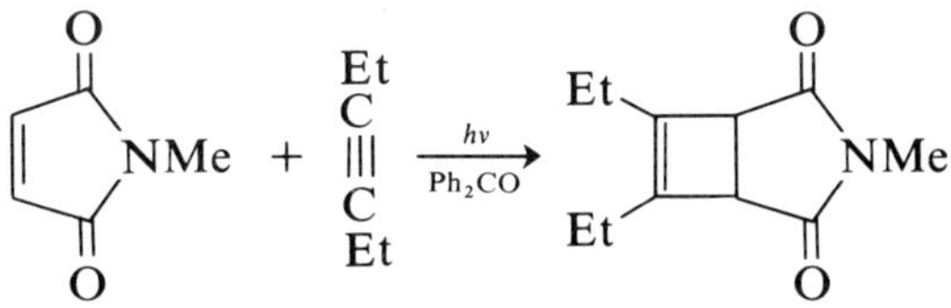

Several interesting reactions between quinones and conjugated dienes with the formation of a variety of ring systems have been reported [59]. Specific excitation of the $n \rightarrow \pi^*$ transition of benzoquinone (34) in the presence of 1,3-dienes gave spirodihydropyran derivatives (35). A macrocyclic ether (36) was also formed on irradiation with buta-1,3-diene.

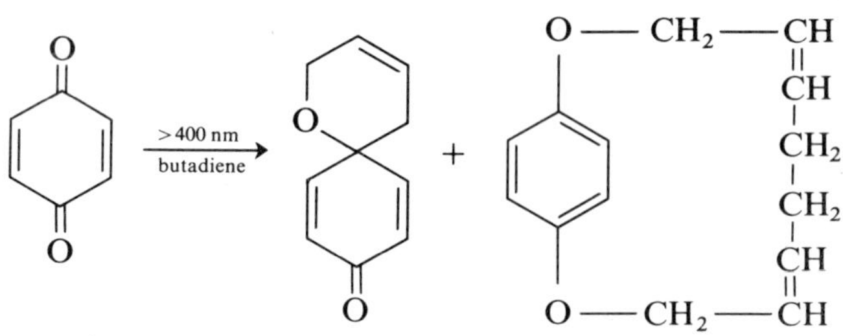

(34) (35) (36)

Additions of non-conjugated olefins and 1,3-dienes to duroquinone all gave cyclobutane derivatives (37); however, chloranil with cyclo-octene formed a 1,2-adduct (38) and a spiro-oxetan.

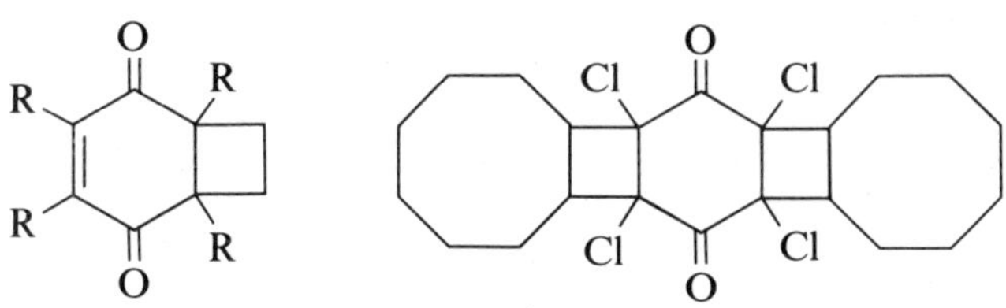

(37) (38)

Barltrop has correlated the nature of the product formed with the electrophilicity of a particular centre in an excited quinone molecule. The n,π^*

excited state results in an electron-deficient oxygen atom and hence addition to the carbonyl group, whereas $\pi \rightarrow \pi^*$ excitation leads to an electron-deficient ring carbon atom and hence cyclobutane formation. Variation of the substituents varies the nature of the lowest excited states, and hence the ring system formed.

o-Quinones form dioxans by addition to olefins [60]. This would appear to be a useful route to the dioxan ring system.

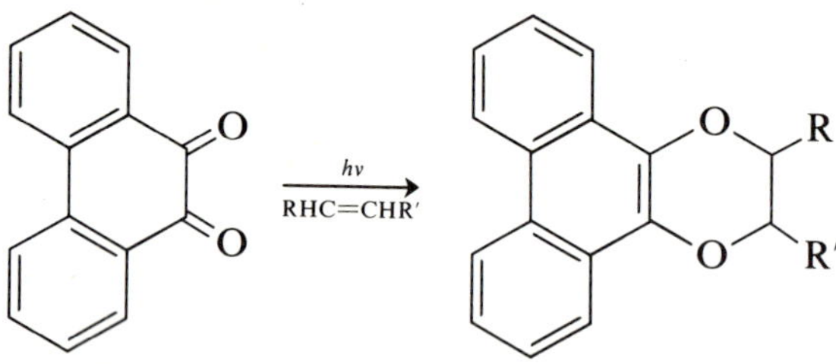

The photochemistry of benzene is of considerable current interest [61]. The reactions may be conveniently divided into isomerization and addition reactions, both of which are of considerable synthetic utility. The isomerization reactions will be reviewed later on in this chapter.

The addition reactions are usually with an ethylenic or acetylenic addend, e.g., maleic anhydride and benzene form the stereospecific 2:1 adduct (39) [62]. Substituted benzenes react in a similar manner but to a lesser extent, as a result of steric and electronic factors. Condensed aromatic compounds

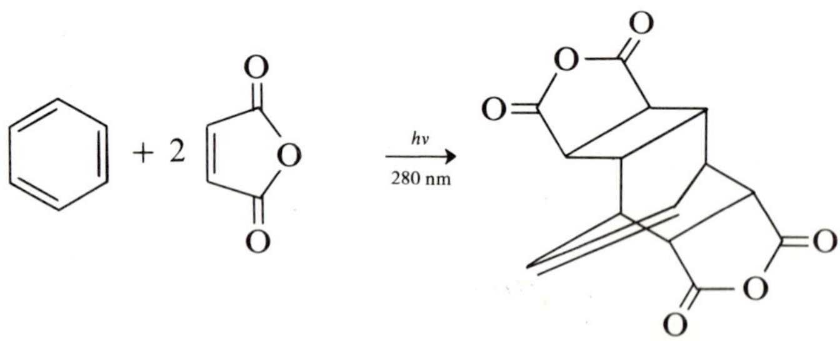

(39)

form only 1:1 adducts with maleic anhydride. When the site of addition has olefinic character, such as the 9,10-bond in phenanthrene, cyclobutane derivatives are formed. In other cases the normal Diels–Alder adducts are preferred [63].

Benzene will readily form 1,2-, 1,3-, and 1,4-adducts. Cyclobutane adducts have been isolated from the irradiation of benzene and olefins [64]. The addition of acetylenes to benzene yielded cyclo-octatetraenes in preparative quantities [65].

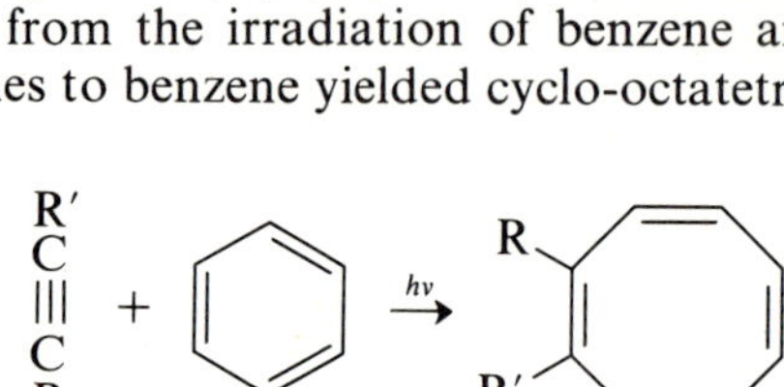

The photochemical 1,3-cycloadditions of olefins to benzene have made a novel ring system available [66].

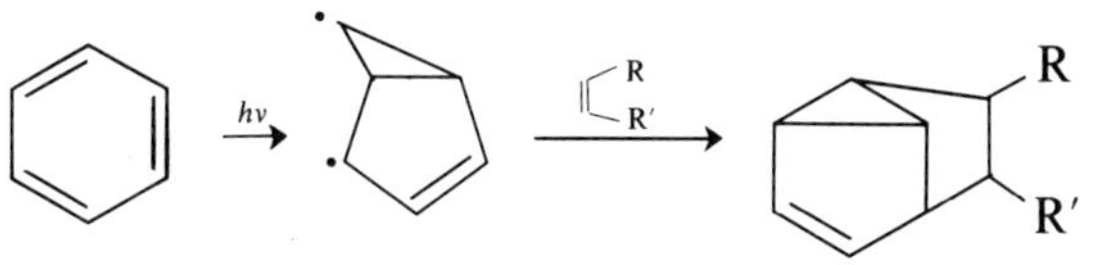

The 1,4-addition reactions of benzene are of interest. Butadiene adds to benzene [67].

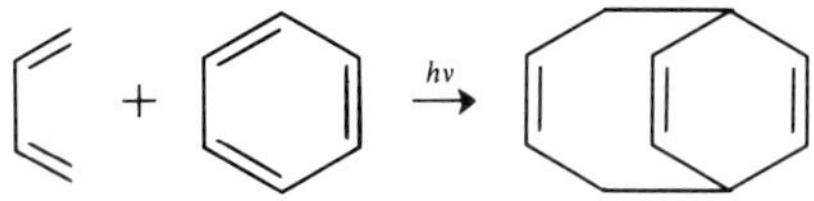

The reductive amination reactions form 1,4-adducts [61].

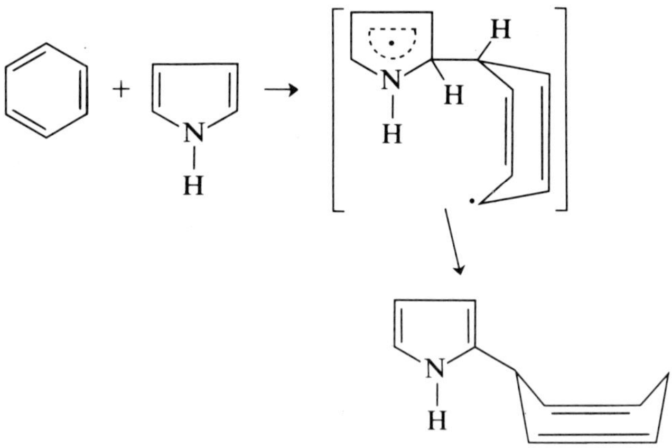

Other photoaddition reactions of synthetic interest include the addition of carbonyls [68], acetals [69], alcohols [70], ethers [71], amines [72], and amides to olefins. Mainly Markovnikov-type addition occurs in these reactions.

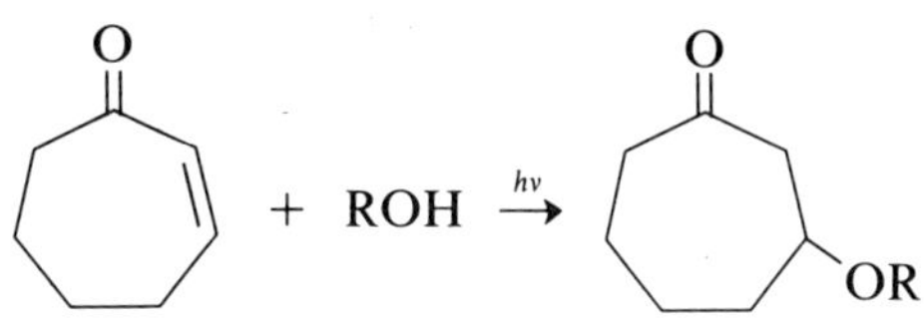

Amines, and even ammonia, readily add to olefins to give adducts [73], and add to styrenes in an anti-Markovnikov manner as expected [72].

$$R'R''C{:}CH_2 + ClNHCO_2Et \xrightarrow{h\nu} \underset{\text{mainly}}{R'R''CCl{-}CH_2NHCO_2Et} + R'R''C(NHCO_2Et)CH_2Cl$$

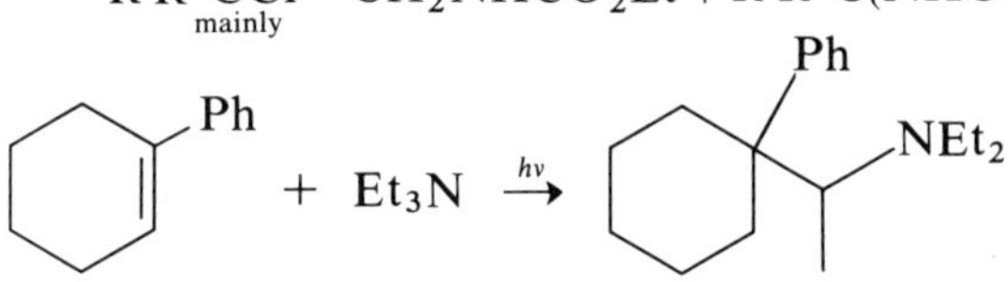

Elad has made several important contributions to this field, having studied the light-induced additions of acetals to olefins [69] and the amidation of olefins and $\alpha\beta$-unsaturated ketones [74]. The photoamidation of acetylenic compounds has also been reported [75]. The formation of 2:2 adducts by photochemical addition of formamide to terminal acetylenes appears to have synthetic possibilities [75].

$$\mathrm{HCONH_2} \xrightarrow[\text{Sens.}]{h\nu} \cdot\mathrm{CONH_2} \xrightarrow{\mathrm{RC{\equiv}CH}} \mathrm{R\dot{C}{=}CHCONH_2} \xrightarrow{\mathrm{R'H}}$$

$$\mathrm{RCH{=}CHCONH_2} \xrightarrow{\cdot\mathrm{CONH_2}} \underset{\underset{\textstyle (A)}{\mathrm{CONH_2}}}{\mathrm{RCH\dot{C}HCONH_2}} \xrightarrow{\mathrm{R'H}} \underset{\mathrm{CONH_2}}{\mathrm{RCHCH_2CONH_2}}$$

$$\mathrm{(A)} \xrightarrow[\text{(ii) R'H}]{\text{(i) RC}\equiv\text{CH}} \underset{\mathrm{H_2NOC}\quad\ \mathrm{CH{=}CHR}}{\mathrm{RCH{-}CHCONH_2}}$$

The addition of polyhalogenomethanes to olefins forms adducts in high yields and has been applied to organic synthesis [76].

The photoaddition of nitrobenzenes to alkenes is worthy of note, as it gave the first members of the class of cyclic derivatives of trivalent nitrogen containing two N—O bonds [77].

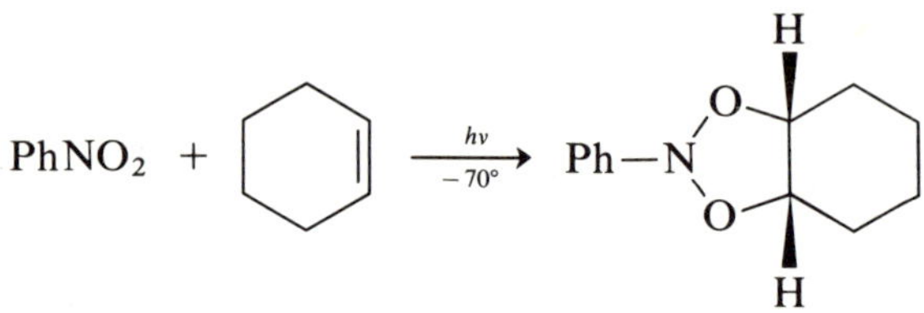

The synthesis of C-2 alkylpyrimidines by the photoinduced methylation of pyrimidines has been reported [78].

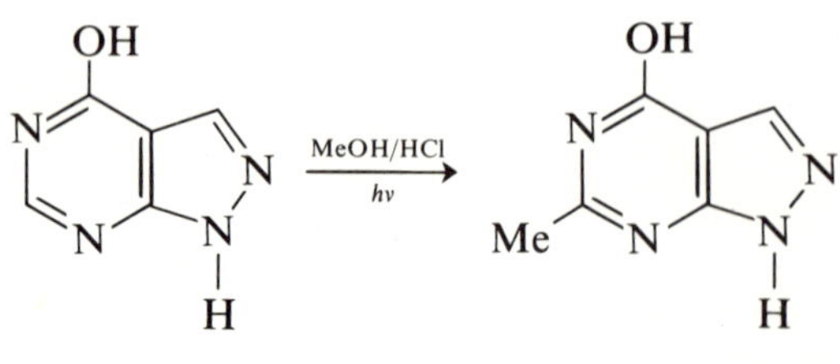

(67%)

Dimerization results in the formation of cyclobutanes (40), Diels–Alder-type adducts (41), and 1,4–1,4 addition products (42) [31]. The ease of formation of dimers is also a great nuisance in a photochemical synthesis, as dimers are often readily formed in any photochemical reaction involving double bonds. Dilute solutions are used to minimize dimerization, but this reduces the quantities of materials that can be handled easily.

The actual products isolated depend on the particular sensitizer (S*), if any, used. For example in the photosensitized dimerization of butadiene, which does give important products [79],

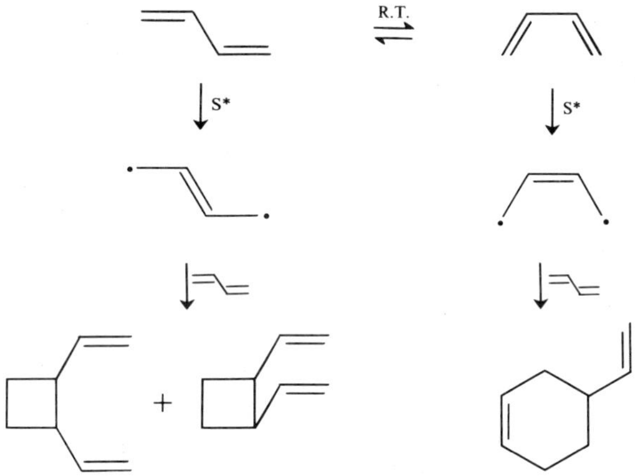

(40) (41)

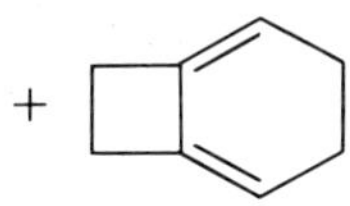

(42)

the *cisoid* diene requires less energy to convert it to the triplet state than does the *transoid* diene. Hence a sensitizer with the higher energy gives a higher percentage of products derived from the *transoid* conformer.

Woodward and Hoffmann have derived some simple selection rules governing the terminal cycloaddition of an m-π-electron system to an n-π-electron molecule [80]. The rules are:

$m+n$	Mode of thermal addition	Mode of photochemical addition
$4q$	*trans-cis* *cis-trans*	*cis-cis* *trans-trans*
$4q+2$	*cis-cis* *trans-trans*	*cis-trans* *trans-cis*

The *trans-trans* and *cis-trans* cycloadditions appear sterically difficult and are very rare. Thus photochemical additions are usually between compounds possessing a total number of π-electrons which is a multiple of four, and the stereochemistry of the addition is *cis-cis*. There are very few examples of photochemical *cis-trans* addition in the case of $4q+2$-π-electron systems [81].

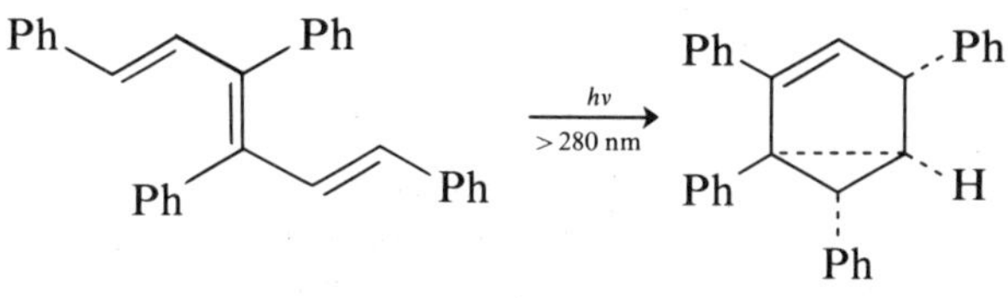

(87%)

6.3 PHOTOCYCLIZATION REACTIONS

These reactions are of considerable importance in the preparation of fused-ring aromatic and heterocyclic compounds. Recently photocyclization reactions have been applied, with considerable success, to the synthesis of natural products.

The photocyclization of anthrone and bianthrone derivatives has been known for a long time and has been employed extensively in synthesis [82, 83]. Thus the *Hypericum* pigment, hypericin (44), may be prepared from the protohypericin derivative (43) [82].

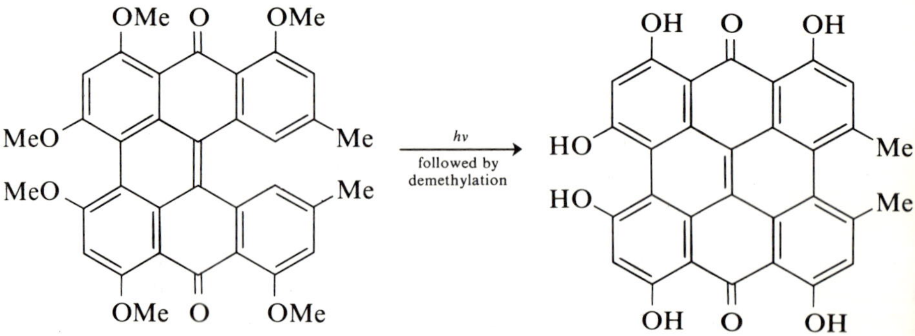

(43) (44)

The photocyclization of conjugated trienes closely parallels the photocyclizations of the anthrone systems and is of some importance in the vitamin D field. Thus irradiation of precalciferol (45) gave ergosterol (46) [84].

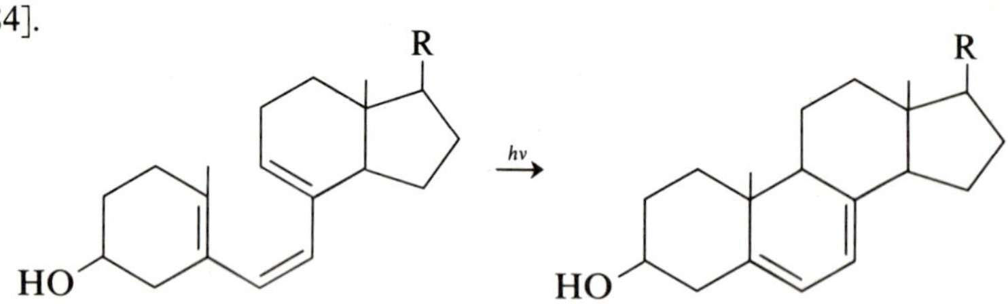

(45) (46)

Several simple hexatrienes have been photocyclized. A low yield of 1,2-dimethyl-$\Delta^{3,5}$-cyclohexadiene (48) was obtained from 1,6-dimethyl-1,3,5-hexatriene (47). An equilibrium was set up between the two compounds,

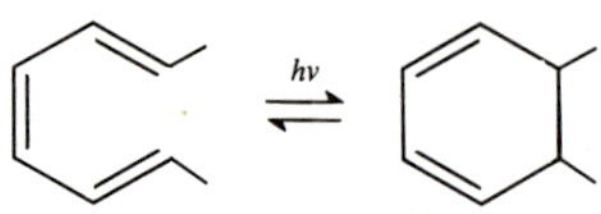

(47) (48)

which lay mainly to the left [85]. Chrysene and 1-phenylnaphthalene can be readily prepared by the photocyclization of 1,6-diphenylhexa-1,3,5-triene and 1,4-diphenylbuta-1,3-diene respectively [86]. There are many other examples of the photocyclization of conjugated trienes in synthesis [87–89].

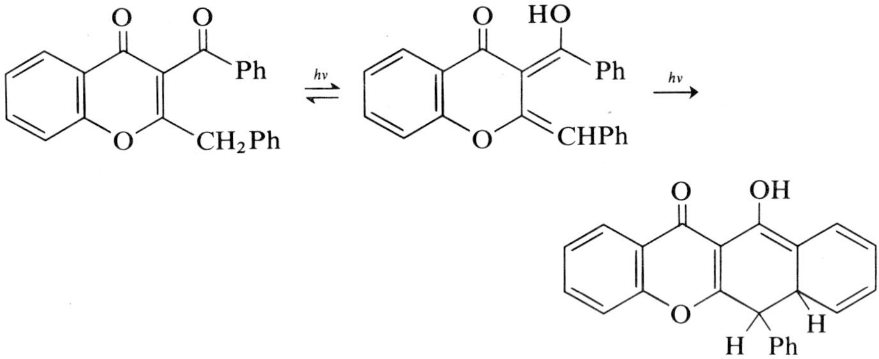

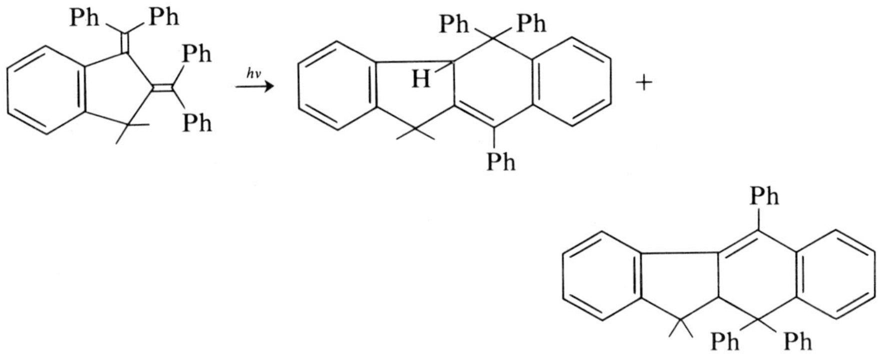

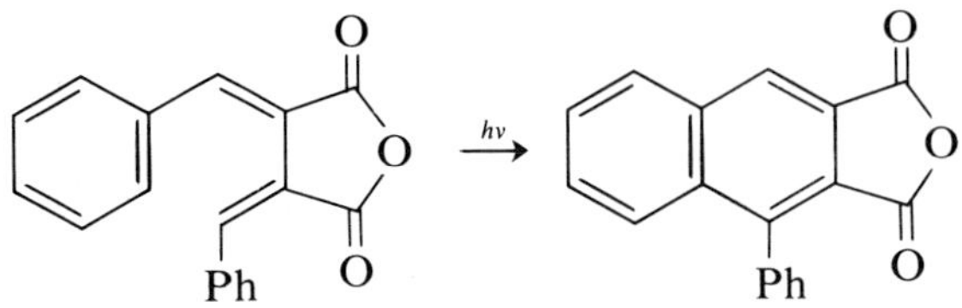

Simple 1,3-dienes also undergo photocyclization [90, 91].

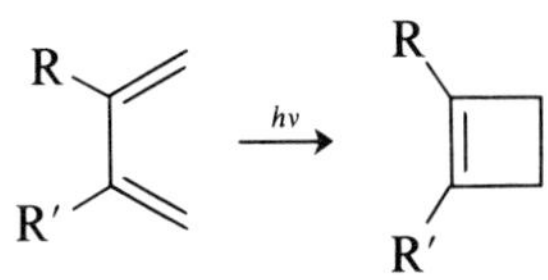

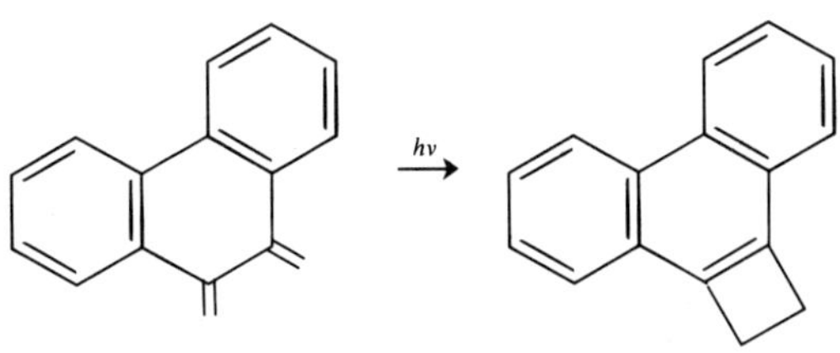

Stilbene (49) possesses the conjugated triene system, and is readily photocyclized to phenanthrene (51) [92]. This reaction provides an easy method of preparing aromatic and heterocyclic ring systems in reasonable yields. However, it is only recently that the full potential of this reaction has been exploited. The mechanism of the photocyclization reaction has been extensively studied [93, 94]. Cyclization takes place via the excited singlet state of *cis*-stilbene (49) to give the dihydro intermediate (50), which undergoes aromatization in a dark, oxidative reaction to phenanthrene (51).

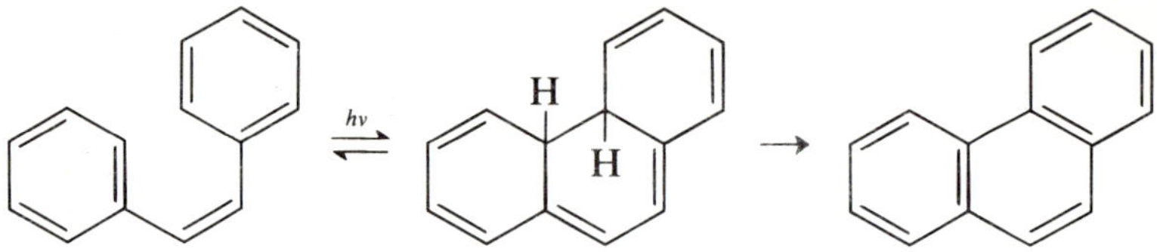

(49) (50) (51)

The scope of this photochemical preparation of phenanthrenes has been explored in detail by Mallory and his co-workers [95]. The method fails only for stilbenes having acetyl, dimethylamino, or nitro substituents. Iodo substituents are lost because of the easy photolysis of C—I bonds. This reaction has been utilized in the synthesis of aristolochic acid (53), which involved the photocyclization of an iodosubstituted nitro stilbene (52) which would not photocyclize under the normal conditions. This is an example of a photochemical Ullmann reaction [96] and may be used to prepare other ring systems. Thus 6,7-dihydro-5*H*-dibenz[*c*,*e*]azepine (54) has been prepared in a similar way in good yield [97].

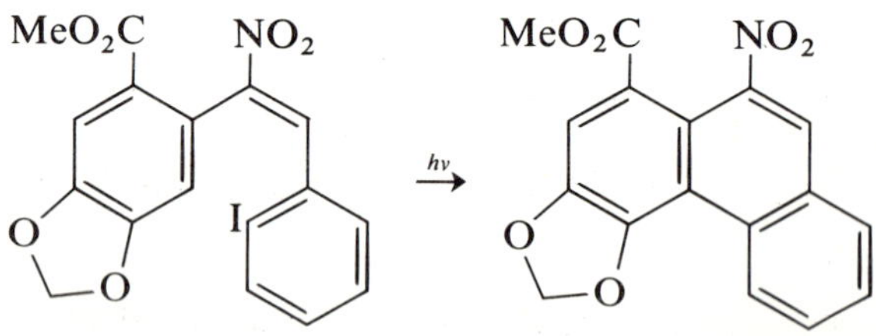

(52) (53)

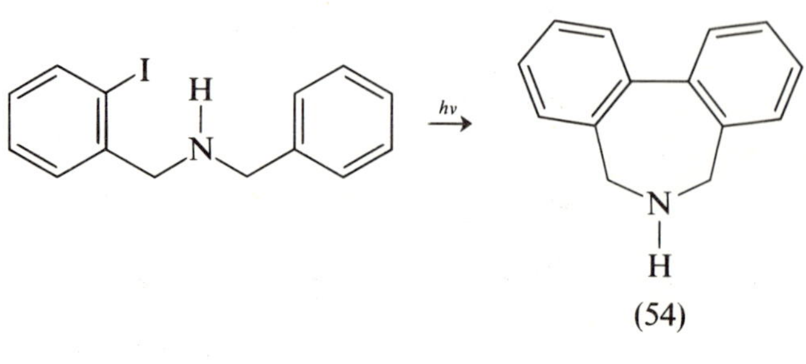

(54)

The most satisfactory conditions for the normal photocyclization reactions involve the use of dilute (0.01 molar) solutions of the stilbene in cyclohexane to prevent the formation of dimers. Excellent yields are obtained. The presence of an oxidizing agent is necessary. Oxygen dissolved in the solvent, iodine, or ferric chloride are the most common oxidants. The use of copper halides as catalysts in the photocyclization of some α,α′-disubstituted stilbenes has been reported [98]. Irradiation of 4,4′-dimethoxy-α,α′-dimethyl stilbene gave only 0·6 % 3,6-dimethoxy-9,10-dimethylphenanthene on irradiation in ethanol in the presence of iodine. Addition of cupric chloride or bromide improved the yield to 54 %. Irradiation of the stilbene in the presence of iodine in cyclohexane solution gave 11 % of the phenanthrene and also 9 % *p*-methoxyacetophenone. The formation of the acetophenone is the first photo-oxidative cleavage to be reported in these systems. Levi and Orchin have used selenium radicals in the photocyclization aromatization of several sterically hindered stilbene analogues [99]. Stilbenes substituted at the point of cyclization have been shown to eject the substituent, and substituent migration has not been observed [95, 100, 101].

Phenanthrene has also been isolated from the photolysis of *cis*-stilbene sulphide [102].

9,10-Dihydrophenanthrenes may be prepared by the degassed irradiation of stilbenes with electron-withdrawing substituents in the α,α′-positions [103].

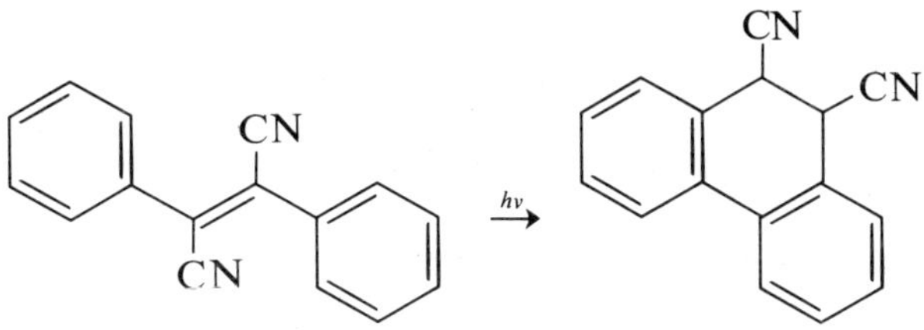

Photocyclization reactions also occur with the styrylnaphthalenes and provide excellent syntheses of chrysene and benzo[*c*]phenanthrene derivatives [104, 105]. Cyclization in the case of 2-styrylnaphthalene (55) is always to the α-position of the naphthalene nucleus.

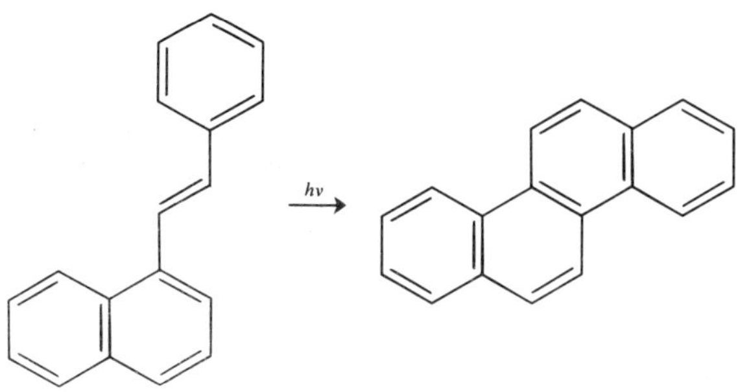

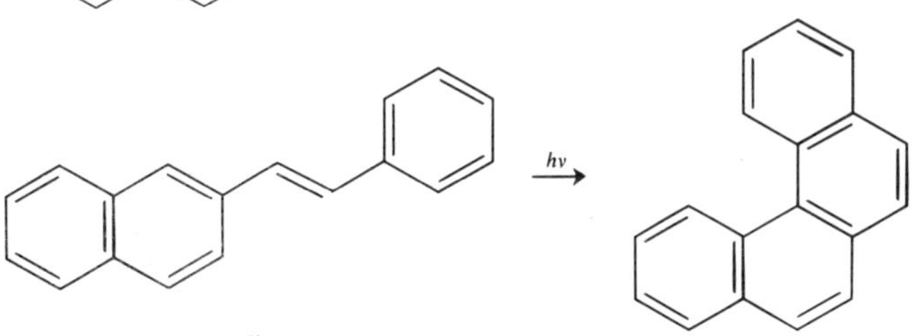

(55)

Benz[*a*]anthracene, benzo[*g*]chrysene, and benzo[*g,h,i*]perylene have all been prepared in this manner [106]. The syntheses of hepta-, octa-, and nonahelicene from the corresponding symmetrical stilbene analogues have been accomplished [107]. These are very simple reactions and provide the method of choice for the preparation of these complex molecules.

1,8-Distyrylnaphthalene (56) did not photocyclize as expected but gave the novel ring system (57) [108].

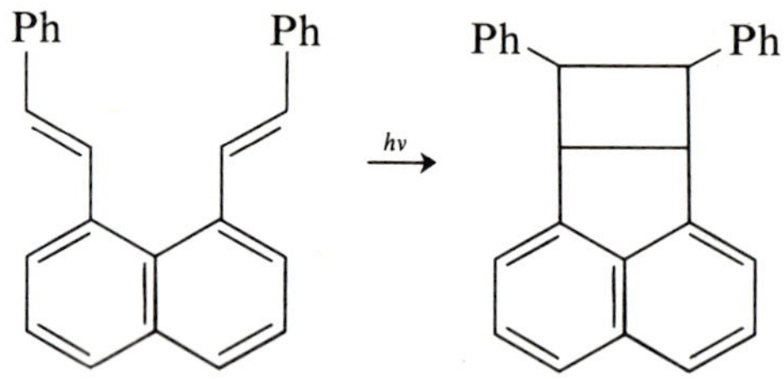

(56) (57)

Styrylpyridines (stilbazoles) photocyclize in a similar manner to the stilbenes and so provide an excellent route to the relatively inaccessible azaphenanthrenes [109]. Thus benzo[*f*]quinoline (58), benz[*f*]isoquinoline (59), and benz[*h*]isoquinoline (60) may be prepared by the photochemical cyclodehydrogenation of 2-, 3-, and 4-stilbazole respectively in cyclohexane. 3-Styrylpyridine has also been reported to photocyclize to benzo[*h*]quinoline (61) [110]. 1- and 3-cyclohexylbenz[*h*]isoquinoline, (62) and (63), have been isolated in the course of the photocyclodehydrogenation of 4-stilbazole to benz[*h*]isoquinoline (60) in cyclohexane [111].

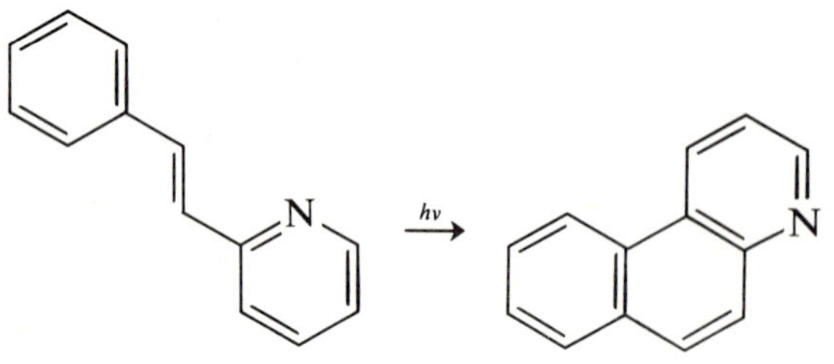

(58)

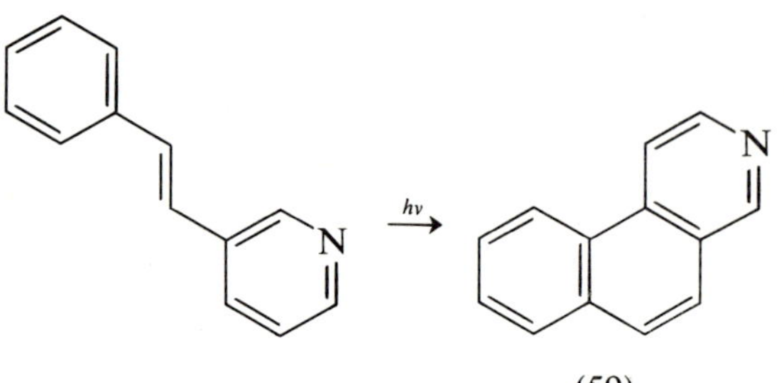

(59)

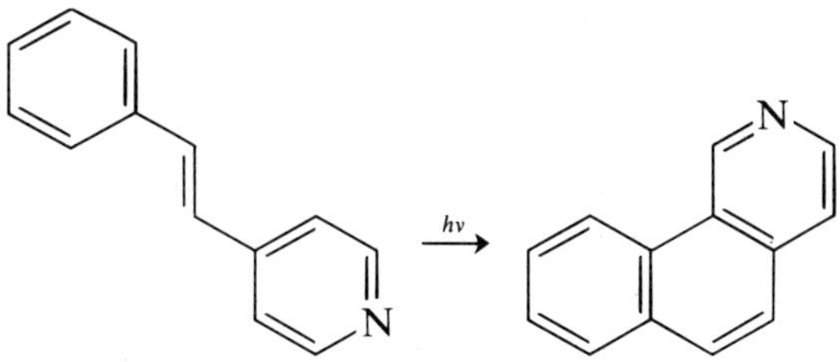

(60)

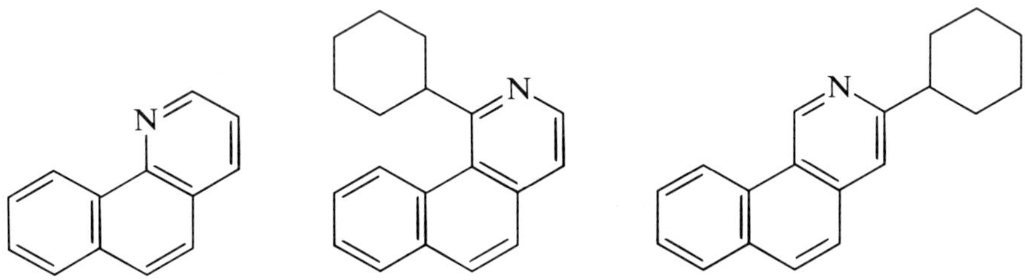

(61) (62) (63)

Benzo[*f*]quinazolines may be prepared by the photocyclodehydrogenation of 4-styrylpyrimidines [112].

Phenanthroline derivatives are readily available via the photocyclization of the dipyridylethylenes [113]. Similarly, the styrylquinolines and styrylisoquinolines photocyclize in the expected manner [114].

A route to the benzo[*a*]quinolizinium salt (65) has been developed by Doolittle and Bradsher via the photocyclization of 1-styrylpyridinium salts (64) [115].

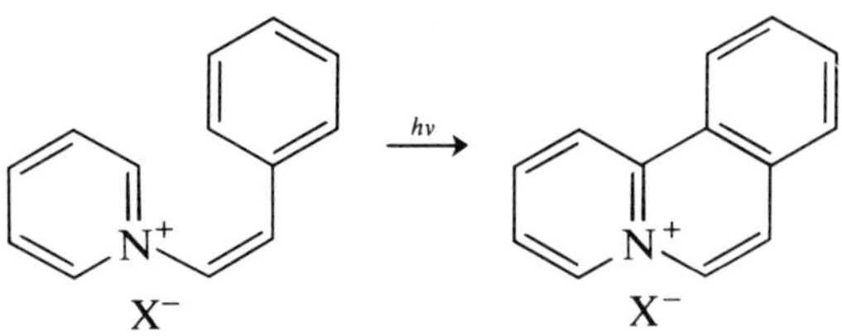

(64) (65)

Benzo[*c*]cinnolines (67) and phenanthridines (69) may be prepared by the photocyclization of azobenzenes (66) and benzalanilines (68) respectively. However, conditions differ from those for stilbene cyclizations [116, 117].

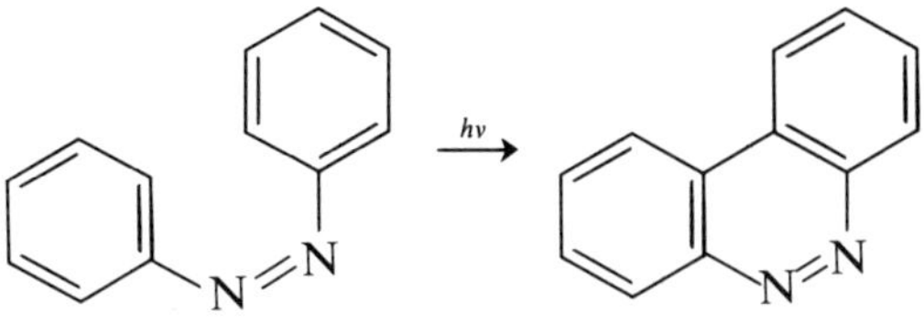

(66) (67)

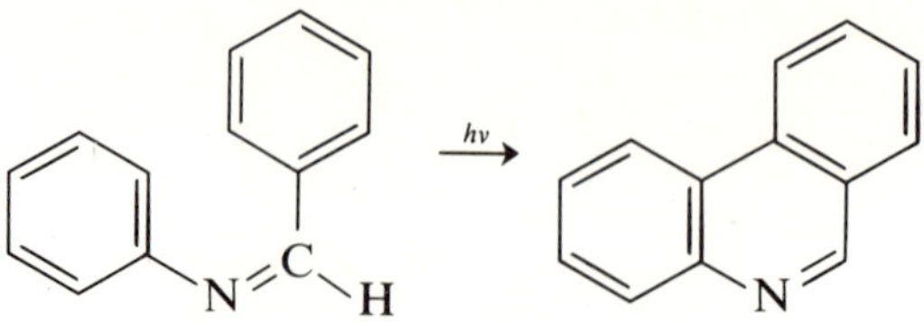

(68) (69)

The azobenzenes and the benzalanilines both photocyclize in concentrated sulphuric acid and also yield reduced compounds in the reaction; hydrazobenzene (70) and benzylaniline (71), respectively. The mechanisms of both reactions are closely related.

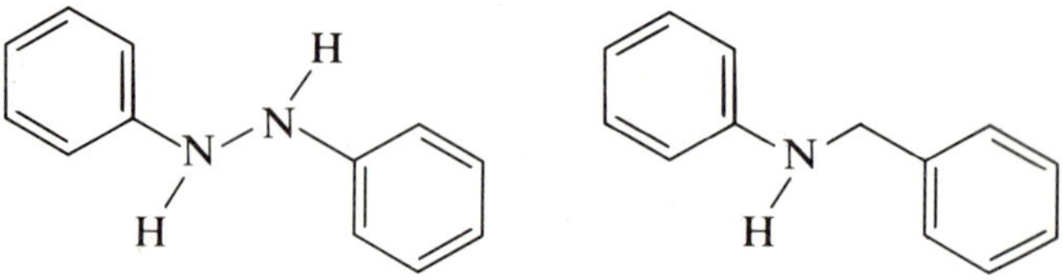

(70) (71)

2,4,6-Trimethylazobenzene gave some 1,2,4-trimethylbenzo[*c*]cinnoline on irradiation. This is the only example of substituent migration that has been reported [118].

The synthetic uses of these reactions are exemplified by the synthesis of calycanine (72) [119].

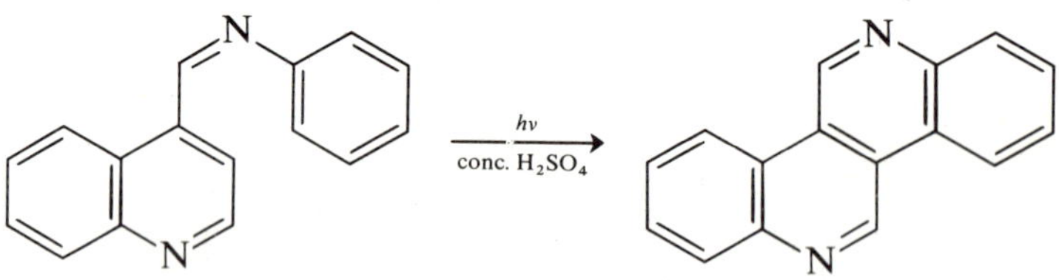

(72)

An interesting reaction of benzalanilines substituted by certain *ortho*-substituents (73) has been reported [120].

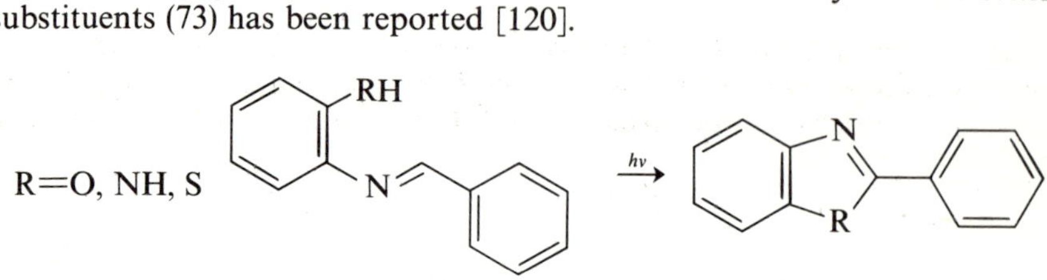

(73)

Some aporphines (74) may be synthesised by utilizing photocyclization reactions [121].

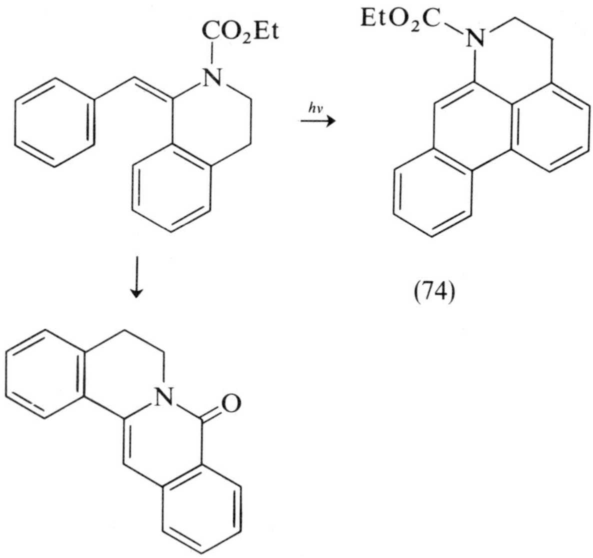

(74)

These reactions have also been utilized in the synthesis of protoberberine alkaloids (e.g., 75) [122].

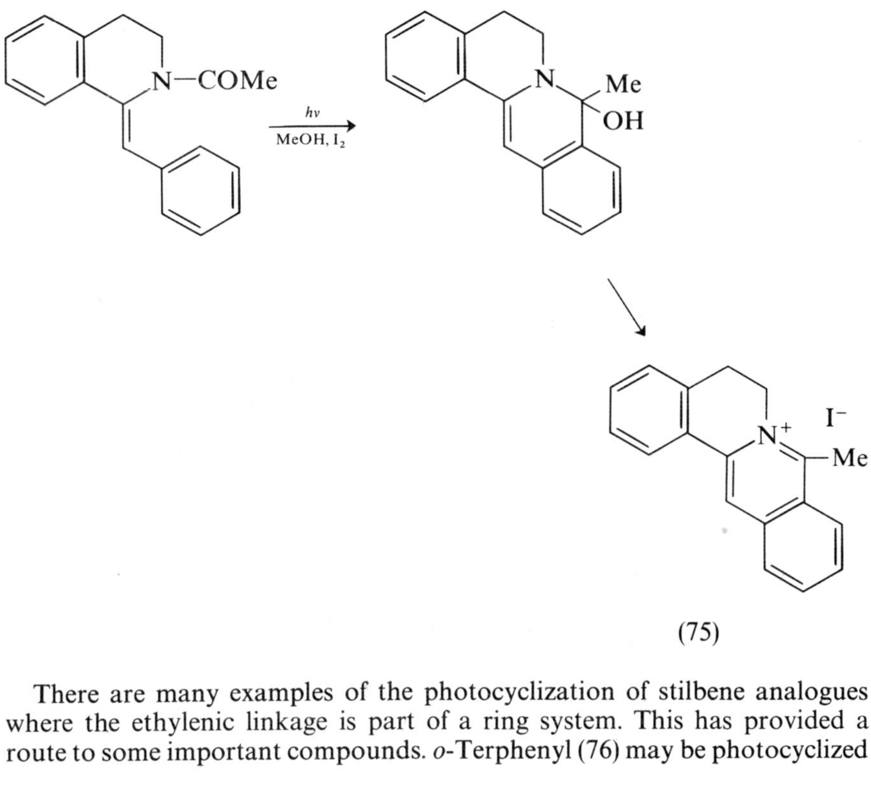

(75)

There are many examples of the photocyclization of stilbene analogues where the ethylenic linkage is part of a ring system. This has provided a route to some important compounds. *o*-Terphenyl (76) may be photocyclized to triphenylene (77) in the presence of iodine [123]. Similarly, *o,p*-quaterphenyl gave 2-phenyltriphenylene.

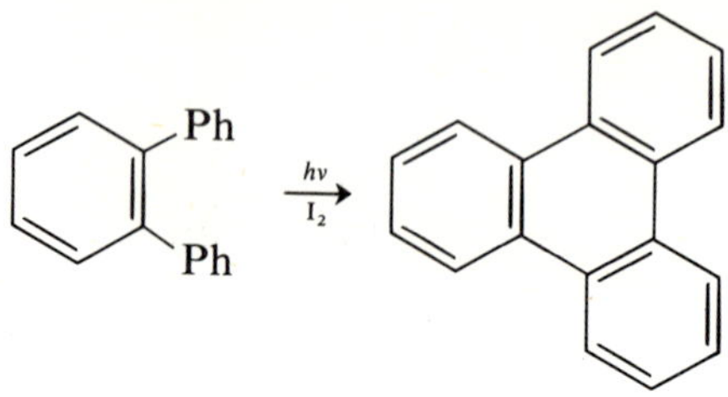

(76) (77)

Cyclization also occurred on irradiation of a tetraphenylcyclopentadienone (78) [124], but only when a protic solvent was used.

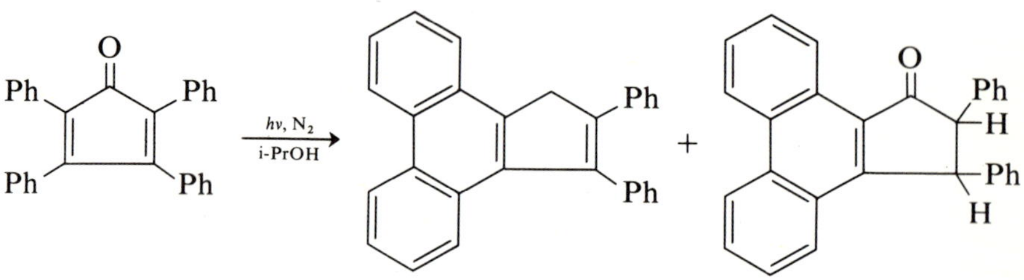

(78)

2,3-Diphenylthiophen (79), 2,3-diphenylfuran, and meso-ionic 1,3,4-thiadiazoles (80) all photocyclize as expected [125–127].

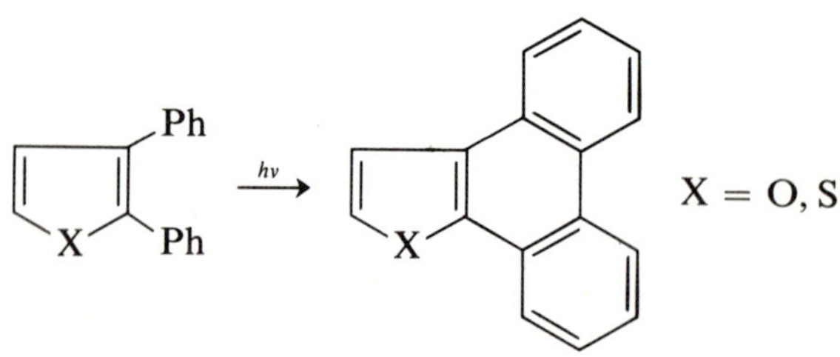

(79)

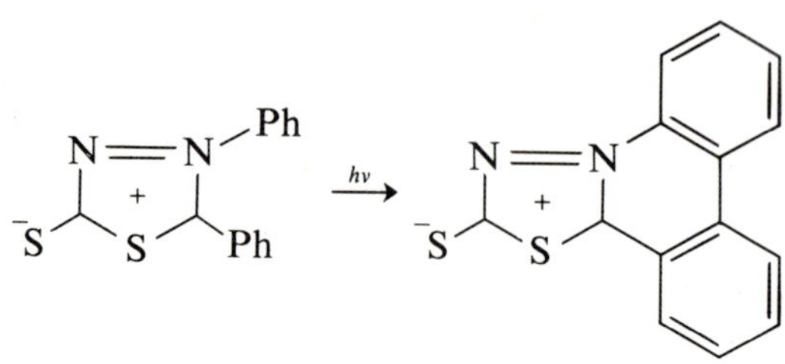

(80)

Diphenylcyclopropenone does not photocyclize but ring-opens to give tolan. Cyclobutanophenanthrene has been detected on irradiation of 1,2-diphenylcyclobutene, but this does not seem to be a reaction capable of synthetic use in view of the low yields obtained [128].

Styrylthiophens, styrylfurans (81), dithienylethylenes, and thienylfurylethylenes (83) all photocyclize as expected to give their respective products in good yields [129]. This photocyclization presents an attractive synthetic route to these heterocyclic ring systems.

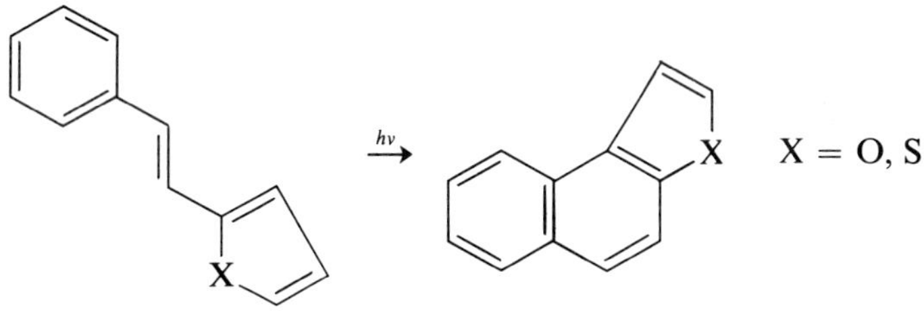

(81) (82)

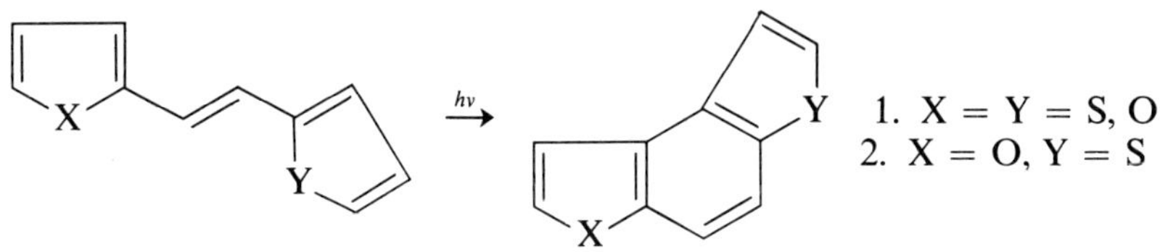

(83) (84)

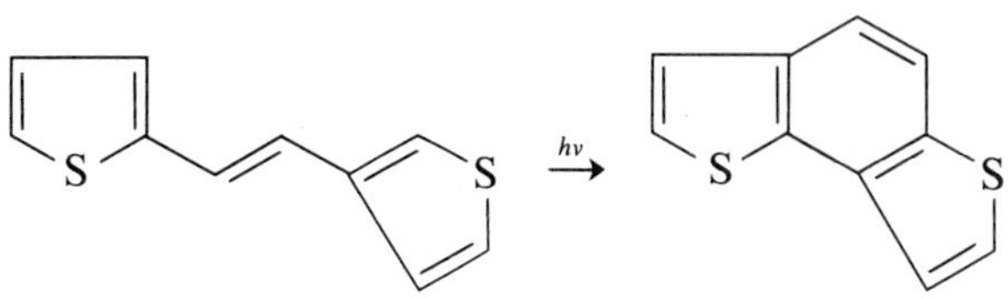

(85) (86)

Carbazoles may be prepared by the photocyclization of diphenylamines. Thus 4-methoxy-4′-methyldiphenylamine (87) gave glycozoline (88) on irradiation [130]. Similarly, decachlorodiphenyl sulphide (89) gave the corre-

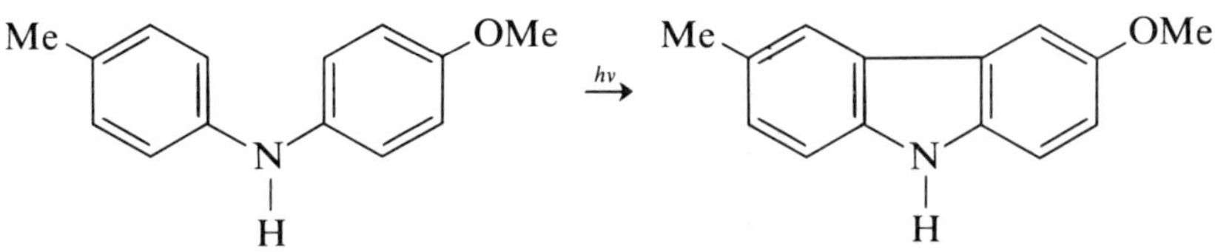

(87) (88)

sponding photocyclized derivative (90) [131]. Clark has reported the photocyclization of some anilinopyridines to carbolines [132].

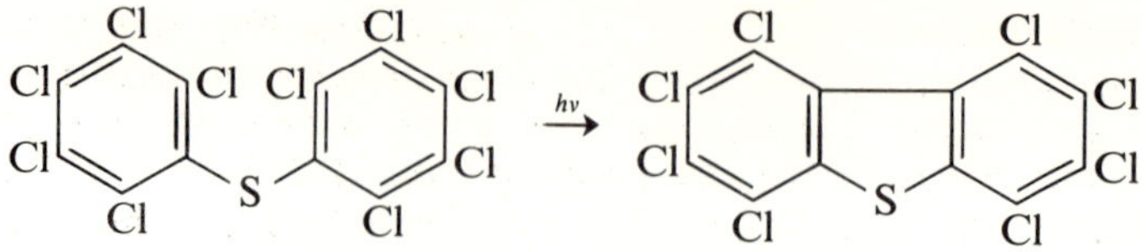

(89) (90)

Other interesting uses of photochemical cyclizations in synthesis include the irradiation of methanolic solutions of the 1,19-diunsubstituted 1,19-dideoxybiladiene-*ac*-dihydrobromides (91) in the presence of base and nickel or cobalt ions to give the corresponding corrole metal complexes (92) [133].

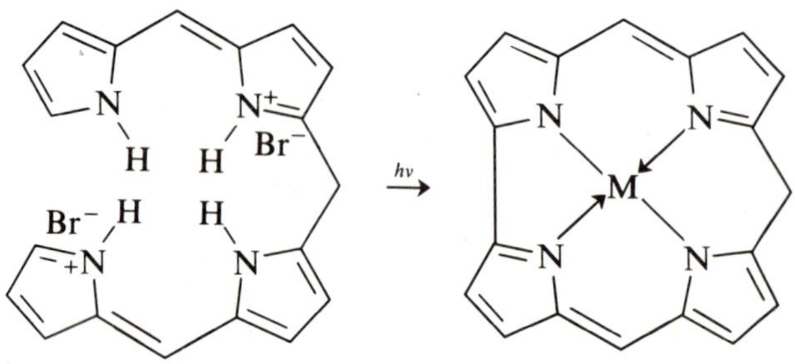

(91) (92)

Irradiation of 2,8-di(phenylethynyl)-biphenyl (93) gave the phenanthrene-like compound (94) [134].

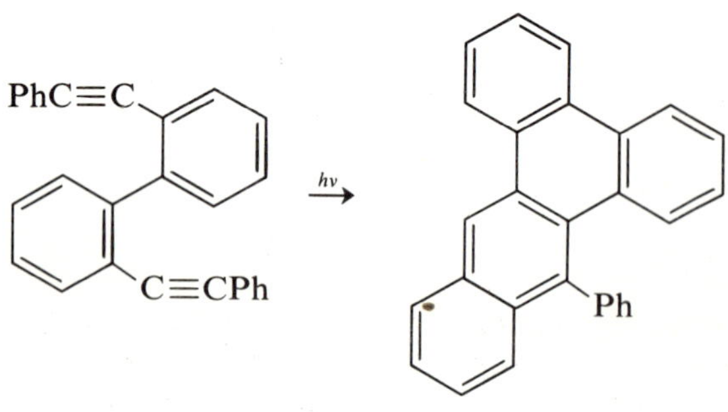

(93) (94)

2-Allylphenols (95) photocyclize [135].

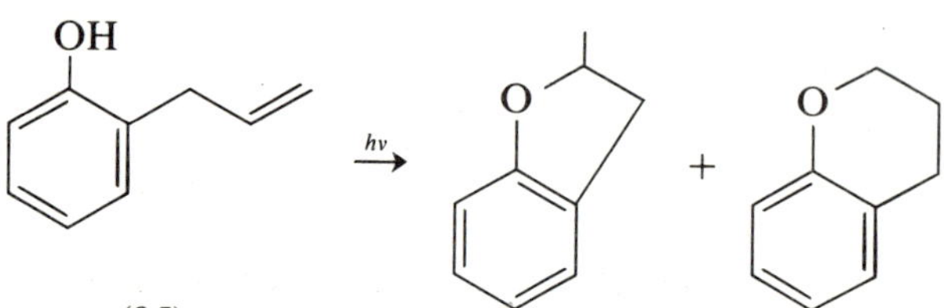

(95)

Many natural products have been made by utilizing a photocyclization step, as we have seen. Other examples include photostreptovarone (96)—a phenalenone—which has been prepared photochemically [136].

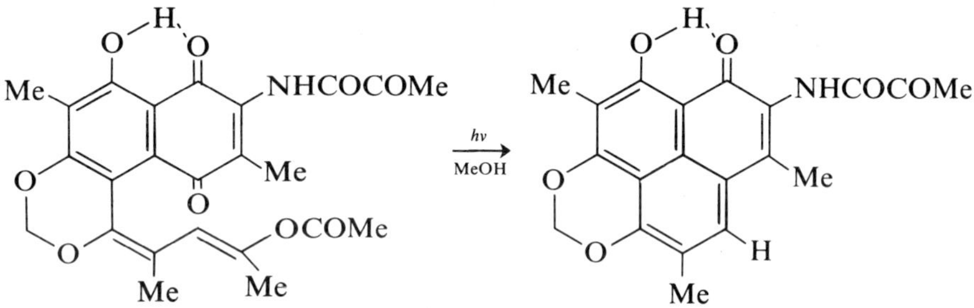

(96)

Irradiation of 3,4-dihydroxycinnamic acid (97), in the presence of oxygen, gave 6,7-dihydroxycoumarin (98) [137].

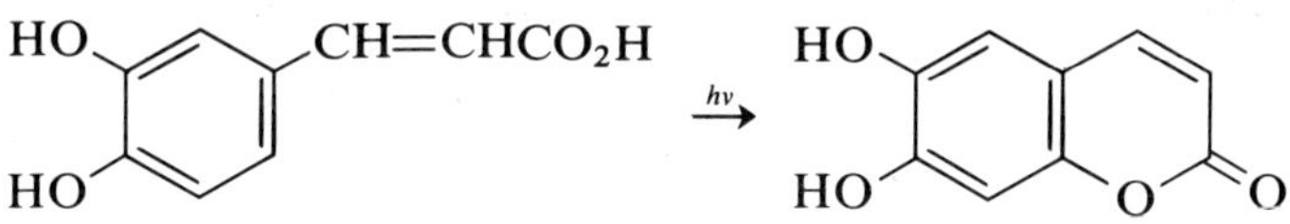

(97) (98)

Swenton has reported the photocyclization of 2-biphenylisocyanate (99) to phenanthridone (100) [138].

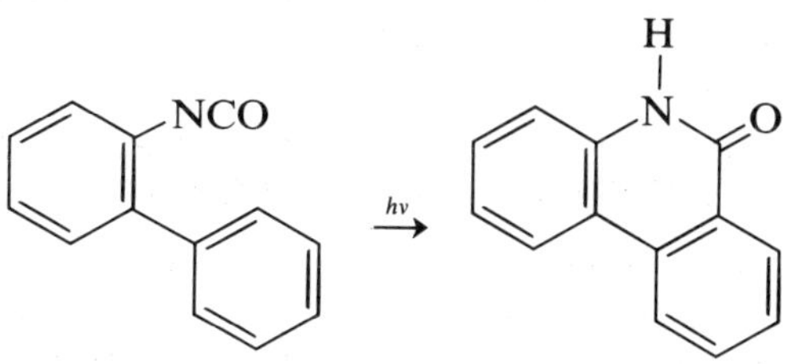

(99) (100)

Several compounds which possess the stilbene chromophore failed to photocyclize. Thus 9,10-diphenylphenanthrene, 2,3,4,5-tetraphenylpyrrole, diphenyltriazine (101), and amarone (102) did not cyclize [139].

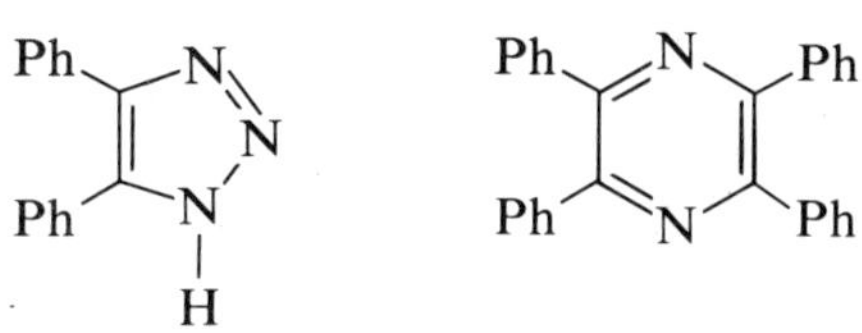

(101) (102)

Several interesting benzofuran syntheses involving photocyclization have been reported [140–142].

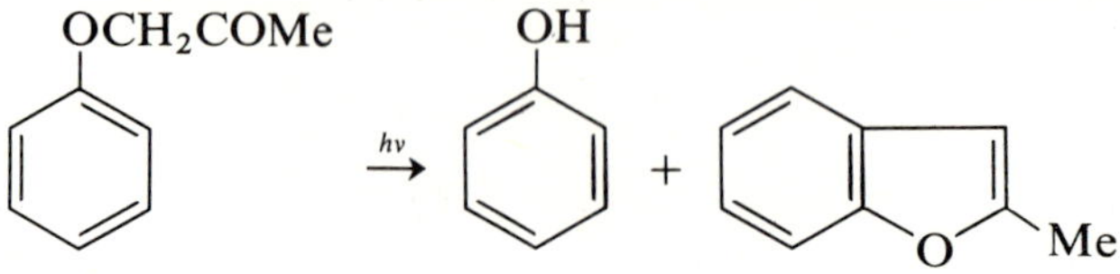

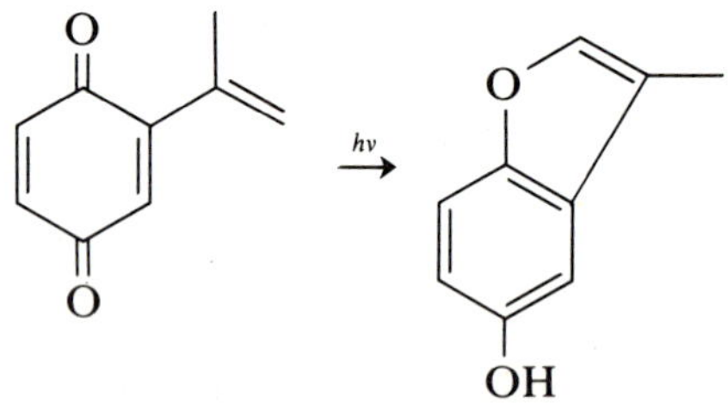

The photocyclization of β-dialkylamino-ketones (103) would appear to be of synthetic use, as it affords a route to azetidines (104) bearing functional groups on ring carbon [143].

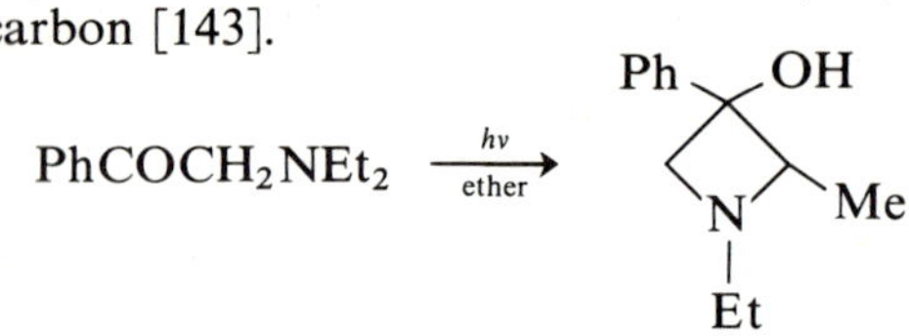

(103) (104)

The photochemical conversion of $3H$-pyrazoles to cyclopropenes appears to provide a convenient route to these compounds [144].

6.4 INTRAMOLECULAR ELECTROCYCLIC TRANSFORMATIONS

Conjugated olefinic systems readily undergo valence-bond tautomerism, and these transformations have made many novel ring systems readily available. The Woodward–Hoffmann theory of electrocyclic reactions will be discussed first, as this enables the prediction of the stereochemistry of the products of these reactions [80].

Consideration of the cyclization of an acyclic olefin containing $k\pi$-electrons (105) to the cyclic olefin (106) shows that the $p\pi$ orbitals on the terminal carbon atoms which lie above and below the plane of the carbon

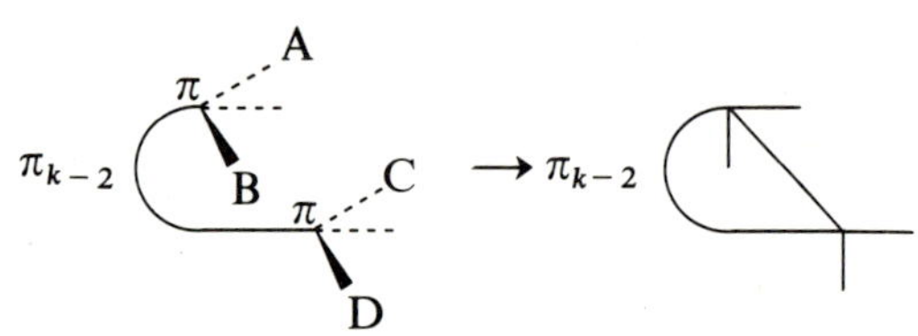

(105) (106)

skeleton in (105) must rotate to lie in the plane of the carbon skeleton in order that they may overlap and form the new bond in (106).

There are two ways in which this can occur:

1. Conrotatory mode;

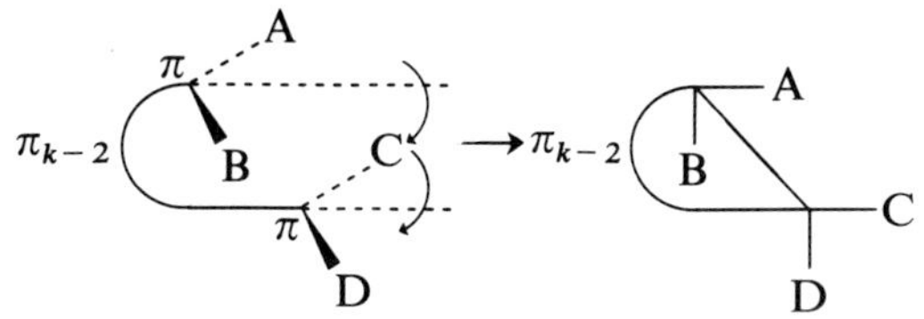

2. Disrotatory mode.

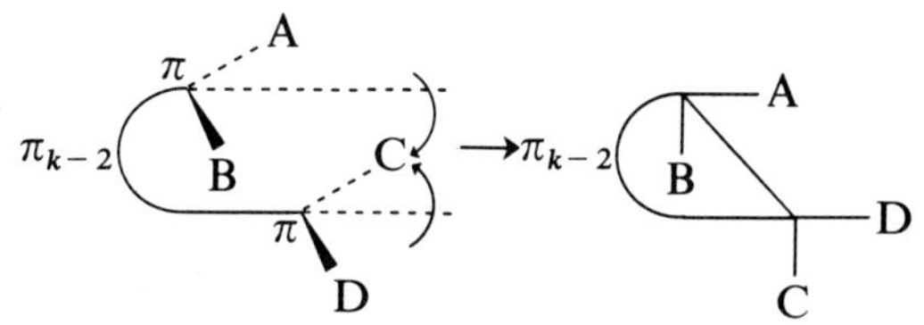

Consideration of the highest singly occupied molecular orbitals of an excited butadiene molecule (107) and hextriene molecule (108) shows that a disrotatory cyclization is expected in the butadiene case and a conrotatory mode in the hexatriene case. This result may be generalized: $4n$-π-electron systems undergo photochemical cyclization/fission in a disrotatory manner and $4n+2$-systems undergo these reactions in a conrotatory manner.

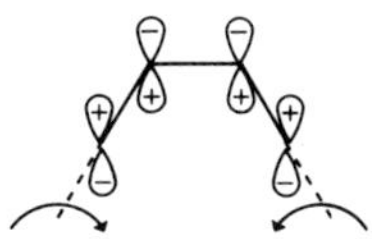

(107)

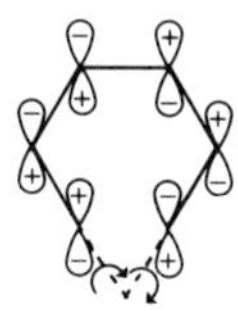

(108)

There are many examples of each type of reaction.

4n [145–147]

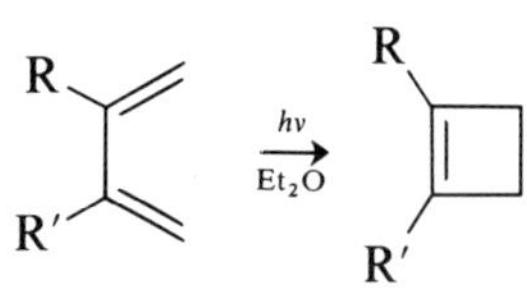

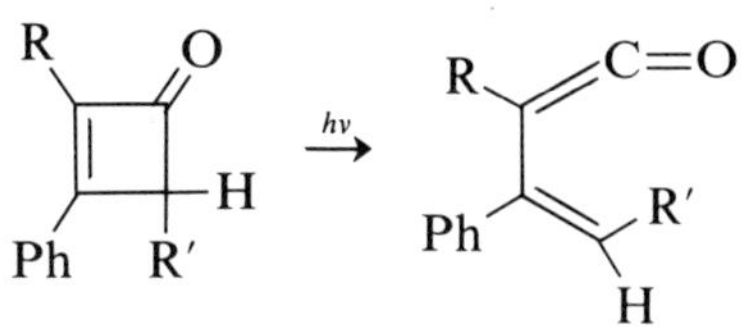

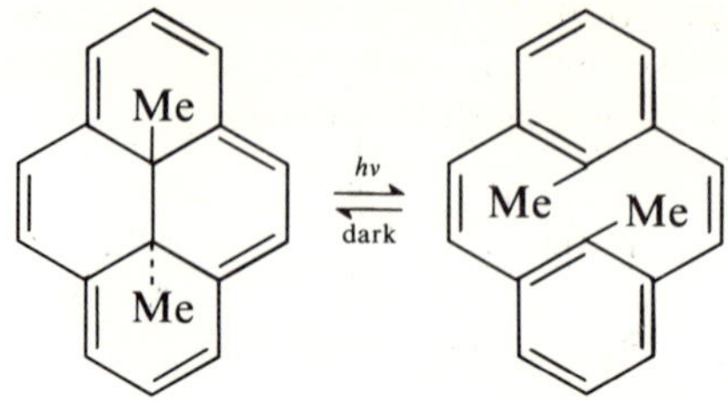

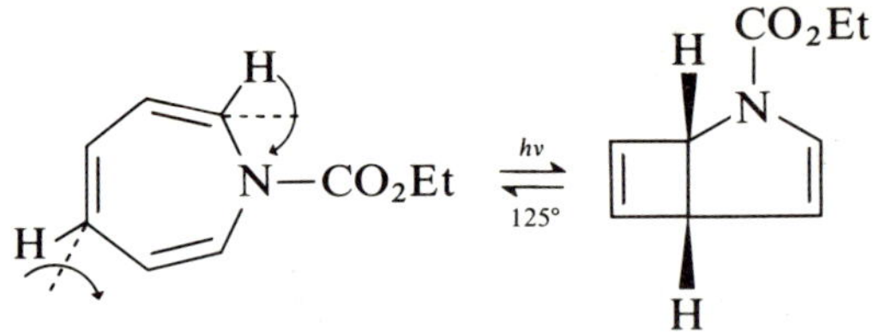

4n+2 [148]

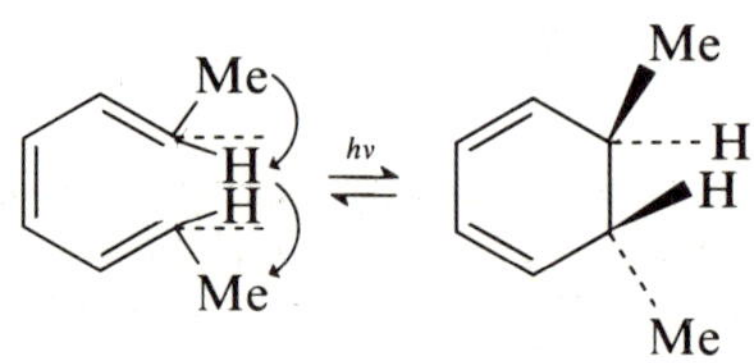

(90–95%) (5–10%)

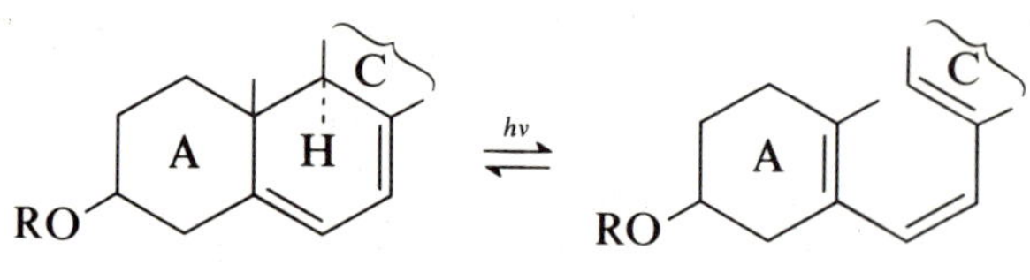

(45–50%) (55–50%)

These equilibria must depend in part on the relative excitation energies and also on steric factors to account for these results.

Many novel ring systems can be prepared involving these valence-bond tautomerization reactions. Bicyclo[1,1,0]butane derivatives are readily available. They have been observed as products in the irradiation of 1,3-butadiene (109) [149] and 3,5-cholestadiene (110) [150]. This ring system has also been postulated as an intermediate in several reactions [151].

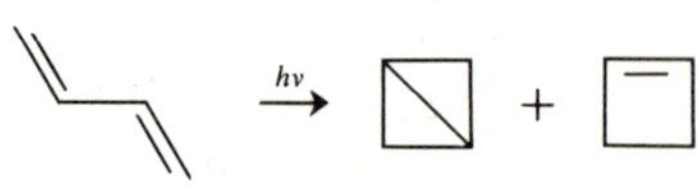

(109)

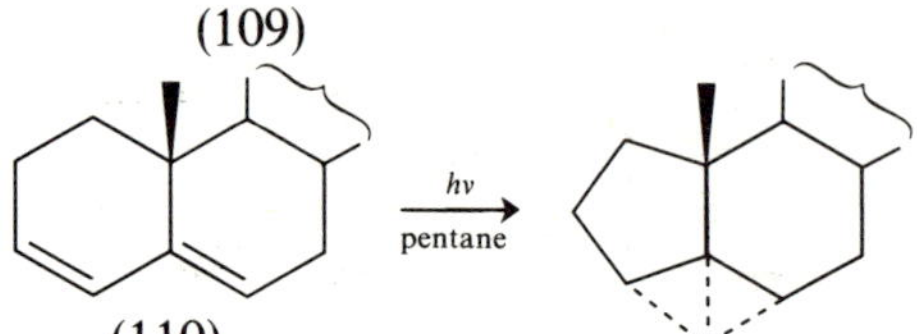

(110)

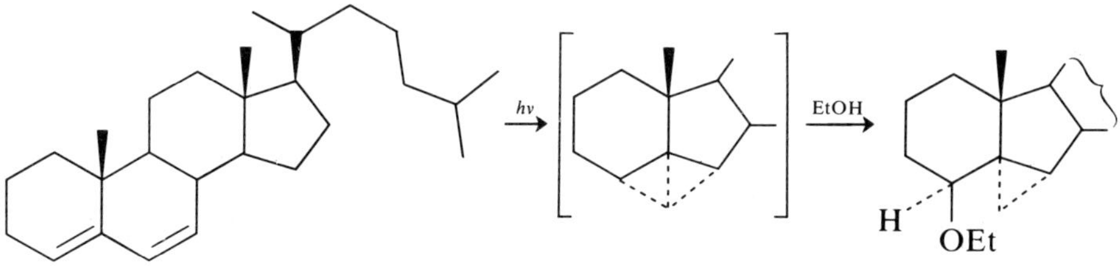

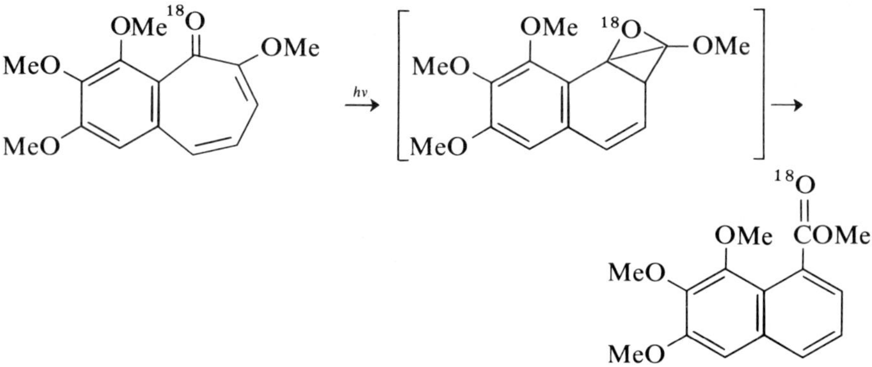

Other interesting ring structures which can be prepared include tricyclo[3,3,0,0,^{2,6}]octane (111) [152] and the bicyclobut[2,1,0]pent-2-ene (112) [153].

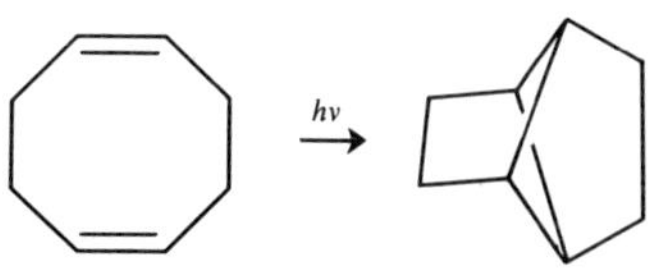

(111)

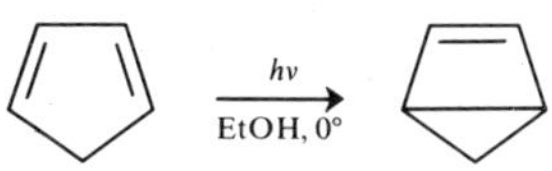

(112)

Such bridging reactions have been reported in many conjugated dienes [154].

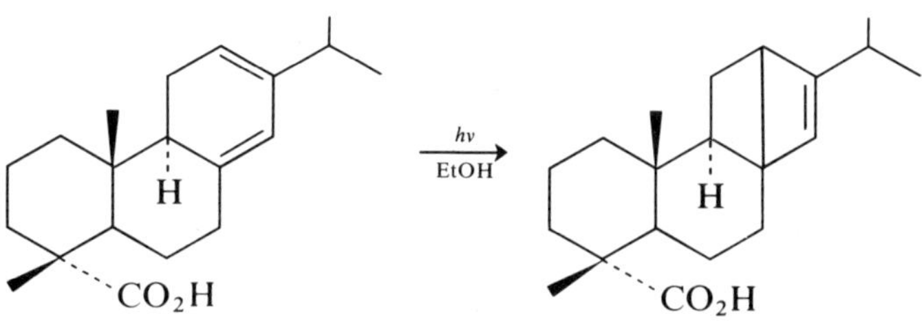

The bicyclo[2,2,0]hexa-2,5-diene ring system and many derivatives can be readily made [155]. This ring system, and prismane have been postulated

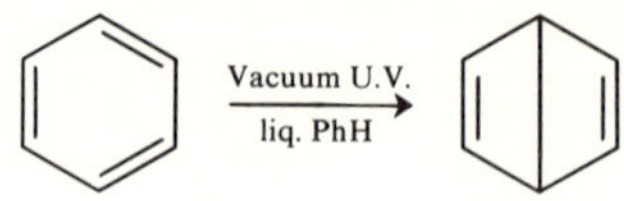

as intermediates in the isomerization of benzene derivatives [156]. Irradiation of mesitylene-1,3,5-$^{14}C_3$ showed that the alkyl groups themselves do not migrate, as the 1,2,4-trimethylbenzene formed was labelled exclusively in the 1-,2- and 4-positions [157]. Similarly, irradiation of benzene-1,3,5-d_3 gave the isomeric trideutero benzenes [158]. The novel prismane structure (114) may be prepared by irradiation of Dewar benzenes (113) [159]. Benzvalene-tricyclo[3,1,0,0^{2,6}]hex-3-ene (115) has also been postulated as an intermediate in these isomerization reactions. It can be isolated from the irradiation of benzenes [160].

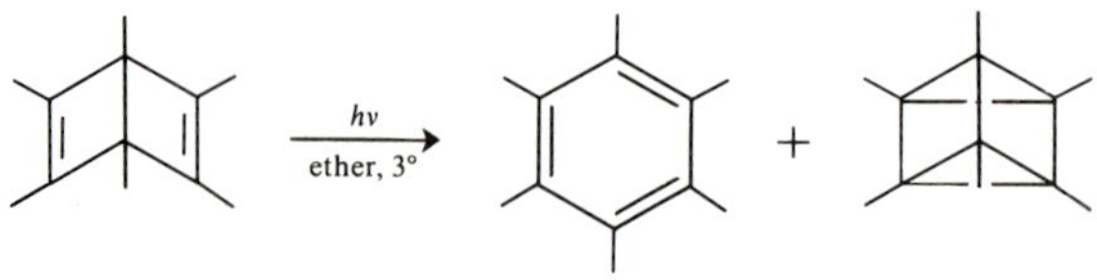

(113) (14%) (5%)
(114)

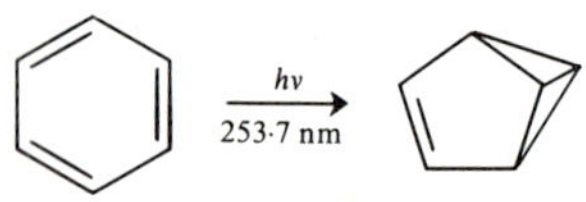

(115)

Heterocyclic analogues can also undergo bridging, and these bicyclic compounds have been postulated as intermediates [161]. These bridging reactions could be of great importance in synthetic chemistry, as they give many novel systems in fair yields and have the advantage that they are simple, one-step reactions. There are many examples of bridging reactions in the photochemistry of seven- and eight-membered rings [162–165].

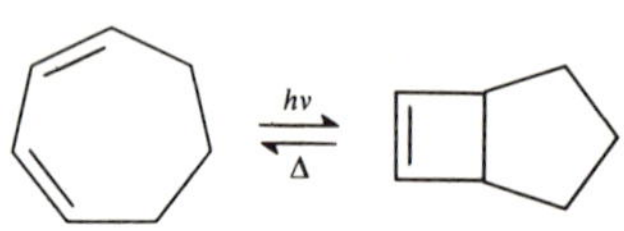

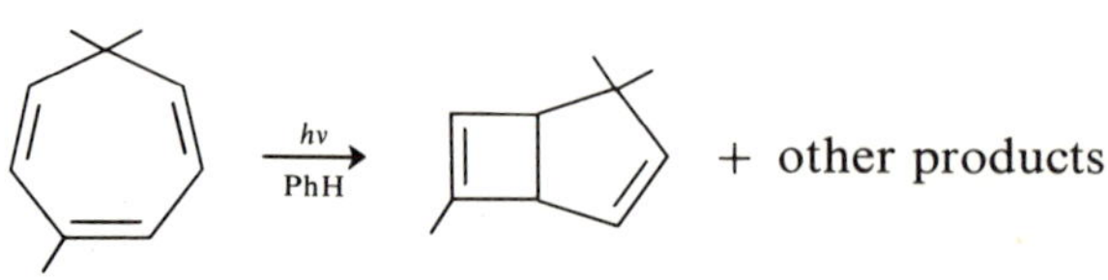

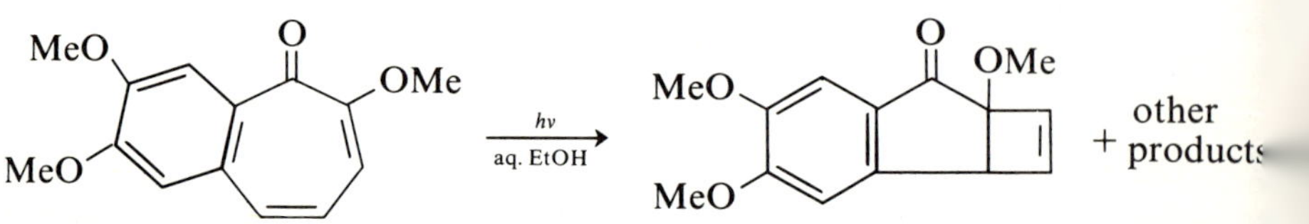

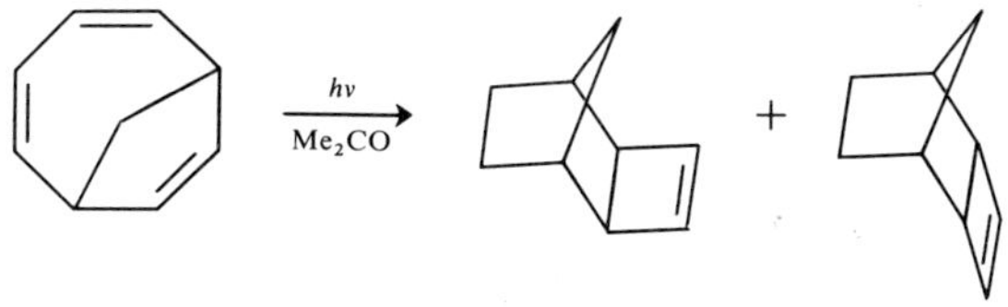

These bond-fission and photocyclization reactions are of immense importance in the vitamin D field [84].

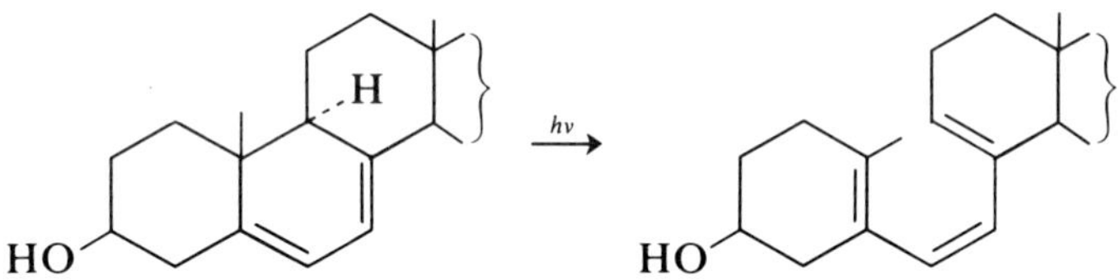

ergosterol previtamin D_2

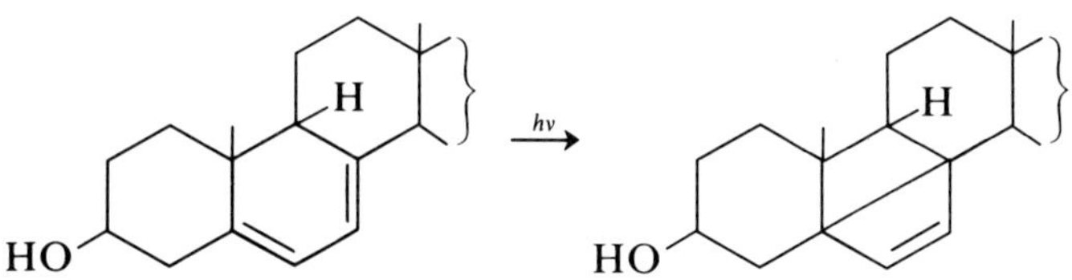

Cross-photocyclization reactions give an excellent pathway to the semibullvalene structure (117) from barrelene (116) [166].

(116) (117)

Internal cycloadditions resulting from valence-bond tautomerism may occur if the olefinic bonds are suitably orientated, as in the synthesis of prismane [159]. There are many other examples [167].

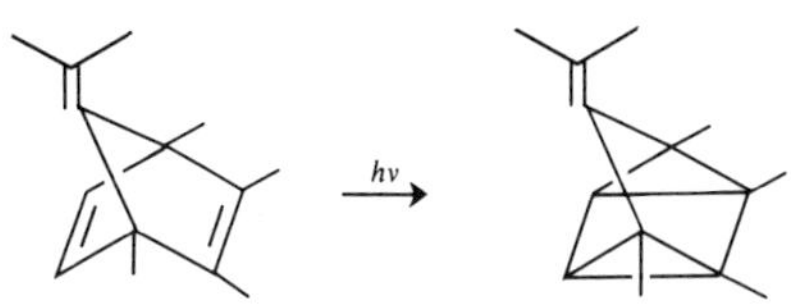

6.5 PHOTOCHEMICAL REARRANGEMENTS

(a) Photochemical Sigmatropic Rearrangements

There are many examples of photochemical sigmatropic rearrangements. Many of these involve hydrogen as the migrating atom. Woodward and Hoffmann define as a sigmatropic change of order $[i,j]$ the migration of a σ-bond, flanked by one or more π-electron systems, to a new position whose termini are $i-1$ and $j-1$ atoms removed from the original bonded loci, in an uncatalysed intramolecular process.

The transition state may be considered as involving a polyolefinic radical and an atom. There are two processes whereby a sigmatropic migration of an atom can occur. The first involves the association of the hydrogen atom with the same face of the π-system of the radical—a suprafacial process—which results in a transition state with a plane of symmetry.

The other mode involves the transfer of the atom from the top face of C_j to the bottom of C_1. Woodward and Hoffmann have called this an antarafacial system [80] and the transition state has a two-fold axis of symmetry. They state that since this rearrangement is concerted, there must be positive orbital overlap throughout the migration of the hydrogen from one terminal carbon to the other, and this process will be controlled by the symmetry of the highest occupied orbital in the polyenyl radical. Consideration of the highest molecular orbitals of the photochemically excited molecules shows that [1,3] and [1,7] hydride shifts would be expected to be suprafacial and [1,5] antarafacial. Similar results pertain to [1,*j*] carbon shifts.

Examples of [1,3] Shifts [168–170]

These reactions generally proceed from a conjugated system to a non-conjugated one, and these are examples of optical pumping (see below, p. 228) and may be of synthetic interest.

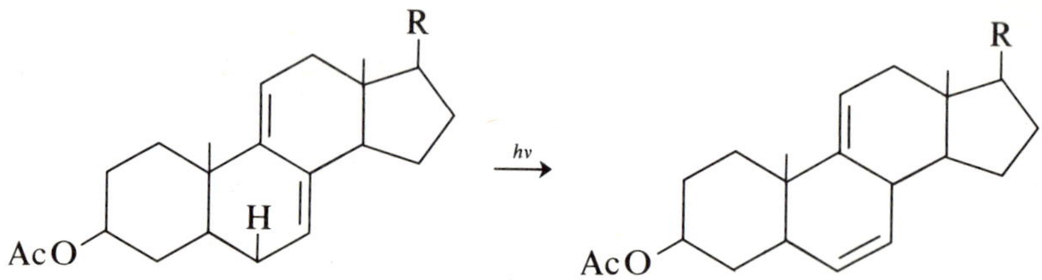

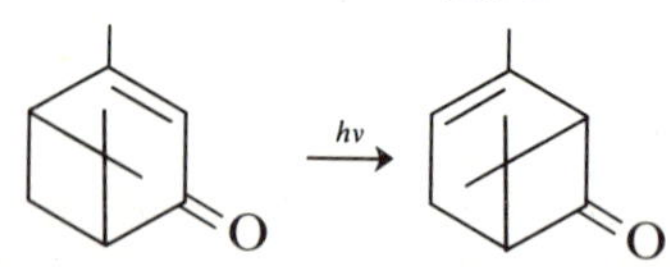

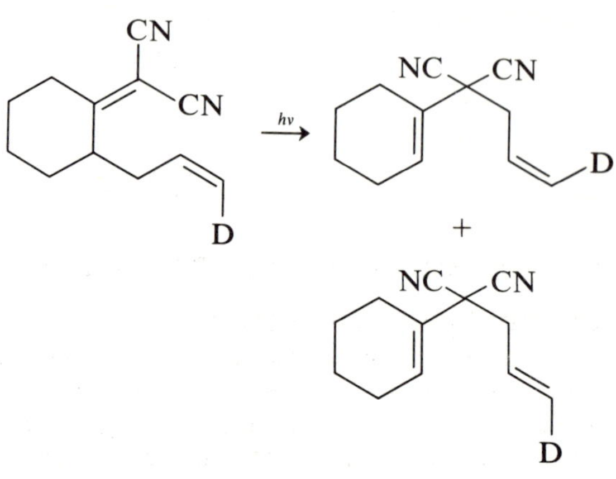

EXAMPLES OF [1,5] SHIFTS [171–173]

The last example may be either a [1,3] or [1,5] substituent migration [173].

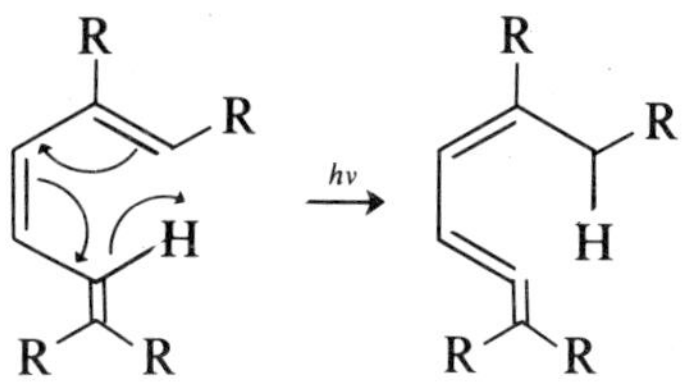

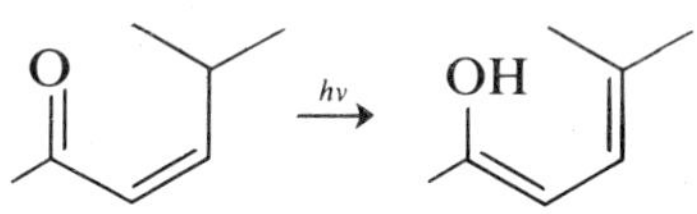

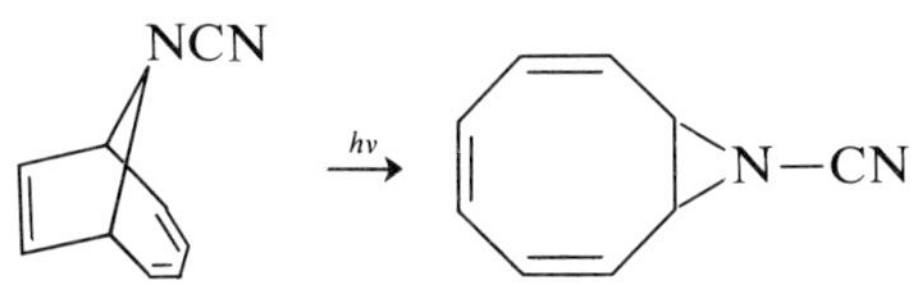

EXAMPLES OF [1,7] SHIFTS [174–175]

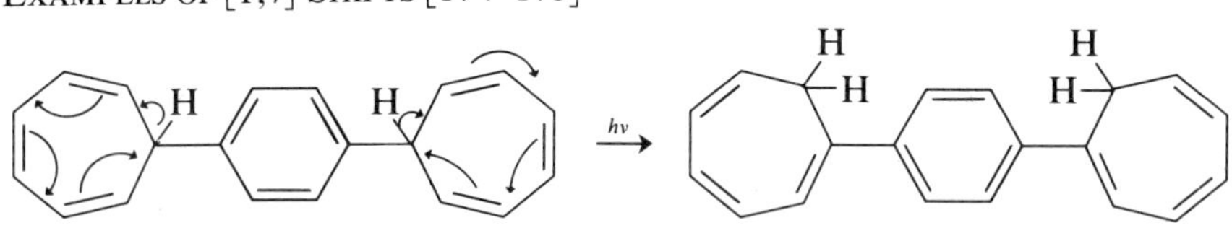

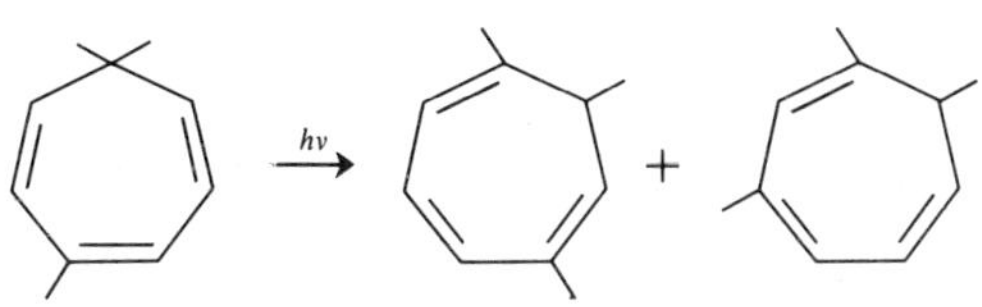

+other products

Woodward and Hoffmann have determined selection rules for sigmatropic reaction of order [i,j].

$i+j$	Stereochemistry of the photochemical transformation
$4q$	Supra-supra antara-antara
$4q+2$	antara-supra supra-antara

These predict the ease of photochemical sigmatropic reactions where $i+j = 4q$ and the absence of examples of the $4q+2$ variety. Hence no photochemical Cope or Claisen rearrangements are observed. However, phenylallyl ethers do rearrange, but the reaction probably proceeds via homolysis or heterolysis of the excited ether to give two stable radicals [176].

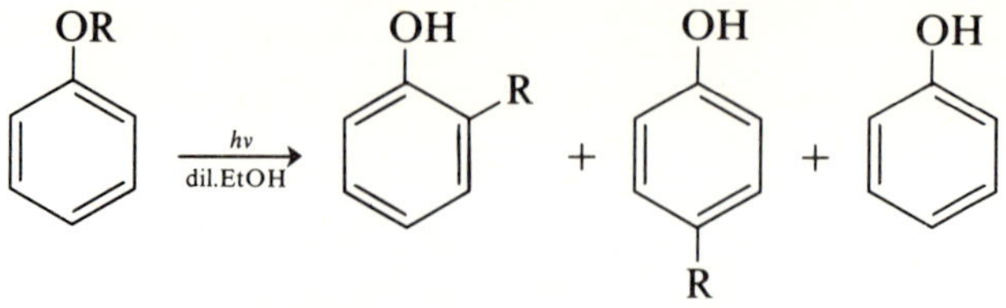

(b) The Photo-Fries Reaction

This reaction is important in the synthesis of *ortho*- and *para*-hydroxy aromatic ketones. The photo-Fries rearrangement has the advantage over the normal Fries rearrangement in that any alkyl groups present are not eliminated, and is, therefore, of considerable importance in the synthesis of compounds containing a number of alkyl substituents. This field has been extensively reviewed by Stenberg [177–179].

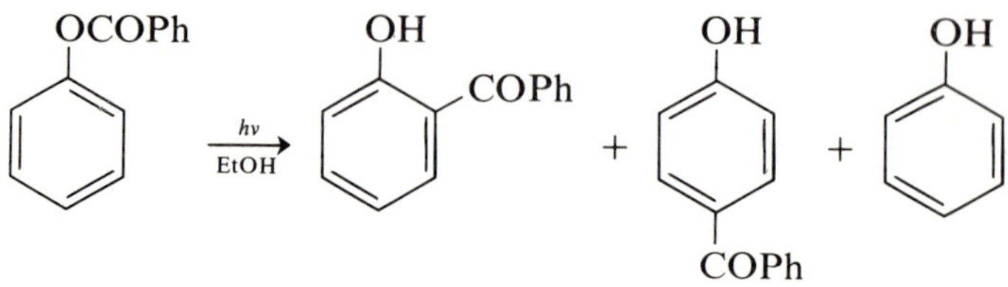

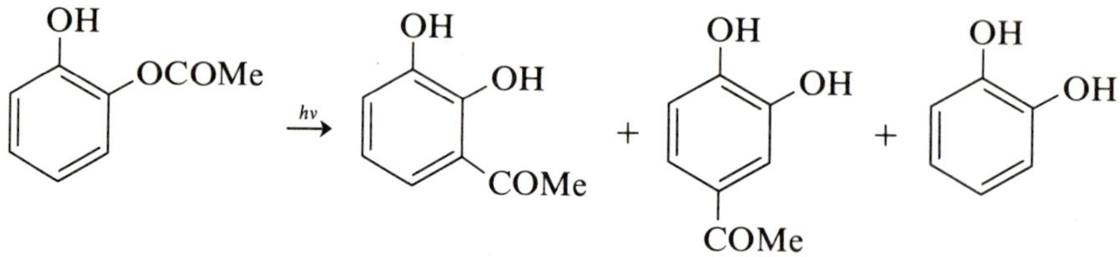

There are some compounds which do not behave as expected under the photo-Fries conditions [180, 181].

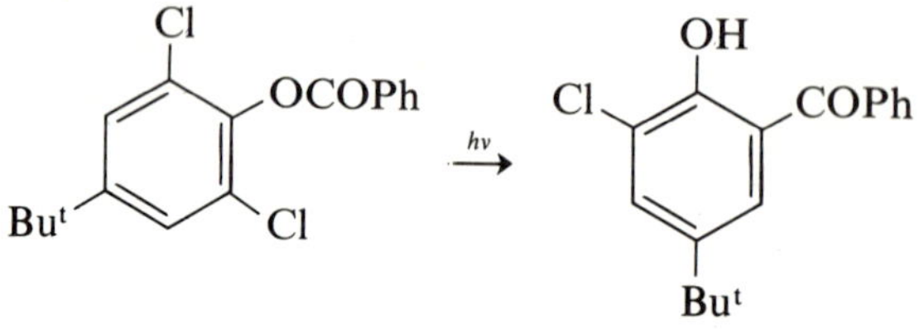

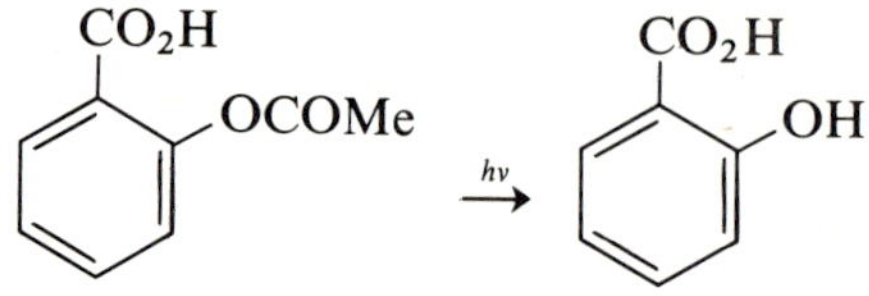

The photo-Fries reaction has some potential in the synthesis of biaryls and alkylaromatics, as these compounds are obtained during the reaction. Even though the yields are low, the simplicity of the reaction recommends itself for the preparation of these compounds rather than the complexity of other routes [182].

$$\text{ArOCOR} \xrightarrow{h\nu} \text{ArR} + \text{CO}_2$$

Anilides will undergo rearrangements similar to those of the aryl esters [183].

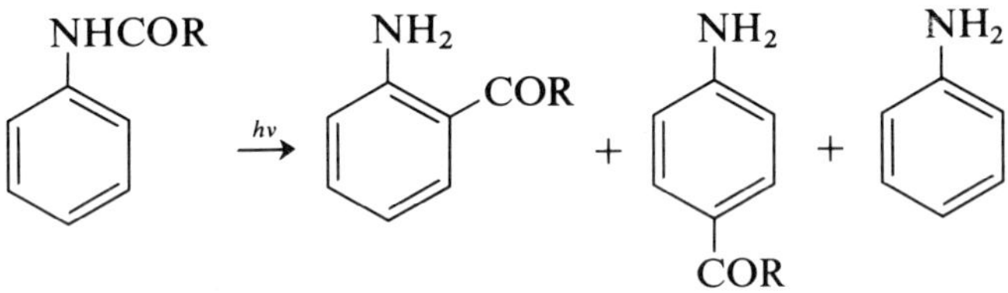

(c) Rearrangement of *N*-Oxides

Examples of both ring-expansion and ring-contraction have been reported in the rearrangement reactions of heterocyclic-*N*-oxides. The former reactions could be of considerable use in the preparation of benz[*d*]-1,3-oxazepines, benz[*f*]-1,3-oxazepines, and related compounds.

The primary photochemical reaction is probably the formation of the oxazirane ring (e.g., (118)), and several compounds containing this system have been isolated [184].

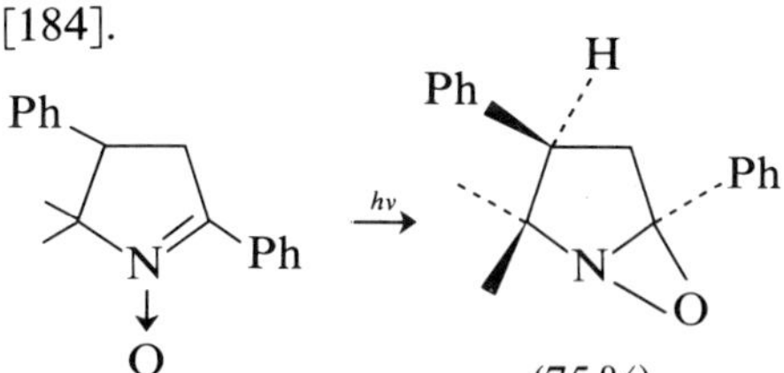

(75%)

(118)

There are many examples of ring-contraction following the irradiation of pyridine-*N*-oxide (119) and its analogues to form pyrroles (120) and similar heterocycles [185]. There is one example of ring-expansion of an

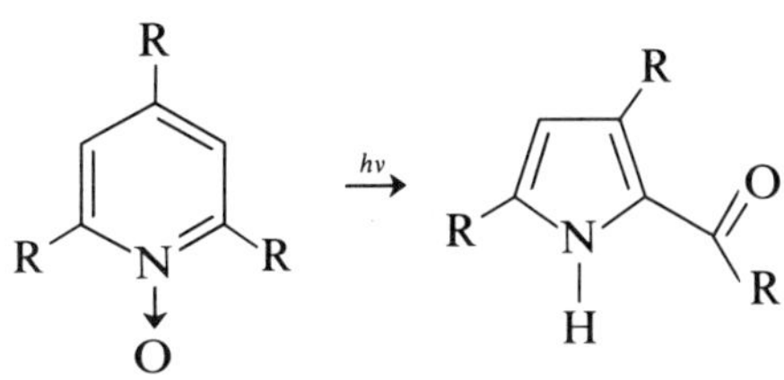

(119) (120)

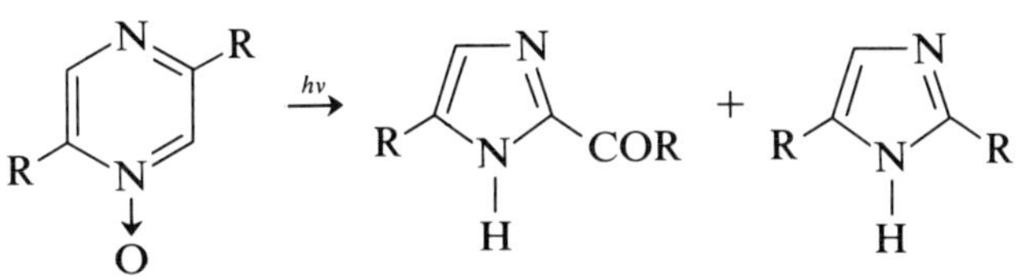

N-substituted pyridine compound, and this gives an excellent route to 1*H*-1,2-diazepines (121) [186].

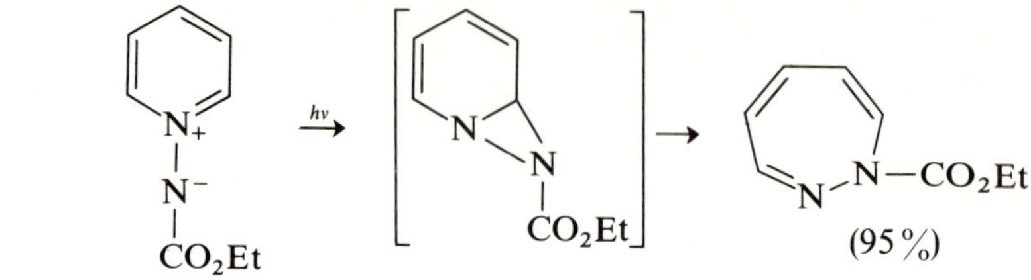

(121)

Irradiation of pyridazine-*N*-oxide (122), however, gave neither ring-expansion nor contraction [187], and probably proceeds via intermediate (123).

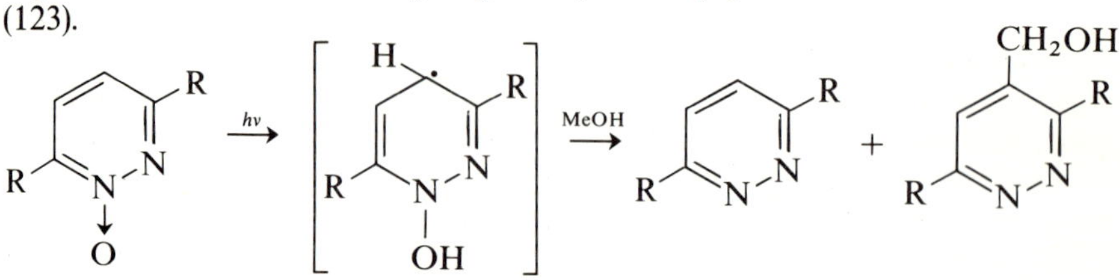

(122) (123)

Quinoline-*N*-oxides (124) react to give different photochemical ring-contraction products. In this case formylindoles (125) and *N*-formyl-2-hydroxy-2,3-dihydroindoles (126) are formed—the latter in fair yield [185, 188–190].

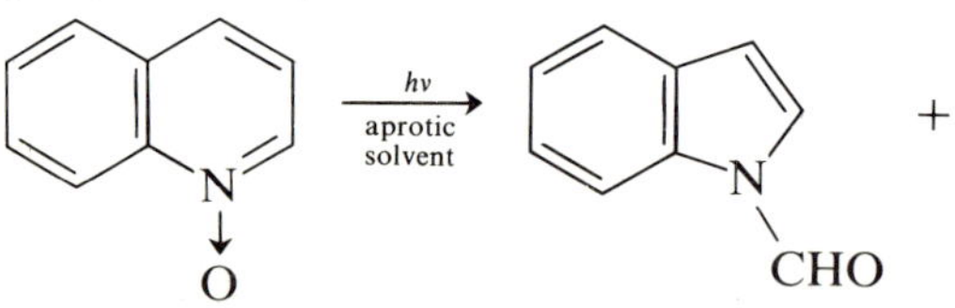

(124) (125)

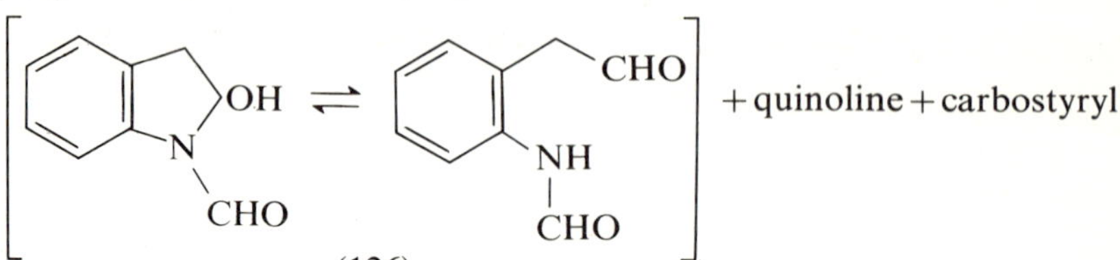

(126)

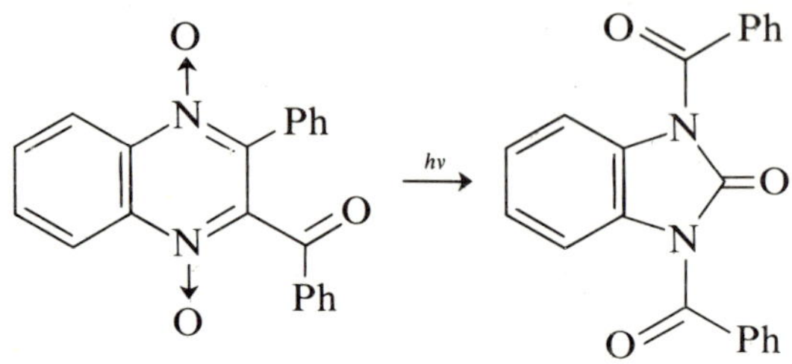

Photolysis of 1,4-diphenylphthalazine-*N*-oxide (127) gave the ring-opened product (128) in 40–60% yield [191].

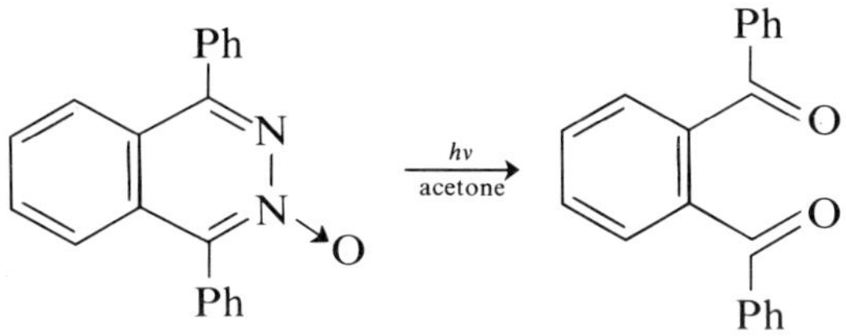

(127) (128)

Benz[*d*]-1,3-oxazepines (130) and benz[*f*]-1,3-oxazepines (132) have only been isolated from the irradiation of quinoline (129) and isoquinoline-*N*-oxides (131) containing a phenyl or a cyano group in the position adjacent to the nitrogen atom, but have been postulated as precursors in the formation of the *N*-acyl indoles [192, 193]. Excellent yields of the oxazepines have been obtained [194]. Irradiation of 4-phenyl-quinazoline-3-oxide (133) gave benz[*f*]-1,3,5-oxadiazepines (134) [195].

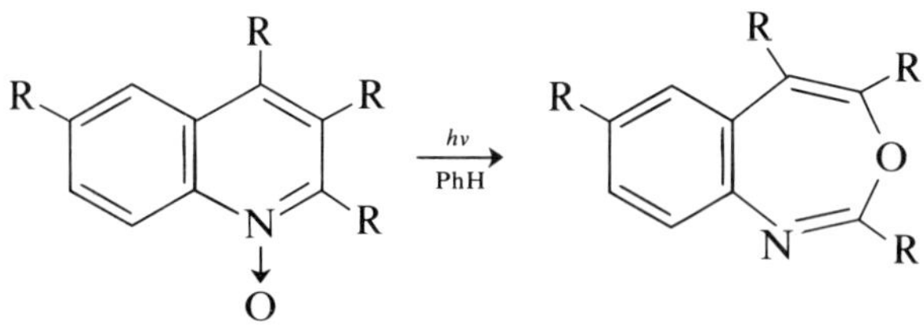

(129) (130)

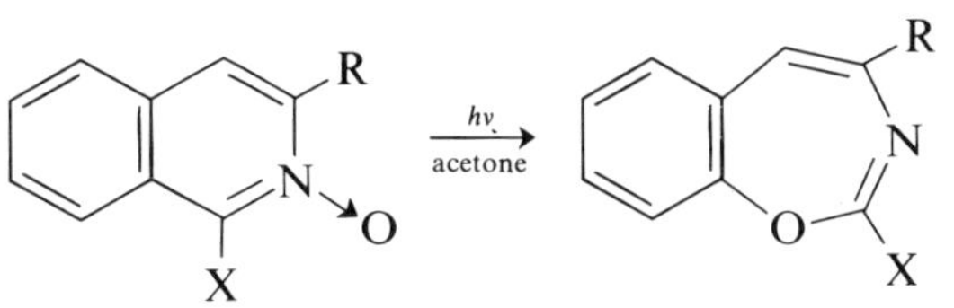

(131) (132)

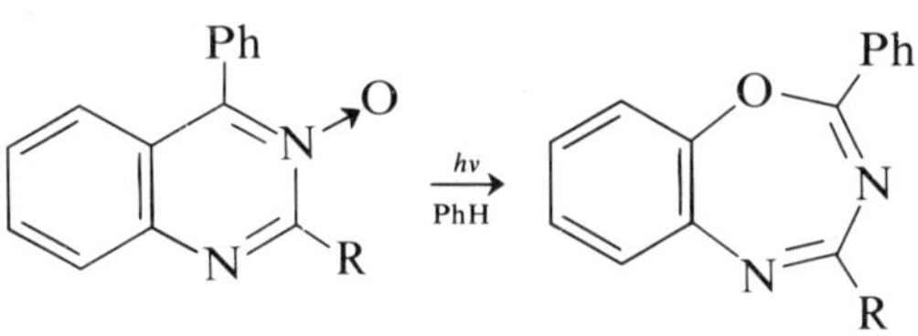

(133) (134)

Other interesting reactions of *N*-oxides have been reported [196].

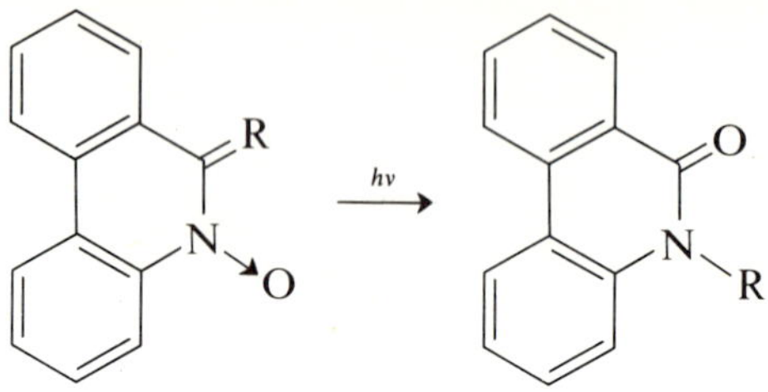

(d) *Cis-trans* Isomerization

This type of reaction is of considerable use in preparing the less stable *cis* olefinic isomers and has been used to prepare the unstable *cis*-azobenzene [197].

Photochemical *cis-trans* isomerization can be effected by direct irradiation or by irradiation in the presence of a sensitizer or a catalyst. It is possible to irradiate at a wavelength at which only one isomer will absorb, by the use of a filter system. This permits the complete conversion to the other isomer, and is known as optical pumping. The less stable *cis* isomers are usually prepared by using this principle.

Bromine, iodine, and iron dodecacarbonyl have all been used as catalysts for isomerization reactions [198]. Sensitizers may be employed to change the *cis*:*trans* ratio in the photostationary state [199].

The report by Hammond and Cole that asymmetric induction could be effected by the use of optically active sensitizers would appear to be of considerable importance. Irradiation of *dl*-*trans*-1,2-diphenylcyclopropane (135) in the presence of an optically active sensitizer gave a photostationary state in which the *trans* component (136) had acquired optical activity [200].

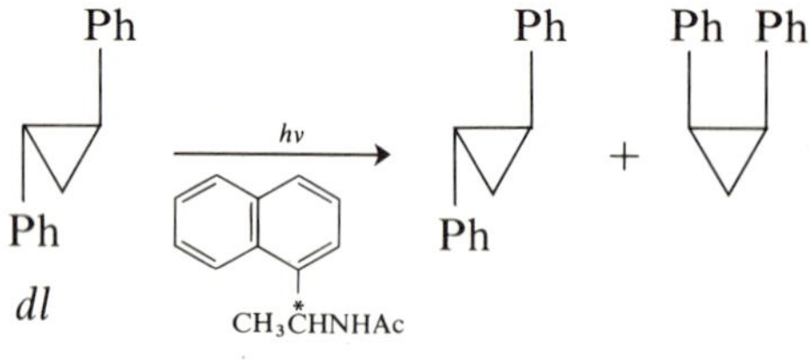

dl

(optically active)

(135) (136)

6.6 AROMATIC SUBSTITUTION REACTIONS

Photochemical aromatic substitution reactions are potentially quite important, as the course of substitution is often different from that in ground-state reactions. Thus, *o*- and *p*-substituted nitrobenzenes could be easily prepared. However, the synthetic possibilities of these reactions do not appear to have been exploited.

There is a large change in electron densities of the benzene rings which occurs upon excitation. The position of greatest electron density in a benzene ring which bears an electron-withdrawing substituent is the *para* position, whereas the *meta* position is the position of greatest electron density in the case of benzene rings substituted by electron-donating groups [201–203].

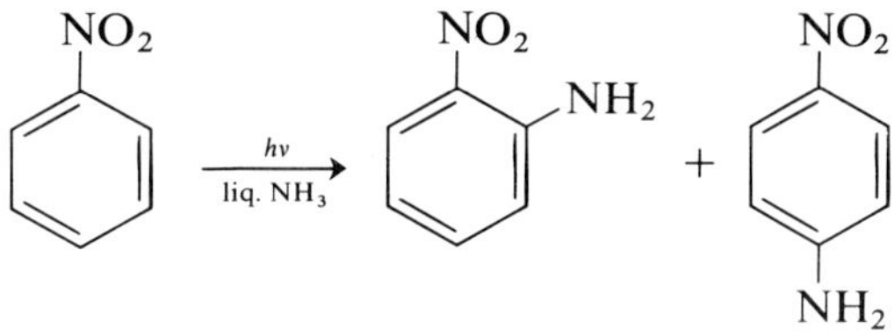

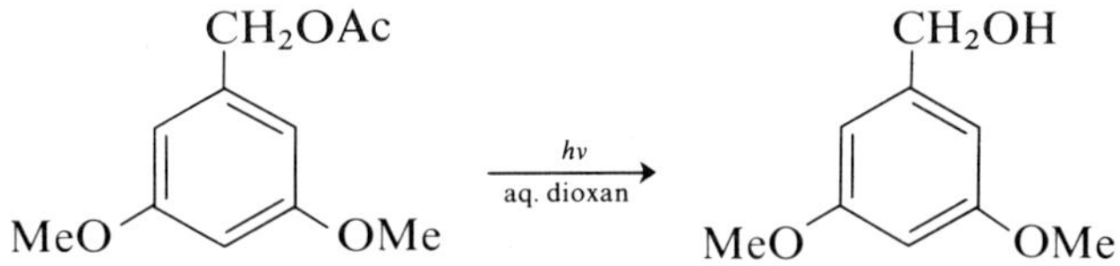

Barton and his co-workers have developed a useful synthesis of aromatic sulphides which involves the photolysis of a 2,4-dinitrobenzene sulphenyl derivative (137) to form an electrophile (138) which then attacks an aromatic moiety present to form a sulphide (139) [204].

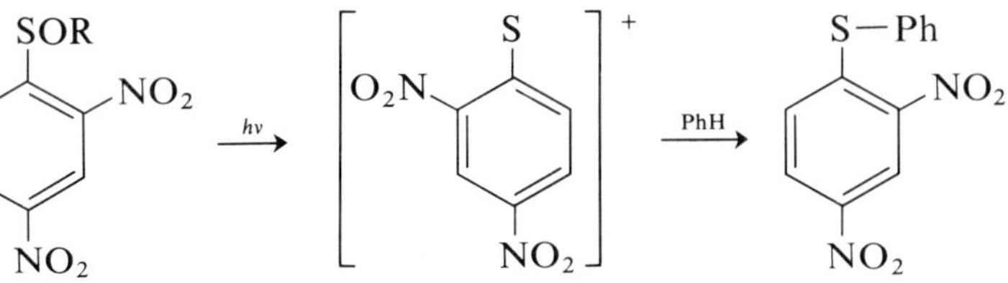

(137) (138) (139)

Solvolysis rates for many systems are affected by the action of light, and this phenomenon might well be of synthetic interest. For example, the solvolysis rate of *m*-nitrophenyl esters is about 300 times that for the *ortho* or *para* isomers [205].

Nucleophilic substitution reactions, like electrophilic substitution reactions, are affected by the action of light. Thus, nucleophilic substitution reactions are aided by *para* withdrawing groups in the ground state and by *meta* withdrawing groups in the excited states [206–208]. Irradiation of

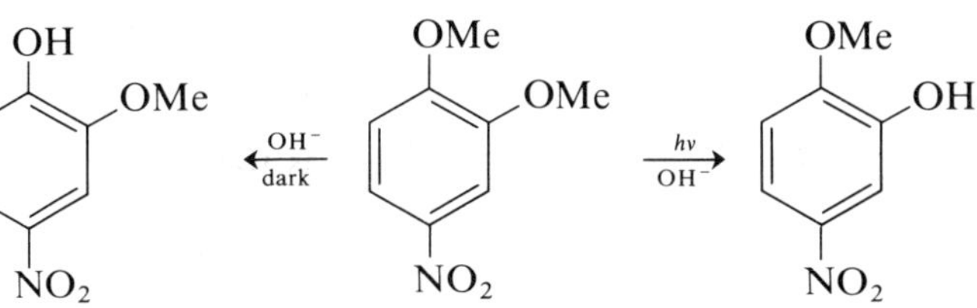

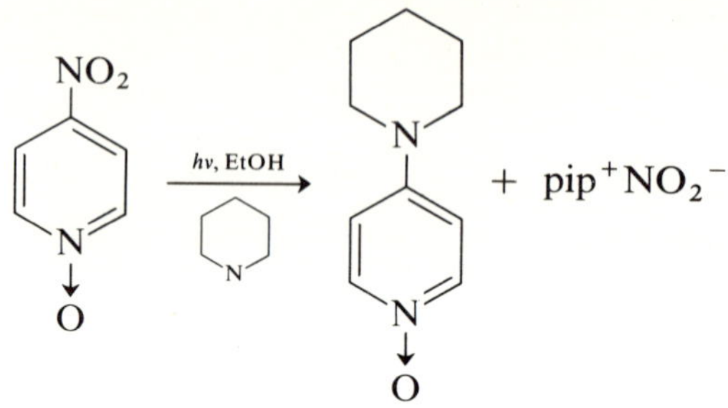

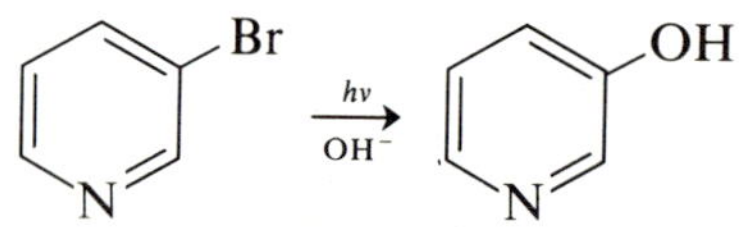

monosubstituted benzenes such as chlorobenzene and anisole in the presence of good nucleophiles leads to a displacement of the substituent by the nucleophiles [209].

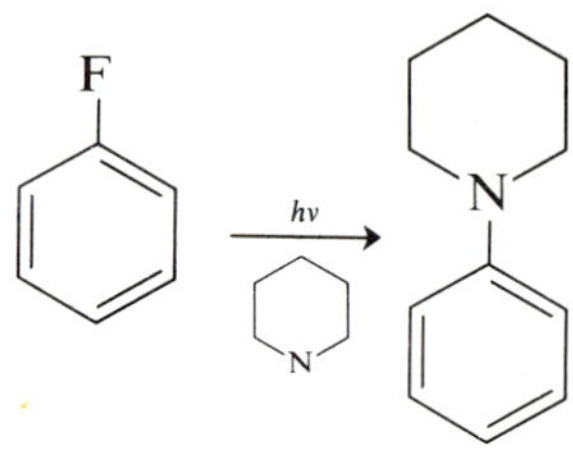

The photolysis of 1,2-di-iodoarenes is an excellent source of arynes which readily undergo addition to give a wide variety of products. The ready availability of a large number of different 1,2-di-iodoaromatic compounds makes these compounds very attractive as general sources for conversion to arynes [210].

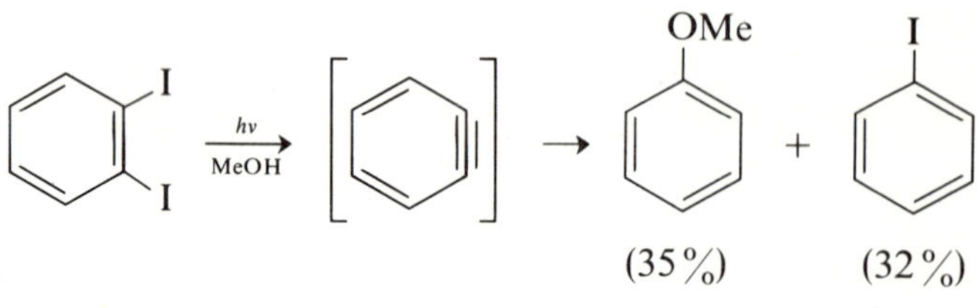

6.7 CARBENES AND NITRENES

The reactions of carbenes [211] and nitrenes [212] are of great importance and utility in organic chemistry. However, the main use of photochemistry in this field is in the production of the carbenes and nitrenes, and so little will be said of the subsequent dark reactions, as there have been extensive reviews on these subjects.

Carbenes may be produced photochemically or thermally in the decomposition of a number of compounds [213].

$$CH_2O \xrightarrow{h\nu} \dot{\underset{\cdot}{C}}H_2 + CO$$

$$CH_2N_2 \xrightarrow{h\nu} \dot{\underset{\cdot}{C}}H_2 + N_2$$

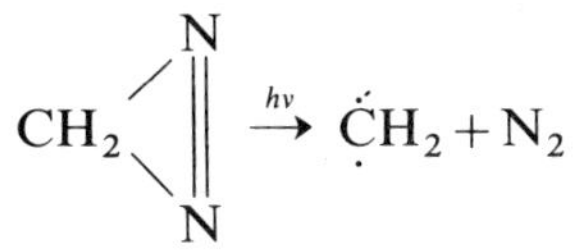

Carbenes undergo a large variety of useful reactions—addition, insertion into C—H bonds, and hydrogen abstraction reactions. Diphenylketene gives diphenyl methylated derivatives on irradiation in organic solvents (SH) such as toluene, cyclohexene, and diethyl ether.

$$Ph_2C{=}C{=}O \xrightarrow{h\nu} Ph\dot{\underset{\cdot}{C}}Ph \xrightarrow{SH} SCHPh_2$$

Addition reactions include the addition to olefins [214] and the 1,3:1′,2′-addition of trifluoroacetylethoxy carbonyl carbene (140) to acetonitrile to form substituted oxazoles (141) [215].

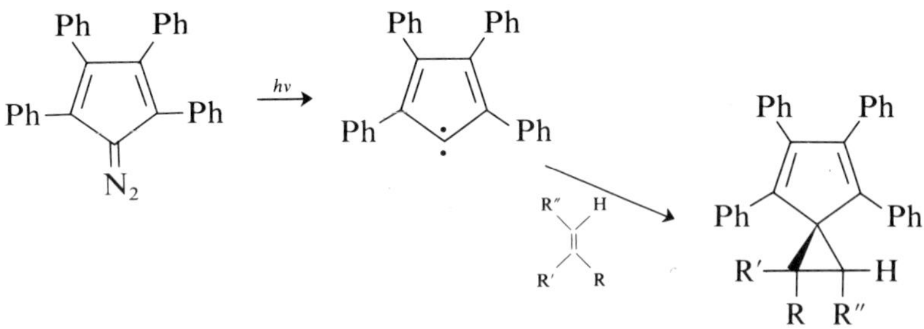

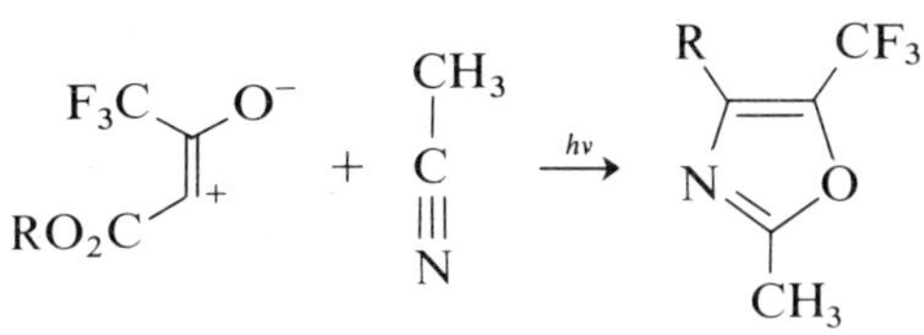

(140) (141)

An example of hydrogen abstraction is afforded by the photolysis of diazo-acetophenone in alcohols. Two products are formed; the ester resulting from a Wolff rearrangement and ketene-solvent addition, and acetophenone formed by hydrogen-abstraction from the solvent [216].

$$Ph{-}\underset{\underset{O}{\|}}{C}{-}CHN_2 \xrightarrow[ROH]{h\nu} PhCH_2CO_2R + PhCOCH_3$$

Reactions of methylene depend to some extent on the way it has been produced. Methylene produced by the photolysis of diazomethane is more reactive and less discriminating in its reactions than is methylene from the photolysis of ketones. Methylene can be formed by triplet-energy transfer from benzophenone, but it is less reactive and will not insert into unactivated C—H bonds [217].

Nitrenes can also be prepared thermally or photochemically from azides and may react in several ways. Two nitrene molecules can recombine to form azo compounds. For example, photolysis of 2-azidobiphenyl (142) gave azo-2-biphenyl (143) [218] and provides an excellent synthesis of this compound.

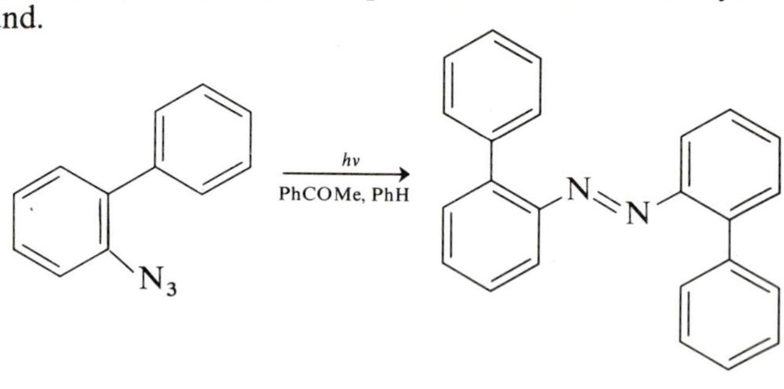

(142) (143)

Hydrogen abstraction from the carbon α to the nitrogen leads to the formation of imines and has considerable synthetic potential [219]. Hydrogen abstraction from the 4- or 5-positions followed by ring-closure results in the formation of pyrrolidines and piperidines.

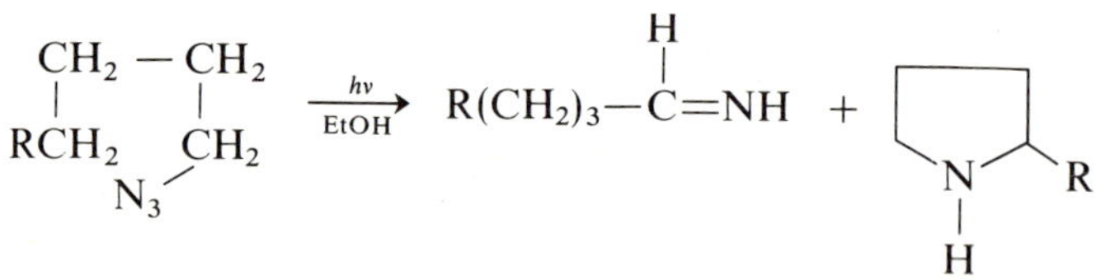

(12 %) (22 %)

The ability of nitrenes to abstract hydrogen atoms from other molecules to form amines has been used in the elegant functionalization of tricyclo-[3,3,0,0^{2,6}]octane. These insertion reactions open the way to an extensive study of the chemistry of C_1- and C_3-substituted tricyclo[3,3,0,0^{2,6}]-octanes [220].

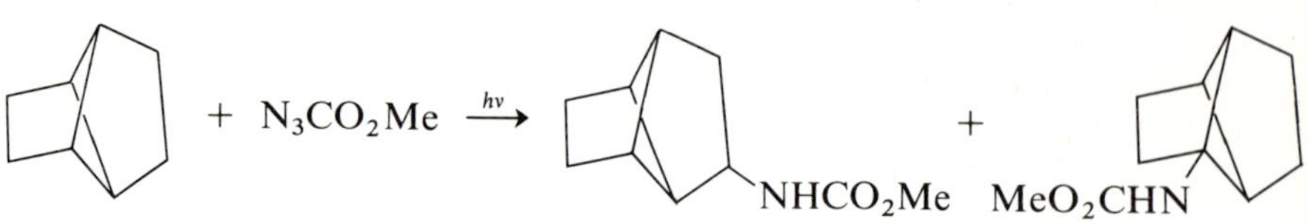

(20 %) (6 %)

Nitrene insertion into double bonds is of considerable use. This is exemplified by the addition of carbethoxy nitrene (144) to 1,3-dienes (145) [221]. Another example is the preparation of azirines from vinyl azides [222].

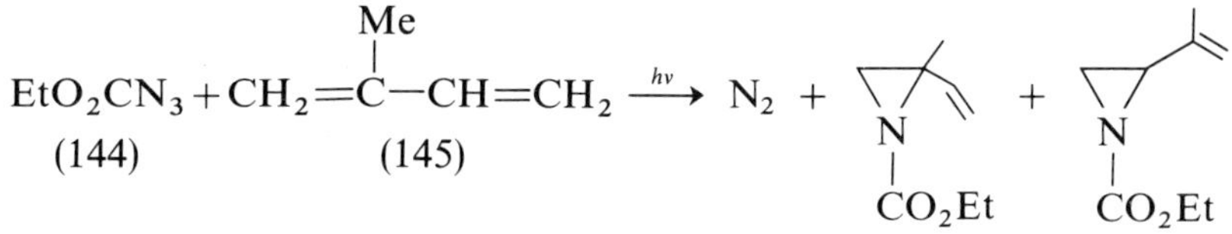

(144) (145)

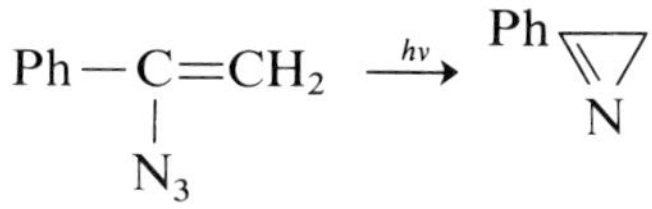

Photolysis of acyl azides leads to the expected Curtius rearrangement. Irradiation of 2,3-diphenylcycloprop-2-ene-carboxylic acid azide (146) in an inert solvent at 70° gave the expected urethane via the isocyanate [223]. The product obtained obviously will depend on the medium used for the reaction. Irradiation of (146) at $-30°$ in ether gave the expected urethane and an amide (147) formed by hydrogen abstraction from the solvent [223].

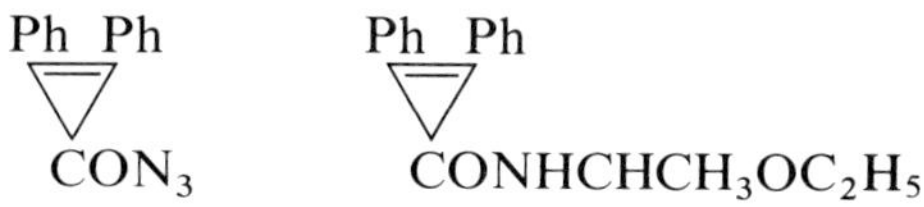

(146) (147)

A recently reported preparation of azepines utilizes ring-enlargement of phenyl azide (148) [224].

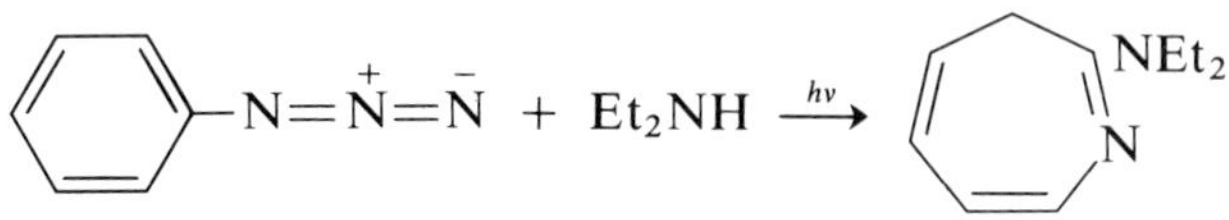

(148)

REFERENCES

[1] Akhtar, M. in *Advances in Photochemistry* (ed. W. A. Noyes, Jr., G. S. Hammond and J. N. Pitts, Jr.), Vol. 2, p. 263, Interscience, New York (1964).

[2] Nussbaum, A. L. and Robinson, C. H. *Tetrahedron* **17**, 35 (1962).

[3] Barton, D. H. R., Beaton, J. M., Geller, L. E. and Pechet, M. M. *J. Am. chem. Soc.* **83**, 4076 (1961).

[4] (a) Barton, D. H. R., Lier, E. F. and McGhie, J. F. *J. chem. Soc.* (*C*) 1031 (1968).

(b) Barton, D. H. R., Kumari, D., Welzel, P., Danks, L. J. and McGhie, J. F. *J. chem. Soc.* (*C*) 332 (1968).

[5] Akhtar, M. and Pechet, M. M. *J. Am. chem. Soc.* **86**, 265 (1964).

[6] Kabasakalian, P., Townley, E. R. and Yudis, M. D. *J. Am. chem. Soc.* **84**, 2716 (1962).

[7] Akhtar, M., Barton, D. H. R. and Sammes, P. G. *J. Am. chem. Soc.* **86**, 3394 (1964).

[8] Barton, D. H. R., Beaton, J. M., Geller, L. E. and Pechet, M. M. *J. Am. chem. Soc.* **82**, 2640 (1960).

[9] Nussbaum, A. L., Carlon, F. E., Oliveto, E. P., Townley, E., Kabasakalian, P. and Barton, D. H. R., *J. Am. chem. Soc.* **82**, 2973 (1960).
[10] Barton, D. H. R. and Beaton, J. M. *J. Am. chem. Soc.* **83**, 4083 (1961).
[11] Barton, D. H. R. and Beaton, J. M. *J. Am. chem. Soc.* **84**, 199 (1962).
[12] Akhtar, M. and Barton, D. H. R. *J. Am. chem. Soc.* **84**, 1496 (1962).
[13] Reimann, H., Capomaggi, A. S., Strauss, T., Oliveto, E. P. and Barton, D. H. R. *J. Am. chem. Soc.* **83**, 4481 (1961).
[14] Kabasakalian, P. and Townley, E. R. *J. Am. chem. Soc.* **84**, 2711 (1962).
[15] Suginome, H., Sato, N. and Masamune, T. *Tetrahedron Lett.* 1557 (1967).
[16] Kabasakalian, P., Townley, E. R. and Yudis, M. D. *J. Am. chem. Soc.* **84**, 2718 (1962).
[17] Robinson, C. H., Gnoj, O., Mitchell, A., Oliveto, E. P. and Barton, D. H. R. *Tetrahedron* **21**, 743 (1965).
[18] Nussbaum, A. L., Robinson, C. H., Oliveto, E. P. and Barton, D. H. R. *J. Am. chem. Soc.* **83**, 2400 (1961).
[19] Suginome, H., Murakami, M. and Masamune, T. *Chem. Commun.* 343 (1966).
[20] Nussbaum, A. L., Wayne, R., Yuan, E., Zagneetko, O. and Oliveto, E. P. *J. Am. chem. Soc.* **84**, 1070 (1962).
[21] Barton, D. H. R., Faro, H. P., Serebryakov, E. P. and Woolsey, N. F. *J. chem. Soc.* 2438 (1965).
[22] Dilling, W. L. *Chem. Rev.* **66**, 373 (1966).
[23] Chapman, O. L. (Ed.), *Organic Photochemistry*, Vol. 1, Arnold, London (1967).
[24] Chieffi, G. and Paterno, E. *Gazz. chim. ital.* **39**, 341 (1909).
[25] Büchi, G., Inman, C. G. and Lipinsky, E. S. *J. Am. chem. Soc.* **76**, 4327 (1954).
[26] Arnold, D. R., Hinman, R. L. and Glick, A. H. *Tetrahedron Lett.* 1425 (1964); Arnold, D. R., Dalton, J. C., Glick, A., Turro, N. J. and Wriede, P. *J. Am. chem. Soc.* **89**, 3950 (1967).
[27] Odaira, Y., Tominaga, T. and Tsutsumi, S. *Bull. chem. Soc. Japan* **40**, 2451 (1967); Hara, M., Odaira, Y. and Tsutsumi, S. *Tetrahedron Lett.* 2981 (1967); Odaira, Y., Shimodaira, T. and Tsutsumi, S. *Chem. Commun.* 757 (1967).
[28] Büchi, G., Kofron, J. T., Koller, E. and Rosenthal, D. *J. Am. chem. Soc.* **78**, 876 (1956).
[29] Arnold, D. R. and Glick, A. H. *Chem. Commun.* 813 (1966).
[30] Jorgenson, M. J., Murov, S., Nussim, M. and Yang, N. C. *Tetrahedron Lett.* 3657 (1964).
[31] Kan, R. O. *Organic Photochemistry*, p. 200, McGraw-Hill, London (1966).
[32] Steinmetz, R., Hartmann, W. and Schenck, G. O. *Chem. Ber.* **98**, 3854 (1965).
[33] Evanega, G. R. and Whipple, E. B. *Tetrahedron Lett.* 2163 (1967).
[34] Ogata, M., Watanabe, H. and Kanō, H. *Tetrahedron Lett.* 533 (1967).
[35] Toki, S. and Sakurai, H. *Tetrahedron Lett.* 4119 (1967).
[36] Shigemitsu, Y., Odaira, Y. and Tsutsumi, S. *Tetrahedron Lett.* 55 (1967).
[37] Bryce-Smith, D. and Gilbert, A. *Tetrahedron Lett.* 3471 (1964).
[38] Bryce-Smith, D., Gilbert, A. and Johnson, M. G. *J. chem. Soc.* (*C*) 383 (1967).
[39] Bryce-Smith, D., Fray, G. I. and Gilbert, A. *Tetrahedron Lett.* 2137 (1964).
[40] Kaiser, E. T. and Wulfers, T. F. *J. Am. chem. Soc.* **86**, 1897 (1964).
[41] Omote, Y., Sugiyama, N., Yamada, K. and Yoshioka, M. *J. org. Chem.* **32**, 3676 (1967).
[42] Criegee, R. and Furrer, H. *Chem. Ber.* **97**, 2949 (1964).
[43] Sunder-Plassmann, P., Nelson, P. H., Durham, L., Edwards, J. A. and Fried, J. H. *Tetrahedron Lett.* 653 (1967).
[44] Huang, C. W. and McCullough, J. J. *Chem. Commun.* 815 (1967).
[45] Lange, G. L., de Mayo, P., Challand, B. D. and Kornis, G. *Chem. Commun.* 704 (1967).

[46] WIESNER, H., JIRKOVSKY, I., FISHMAN, M. and WILLIAMS, C. A. J. *Tetrahedron Lett.* 1523 (1967).
[47] YAMADA, Y., UDA, H. and NAKANISHI, K. *Chem. Commun.* 423 (1966).
[48] HIKINO, H. and DE MAYO, P. *J. Am. chem. Soc.* **86**, 3582 (1964).
[49] NOZAKI, H., KURITA, M., MORI, T. and NOYORI, R. *Tetrahedron* **24**, 1821 (1968).
[50] COREY, E. J., BASS, J. D., LE MAHIEU, R. and MITRA, R. B. *J. Am. chem. Soc.* **86**, 5570 (1964).
[51] EATON, P. E. and LIN, K. *J. Am. chem. Soc.* **87**, 2052 (1965).
[52] CHAPMAN, O. L., GRISWOLD, A. A., HOGANSON, E., LENZ, G. and REASONER, J. *Pure appl. Chem.* **9**, 585 (1965).
[53] CARLOUGH, K. H. and SRINIVASAN, R. *J. Am. chem. Soc.* **89**, 4932 (1967).
[54] COLE, T. W. and EATON, P. E. *J. Am. chem. Soc.* **86**, 962 (1964); BARBORAK, J. C. and PETTIT, R. *J. Am. chem. Soc.* **89**, 3080 (1967); JONES, M. and SCOTT, L. T. *J. Am. chem. Soc.* **89**, 150 (1967).
[55] BROWN, M. *J. org. Chem.* **33**, 162 (1968).
[56] DE MAYO, P., REID, S. T. and YIP, R. W. *Can. J. Chem.* **42**, 2828 (1964); BARLTROP, J. A. and ROBSON, P. *Tetrahedron Lett.* 597 (1963).
[57] METZNER, W., PARTALE, H. and KRAUCH, C. H. *Chem. Ber.* **100**, 3156 (1967); CHILDS, R. F. and JOHNSON, A. W. *J. chem. Soc.* (*C*) 874 (1967).
[58] SCHARF, H.-D. *Tetrahedron Lett.* 4231 (1967).
[59] BARLTROP, J. A. and HESP, B. *J. chem. Soc.* (*C*) 1625 (1967).
[60] BRUCE, J. M. *Q. Rev. chem. Soc.* 405 (1967) and references therein.
[61] BELLAS, M., BRYCE-SMITH, D. and GILBERT, A. *Chem. Commun.* 263 (1967) and references therein.
[62] BRYCE-SMITH, D., FRAY, G. I. and VICKERY, B. *J. chem. Soc.* (*C*) 390 (1967).
[63] BRYCE-SMITH, D. and VICKERY, B. *Chemy Ind.* 429 (1961); SCHENCK, G. O., KUHLS, J., MANNSFELD, S. P. and KRAUCH, C. H. *Chem. Ber.* **96**, 813 (1963).
[64] JOB, B. E. and LITTLEHAILES, J. D. *J. chem. Soc.* (*C*) 886 (1968).
[65] ANET, F. A. L. and GREGOROVICH, B. *Tetrahedron Lett.* 5961 (1966).
[66] BRYCE-SMITH, D., GILBERT, A. and ORGER, B. H. *Chem. Commun.* 512 (1966).
[67] KRAFT, K. and KOLTZENBURG, G. *Tetrahedron Lett.* 4357 (1967).
[68] DE MAYO, P., STOTHERS, J. B. and TEMPLETON, W. *Can. J. Chem.* **39**, 488 (1961).
[69] ELAD, D. and ROSENTHAL, I. *Chem. Commun.* 684 (1966).
[70] NOZAKI, H., KURITA, M. and NOYORI, R. *Tetrahedron Lett.* 2025 (1968); BLADON, P. and WILLIAMS, I. A. *J. chem. Soc.* (*C*) 2032 (1967).
[71] ELAD, D. and YOUSSEFYEH, R. D. *J. org. Chem.* **29**, 2031 (1964).
[72] COOKSON, R. C., HUDEC, J. and MIRZA, N. A. *Chem. Commun.* 180 (1968); COOKSON, R. C. *Q. Rev. chem. Soc.* **22**, 423 (1968).
[73] SCHRAGE, K. *Tetrahedron Lett.* 5795 (1966).
[74] ROKACH, J. and ELAD, D. *J. org. Chem.* **31**, 4210 (1966).
[75] FRIEDMAN, G. and KOMEN, A. *Tetrahedron Lett.* 3357 (1968).
[76] MAROLEWSKI, T. and YANG, N. C. *Chem. Commun.* 1225 (1967).
[77] CHARLTON, J. L. and DE MAYO, P. *Can. J. Chem.* **46**, 1041 (1968).
[78] OCHIAI, M. and MORITA, K. *Tetrahedron Lett.* 2349 (1967).
[79] FISCHER, A., HAMMOND, G. S. and TURRO, N. J. *J. Am. chem. Soc.* **83**, 4674 (1961).
[80] WOODWARD, R. B. and HOFFMANN, R. *J. Am. chem. Soc.* **87**, 395, 2511 (1965); HOFFMANN, R. and WOODWARD, R. B. *J. Am. chem. Soc.* **87**, 2046, 4388, 4389 (1965); *Acc. chem. Res.* **1**, 17 (1968).
[81] DAUBEN, W. G. and SMITH, J. H. *J. org. Chem.* **32**, 3244 (1967).
[82] BROCKMANN, H. *Proc. chem. Soc.* 304 (1957).
[83] SCHÖNBERG, R. and SODTKE, U. *Tetrahedron Lett.* 4977 (1967).
[84] HAVINGA, E. and SCHLATMANN, J. L. M. A. *Tetrahedron* **16**, 146 (1961).
[85] FONKEN, G. J. *Tetrahedron Lett.* 549 (1962).
[86] FONKEN, G. J. *Chemy Ind.* 1327 (1962).

[87] HENDERSON, W. A. and ULLMAN, E. F. *J. Am. chem. Soc.* **87**, 5424 (1965).
[88] HELLER, H. G., AULD, D. and SALISBURY, K. *J. chem. Soc.* (*C*) 682 (1967).
[89] WEINSTEIN, B. and BRATTESANI, D. N. *Chemy Ind.* 1292 (1967).
[90] SRINIVASAN, R. *J. Am. chem. Soc.* **84**, 4141 (1962).
[91] LAMOLA, A. A., HAMMOND, G. S. and MALLORY, F. B. *Photochem. Photobiol.* **4**, 259 (1965).
[92] PARKER, C. O. and SPOERRI, P. E. *Nature, Lond.* **166**, 603 (1950).
[93] MUSZKAT, K. A. and FISCHER, E. *J. chem. Soc.* (*B*) 662 (1967).
[94] BLACKBURN, E. V., LOADER, C. E. and TIMMONS, C. J. *J. chem. Soc.* (*C*) 163 (1970).
[95] MALLORY, F. B., WOOD, C. S., GORDON, J. T., LINDQUIST, L. C. and SAVITZ, M. L. *J. Am. chem. Soc.* **84**, 4361 (1962); MALLORY, F. B., GORDON, J. T. and WOOD, C. S. *J. Am. chem. Soc.* **85**, 828 (1963); MALLORY, F. B., WOOD, C. S. and GORDON, J. T. *J. Am. chem. Soc.* **86**, 3094 (1964); WOOD, C. S. and MALLORY, F. B. *J. org. Chem.* **29**, 3373 (1964).
[96] KUPCHAN, S. M. and WORMSER, H. C. *Tetrahedron Lett.* 359 (1965).
[97] JEFFS, P. W. and HANSEN, J. F. *J. Am. chem. Soc.* **89**, 2798 (1967).
[98] COLLINS, D. J. and HOBBS, J. J. *Aust. J. Chem.* **20**, 1905 (1967).
[99] LEVI, E. K. J. and ORCHIN, M. *J. org. Chem.* **31**, 4302 (1966).
[100] BLACKBURN, E. V., LOADER, C. E. and TIMMONS, C. J. *J. chem. Soc.* (*C*) 1576 (1968).
[101] SERVIS, K. L. and FANG, K. N. *Tetrahedron Lett.* 967 (1968).
[102] GOTO, Y., HATA, K., HAYASHI, S., SATO, T. and TOHYAMA, T. *Bull. chem. Soc. Japan* **40**, 2975 (1967).
[103] SARGENT, M. V. and TIMMONS, C. J. *J. chem. Soc.* 5544 (1964).
[104] CARRUTHERS, W. and STEWART, H. N. M. *J. chem. Soc.* (*C*) 560 (1967).
[105] SCHOLZ, M., MÜHLSTÄDT, M. and DIETZ, F. *Tetrahedron Lett.* 665 (1967).
[106] CARRUTHERS, W. *J. chem. Soc.* (*C*) 1525 (1967).
[107] MARTIN, R. H., FLAMMANG-BARBIEUX, M., COSYN, J. P. and GELBCKE, M. *Tetrahedron Lett.* 3507 (1968).
[108] HOULTON, P. R. and KEMP, W. *Tetrahedron Lett.* 1045 (1968).
[109] LOADER, C. E. and TIMMONS, C. J. *J. chem. Soc.* (*C*) 1078 (1966).
[110] BORTOLUS, P., CAUZZO, G. and GALIAZZO, G. *Tetrahedron Lett.* 3717 (1966).
[111] LOADER, C. E. and TIMMONS, C. J. *J. chem. Soc.* (*C*) 1457 (1967).
[112] LOADER, C. E. and TIMMONS, C. J. *J. chem. Soc.* (*C*) 1343 (1967).
[113] PERKAMPUS, H. H. and KASSEBEER, G. *Annl. Chem.* (*Liebigs*) **696**, 1 (1966).
[114] DYKE, S. F. and SAINSBURY, M. *Tetrahedron* **23**, 3161 (1967).
[115] BRADSHER, C. K. and DOOLITTLE, R. E. *J. org. Chem.* **31**, 2616 (1966).
[116] JOSHUA, C. P. and LEWIS, G. E. *Aust. J. Chem.* **20**, 2229 (1967).
[117] CAVA, M. P. and SCHLESSINGER, R. H. *Tetrahedron Lett.* 2109 (1964).
[118] BADGER, G. M., DREWER, R. J. and LEWIS, G. E. *Aust. J. Chem.* **17**, 1036 (1964).
[119] CLARK, V. M. and COX, A. *Tetrahedron* **22**, 3421 (1966).
[120] GRELLMANN, K. H. and TAUER, E. *Tetrahedron Lett.* 1909 (1967).
[121] CAVA, M. P., HAULICEK, S. C., LINDERT, A. and SPANGLER, R. J. *Tetrahedron Lett.* 2937 (1966); YANG, N. C., LENZ, G. R. and SHANI, A. *Tetrahedron Lett.* 2941 (1966).
[122] LENZ, G. R. and YANG, N. C. *Chem. Commun.* 1136 (1967).
[123] KHARASCH, N., ALSTON, T. G., LEWIS, H. B. and WOLF, W. *Chem. Commun.* 242 (1965).
[124] TOSHIMA, N. and MORITANI, I. *Tetrahedron Lett.* 357 (1967).
[125] BUTER, J., KELLOGG, R. M., VAN DRIEL, H. and WYNBERG, H. *J. Am. chem. Soc.* **89**, 3487 (1967).
[126] PADWA, A. and HARTMAN, R. *J. Am. chem. Soc.* **88**, 3759 (1966).

[127] MORIARTY, R. M., KLIEGMAN, J. M. and DESAI, R. B. *Chem. Commun.* 1255 (1967).
[128] WHITE, E. H. and ANHALT, J. P. *Tetrahedron Lett.* 3937 (1965).
[129] CARRUTHERS, W. and STEWART, H. N. M. *J. chem. Soc.* 6221 (1965); LOADER, C. E. and TIMMONS, C. J. *J. chem. Soc.* (*C*) 1677 (1967); GROEN, M. B., KELLOGG, R. M. and WYNBERG, H. *J. org. Chem.* **32**, 3093 (1967).
[130] CARRUTHERS, W. *Chem. Commun.* 272 (1966).
[131] KHARASCH, N. and ARIYAN, Z. S. *Chemy Ind.* 302 (1965).
[132] CLARK, V. M., COX, A. and HERBERT, E. J. *J. chem. Soc.* (*C*) 831 (1968).
[133] JOHNSON, A. W. *Chemy Br.* **3**, 253 (1967).
[134] WHITE, E. H. and SIEBER, A. A. F. *Tetrahedron Lett.* 2713 (1967).
[135] FRÁTER, GY. and SCHMID, H. *Helv. chim. Acta* **50**, 255 (1967).
[136] RINEHART, K. L. and SCHACHT, R. J. *J. Am. chem. Soc.* **89**, 2239 (1967).
[137] KAGAN, J. *J. Am. chem. Soc.* **88**, 2617 (1966).
[138] SWENTON, J. S. *Tetrahedron Lett.* 2855 (1967).
[139] BLACKBURN, E. V. and TIMMONS, C. J., unpublished results.
[140] HILL, J. *Chem. Commun.* 260 (1966).
[141] BRUCE, J. M. and KNOWLES, P. *J. chem. Soc.* (*C*) 1627 (1966).
[142] MILLER, B. *Chem. Commun.* 327 (1966).
[143] CLASEN, R. A. and SEARLES, S. *Chem. Commun.* 289 (1966).
[144] CLOSS, G. L., BÖLL, W. A., HEYN, H. and DEV, V. *J. Am. chem. Soc.* **90**, 173 (1968).
[145] BALDWIN, J. E. and MCDANIEL, M. C. *J. Amer. chem. Soc.* **89**, 1537 (1967).
[146] BLATTMANN, H.-R., BOEKELHEIDE, V., HEILBRONNER, E., MEUCHE, D. and MOLYNEUX, R. J. *J. Am. chem. Soc.* **87**, 130 (1965).
[147] PAQUETTE, L. A. and BARNETT, J. H. *J. Amer. chem. Soc.* **88**, 1718 (1966).
[148] AUTREY, R. L., BARTON, D. H. R., GANGULY, A. K. and REUSCH, W. H. *J. chem. Soc.* 3313 (1961).
[149] SRINIVASAN, R. *J. Am. chem. Soc.* **85**, 4045 (1963).
[150] DAUBEN, W. G. and WILLEY, F. G. *Tetrahedron Lett.* 893 (1962).
[151] GASSMAN, P. G. and HYMANS, W. E. *Chem. Commun.* 795 (1967); CHAPMAN, O. L. and MURPHY, T. J. *J. Am. chem. Soc.* **89,** 3476 (1967).
[152] HALLER, I. and SRINIVASAN, R. *J. Am. chem. Soc.* **88**, 5084 (1966).
[153] BRAUMAN, J. I., ELLIS, L. E. and VAN TAMELEN, E. E. *J. Am. chem. Soc.* **88**, 846 (1966).
[154] COATES, R. M. and DAUBEN, W. G. *J. Am. chem. Soc.* **86**, 2490 (1964).
[155] WARD, H. R. and WISHNOK, J. S. *J. Am. chem. Soc.* **90**, 1085 (1968).
[156] ARNETT, E. M. and BOLLINGER, J. M. *Tetrahedron Lett.* 3803 (1964).
[157] WILZBACH, K. E., KAPLAN, L., BROWN, W. G. and YANG, S. S. *J. Am. chem. Soc.* **87**, 675 (1965).
[158] WILZBACH, K. E., HARKNESS, A. L. and KAPLAN, L. *J. Am. chem. Soc.* **90**, 1116 (1968).
[159] ASKANI, R., CRIEGEE, R., GRÜNER, H. and SHÄFER, W. *Angew. Chem. int. Edn.* **6**, 78 (1967).
[160] WILZBACH, K. E., RITSCHER, J. S. and KAPLAN, L. *J. Am. chem. Soc.* **89**, 1031 (1967).
[161] PADWA, A. and HARTMAN, R. *J. Am. chem. Soc.* **86**, 4212 (1964).
[162] JONES, L. B. and JONES, V. K. *J. Am. chem. Soc.* **89**, 1880 (1967) and references therein.
[163] DAUBEN, W. G. and CARGILL, R. L. *Tetrahedron* **12**, 186 (1961).
[164] FORBES, E. J. and GRIFFITHS, J. *J. chem. Soc.* (*C*) 2072 (1966).
[165] CANNELL, L. G. *Tetrahedron Lett.* 5967 (1966).
[166] ZIMMERMAN, H. E. and GRUNEWALD, G. L. *J. Am. chem. Soc.* **88**, 183 (1966).
[167] PRINZBACH, H. and RIVIER, J. *Tetrahedron Lett.* 3713 (1967).

[168] DAUBEN, W. G. and WIPKE, W. T. *Pure appl. Chem.* **9**, 539 (1964).
[169] HURST, J. J. and WHITHAM, G. H. *J. chem. Soc.* 2864 (1960).
[170] BROWN, R. F. C., COOKSON, R. C. and HUDEC, J. *Chem. Commun.* 823 (1967); *Tetrahedron* **24**, 3955 (1968).
[171] SRINIVASAN, R. *J. Am. chem. Soc.* **84**, 3982 (1962).
[172] YANG, N. C. and JORGENSON, M. J. *Tetrahedron Lett.* 1203 (1964).
[173] ANASTASSIOU, A. G. and CELLURA, R. P. *Chem. Commun.* 762 (1967).
[174] MURRAY, R. W. and KAPLAN, M. L. *J. Am. chem. Soc.* **88**, 3527 (1966).
[175] JONES, L. B. and JONES, V. K. *J. Am. chem. Soc.* **89**, 1880 (1967); **90**, 1540 (1968).
[176] KELLY, D. P., PINHEY, J. T. and RIGBY, R. D. G. *Tetrahedron Lett.* 5953 (1966).
[177] STENBERG, V. I. in *Organic Photochemistry* (ed. O. L. Chapman), p. 127, Arnold, London (1967).
[178] ANDERSON, J. C. and REESE, C. B. *J. chem. Soc.* 1781 (1963).
[179] OBARA, H. and TAKAHASHI, H. *Bull. chem. Soc. Japan* **40**, 1012 (1967).
[180] KOBSA, H. *J. org. Chem.* **27**, 2293 (1962).
[181] REID, S. T. and SUBRAMANIAN, G. *Chem. Commun.* 245 (1966).
[182] FINNEGAN, R. A. and KNUTSON, D. *Chem. Commun.* 172 (1966); *J. Am. chem. Soc.* **89**, 1970 (1967).
[183] ELAD, D. *Tetrahedron Lett.* 873 (1963).
[184] KAMINSKY, L. S. and LAMCHEN, M. *J. chem. Soc.* (*C*) 2295 (1966); BAPAT, J. B. and BLACK, D. ST. C. *Chem. Commun.* 73 (1967).
[185] STREITH, J., DARRAH, H. K. and WEIL, M. *Tetrahedron Lett.* 5555 (1966) and references therein; IKEKAWA, N., HONMA, Y. and KENKYUSHO, R. *Tetrahedron Lett.* 1197 (1967).
[186] STREITH, J. and CASSAL, J. M. *Angew Chem.* **80**, 117 (1968).
[187] OGATA, M. and KANŌ, K. *Chem. Commun.* 1176 (1967).
[188] BUCHARDT, O., BECHER, J. and LOHSE, C. *Acta chem. scand.* **20**, 2467 (1966).
[189] HADDADIN, M. J. and ISSIDORIDES, C. H. *Tetrahedron Lett.* 753 (1967).
[190] KANEKO, C., YAMADA, S., YOKOE, I. and ISHIKAWA, M. *Tetrahedron Lett.* 1873 (1967).
[191] BUCHARDT, O. *Tetrahedron Lett.* 1911 (1968).
[192] BUCHARDT, O., JENSEN, B. and KJØLLER-LARSEN, I. *Acta chem. scand.* **21**, 1841 (1967).
[193] BUCHARDT, O., LOHSE, C., DUFFIELD, A. M. and DJERASSI, C. *Tetrahedron Lett.* 2741 (1967).
[194] BUCHARDT, O. *Tetrahedron Lett.* 6221 (1966).
[195] KANEKO, C. and YAMADA, S. *Tetrahedron Lett.* 5233 (1967).
[196] TAYLOR, E. C. and SPENCE, G. G. *Chem. Commun.* 767 (1966).
[197] FISCHER, E. *J. Am. chem. Soc.* **90**, 796 (1968).
[198] BACK, M. H. and CVETANOVIĆ, R. J. *Can. J. Chem.* **41**, 1396 (1963); CARR, M. D., KANE, V. V. and WHITING, M. C. *Proc. chem. Soc.* 408 (1964).
[199] SEARLE, R., WILLIAMS, J. L. R., DE MEYER, D. E. and DOTY, J. C. *Chem. Commun.* 1165 (1967).
[200] COLE, R. S. and HAMMOND, G. S. *J. Am. chem. Soc.* **87**, 3256 (1965).
[201] ZIMMERMAN, H. E. and SANDEL, V. R. *J. Am. chem. Soc.* **85**, 915 (1963).
[202] VAN VLIET, A., KRONENBERG, M. E. and HAVINGA, E. *Tetrahedron Lett.* 5957 (1966).
[203] ZIMMERMAN, H. E. in *Advances in Photochemistry* (ed. W. A. Noyes, Jr., G. S. Hammond and J. N. Pitts, Jr.), Vol. 1, p. 183, Wiley, London (1963).
[204] BARTON, D. H. R., NAKANO, T. and SAMMES, P. G. *J. chem. Soc.* (*C*) 322 (1968).
[205] HAVINGA, E., DE JONGH, R. O. and DORST, W. *Recl. Trav. chim. Pays-Bas Belg.* **75**, 378 (1956).
[206] HAVINGA, E. and DE JONGH, R. O. *Bull. Socs. chim. belg.* **71**, 803 (1962).

[207] JOHNSON, R. M. and REESE, C. W. *Proc. chem. Soc.* 213 (1964).
[208] VAN DER STEGEN, G. H. D., POZIOMEK, E. J., KRONENBERG, M. E. and HAVINGA, E. *Tetrahedron Lett.* 6371 (1966).
[209] BARLTROP, J. A., BUNCE, N. J. and THOMSON, A. *J. chem. Soc.* (*C*) 1142 (1967).
[210] KHARASCH, N. and SHARMA, R. K. *Chem. Commun.* 492 (1967).
[211] KIRMSE, W, *Carbene Chemistry*, Academic Press, London (1964); HINE, J. *Divalent Carbon*, Ronald Press, New York (1964).
[212] SMITH, P. A. S. *The Chemistry of Open Chain Organic Nitrogen Compounds*, Benjamin, New York (1965).
[213] MOSS, R. A. *Tetrahedron Lett.* 4905 (1967); CIGANEK, E. *J. Am. chem. Soc.* **89**, 1458 (1967); NOZAKI, H., NAKANO, M. and KONDÔ, K. *Tetrahedron* **22**, 477 (1966).
[214] DÜRR, H. and SCHEPPERS, G. *Chem. Ber.* **100**, 3236 (1967).
[215] DWORSCHAK, H. and WEYGAND, F. *Chem. Ber.* **101**, 302 (1968).
[216] PADWA, A. and LAYTON, R. *Tetrahedron Lett.* 2167 (1965).
[217] KOPECKY, K. R., HAMMOND, G. S. and LEERMAKERS, P. A. *J. Am. chem. Soc.* **84**, 1015 (1962).
[218] SWENTON, J. S. *Tetrahedron Lett.* 3421 (1968).
[219] BARTON, D. H. R. and MORGAN, L. R. *J. chem. Soc.* 622 (1962); BARTON, D. H. R. and STARRATT, A. N. *J. chem. Soc.* 2444 (1965).
[220] MEINWALD, J. and AUE, D. H. *Tetrahedron Lett.* 2317 (1967).
[221] LWOWSKI, W., MISHRA, A. and RICE, S. N. *J. org. Chem.* **33**, 481 (1968).
[222] WOERNER, F. P., REIMLINGER, H. and ARNOLD, D. R. *Angew. Chem.* **80**, 119 (1968).
[223] CASTELLUCCI, N. C., KATO, M., MASAMUNE, S. and ZENDA, H. *Chem. Commun.* 473 (1967).
[224] DOERING, W. VON E. and ODUM, R. A. *Tetrahedron* **22**, 81 (1966).

7

Syntheses of Aromatic Compounds

J. W. Barton, *Bristol University*

The term 'aromatic' covers an ever-widening spectrum of organic compounds [1]. In order to place some limitation on the present discussion it is proposed to deal only with syntheses involving benzenoid compounds or involving species closely related to these.

RING SYNTHESES

The few methods available for the direct construction of benzene rings have been confined to the synthesis of particular types of compound; for instance, the condensation of ketones with the sodium salt of nitromalondialdehyde leads to nitrophenols of type (1).

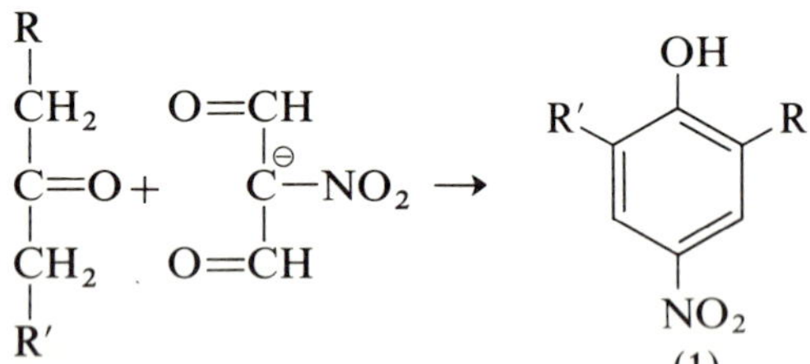

(1) [2]

These older methods are now supplemented by several new ones based on diene additions. The reaction of *trans*, *trans*-1,4-diacetoxybuta-1,3-diene with suitably activated double bonds gives adducts (2) which aromatize easily by the loss of acetic acid [3].

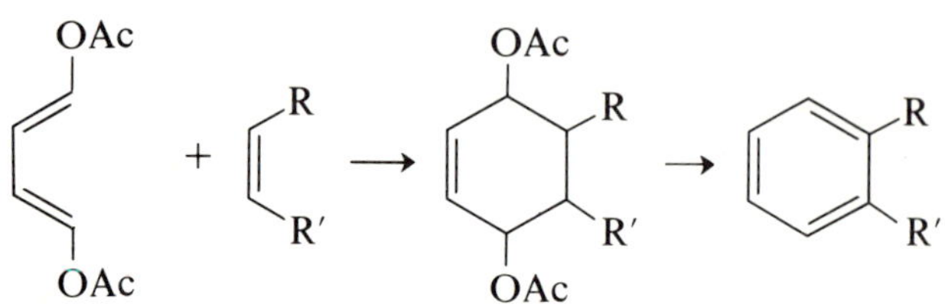

(2)

The method is especially useful for the annellation of benzene rings to quinones. In a closely related procedure, 1-acetoxybuta-1,3-diene derivatives (3), prepared by the acetylation of $\alpha\beta$-unsaturated aldehydes, are reacted with maleic anhydride to give adducts (4). These eliminate acetic acid and are dehydrogenated to give phthalic anhydrides when boiled in pyridine [4].

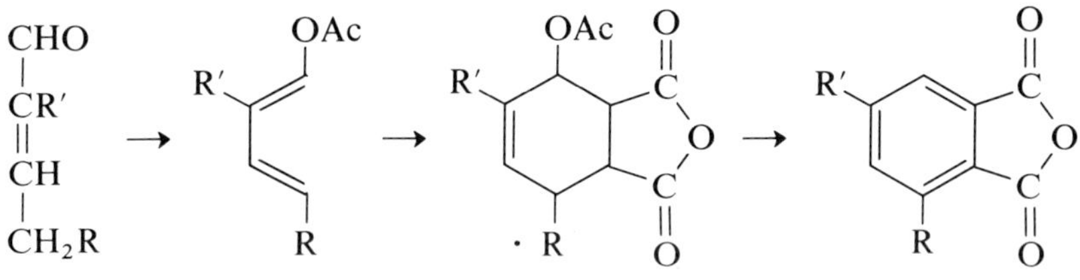

(3) (4)

The use of acetylenes as dienophiles in this sequence obviates the dehydrogenation step, but the yields reported [3] are low.

Additions of enamines, such as (5), to ethyl penta-2,4-dienoate and related esters give cyclohexadienes (6) on elimination of the amine moiety. These may be aromatized by mild oxidation, or if the reaction is conducted at higher temperature the benzene derivative is obtained directly [5].

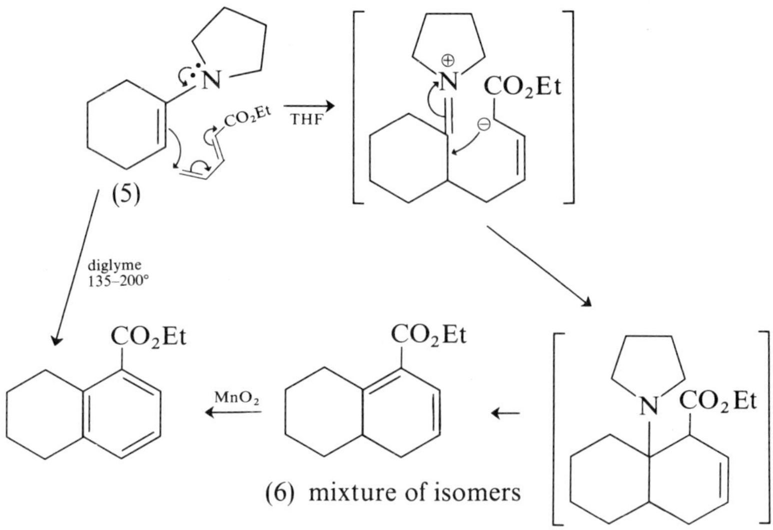

(5)

(6) mixture of isomers

Comparable results are obtained with 5- and 7-membered cyclic enamines.

Simple derivatives of α-pyrone such as (7) are fairly reactive dienes, giving adducts (8) which readily lose the lactone bridge to give benzene derivatives [6]. Such reactions have received little attention, but the publication of an improved method for the preparation of the parent α-pyrone [7] should prompt further study.

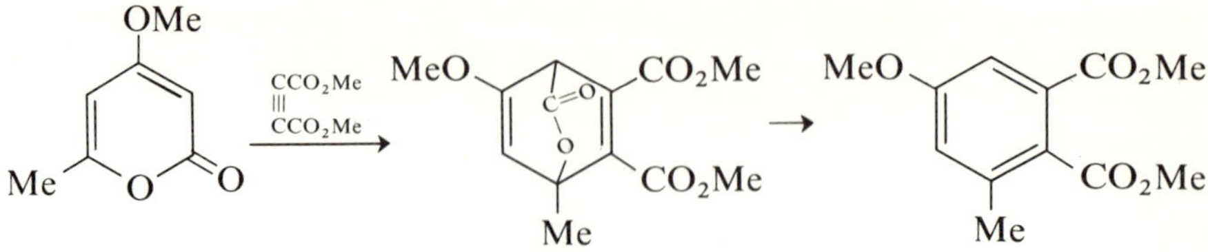

(7) (8)

In similar fashion cyclopentadienone derivatives like tetracyclone (9), an efficient aryne trap, and their ketals, such as (10), have been used as dienes. The carbonyl and ketal bridges in the resulting adducts are easily removed, if they are not lost spontaneously.

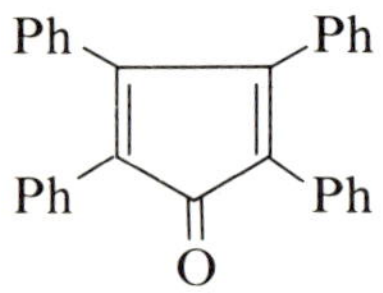

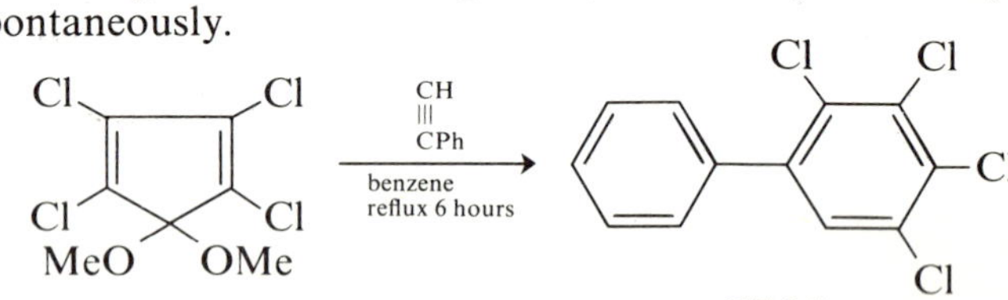

(17%)

(9) (10) [8(a)]

Alkyl and arylbenzenes may be constructed by using transition metal compounds to bring about the trimerization of acetylene derivatives. Following the work of Reppe, it has been found that a number of metal carbonyls and complexes, e.g., (11), bring about the trimerization of diphenylacetylene to hexaphenylbenzene in good yield [9].

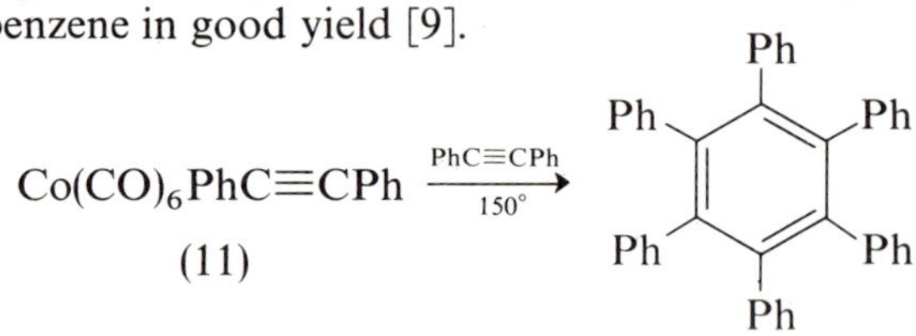

(11)

(70%)

The method has been extended to produce mixed trimers, notably sterically crowded ones such as 1,2-di-t-butylbenzene, formed from acetylene and the complex from di-t-butylacetylene and dicobalt octacarbonyl [10], after conventional methods of synthesis had failed.

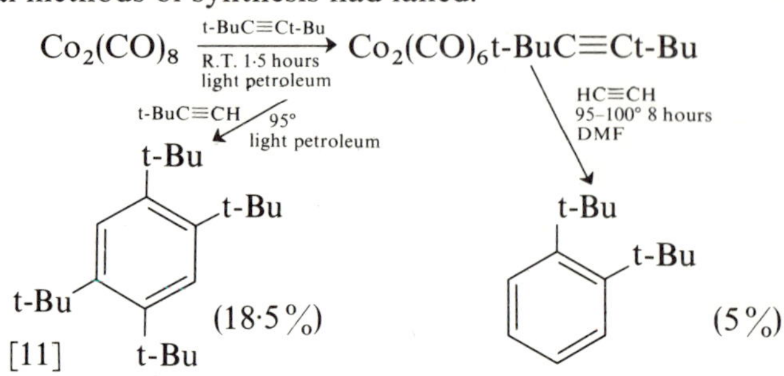

(18·5%) (5%)

[11] [10]

See also [12, 13].

Similarly, the reaction of dimethylacetylene with triphenylchromium tetrahydrofuranate (12) gives hexamethylbenzene, but in this case the phenyl groups in the complex participate, with the formation of some 1,2,3,4-tetramethylnaphthalene, particularly if the amount of dimethylacetylene is limited [14].

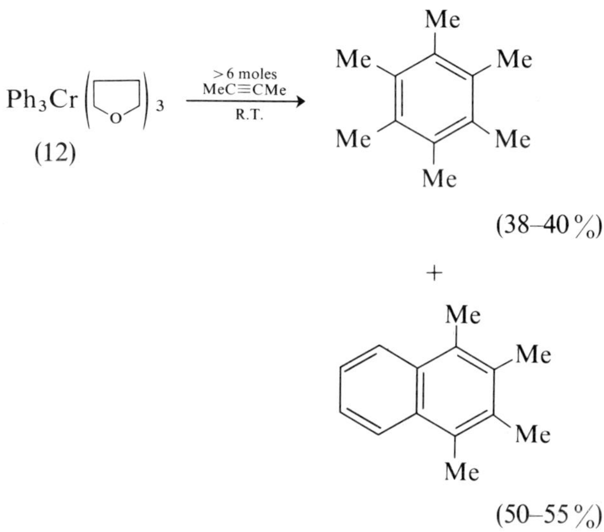

The method may be extended for the annellation of further rings on to polycyclic aromatic compounds. The tri(2-naphthyl) complex gives both 1,2,3,4-tetramethylanthracene and 1,2,3,4-tetramethylphenanthrene, the latter predominating.

More recently it has been reported that palladium metal catalyses the trimerization of dimethyl acetylenedicarboxylate to give almost quantitative yields of hexamethylmellitate [15]; however, this reaction appears to be very limited in scope.

SYNTHESES WITH REACTIVE INTERMEDIATES

Since the existence of 1,2-dehydrobenzene (13)† or 'benzyne' as a reactive intermediate was demonstrated in 1953 [16], a vast amount of information has been gathered concerning this species (for reviews see [17]), and some reactions of it are found to have considerable synthetic utility. There are two main types of reaction available for the generation of benzynes: (a) by elimination reactions from monosubstituted or from 1,2-disubstituted benzene derivatives; and (b) by the fragmentation of benzoheterocycles or of salts like benzenediazonium-2-carboxylate and diphenyliodonium-2-carboxylate. Reactions of type (a) do not have general applicability in synthesis because the method of generation of the benzyne usually involves the use of a nucleophile which readily reacts further with it. For example,

† Structure (13) is used as being the most convenient, but not necessarily the best, representation of this species (see [17]).

the reaction of 2-bromoiodobenzene with lithium gives triphenylene (14) [18] by a stepwise reaction, not by trimerization [19].

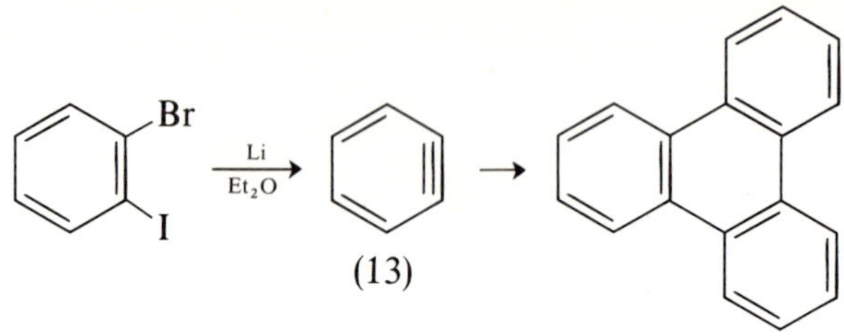

(13)

(53–59 %)

(14)

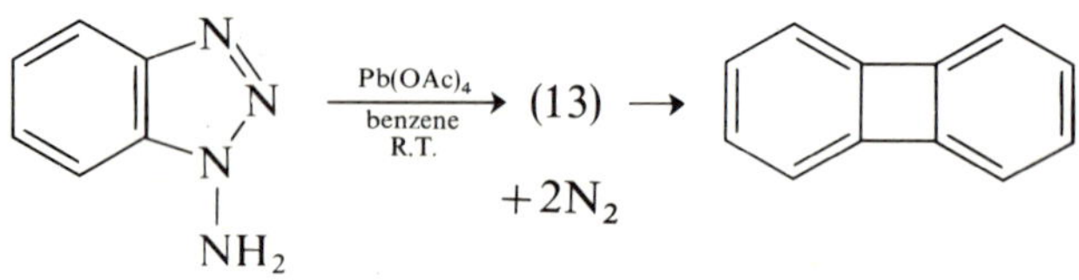

(15) (16)

Only low yields of the benzyne dimer, biphenylene (16), are obtainable from this type of reaction [20] and it is better to employ one of type (b), such as the oxidation of 1-aminobenzotriazole (15) with lead tetra-acetate [21]. The latter reaction has been extended to the synthesis of poly-cyclic biphenylenes which are otherwise inaccessible [22].

In some instances reactions of type (a) are preferred. The generation of a carbanion and an 'yne' function simultaneously results in reaction between the two, and the process may be employed for the arylation of ketones, esters, etc., as in the synthesis of ketone (17) [23].

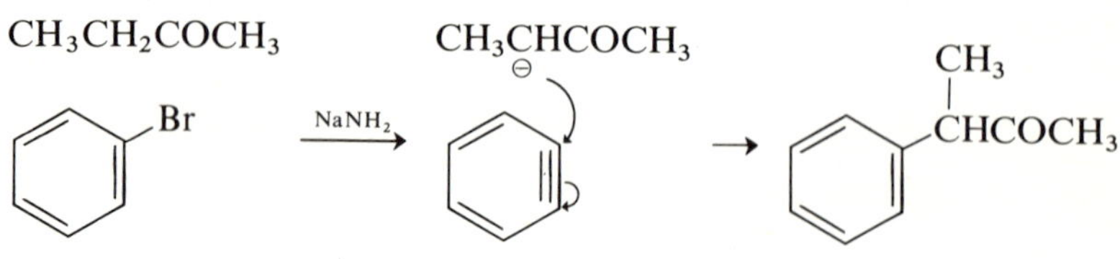

(75 %)

(17)

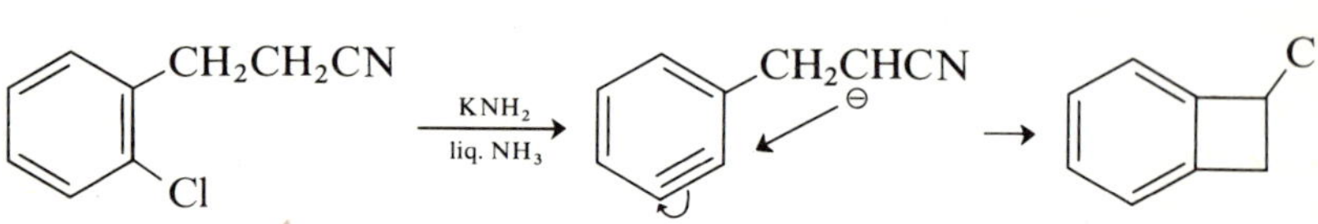

(61 %)

(18) (19)

This method is effective for the cyclization of sidechains of four to seven carbon atoms, e.g., (18) → (19), providing that a good carbanion-stabilizing group is present [24]. The corresponding *m*-halogenocompounds give lower yields of the same products. The same reaction may be used to effect cyclizations with a heteroatom in a side chain serving as the nucleophile, e.g., (20) → (21) [25]. In this manner a large number of benzoheterocycles has been synthesized, some with rings having as many as sixteen and seventeen atoms [26].

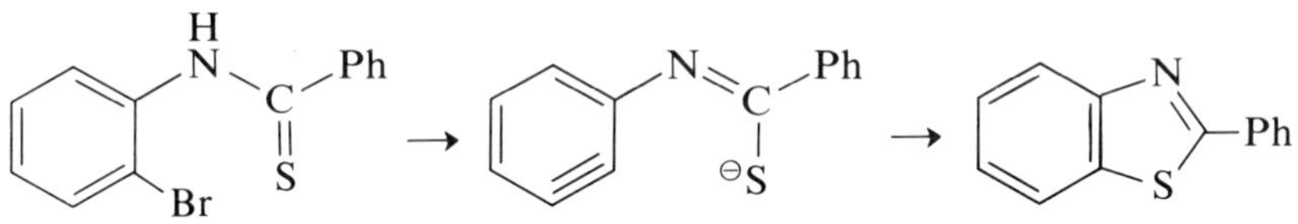

(90%)

(20) (21)

When used as dienophiles, benzynes are best generated by methods of type (b). Many such reactions are recorded; for example, the decomposition of benzenediazonium-2-carboxylate (22) in the presence of furan gives the endoxide (23) [27], while the analogous reaction with anthracenes as dienes is by far the most convenient route to triptycene (24) and its derivatives [28], and supersedes the reaction of type (a) [29].

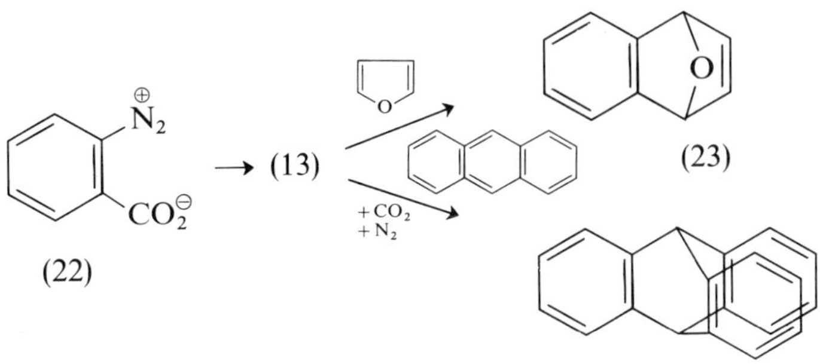

(22) (23) (24)

Direct diene syntheses of aromatic rings (see above) by reactions of benzyne with 1-acetoxybuta-1,3-diene [3], α-pyrone [30] and phenylacetylenes [31] are also known.

The marked electrophilic character of benzynes allows 1,2-additions to occur with suitably reactive double bonds and other electron-rich systems. The decomposition of benzenediazonium-2-carboxylate (22) in the presence of vinyl esters and ethers [32] provides a direct route to benzocylobutene derivatives (25), although an attempt to extend the method using 1-ethoxyvinyl acetate gave ethyl 2-acetylphenylacetate as the main product (25%). The same behaviour is observed with ethoxyacetylene, which gives a moderate yield of 2-ethoxyphenylacetylene [31].

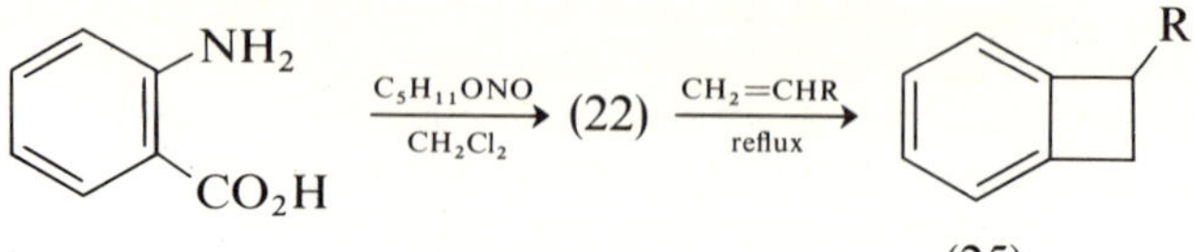

(25)
R = OEt (40 %)
R = OAc (45 %)

Similar reactions with aromatic nitrosocompounds lead to *N*-phenylcarbazoles [33], while 1,3-additions take place with azides and diazomethane derivatives, forming benzotriazoles [34] and indazoles [35] respectively.

Quinone Methides (or Methanes)

The dimethides of *ortho* and *para* benzoquinone, usually referred to as *o*- (26) and *p*-xylylene (27), have received a lot of attention in recent years [36]. The *para* compound (27) is relatively stable in solution at low temperature, while *o*-xylylene is known only as a transient intermediate or in combination with transition metals [37]. For use as a diene it is conveniently generated by the action of zinc on $\alpha\alpha'$-dibromo-*o*-xylene, as in the synthesis of 2-cyano-1,2,3,4-tetrahydronaphthalene (28) [38].

(27)

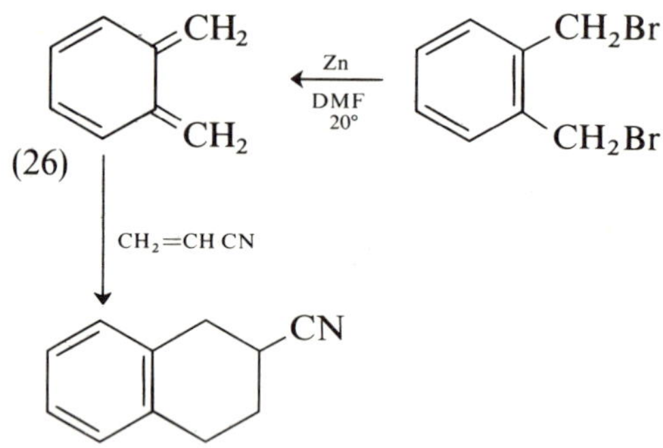

(28) (80 %)

FUNCTIONAL GROUPS

In synthesis, it is usual to introduce the desired functional groups into aromatic nuclei by replacement of hydrogen or some other substituent. These replacement processes are usually electrophilic or nucleophilic substitutions, or less commonly homolytic ones. The general preparative aspects of electrophilic [39] and nucleophilic [40] substitution have been reviewed, and the present discussion is confined to areas where developments have taken place.

Halides

Direct iodination by classical methods, more difficult to achieve and less well investigated than chlorination and bromination, has been further studied [41]. Other useful iodination procedures have been developed, using iodine in conjunction with peracetic acid [42] and with silver trifluoroacetate [43], in which the attacking reagent is considered to be the acyl hypoiodite. Halogenation methods involving halogen cations or closely related species [44] promise to be more generally applicable to substitution in relatively unreactive nuclei. For example, benzoic acid is readily iodinated by a mixture of iodine and silver sulphate in strong sulphuric acid [45]. Chlorination [46] and bromination [47] may be accomplished by similar means.

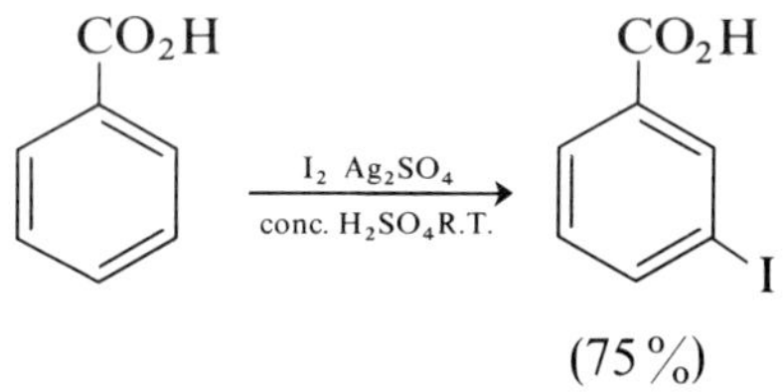

[45(a)]

In this type of reaction, protonation of the substrate may occur to influence the position of substitution (see [47(b)]). More recently it has been shown that mixtures of *N*-chloro- and *N*-bromo-succinimide with 50% aqueous sulphuric acid will halogenate aromatics [48]. The yields obtained are comparable with those from the silver salt method.

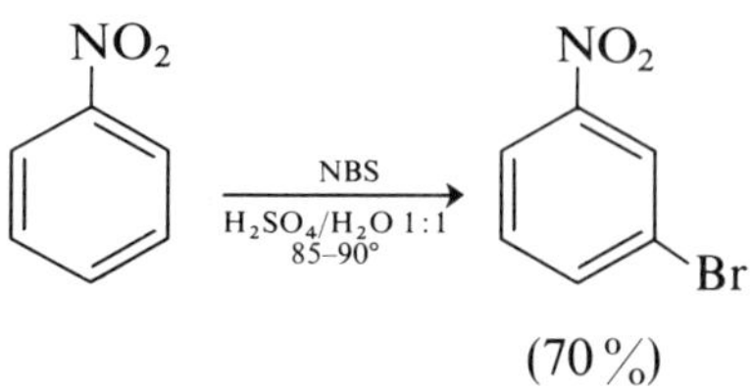

[48]

Studies have been made of the halogenation of aromatic [49(a–c)] and heteroaromatic [49(d, e)] compounds in the presence of relatively large amounts of aluminium chloride, the so-called 'catalyst swamping effect'. Some of these reactions are useful in synthesis; for example, acetophenone is brominated at position 3-, side-chain attack being completely suppressed.

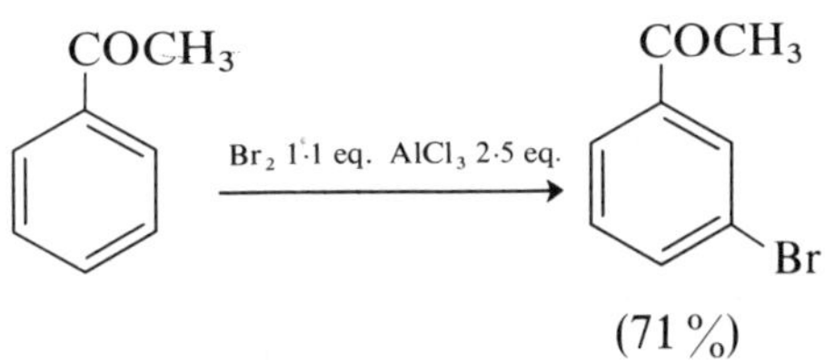

[49(a)]

In addition to methods for the direct introduction of halogen into aromatics there are the indirect ones, the commonest being the Sandmeyer reaction (for developments see the section on amines). The mercuration of aromatic compounds is a useful synthetic reaction because the resulting mercurials are readily converted into halides by treatment with halogens under very mild conditions [50]. In favourable cases, conversions into other compounds, such as nitroso, nitro and carbonyl, are also possible. Except in the case of amines and phenols, mercuration has not been extensively investigated until recently, when isomer distributions and kinetic data for a number of hydrocarbons have been forthcoming [51]. In the absence of solvents and in acetic acid the reaction is at least partly homolytic and isomer distributions tend to be random. However, when it is catalysed by strong acids or their salts it proceeds more rapidly, and the isomer distributions are nearer to those expected in electrophilic substitution. The reaction is reversible; hence build-up of certain isomers may occur if the reaction time is prolonged, or if the temperature is varied. By control of the reaction conditions either of the two monoiodobiphenylenes may be obtained as the main product via a perchloric-acid-catalysed mercuration reaction [52].

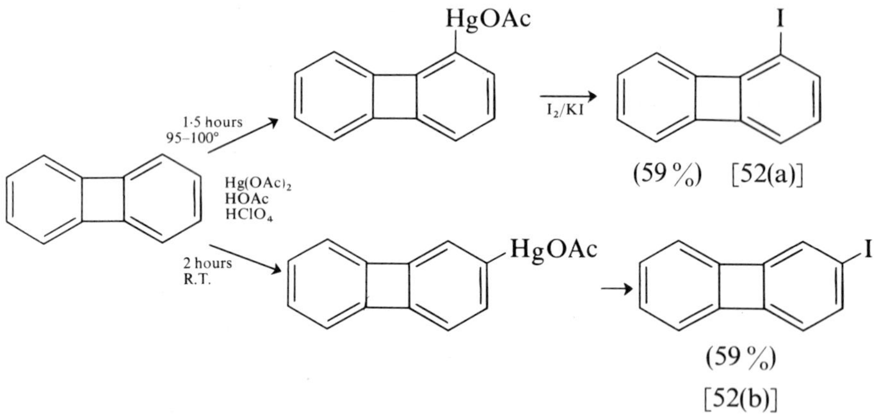

Conversions of carboxylic acids into aryl halides have been achieved by using the Hunsdiecker and related reactions [53], but the yields obtained are poor unless electron-withdrawing substituents are present in the nucleus. Recently it has been reported [54(a)] that aroyl chlorides may be decarbonylated in very high yields by using tris(triphenylphosphine)rhodium chloride as a catalyst. Arylsulphonyl chlorides undergo an analogous reaction [54(b)] under more strenuous conditions.

Nucleophilic replacement of halogen in aryl halides is difficult to achieve unless the halogen atom is activated by electron-withdrawing groups in the nucleus. Recent studies of the use of dipolar aprotic solvents [55] and of copper-promoted reactions [56] have helped to extend the scope of such reactions in synthesis. The Rosenmund–von Braun nitrile synthesis is the best-known example of this type of reaction. In its original form it consisted of heating aryl bromides with cuprous cyanide and a limited amount of pyridine or quinoline as a promoter at temperatures in the range 130–200°. Aryl chlorides required longer reaction times and higher temperatures, so that sealed systems were often required. Less strenuous conditions are necessary if solvents such as *N*,*N*-dimethylformamide [57], *N*-methyl-pyrrolidone [58], or dimethyl sulphoxide [59] are used, and the isolation procedures are made easier.

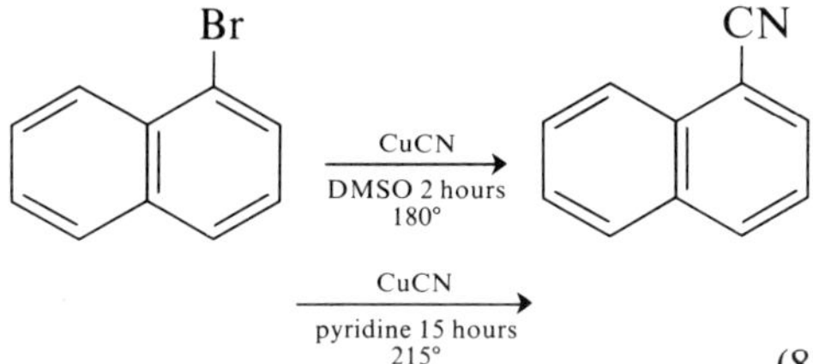

(99 %) [59]

(82–90 %) [60]

Conversions of aryl bromides or iodides into chlorides are readily effected in the same manner [59]. Conditions have been worked out for replacements by other nucleophiles [61] in reactions promoted by cuprous oxide or cuprous halides. Heterocyclic bases also give good results as media for some of these reactions, as exemplified by the preparation of 9-methoxy-phenanthrene and 9-methoxyanthracene from the corresponding bromo compounds [61(b)].

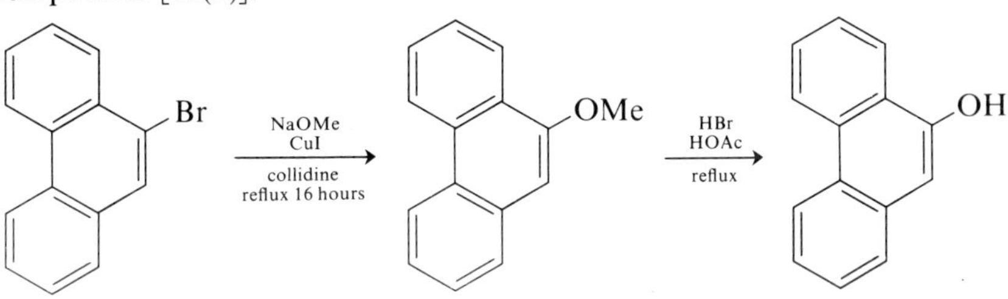

(Quantitative) (85 %) [61(b)]

In such cases, replacement of halogen by alkoxyl followed by dealkylation is the preferred sequence for the conversion of halide into phenol. The main alternative methods, which involve conversion of the halide into an organic magnesium or lithium derivative followed by direct or indirect oxidation of this, e.g., (29) → (30) [62], do not give uniformly good results.

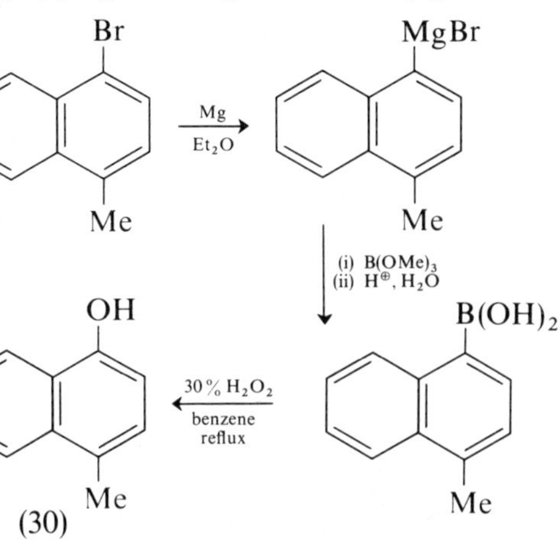

(29) (30)

(33 % overall) [62(b)]

The replacement of halogen using copper acetylides is useful for the introduction of acetylenic side chains [63].

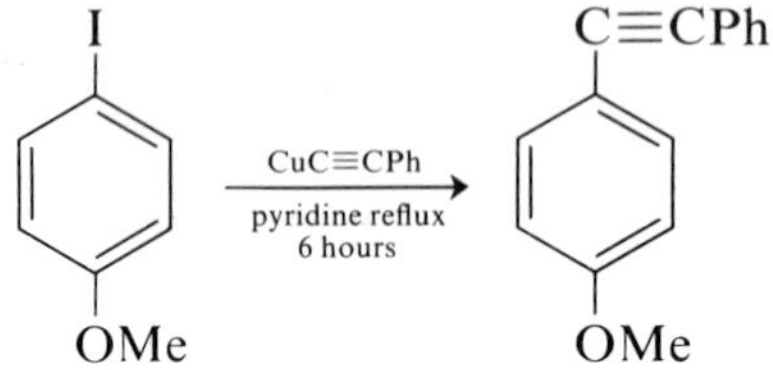

(83 %) [63]

Certain *ortho*-substituted compounds react further; for instance, *o*-hydroxyhalides give rise to benzofurans.

Diaryliodonium salts, which are readily obtainable by a variety of methods [64], may find some use in synthesis, as they are attacked by nucleophiles very much more easily than are aryl iodides, sometimes under exceptionally mild conditions, as in the conversion of 2-nitrodiphenyliodonium bromide (31) into a mixture of iodobenzene and *o*-dinitrobenzene.

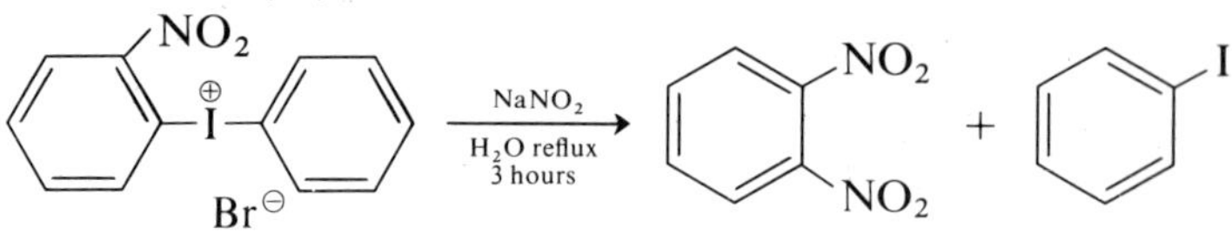

(31) (84 %) [65]

Diphenyliodonium bromide has been used for the *O*-phenylation of hydroxy compounds [66] and for the *C*-phenylation of indandiones [67].

Amino Compounds

Aromatic amines are commonly obtained by the reduction of nitro compounds, by the Beckmann rearrangement of oximes [68], or from carboxylic acids and their derivatives by the Hofmann degradation and related reactions [69]. The Bucherer reaction of phenols [70] is more or less limited to bi- and polycyclic compounds, but a sequence (32) → (33) applicable to other phenols has been reported [71].

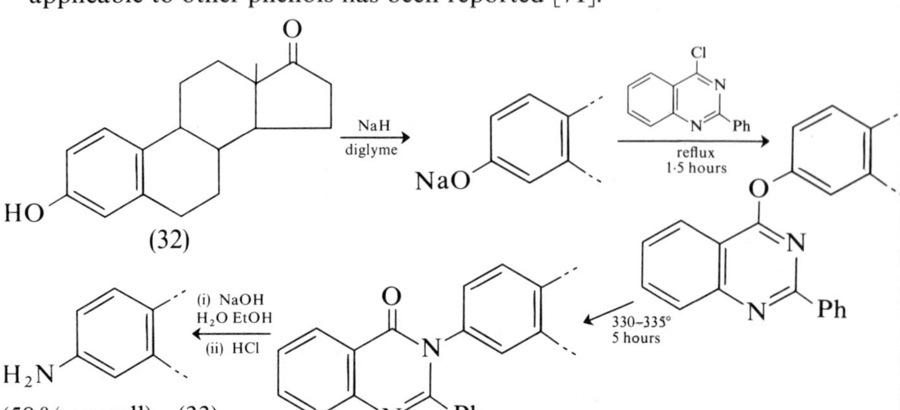

(32)

(58 % overall) (33)

For the reduction of nitro compounds the recent tendency has been to replace catalytic hydrogenation by reagent/catalyst combinations, several of which are found to be effective. The method due to H. C. and C. A. Brown of generating hydrogen together with the catalyst *in situ* is a direct alternative to quantitative hydrogenation at atmospheric pressure [72]. Systems using palladium/carbon as catalyst with sodium borohydride [73] or α-phellandrene [74] as the hydrogen-donor are reported to give good results with nitro compounds, and catalytic hydrazine reductions have been widely used [75]. The latter give yields equal to or better than the corresponding hydrogenation and require no special equipment. Halogen substituents on an aromatic nucleus are usually retained [76].

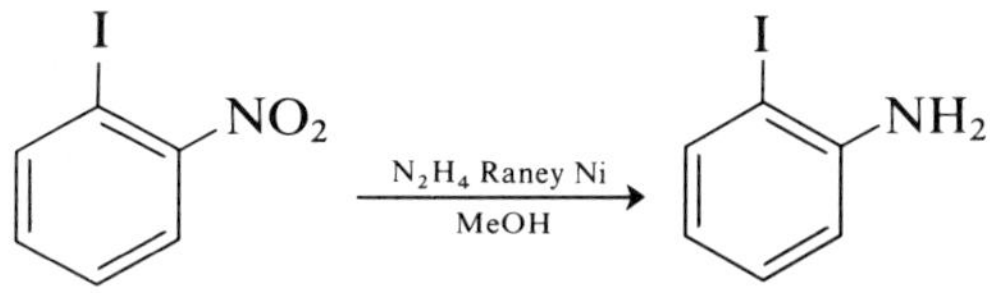

(64 %) [76]

Much attention has been given to the direct amination of benzenoid compounds, but a generally useful method has yet to be found. Hydrazoic acid [77] in the presence of Friedel–Crafts catalysts gives acceptable yields of amines with fairly reactive hydrocarbons such as toluene; with hydroxylamine-*o*-sulphonic acid [78] and other hydroxylamine derivatives [79] the yields are lower. The isomer distributions recorded indicate that the reactions are less selective than normal electrophilic substitutions.

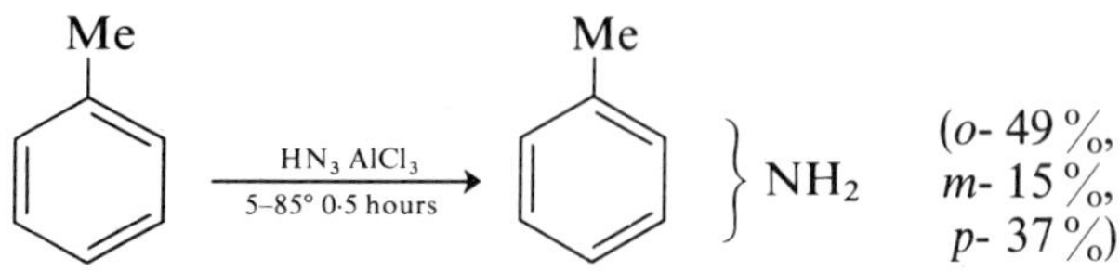

(65 %) [77(a)]

The corresponding dialkylamination takes place more readily [80] and acylaminations with hydroxamic acids have also been reported [81].

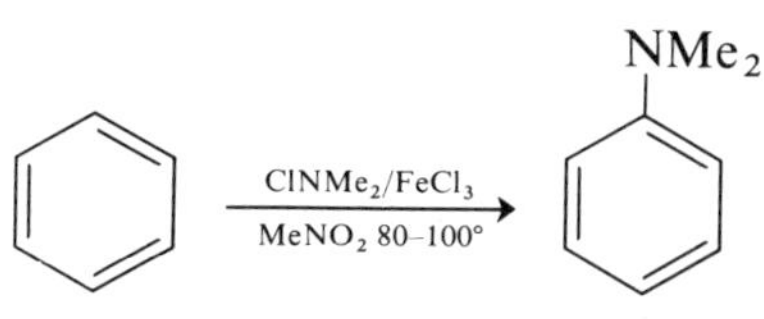

(90 %) [80]

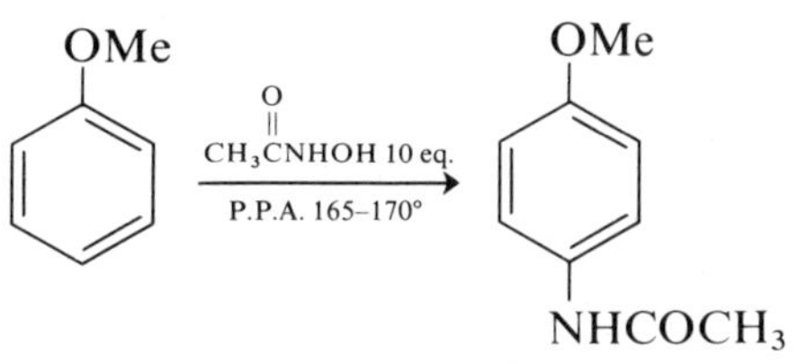

(57 %) [81]

It may be noted that hydroxylamine-*O*-sulphonic acid has also been used for *N*- and *O*-amination in basic solution in the same way that dimethyl sulphate is used for methylation. Aromatic amines give arylhydrazines with this reagent [82] while phenol gives the unstable *O*-phenylhydroxylamine [83].

In connection with the reactions of amines, nitrosation, diazotization, and deamination [84] have been reviewed: the uses of diazonium compounds [85] in synthesis continue to be extended. An alternative to the Sandmeyer reaction which obviates the use of copper salts consists of the aprotic diazotization of an arylamine in bromoform solution, giving the aryl bromide directly [86]. It is less satisfactory for chlorides.

It has been known for some time that diazonium salts may be used to arylate oximes [87] which then, on hydrolysis, give aldehydes and ketones.

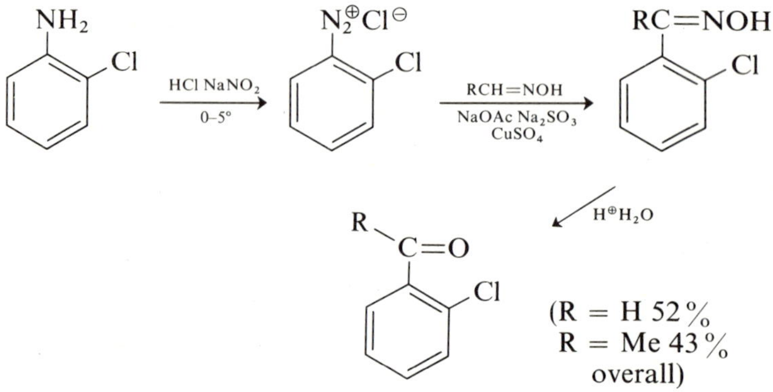

More recently, a method for the conversion of aromatic amines into carboxylic acids and esters has become available [88].

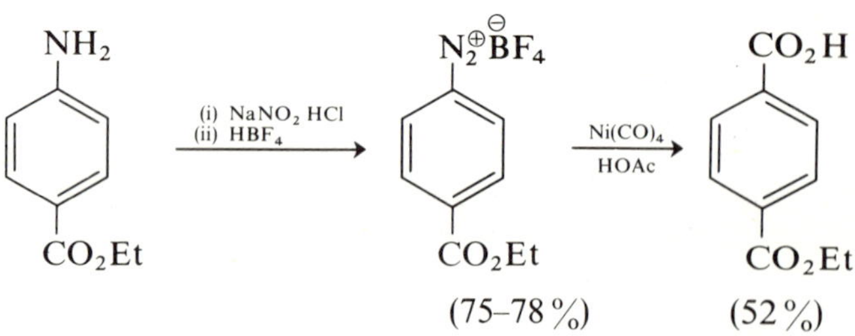

A mixture of the diazonium fluoroborate with nickel carbonyl is decomposed by the addition of acetic acid, giving the aromatic acid; the use of ethanol in place of acetic acid gives the corresponding ester. Although the yields are variable, the method appears to offer some advantage over the sequence $ArNH_2 \rightarrow ArN_2{}^+X^- \rightarrow ArCN \rightarrow ArCO_2H$ if a compound has other groups which are susceptible to hydrolysis.

In cases where a nitro compound is not available by direct nitration, it has been prepared from the corresponding amine via the diazonium salt or by direct oxidation with Caro's acid, which only gives good results with nitroamines. Newer oxidation procedures using peracetic [89] and permaleic [90] acids give high yields of nitro compounds; thus the oxidation method is now superior to that employing diazonium salts.

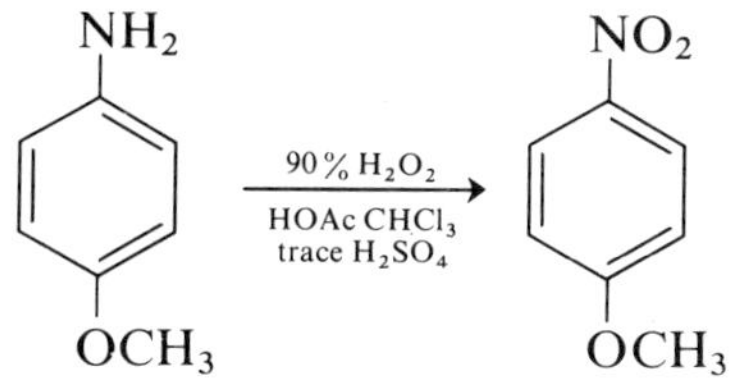

(82%) [89(a)]

Phenols

There is as yet no satisfactory general method for the direct hydroxylation of aromatic compounds. Only a few reactive hydrocarbons like mesitylene undergo electrophilic hydroxylation with peroxytrifluoroacetic acid [91]. Thus the introduction of a hydroxyl group is normally carried out via the corresponding amine or sulphonic acid, or a carbonyl compound may be oxidized by the Baeyer–Villiger method [92]. However, if a compound has a hydrogen which is capable of undergoing metal exchange (lithiation) [93], then a hydroxyl group may be introduced by methods analogous to those used for converting halides into phenols (see above, ref. [62]). For example, 2,3′,5′-trimethoxybiphenyl has been hydroxylated at position 4′ to give (34) [94(a)].

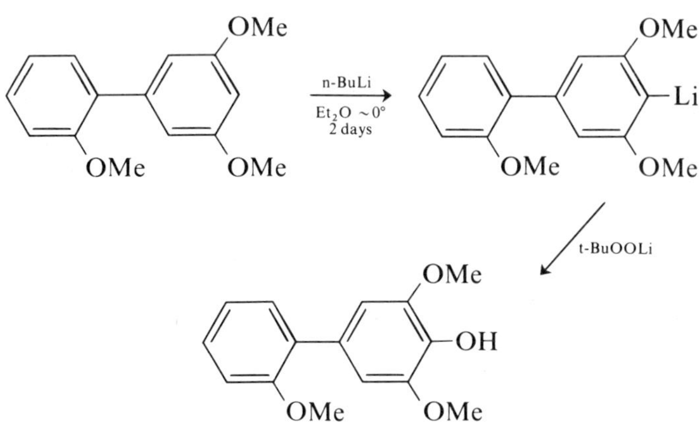

(34) (53% overall)

The discovery that lithiation reactions are catalysed by tertiary amines of low steric requirement [95] increases the scope of such procedures.

Carbonyl Compounds

Acylation and related reactions, Fries migration, etc., have been dealt with in a recent treatise on the Friedel-Crafts reaction [96], and new methods continue to appear. For example, the direct formylation of aromatics with dichloromethyl ethers, easily prepared from alkyl formates and phosphorus pentachloride, gives uniformly good results [97] and does not require such reactive substrates as do the Gattermann [98] and Vilsmeier [99] formylation reactions.

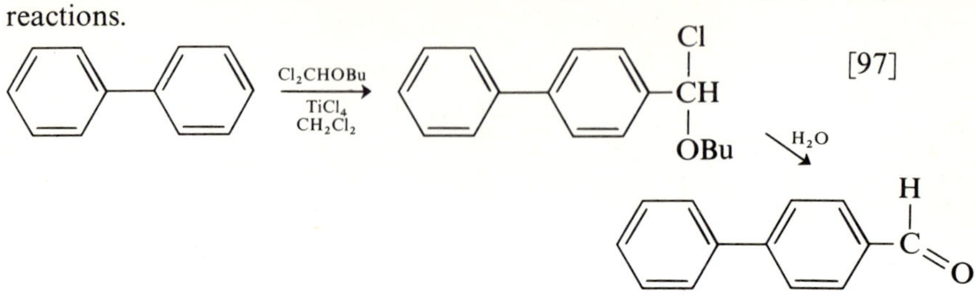

[97]

(64 %)

An extension of this reaction using dichloromethylenedioxybenzene (35), prepared from catachol carbonate and phosphorus pentachloride, gives good yields of carboxylic acids [100].

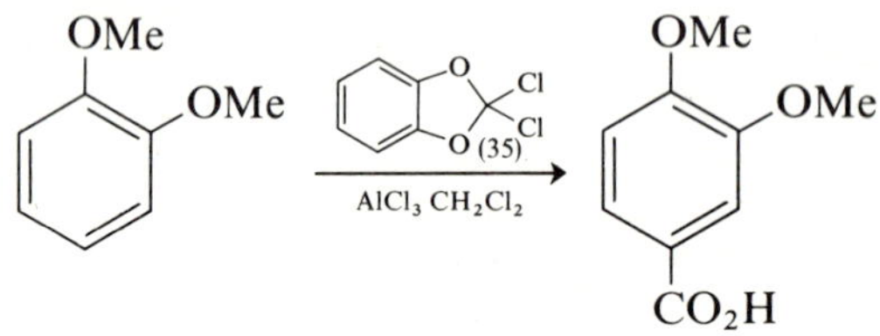

(88 %) [100]

Recent indirect methods for the preparation of aromatic aldehydes include the reduction of nitriles with Raney nickel alloy in aqueous formic acid [101] and the aprotic diazotization of benzylamines in dimethyl sulphoxide [102], the latter reaction being similar to the earlier one employing benzyl halides with pyridine *N*-oxide [103].

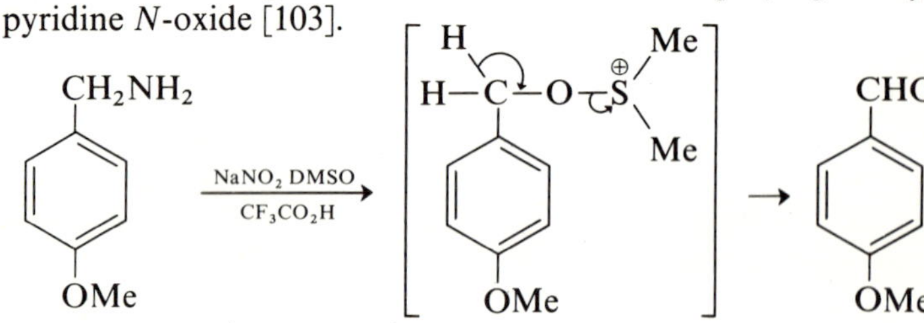

[102]

(65 %)

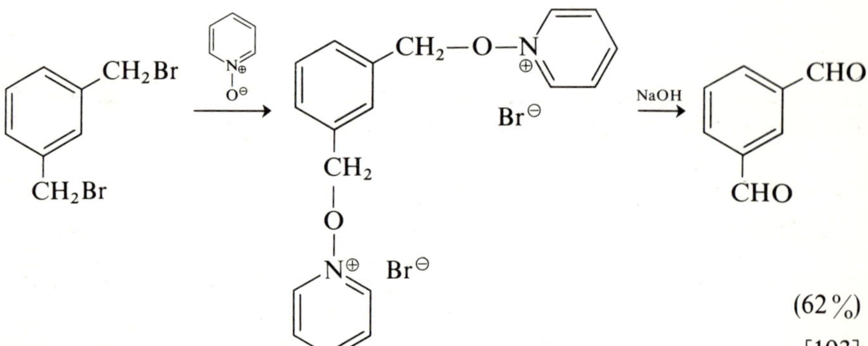

(62 %)

[103]

Alkylation

For reviews of electrophilic alkylation, halomethylation, etc., see ref. [96].

Solutions containing the carbanion from dimethyl sulphoxide, generated by using potassium t-butoxide or sodium hydride, are found to attack the more reactive aromatic hydrocarbons, the overall reaction being one of methylation [104].

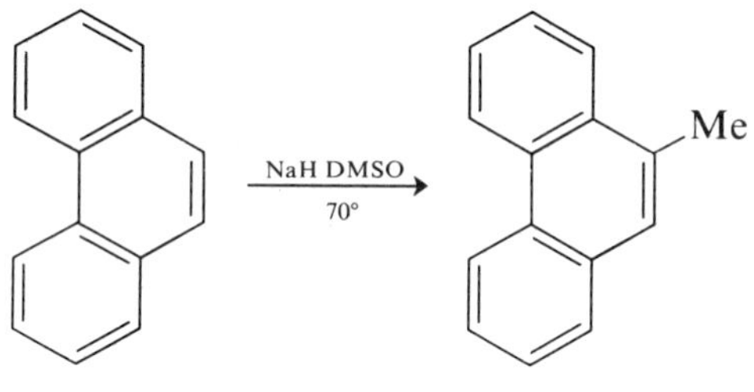

(76 %) [104(b)]

In similar fashion dimethylsulphoxonium methylide has given low yields of methylated nitro compounds [105].

Biaryl Formation

The use of a dipolar aprotic solvent, usually *N*,*N*-dimethylformamide, as a diluent in the Ullmann reaction [106] makes isolation procedures much easier and often results in increased yields at lower temperatures. This is probably due mainly to dissolution of the copper halide so that the copper surface remains clean and active throughout the reaction.

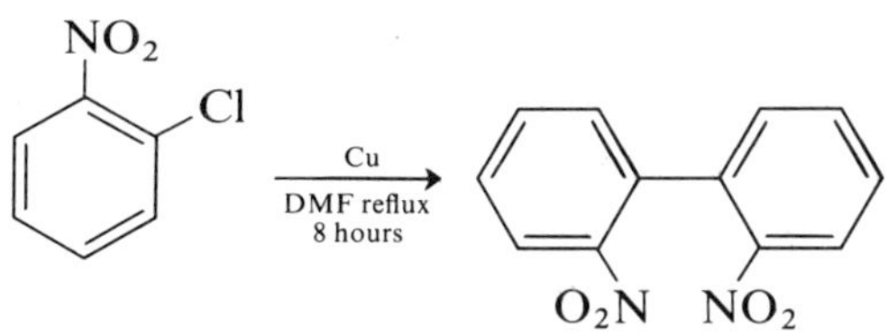

[107(a)]

(80 %)

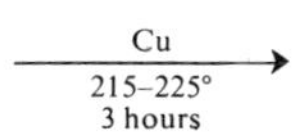

(52–61 %) [107(b)]

It is important that the diluent used should be reasonably dry; otherwise dehalogenation of the starting material may be the chief reaction. For example, 10,10′-dinitro-9,9′-biphenanthryl (36) is formed by the reaction of 9-bromo-10-nitrophenanthrene with copper in the absence of a diluent [108]. If, however, the reaction is conducted in undried *N*,*N*-dimethylformamide the product is 9-nitrophenanthrene; this is in fact the best route to the latter compound.

A recent study of the copper-catalysed decarboxylation of aromatic acids in quinoline indicates that this reaction and the Ullmann biaryl synthesis probably proceed via a common intermediate, which may be an aryl copper [109]. Two useful extensions of the Ullmann reaction have emerged: (a) that unsymmetrical biaryls are conveniently prepared by heating an aromatic acid with an aryl halide and cuprous oxide in quinoline; and (b) that

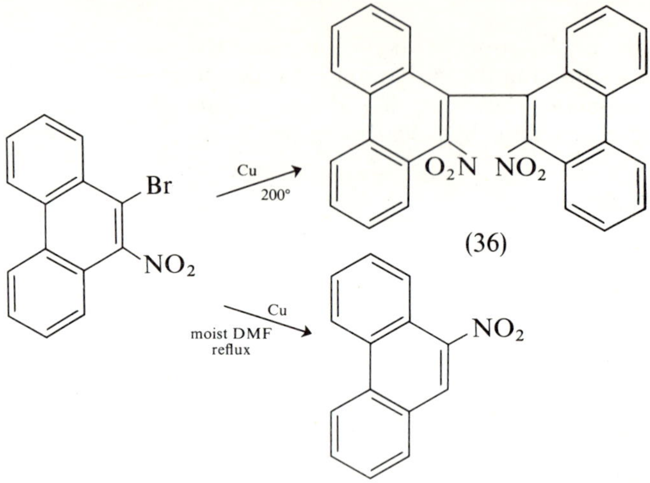

(36)

[108]

in favourable cases the reactive intermediate may be generated by removal of hydrogen from an aromatic ring, e.g., by heating *m*-dinitrobenzene with cuprous oxide in quinoline.

(a)

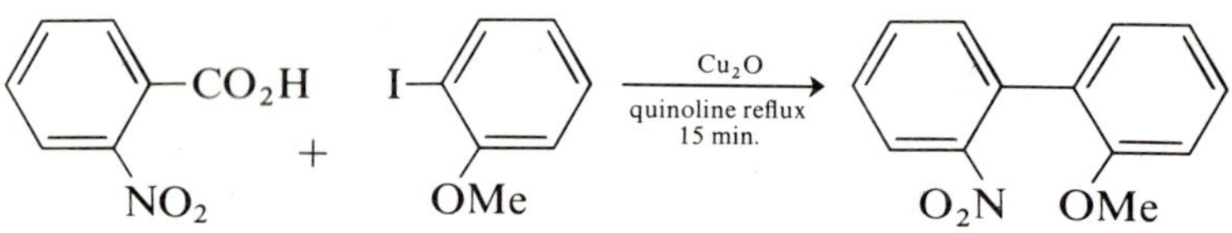

(50 %) [109(a)]

(b)

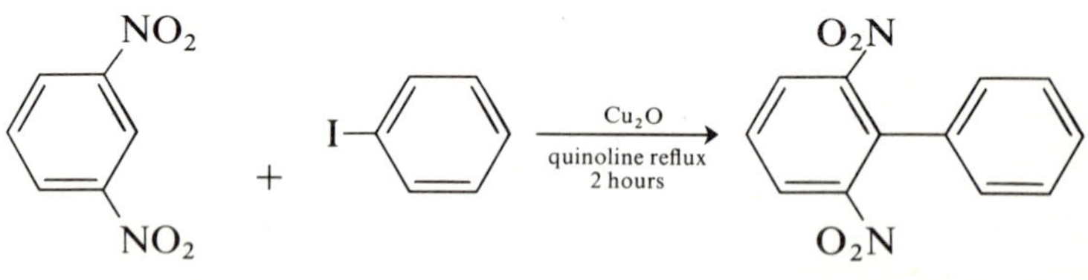

(41 %) [109(b)]

For biaryl formation the Gomberg arylation reaction [110] is the main alternative to the Ullmann and lesser-known Busch [111] coupling reactions of halides, particularly for the preparation of unsymmetrical biaryls. Useful modifications of the standard procedures have been developed, involving the aprotic diazotization of an aromatic amine which is dissolved in the compound to be arylated [112].

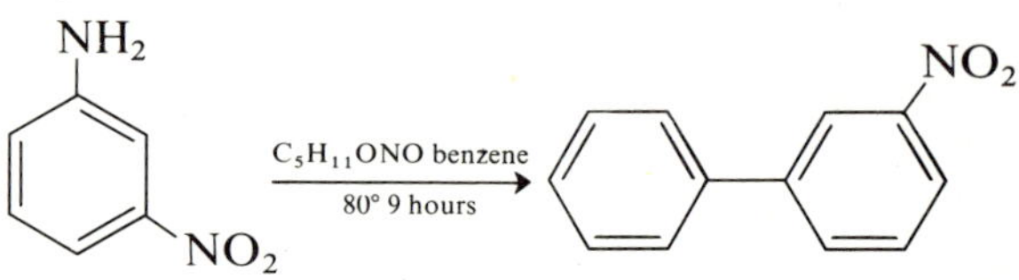

(50–60 %) [112(b)]

It appears that these reactions are not straightforward aprotic diazotizations (cf. [112(c)]), as the primary product, isolable in some cases, is the diazoamino compound. In the above example a reaction time of one hour gives 3,3′-dinitrodiazoaminobenzene (71 %) together with biaryl (16 %) [113].

REFERENCES

[1] Baker, W. The Chemical Society Pedler Lecture, *Chem. Br.* **1**, 191, 250 (1965).
[2] Fanta, P. E. and Stein, R. A. *Chem. Rev.* **60**, 261 (1960).
[3] Hill, R. K. and Carlson, R. M. *Tetrahedron Lett.* 1157 (1964).
[4] Blanc, P. *Helv. chim. Acta* **44**, 1 (1961).
[5] Danishefsky, S. and Cunningham, R. *J. org. Chem.* **30**, 3676 (1965).
[6] Bu'Lock, J. D. and Smith, H. G. *J. chem. Soc.* 502 (1960).
[7] Zimmerman, H. E., Grunewald, G. L. and Paufler, R. M. *Org. Synth.* **46**, 101 (1966).
[8] (a) McBee, E. T., Diveley, W. R. and Burch, J. E. *J. Am. chem. Soc.* **77**, 385 (1955).
(b) Eaton, P. E. and Hudson, R. A. *J. Am. chem. Soc.* **87**, 2769 (1965).
[9] Hübel, W. and Hoozgand, C. *Chem. Ber.* **93**, 103 (1960).
[10] Arnett, E. M. and Strem, M. E. *Chemy. Ind.* 2008 (1961).
[11] Arnett, E. M., Strem, M. E. and Friedel, R. A. *Tetrahedron Lett.* 658 (1961).
[12] Hübel, W. and Hoozgand, C. *Tetrahedron Lett.* 637 (1961).
[13] Kruerke, U., Hoozgand, C. and Hübel, W. *Chem. Ber.* **94**, 2817 (1961).
[14] Herwig, W., Metlesics, W. and Zeiss, H. H. *J. Am. chem. Soc.* **81**, 6203 (1959).
[15] Bryce-Smith, D. *Chemy Ind.* 239 (1964).
[16] Roberts, J. D., Simmons, H. E., Carlsmith, L. A. and Vaughan, C. W. *J. Am. chem. Soc.* **75**, 3290 (1953).
[17] (a) Bunnett, J. F. *J. chem. Educ.* **38**, 278 (1961); (b) Heaney, H. *Chem. Rev.* **62**, 81 (1962); (c) Wittig, G. *Angew. Chem. Int. Ed.* **4**, 731 (1965).
[18] Heaney, H. and Millar, I. T. *Org. Synth.* **40**, 105 (1960).
[19] Bartle, K. D., Heaney, H., Jones, D. W. and Lees, P. *Tetrahedron* **21**, 3289 (1965).
[20] Wittig, G. and Pohmer, L. *Chem. Ber.* **89**, 1334 (1956).
[21] Campbell, C. D. and Rees, C. W. *Chem. Commun.* 192 (1965).
[22] Barton, J. W. and Jones, S. A. *J. chem. Soc.* (*C*) 1276 (1967).
[23] Leake, W. W. and Levine, R. *J. Am. chem. Soc.* **81**, 1169 (1959).
[24] Bunnett, J. F. and Skorcz, J. A. *J. org. Chem.* **27**, 3836 (1962).
[25] Bunnett, J. F. and Hrutfiord, B. F. *J. Am. chem. Soc.* **83**, 1691 (1961).
[26] Huisgen, R., Konig, H. and Lapley, A. R. *Chem. Ber.* **93**, 1496 (1960).
[27] Stiles, M., Miller, R. G. and Burkhardt, U. *J. Am. chem. Soc.* **85**, 1792 (1963).
[28] Friedman, L. and Logullo, F. M. *J. Am. chem. Soc.* **85**, 1549 (1963).
[29] Wittig, G. *Org. Synth.* **39**, 75 (1959).
[30] Wittig, G. and Hoffman, R. W. *Chem. Ber.* **95**, 2718 (1962).
[31] Stiles, M., Burkhardt, U. and Haag, A. *J. org. Chem.* **27**, 4715 (1962).
[32] Wasserman, H. H. and Solodar, J. *J. Am. chem. Soc.* **87**, 4002 (1965).
[33] Steinhoff, G. W. and Henry, M. C. *J. org. Chem.* **29**, 2808 (1964).
[34] Huisgen, R. and Knorr, R., *Naturwissenschaften* **48**, 716 (1961).
[35] Ried, W. and Shön, M. *Angew. Chem. Int. Edn* **3**, 67 (1964).
[36] Tedder, J. M. *A. Rep. Prog. Chem.* **58**, 225 (1961).
[37] Roth, W. R. and Meier, J. D. *Tetrahedron Lett.* 2053 (1967).
[38] Alder, K. and Fremery, M. *Tetrahedron* **14**, 190 (1961).
[39] Norman, R. O. C. and Taylor, R. *Electrophilic Substitution in Benzenoid Compounds*, Elsevier, Amsterdam (1965).

[40] SAUER, J. and HUISGEN, R. *Angew. Chem.* **8**, 294 (1960).
[41] (a) WIRTH, H. O., KONIGSTEIN, O. and KERN, W. *Annln Chem.* (*Liebigs*) **634**, 84 (1960).
(b) SUZUKI, H., NAKAMURA, K. and GOTO, R. *Bull. chem. Soc. Japan* **39**, 128 (1966).
[42] (a) HASZELDINE, R. N. and SHARPE, A. G. *J. chem. Soc.* 993 (1952).
(b) JANSSEN, D. E. and WILSON, C. V. *Org. Synth.* **36**, 46 (1956).
[43] OGATA, Y. and NAKAJIMA, K. *Tetrahedron* **20**, 43 (1964).
[44] (a) AROTSKY, J. and SYMONS, M. C. R. *Q. Rev. chem. Soc.* **16**, 282 (1962).
(b) BERLINER, E. *J. chem. Educ.* **43**, 124 (1966).
[45] (a) DERBYSHIRE, D. H. and WATERS, W. A. *J. chem. Soc.* 3694 (1950).
(b) BARKER, I. R. L. and WATERS, W. A. *J. chem. Soc.* 150 (1952).
[46] GORVIN, J. H. *Chemy Ind.* 910 (1951).
[47] (a) DERBYSHIRE, D. H. and WATERS, W. A. *J. chem. Soc.* 573 (1950).
(b) GORVIN, J. H. *J. chem. Soc.* 1237 (1953).
[48] LAMBERT, F. L., ELLIS, W. D. and PARRY, R. J. *J. org. Chem.* **30**, 304 (1965).
[49] PEARSON, D. E. *et al. J. org. Chem.*: (a) **23**, 1412 (1958); (b) **27**, 447 (1962); (c) **28**, 3147 (1963); (d) **26**, 789 (1961); (e) **29**, 329 (1964).
[50] KOBE, K. A. and DOUMANI, T. F. *Ind. Engng Chem.* **33**, 70 (1941).
[51] BROWN, H. C. *et al. J. org. Chem.* **32**, 745, 752 (1967) and preceding papers.
[52] BARTON, J. W., WHITAKER, K. E. and SMITH, C. M. *J. chem. Soc.* (*C*) 28 (1968).
[53] (a) JOHNSON, R. G. and INGHAM, R. K. *Chem. Rev.* **56**, 219 (1956).
(b) WILSON, C. V. in *Organic Reactions*, Vol. IX, pp. 332–387, Wiley, New York (1957).
[54] BLUM, J. *Tetrahedron Lett.*: (a) 1605 (1966); (b) 3041 (1966).
[55] PARKER, A. J. *Adv. org. Chem.* **5**, 1–46 (1965).
[56] BACON, R. G. R. and HILL, H. A. O. *Q. Rev. chem. Soc.* **19**, 95 (1965).
[57] FRIEDMAN, L. and SCHECTER, H. *J. org. Chem.* **26**, 2522 (1961).
[58] NEWMAN, M. S. and BODEN, H. *J. org. Chem.* **26**, 2525 (1961).
[59] BACON, R. G. R. and HILL, H. A. O. *J. chem. Soc.* 1097 (1964).
[60] NEWMAN, M. S. *Org. Synth.* **21**, 89 (1941).
[61] (a) BACON, R. G. R. and HILL, H. A. O. *J. chem. Soc.* 1108 (1964).
(b) BACON, R. G. R. and RENNISON, S. C. *Chemy Ind.* 812 (1966).
[62] (a) HAWTHORNE, M. F. *J. org. Chem.* **22**, 1001 (1957).
(b) YARBORO, T. L. and KARR, C. *J. org. Chem.* **24**, 1141 (1959).
[63] CASTRO, C. E. and STEPHENS, R. D. *J. org. Chem.* **28**, 2163, 3313 (1963).
[64] BERINGER, F. M. *et al. J. Am. chem. Soc.*: (a) **75**, 2705 (1953); (b) **81**, 342 (1959).
[65] BERINGER, F. M. *et al. J. Am. chem. Soc.* **75**, 2708 (1953).
[66] MAYER, W., FIKENTSCHER, R., SCHMIDT, J. and SCHMIDT, O. T. *Chem. Ber.* **93**, 2761 (1960).
[67] BERINGER, F. M., GALLON, S. A. and HUANG, S. J. *J. Am. chem. Soc.* **84**, 2819 (1962).
[68] DONARUMA, I. G. and HELDT, W. Z. *Org. React.* **11**, 1–156 (1960).
[69] SMITH, P. A. S. *Molecular Rearrangements*, Part I, pp. 457–568, Interscience, New York (1963).
[70] DRAKE, N. L. *Org. React.* **1**, 105–128 (1942).
[71] MORROW, D. F. and HOFER, R. M. *J. mednl Chem.* **9**, 249 (1966).
[72] BROWN, H. C. and BROWN, C. A. *J. Am. chem. Soc.* **84**, 1495 (1962).
[73] NEILSON, T., WOOD, H. C. S. and WYLIE, A. G. *J. chem. Soc.* 371 (1962).
[74] PALLAND, R. and ANH-HOA, H. *Compt. rend. Acad. Sci. Paris* **252**, 2896 (1961).
[75] FURST, A., BERLO, R. C. and HOOTON, S. *Chem. Rev.* **65**, 51 (1965).
[76] LEGGETTER, B. F. and BROWN, R. K. *Can. J. Chem.* **38**, 2363 (1960).
[77] (a) KOVACIC, P., RUSSELL, R. L. and BENNETT, R. P. *J. Am. chem. Soc.* **86**, 1588 (1964).
(b) HOOP, G. M. and TEDDER, J. M. *J. chem. Soc.* 4685 (1961).

[78] KOVACIC, P. and BENNETT, R. P. *J. Am. chem. Soc.* **83**, 221 (1961).
[79] KOVACIC, P. and FOOTE, J. L. *J. Am. chem. Soc.* **83**, 743 (1961).
[80] BOCK, H. and KOMPA, K. L. *Angew. Chem. int. Edn* **4**, 783 (1965).
[81] WASSMUNDT, F. W. and PADEGIMAS, S. J. *J. Am. chem. Soc.* **89**, 7131 (1967).
[82] GOSL, R. and MEUWSEN, A. *Chem. Ber.* **92**, 2521 (1959).
[83] BUMGARDENER, C. L. and LILLY, R. L. *Chemy Ind.* 559 (1962).
[84] RIDD, J. H. *Q. Rev. chem. Soc.* **15**, 418 (1961).
[85] (a) ZOLLINGER, H. *Azo and Diazo Chemistry*, Interscience, New York (1961).
(b) BELOV, B. I. and KOZLOV, V. V. *Russ. Chem. Revs* **32**, 59 (1963).
[86] CADOGAN, J. I. G., ROY, D. A. and SMITH, D. M. *J. chem. Soc.* (*C*) 1249 (1966).
[87] BEECH, W. F. *J. chem. Soc.* 1297 (1954).
[88] CLARK, J. C. and COOKSON, R. C. *J. chem. Soc.* 686 (1962).
[89] (a) EMMONS, W. D. *J. Am. chem. Soc.* **79**, 5528 (1957).
(b) HOLMES, R. R. and BAYER, R. P. *J. Am. chem. Soc.* **82**, 3454 (1960).
[90] WHITE, R. W. and EMMONS, W. D. *Tetrahedron* **17**, 37 (1961).
[91] HART, H. and BUCHLER, C. A. *J. org. Chem.* **29**, 2397 (1964).
[92] HASSALL, C. H. *Org. React.* **9**, 73–106 (1957).
[93] GILMAN, H. and MORTON, J. W. *Org. React.* **8**, 258–304 (1954).
[94] (a) NILSSON, M. and NORIN, T. *Acta chem. scand.* **17**, 1157 (1963).
(b) MCOMIE, J. F. W. *et al. J. chem. Soc.* (*C*) 328 (1968).
[95] SCRETTAS, C. G. and EASTHAM, J. F. *J. Am. chem. Soc.* **87**, 3276 (1965).
[96] OLAH, G. A. *Friedel-Crafts and Related Reactions*, Vol. 3, Interscience, New York (1965).
[97] RIECHE, A., GROSS, H. and HOFT, E. *Chem. Ber.* **93**, 88 (1960).
[98] TRUCE, W. E. *Org. React.* **9**, 37–72 (1957).
[99] (a) MINKIN, V. I. and DORFEENKO, G. N. *Russ. Chem. Revs* **29**, 599 (1960).
(b) DE MAHEAS, M. R. *Bull. Soc. chim. Fr.* 1989 (1962).
[100] GROSS, H., RUSCHE, J. and MIRACHE, M. *Chem. Ber.* **96**, 1382 (1963).
[101] VAN ES, T. and STASKUN, B. *J. chem. Soc.* 5775 (1965).
[102] SCHEIT, K. H. and KAMFE, W. *Angew. Chem. Int. Edn* **4**, 787 (1965).
[103] FEELY, W., LEHN, W. L. and BOCKELHEIDE, V. *J. org. Chem.* **22**, 1135 (1957).
[104] (a) ARGABRIGHT, P. A., HOFMANN, J. E. and SCHRIESHEIM, A. *J. org. Chem.* **30**, 3233 (1965).
(b) NOZAKI, H., YAMAMOTO, Y. and NOYORI, R. *Tetrahedron Lett.* 1123 (1966).
(c) RUSSELL, G. A. and WEINER, S. A. *J. org. Chem.* **31**, 248 (1966).
[105] TRAYNELIS, V. J. and MCSWEENEY, J. U. *J. org. Chem.* **31**, 243 (1966).
[106] FANTA, P. E. *Chem. Rev.* **38**, 139 (1946); **64**, 613 (1964).
[107] (a) KORNBLUM, N. and KENDALL, D. L. *J. Am. chem. Soc.* **74**, 5782 (1952).
(b) FUSON, R. C. and CLEVELAND, E. A. *Org. Synth.* **20**, 45 (1940).
[108] BRAITHWAITE, R. S. W. and HOLT, P. F. *J. chem. Soc.* 3025 (1959).
[109] (a) NILSSON, M. *Acta chem. scand.* **20**, 423 (1966).
(b) BJORKLUND, C. and NILSSON, M. *Tetrahedron Lett.* 675 (1966).
[110] BACHMANN, W. E. and HOFFMAN, R. A. *Org. React.* **2**, 224–261 (1944).
[111] BUSCH, M. and WEBER, W. *J. prakt Chem.* **146**, 1 (1936).
[112] (a) ATKINSON, C. M. and SHARPE, C. J. *J. chem. Soc.* 2858 (1959).
(b) HWANG HSU. *Acta chim. sin.* **33**, 171 (1959).
(c) CADOGAN, J. I. G. *J. chem. Soc.* 5257 (1962).
(d) LIDGETT, R. A., LYNCH, E. R. and MCCALL, E. B. *J. chem. Soc.* 3754 (1965).

8

Recent Methods for Forming Heterocyclic Rings

G. W. H. CHEESEMAN, *Queen Elizabeth College, London*

Recent methods are regarded for the purpose of this review as those published in the 1960s. In the following discussion, heterocyclic syntheses employing a common type of reagent have been grouped together, and also those in which a general principle of synthesis is applied. No attempt has been made to summarize more specialized aspects of heterocyclic synthesis, such as the preparation of polypyrroles [1]. The reader is referred to publications such as *Angewandte Chemie*, *Quarterly Reviews*, and *Chemical Reviews*. The *Annual Reports of the Chemical Society* (*London*) are a convenient source for locating leading references. *Organic Syntheses* also contains many valuable bibliographies.

SYNTHESES USING ACETYLENIC INTERMEDIATES

Acetylenic intermediates are widely employed in heterocyclic synthesis [2]. These syntheses involve reaction of the acetylene with a nucleophile, and suitable activation of the carbon–carbon triple bond is achieved with groupings such as CO_2R, CN, and CF_3. Activated acetylenes like dimethyl acetylenedicarboxylate react with a wide range of nitrogen-, oxygen-, sulphur-, and phosphorus-containing nucleophiles. For example, a new pyrrole and furan synthesis involves the reaction of an acetylenic ester with an α-amino- or α-hydroxy-ketone, respectively.

$$C_6H_5CO{\cdot}CH(NH_2)CH_3 + CH_3O_2CC{\equiv}CCO_2CH_3 \rightarrow$$

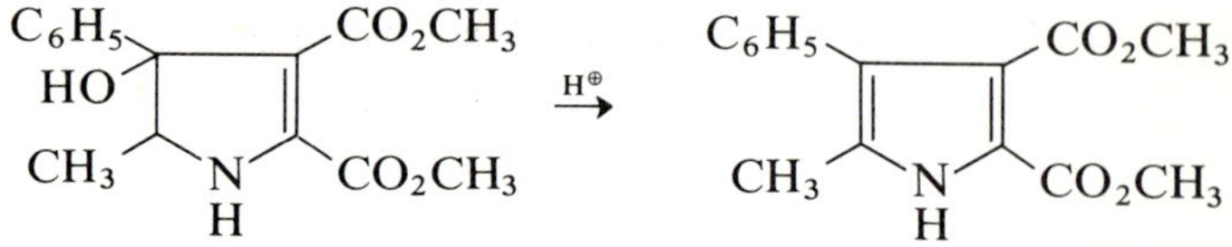

(80%) [3]

$C_6H_5COCH(OH)CH_3 + CH_3O_2CC{\equiv}CCO_2CH_3 \rightarrow$

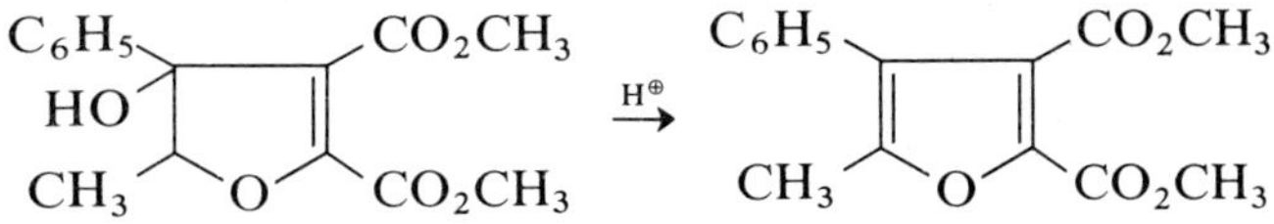

(44 %) [3]

The corresponding reaction of acetylenic esters and α-mercapto carbonyl compounds results in thiophen formation, but in this case an acetylenic ester group is involved in the ring-closure.

$CH_3COCH_2SH + C_2H_5CC{\equiv}CCO_2C_2H_5 \xrightarrow{CH_3CO_2Na}$

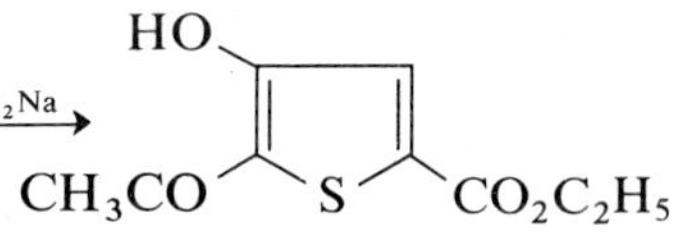

(23 %) [3]

$C_2H_5O_2CCH_2SH + C_2H_5O_2CC{\equiv}CCH(CH_3)_2 \xrightarrow{C_2H_5ONa}$

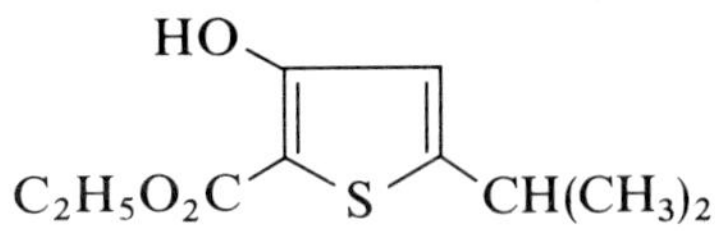

(47 %) [4]

When thiourea is treated with dimethyl acetylenedicarboxylate, the thiazolidone (1) is formed, and treatment of 1,2-diaminopropane with the diethyl ester gives the piperazine (2).

$CS(NH_2)_2 + CH_3O_2CC{\equiv}CCO_2CH_3 \rightarrow$

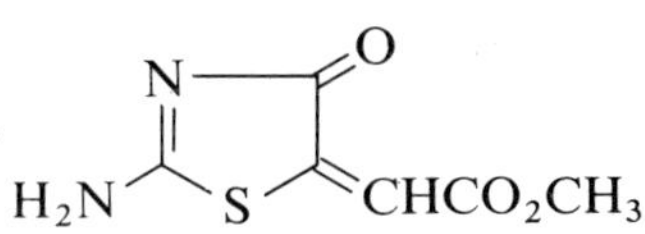

(91 %)

(1) [3]

$CH_3CH(NH_2)CH_2NH_2 + C_2H_5O_2CC{\equiv}CCO_2C_2H_5 \rightarrow$

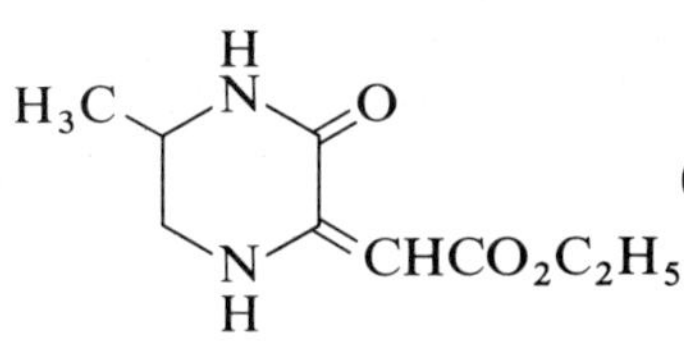

(89 %)

(2) [5]

Six-membered heterocycles are also formed by the reaction of *o*-substituted anilines and acetylenedicarboxylic acid esters [6]. The adduct (4) from 6-aminopiperonal (3) and the dimethyl ester gives the quinoline (5) on treatment with acid [3]. 2-Aminopyridine reacts with propiolic acid via the ring nitrogen atom, and a mixture of the pyridopyrimidone (6) and the *trans*-adduct (7) is obtained [7].

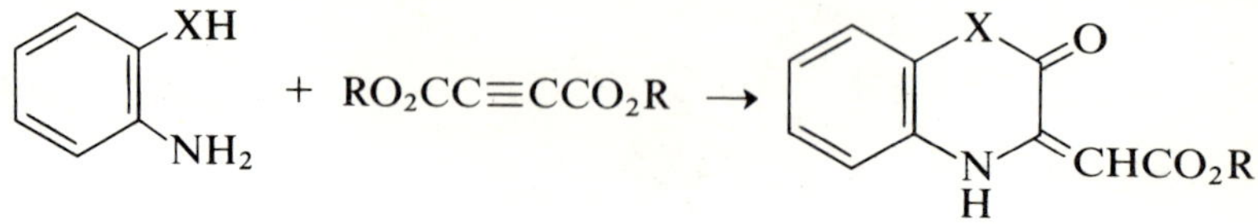

(X = NH, O, S)

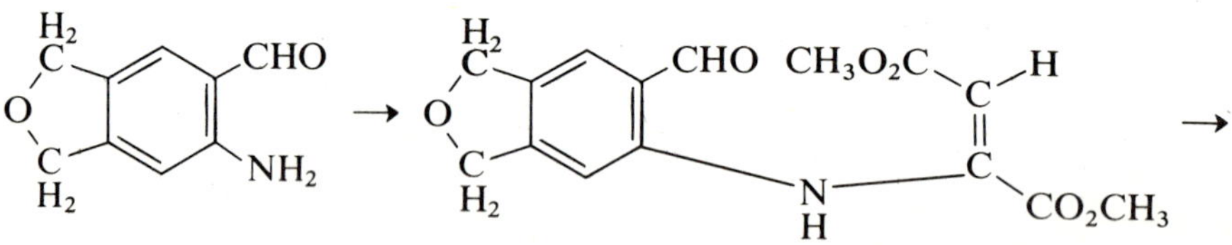

(3) (4)

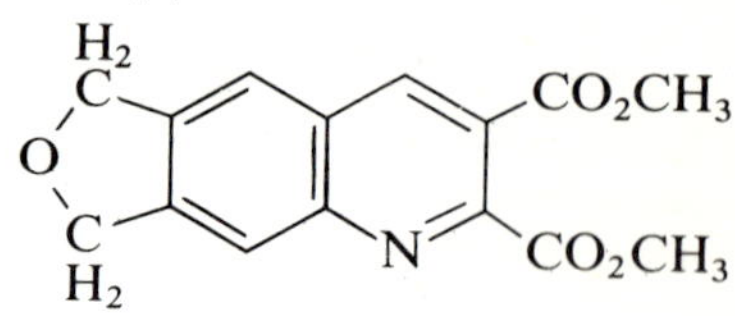

(5)

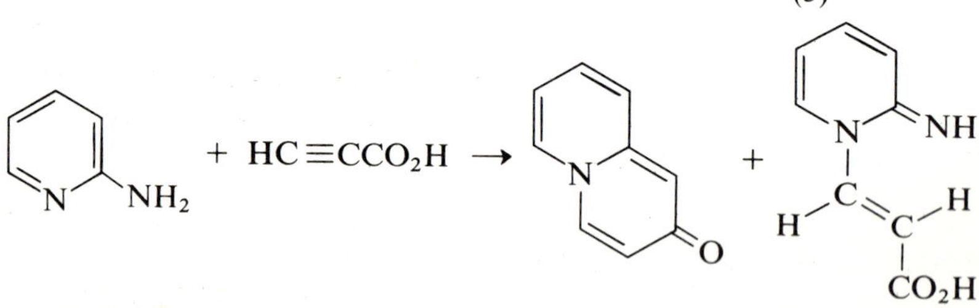

(6) (7)

The reactions of tertiary aromatic amines with acetylenic esters were first carried out by Diels and Alder; the nature of these products has been investigated intensively in recent years, notably by Acheson and his co-workers [8]. Pyridine forms a 1:adduct with dimethyl acetylenedicarboxylate; the 9a*H*-quinolizine (9) initially formed is unstable and rapidly isomerizes to the 4*H*-quinolizinetetracarboxylic ester (10). The quinolizinone (11) is also formed as a by-product in this reaction. Support for this mechanism

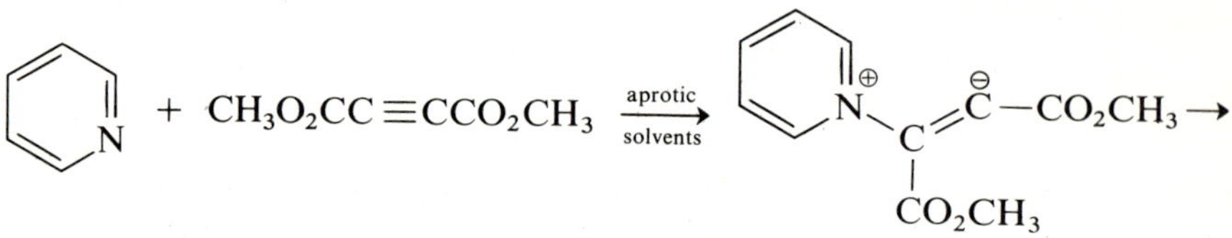

(8a)

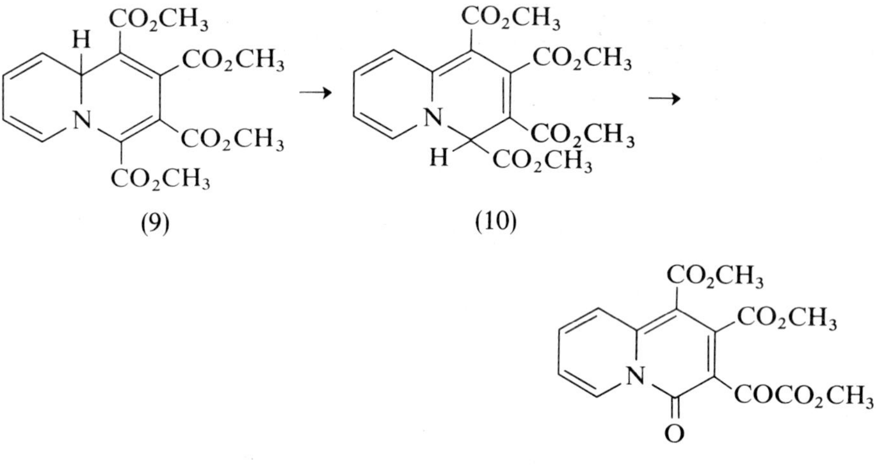

(9) (10) (11)

has been adduced by interception of the intermediate 1:1-adduct (8a) with carbon dioxide; this results in the formation of the unstable isomeric betaines (12) and (13) [9].

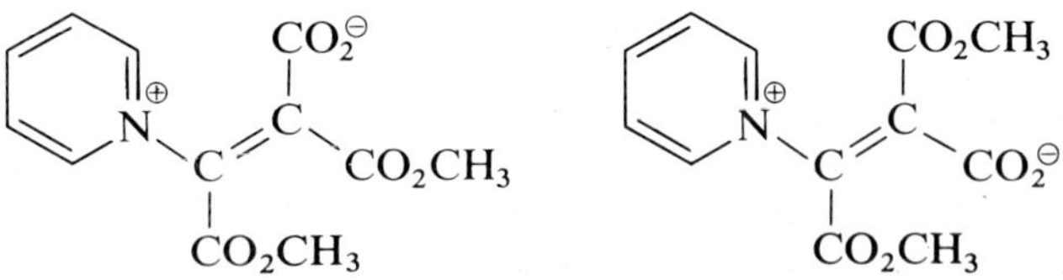

(12) (13)

The 1:1-adduct, classified by Huisgen as a 1,4-dipole (8b), in addition to undergoing ring-closure with a second molecule of dimethyl acetylene-dicarboxylate, also undergoes ring-closure with phenyl isocyanate to give compound (14). 1,4-Dipolar cycloaddition (cf. 1,3-dipolar cycloaddition, discussed in the following section) offers a general principle for heterocyclic synthesis [10].

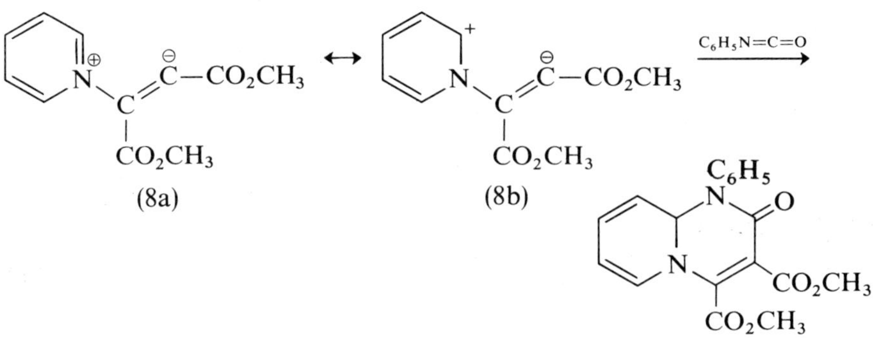

(8a) (8b) (14)

A study of the reaction of methyl α-pyridylacetate (15) with dimethyl acetylenedicarboxylate in t-butanol has shown that initial nucleophilic attack on the pyridine nitrogen to give the betaine (16) is followed by intramolecular cyclization to the 2-quinolizone (17).

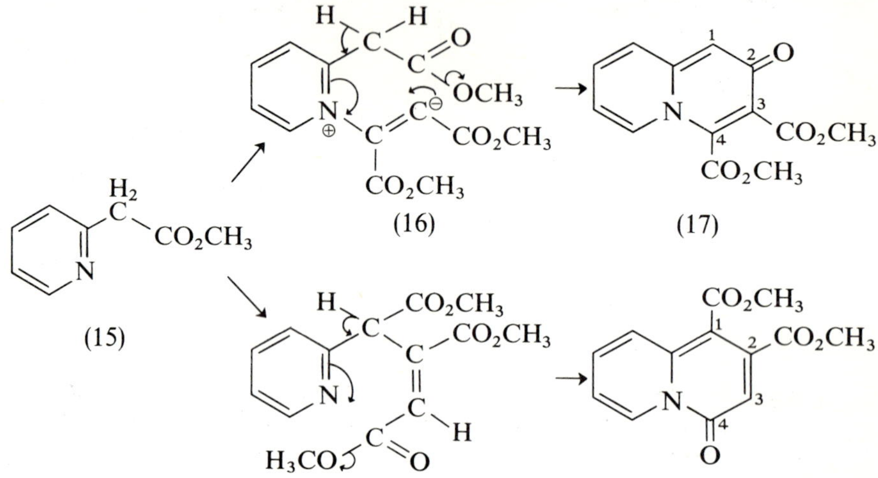

(15) (16) (17) (18) (19)

The alternative reaction pathway leading to the 4-quinolizinone (19) involves initial attack by the acetylenic ester at the α-carbon atom of the pyridyl ester to give (18) followed by a cyclization step. The 4-quinolizone (19) is the major product in non-polar solvents such as benzene [11].

When dimethyl acetylenedicarboxylate reacts with five-membered heterocycles such as pyrrole, the normal reaction is one of substitution, with the introduction of an $\alpha\beta$-dicarbomethoxyvinyl group [$-C(CO_2CH_3){=}CHCO_2CH_3$]. Indolizine (20) reacts smoothly with the

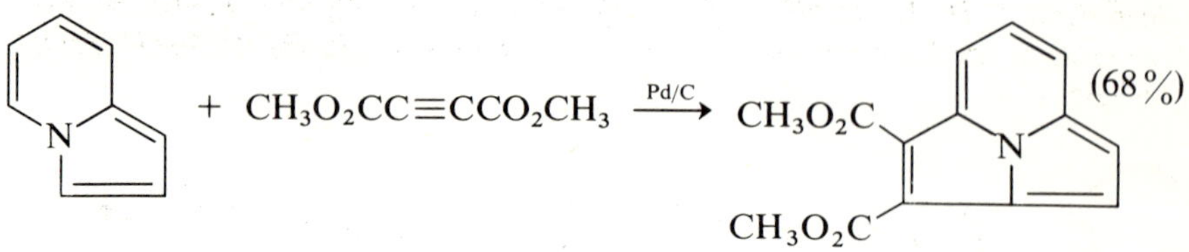

(20) (21)

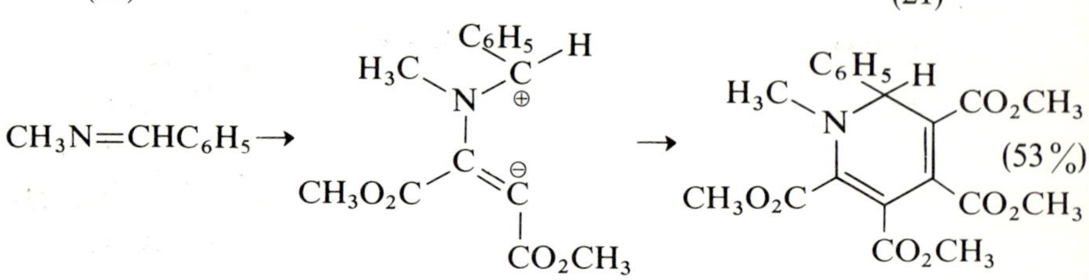

(22) (23) (24)

ester in the presence of a dehydrogenation catalyst to give 1,2-dicarbomethoxycycl[3,2,2]azine (21). In this case initial substitution at position 3 is followed by intramolecular cyclization [12]. Dimethyl acetylenedicarboxylate reacts with imines such as benzylidenemethylamine (22) to form a 2:1 adduct (24). The addition of the second molecule of acetylenic ester to the intermediate 1:1-adduct (23) is a further example of 1,4-dipolar cycloaddition [13].

Heteroaromatic *N*-oxides, in addition to acting as 1,3-dipoles, undergo some unusual reactions with acetylenic esters; for example, 1:1-adducts of type (25) are formed from phenanthridine 5-oxide and its 6-alkyl derivatives. The adduct (R = CH_3), is cyclized on heating to the indolizine (26), which is readily transformed to the parent heterocycle [14].

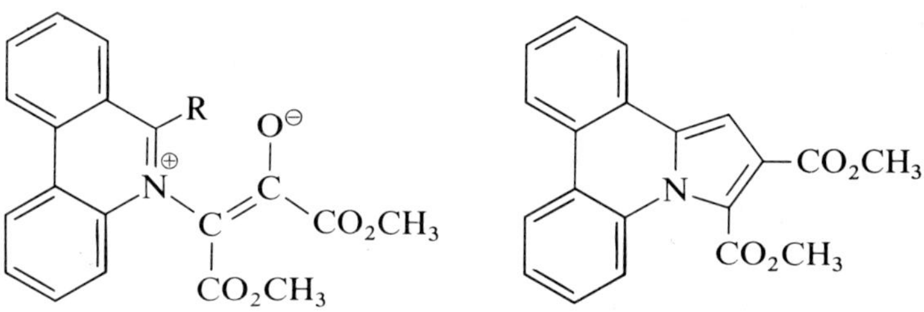

(25) (26)

Hexafluorobut-2-yne (27) reacts with sulphur vapour to give bis-trifluoromethyl-1,2-dithietene (28). The thietene readily dimerizes; with olefins it is converted into dihydro-*p*-thiins (e.g., 29), and with acetylenes to *p*-thiins which decompose thermally with loss of sulphur to the corresponding thiophen (e.g., 30) [15].

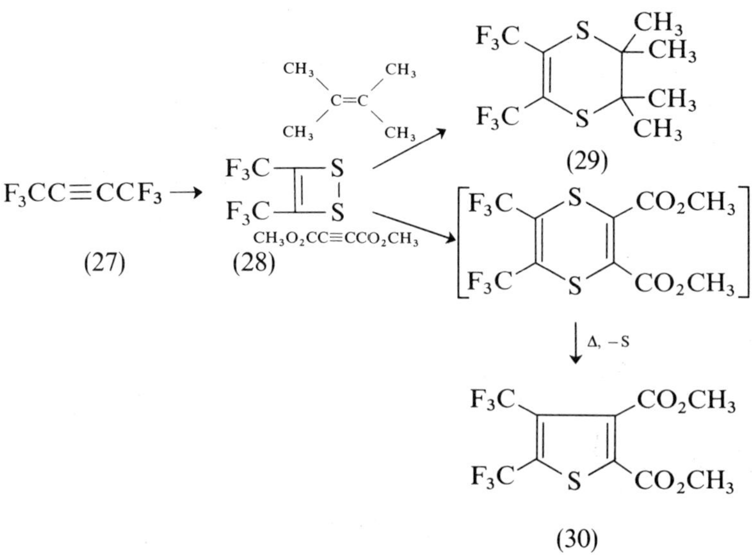

(27) (28) (29) (30)

Further uses of acetylenic intermediates in heterocyclic synthesis are mentioned in the following section on 1,3-dipolar cycloaddition; acetylenes also function as dienophiles in Diels–Alder syntheses.

1,3-DIPOLAR CYCLOADDITION REACTIONS

1,3-Dipolar cycloaddition is an important general principle for the formation of five-membered rings. The scope of this reaction has been investigated, particularly by Huisgen and his co-workers, and in general terms it involves the addition of a 1,3-dipole, $\overset{+}{a}$-b-$\overset{-}{c}$, to a multiple bond which may be, for example, acetylenic, olefinic or carbonyl in nature. For a comprehensive classification of 1,3-dipoles the reader is referred to recent review articles [16]. The following examples are illustrative of some synthetically useful applications of this general method of synthesis.

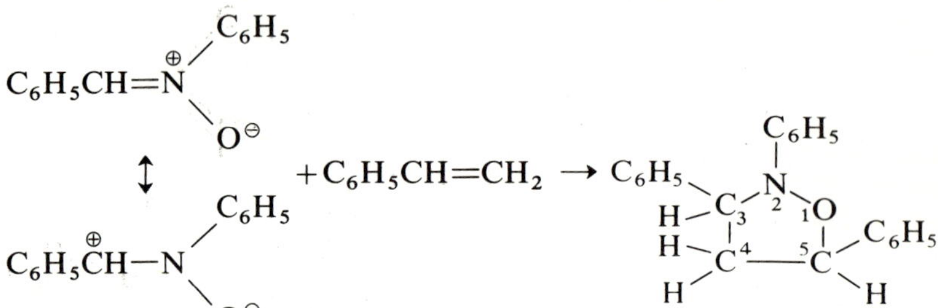

N,α-diphenylnitrone

2,3,5-triphenylisoxazolidine (92–99 %)

[17]

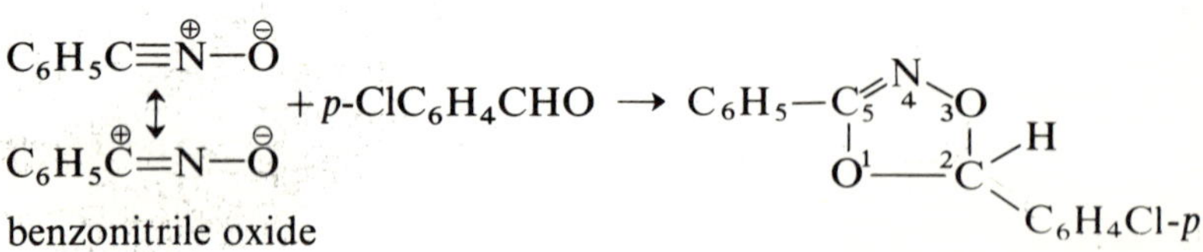

benzonitrile oxide

2-*p*-chlorophenyl-5-phenyl-1,3,4-dioxazole

(84 %)

[18]

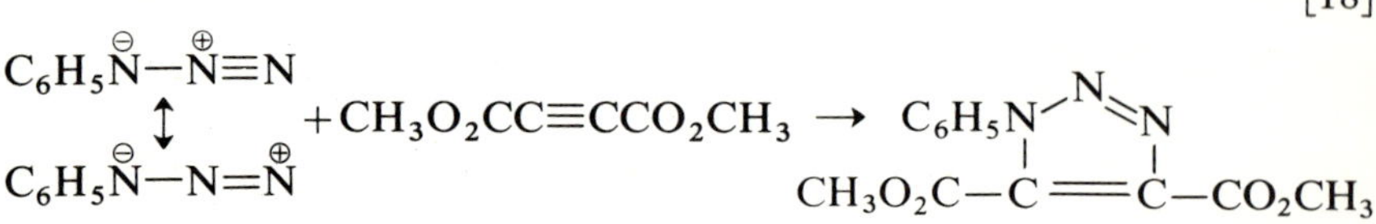

phenylazide

Dimethyl 1-phenyl-1,2,3-triazole 4,5-dicarboxylate

[19]

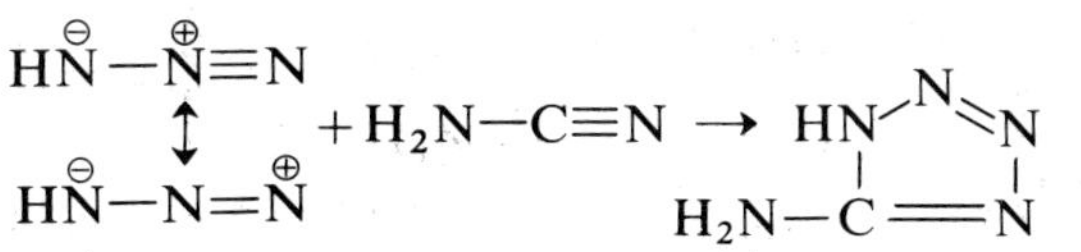

[20]

hydrazoic acid cyanamide 5-aminotetrazole

Benzyne may also function as a 1,3-dipolarophile as illustrated in the following equation:

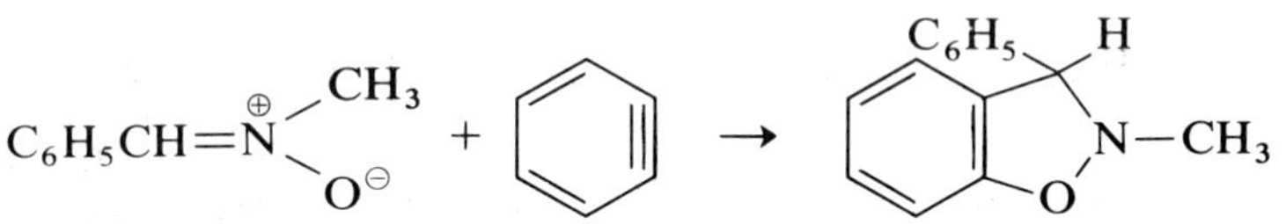

[21]

N-methyl-α-phenylnitrone benzyne generated from benzene-diazonium *o*-carboxylate 2-methyl-3-phenyl-2,3-dihydrobenz[*d*]-isoxazole

An interesting recent development of this method has been the investigation of the cycloaddition reactions of mesionic compounds. For example, *C*-methyl-*N*-phenylsydnone (31) behaves as a 1,3-dipolar reagent and reacts smoothly with dimethyl acetylenedicarboxylate to give the pyrazole (32); an analogous reaction of the mesionic oxazolone (33), gives the pyrrole (34) [22].

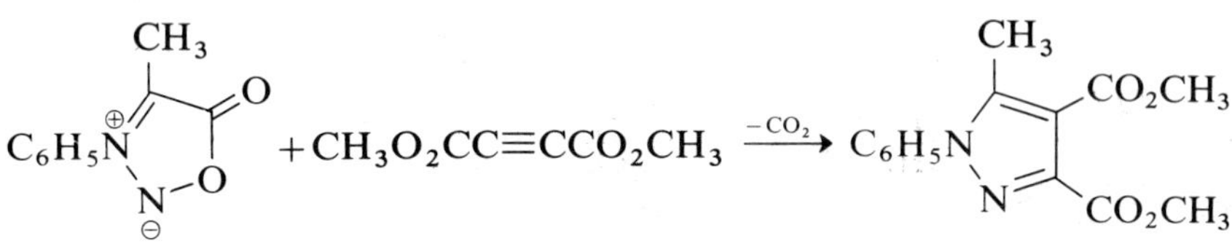

(31) (32) (99 %)

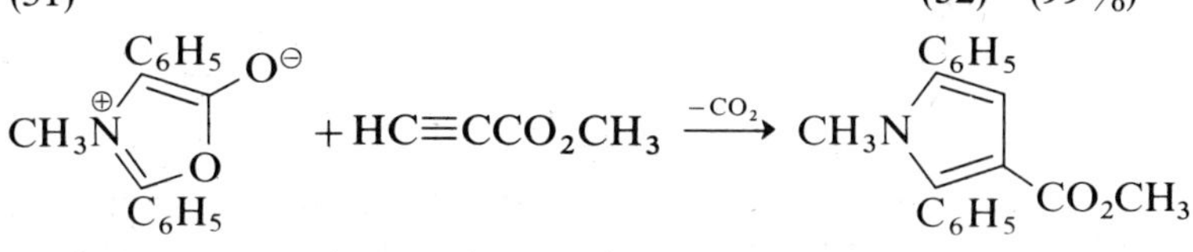

(33) (34) (95 %)

DIELS–ALDER SYNTHESES

There has been an increasing number of applications of the Diels–Alder reaction to the synthesis of heterocyclic ring systems. Dienophiles with heteroatoms include compounds with >C=O; >C=S; —C=N—; —N=N—; —N=O; —N=S=O groups [23, 24]. A new pyridine synthesis is exemplified by the reaction of 2,3-dimethylbutadiene (35) with *N*-(butyloxy-carbonylmethylene)*p*-toluenesulphonamide (36); the adduct (37) on treatment with alkali gives 4,5-dimethylpyridine-2-carboxylic acid (38) [25].

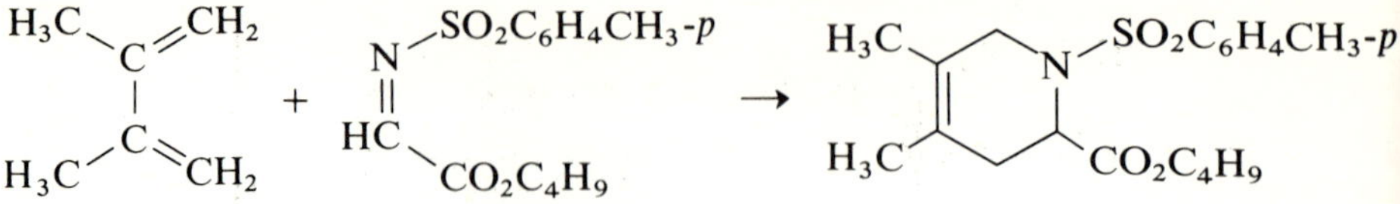

(35) (36) (37)

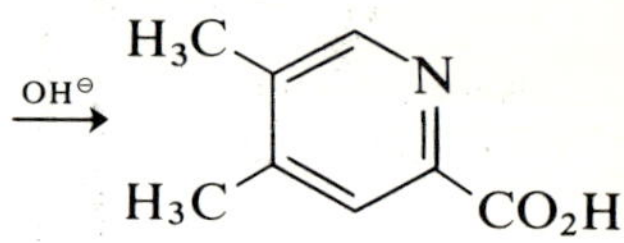

(38)

The highly reactive *cis*-azodienophiles (39) [26] and (40) [27] have been generated as shown, and their use as intermediates in heterocyclic synthesis is illustrated by their reactions with 2,3-dimethylbutadiene.

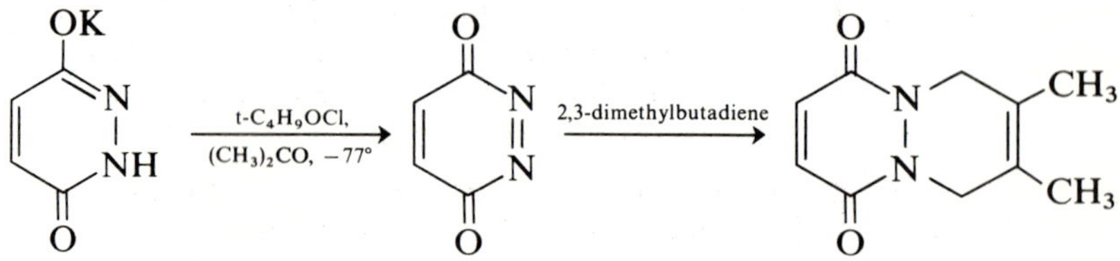

monopotassium salt of maleic hydrazide (39)

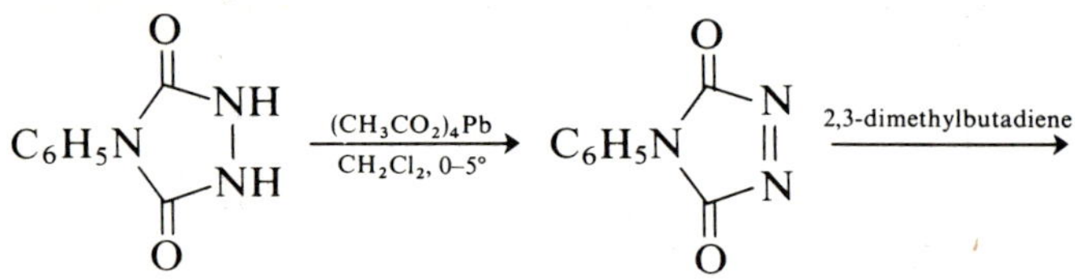

4-phenylurazole (40)

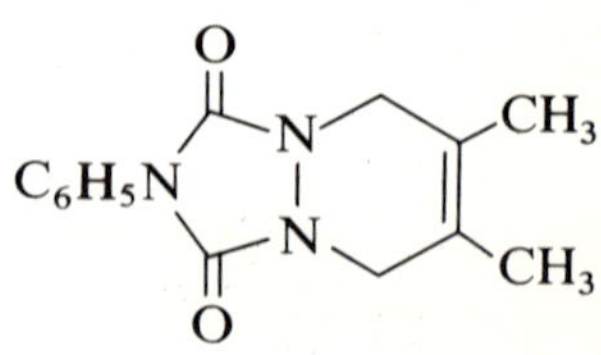

Diels–Alder adducts (43) have been prepared by the addition of nitroso compounds (42) to dienecarboxylic esters (41). Subsequent reaction of these adducts with basic reagents gives pyrrole 2-carboxylic esters (44) [28].

2-Acylpyrroles have been prepared similarly, starting from ketodienes [29].

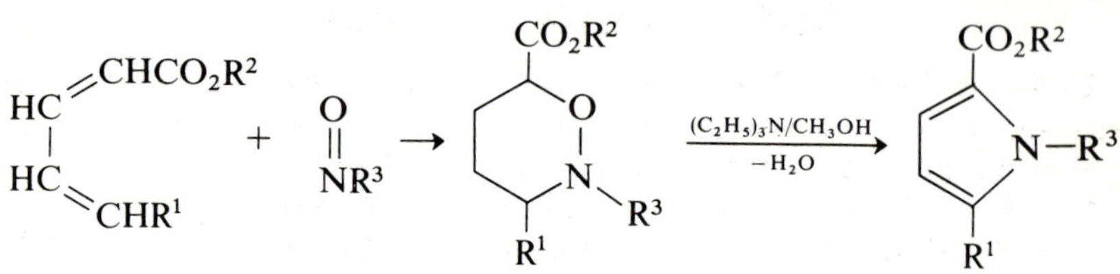

(41) (42) (43) (44)

The alternative application of the Diels–Alder reaction employing dienes with heteroatoms is illustrated by the addition of $\alpha\beta$-unsaturated carbonyl compounds to vinyl ethers [30]. For example, addition of cinnamaldehyde to n-butyl vinyl ether gives the dihydropyran (45) in 60% yield. This is converted in 76% overall yield to 4-phenylpyridine (46) by successive reaction with acid and hydroxylamine. Oxazoles can function as dienes: the adduct (49) from 5-ethoxy-4-methyloxazole (47) and diethyl maleate (48) gives the diethyl pyridinedicarboxylate (50) on treatment with ethanolic hydrogen chloride. Reduction of compound (50) with lithium aluminium hydride yields pyridoxine (51) [31].

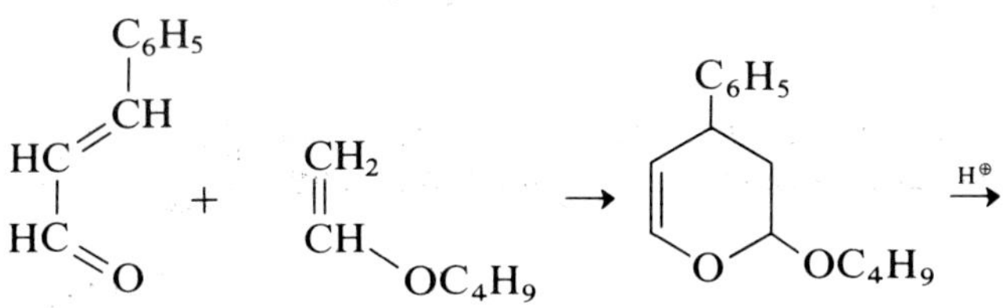

cinnamaldehyde n-butylvinyl ether (45)

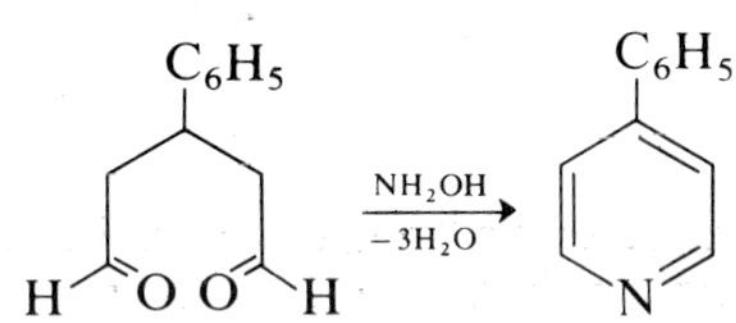

(46)

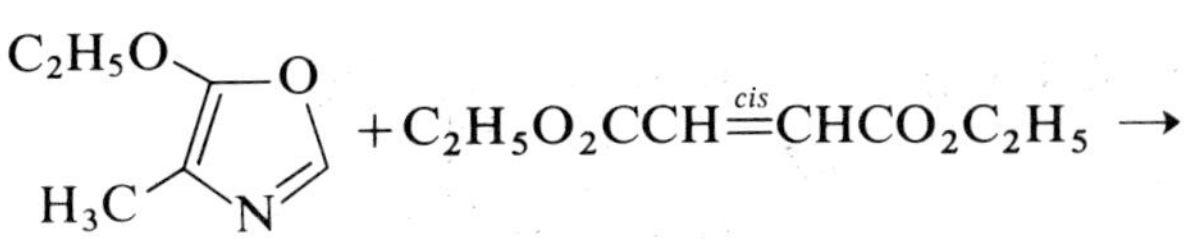

(47) (48)

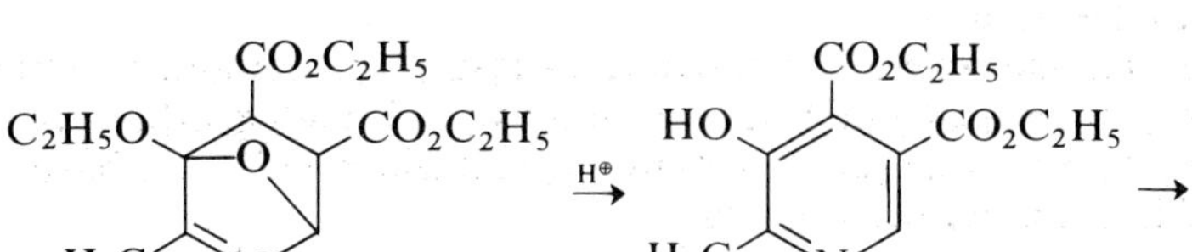

(49) (50)

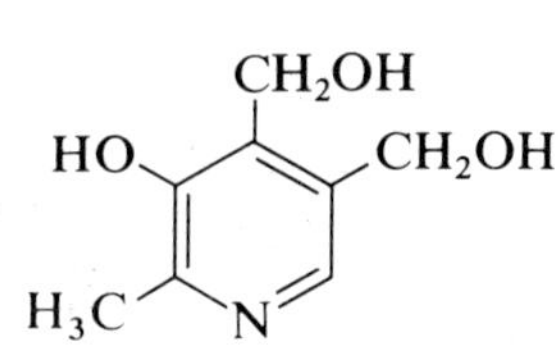

(51)

PHOTOCHEMICAL RING-CLOSURES

These are conveniently grouped into two types, oxidative ring-closures and non-oxidative ring-closures. The oxidative photocyclization of diphenylamines leads to a convenient preparative route to carbazoles, as illustrated by the synthesis of glycozoline (3-methoxy-6-methylcarbazole) [32, 33].

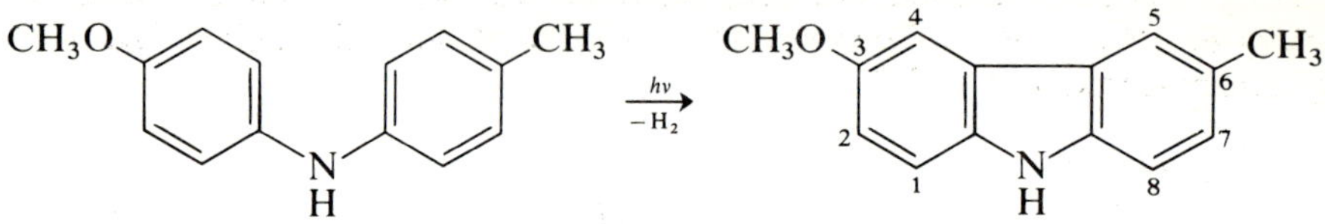

Five-membered rings are also formed on irradiation of the benzylidene derivatives of *o*-phenylenediamine, *o*-aminophenol, and *o*-aminothiophenol [34].

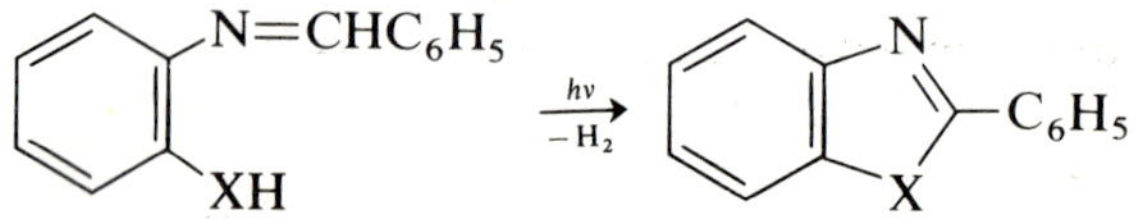

(X = NH, O, S)

Irradiation of benzanilide in the presence of iodine gives phenanthridone [35], and irradiation of 3,4-dihydroxycinnamic acid in the presence of oxygen gives 6,7-dihydroxycoumarin (esculetin) [36].

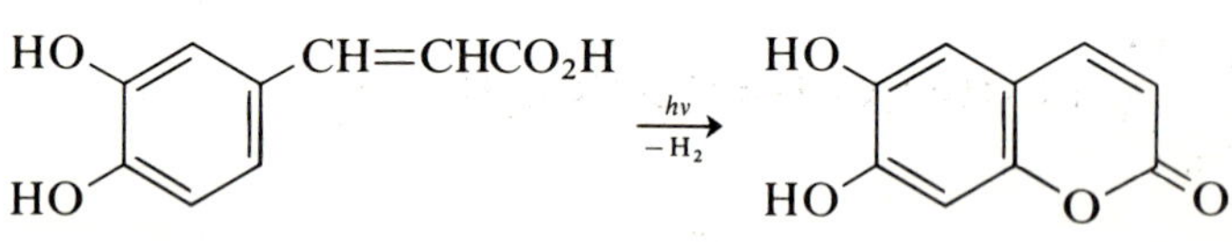

In this case the primary photochemical reaction is the isomerization of the *trans*- to the *cis*-acid.

Two examples of non-oxidative ring-closure are the photocyclization of α-methylacrylic acid anilide (52) to 3,4-dihydro-3-methylcarbostyril (53) [37] and of 2-biphenylisocyanate (54) to phenanthridone (55). Carbazole is also formed in the latter reaction [38].

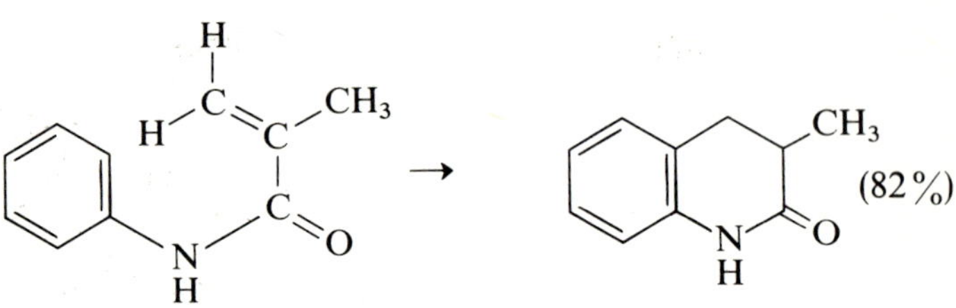

(52) (53)

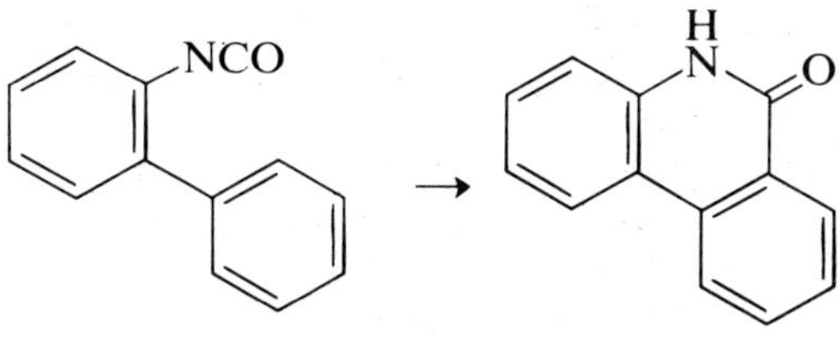

(54) (55)

RING EXPANSION REACTIONS

There are many examples of reactions of this type in the recent literature. Tetramethylpyrroles (57) are conveniently prepared from the readily available dichlorocyclobutene (56) by reaction with ammonia or primary amines [39].

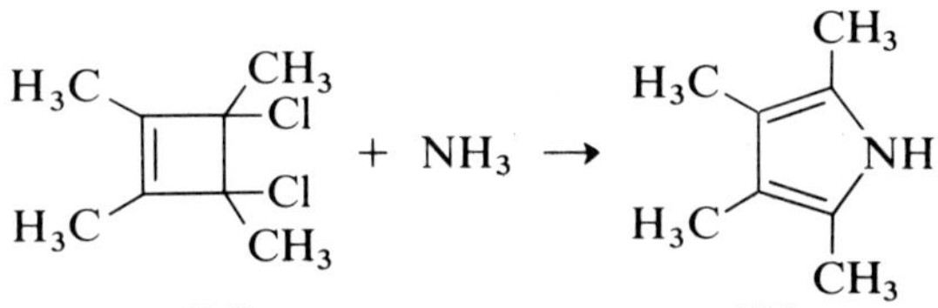

(56) (57)

A new route to Δ^2-thiazolines involves thc treatment of an episulphide with a nitrile in the presence of a strong acid.

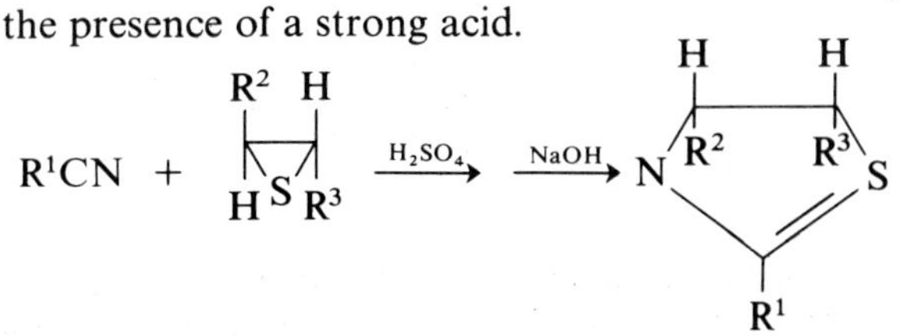

The reaction is stereospecific, since isomeric thiazolines are obtained from *cis*- and *trans*-2-butene episulphides; yields of up to 80% are reported [40]. 1-Aryl-Δ^2-1,2,3-triazolines (59) are obtained in high yields by iodide-ion-promoted isomerization of 1-arylazoaziridines (58) [41]. The latter compounds are prepared from aryldiazonium salts and aziridine.

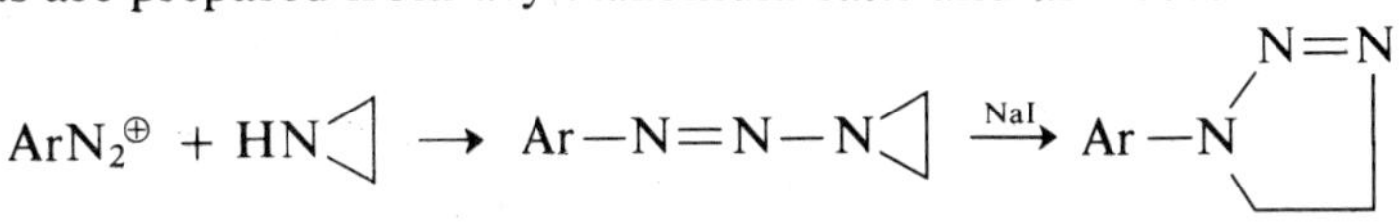

(58) (59)

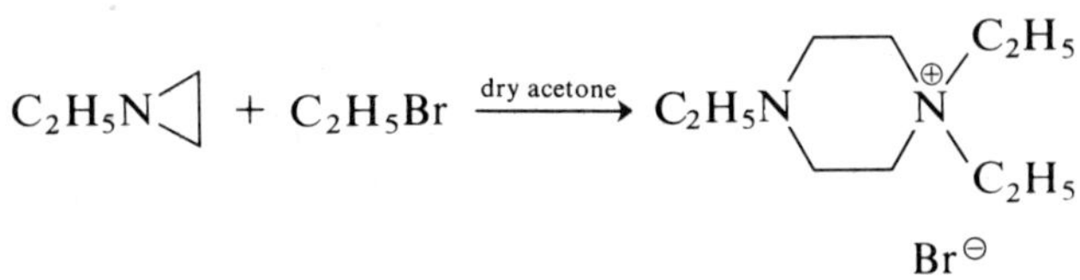

1-ethylaziridine 1,1,4-triethylpiperazinium bromide (99%)

The reaction of various 1-alkylaziridines with alkyl, alkenyl, or benzyl halides gives nearly quantitative yields of the corresponding 1,1,4-trisubstituted piperazine halides [42].

A variety of heterocyclic systems has been synthesized from 3-unsubstituted isoxazolium salts (e.g., 60). Abstraction of the 3-proton by base gives an intermediate α-ketoketenimine (e.g., 61) which can undergo ring-closure reactions with suitable nucleophiles [43].

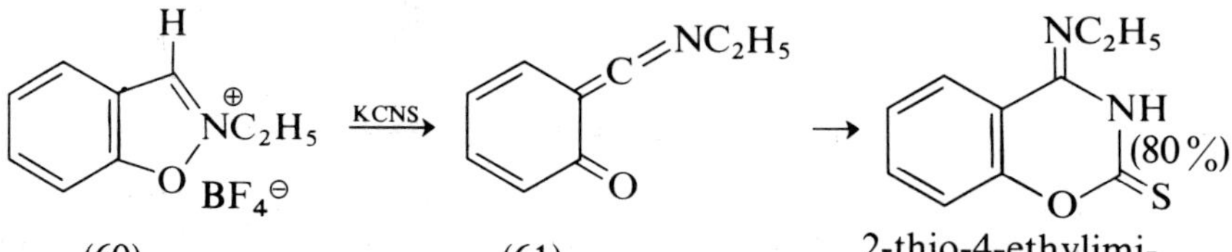

(60) (61) 2-thio-4-ethyliminobenzo[1,3]oxazine-2,4-dione

3-Alkyl-2-hydroxyisocarbostyrils are formed by reaction of 2-alkylindanones with hydrochloric acid and n-butyl nitrite, e.g.,

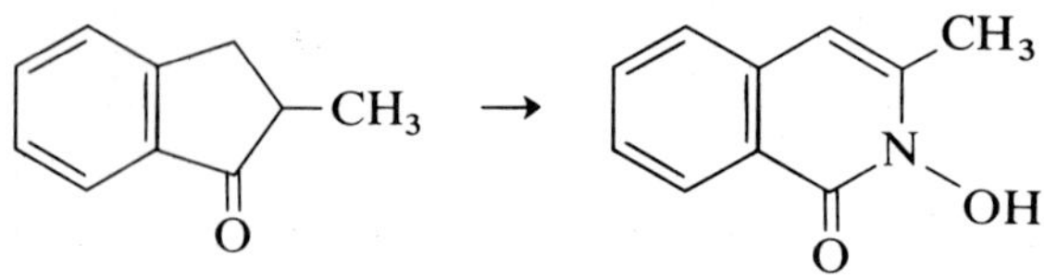

[44]

2-methyl-1-indanone 2-hydroxy-3-methylisocarbostyril (62–69%)

Ring expansion reactions leading to the formation of azepines from benzene or its derivatives are illustrated in the following equations.

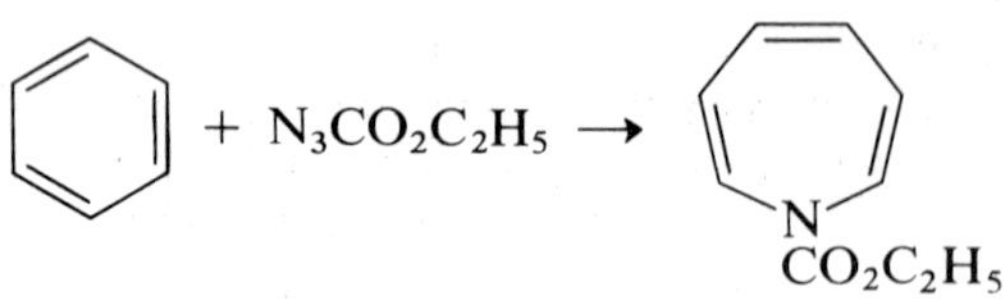

[45]

ethyl azidoformate N-ethoxycarbonylazepine (70%)

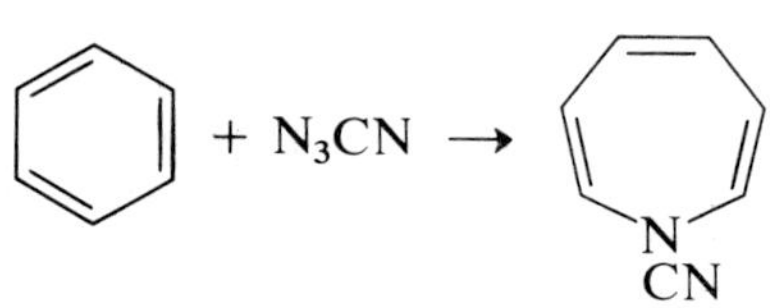

cyanogen azide N-cyanoazepine (70%) [46]

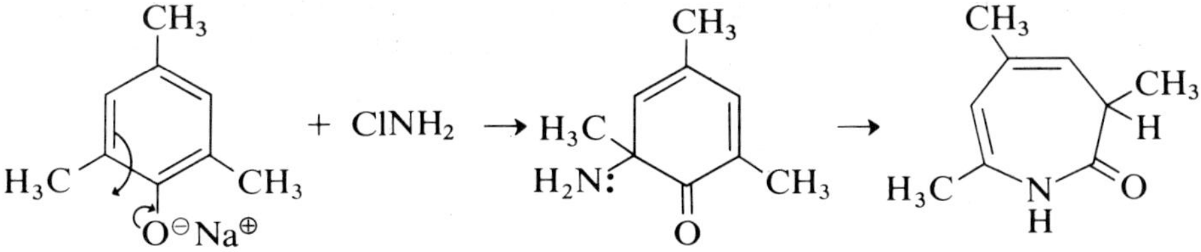

chloramine (62) 1,3-dihydro-3,5,7-trimethyl-2*H*-azepin-2-one (45–53 %) [47]

In the last reaction, the nucleophilic centre in the trimethylphenoxide ion is C-2. Formation of the intermediate (62) is followed by intramolecular nucleophilic attack of NH_2 on the adjacent carbonyl function and ring expansion.

Azepines are also formed by ring expansion of dihydropyridines; for example, the dihydropyridine (63) gives the 3*H*-azepine (64) on boiling with ethanolic sodium ethoxide. Treatment of the *N*-methyldihydropyridine (65) with potassium t-butoxide gives either the 1*H*-azepine (66) or its bicyclic valency tautomer (67), depending on the conditions. In this case formation of a 3*H*-azepine is structurally precluded. Compound (67) isomerizes to compound (66) in carbon tetrachloride [48].

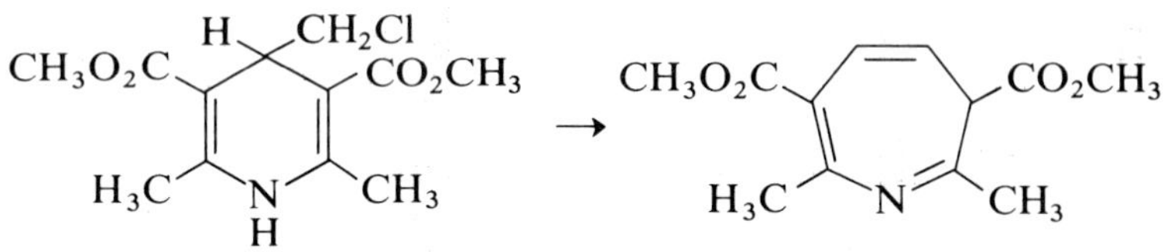

(63) (64)

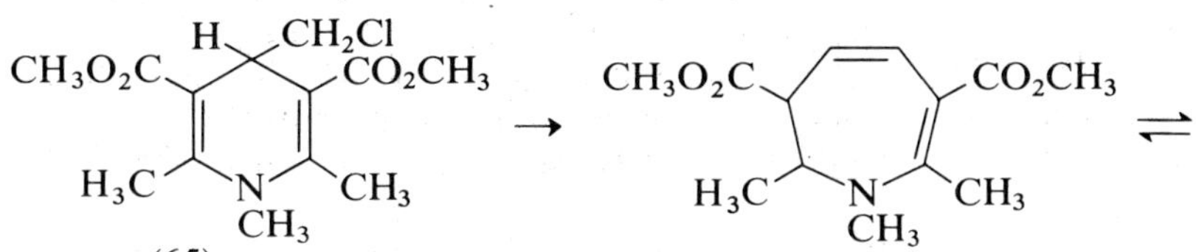

(65) (66)

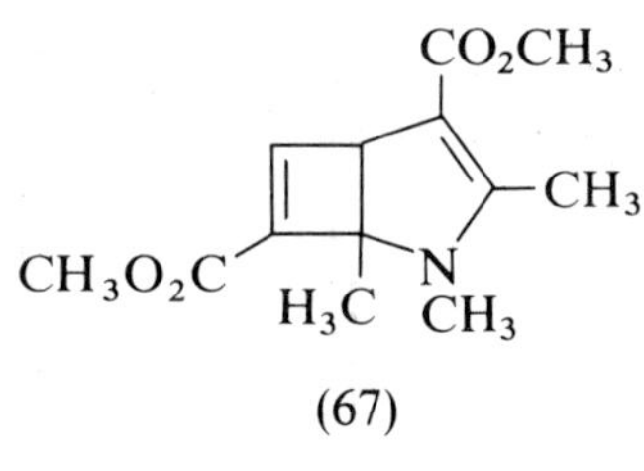

(67)

A ring-expanded product is obtained on treatment of 4-azido-1,2-naphthoquinone with concentrated sulphuric acid at room temperature [49].

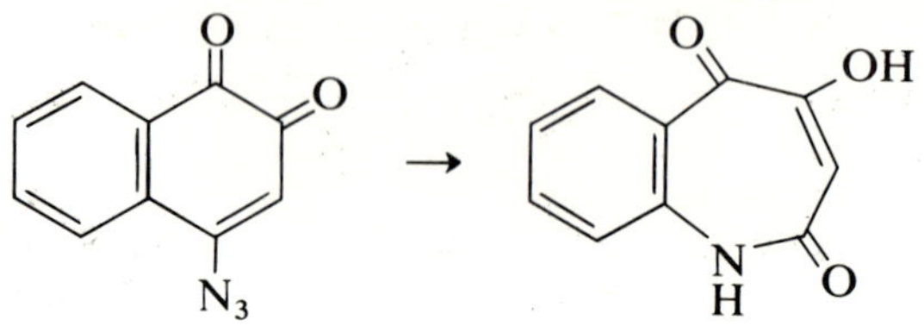

2,5-*H*-4-hydroxy-benzazepine-2,5-dione (82%)

Oxepins (70) are conveniently prepared by dehydrohalogenation of 1,2-epoxy-4,5-dibromocyclohexanes (68). The parent compound (70; R = R′ = H) exists in rapid equilibrium with benzene oxide (69; R = R′ = H), and the proportion of benzene oxide in the equilibrium mixture increases with increasing dielectric constant of the solvent; in iso-octane it is 30%, while in water–methanol (85:15) it is about 90%.

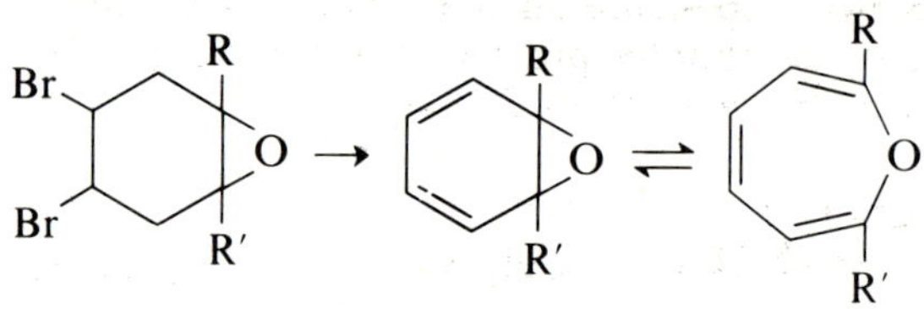

(68) (69) (70)

In the case of 2,7-dimethyloxepin (70; R = R′ = CH_3) it is not possible to detect the oxide in the equilibrium mixture by spectroscopic methods. The equilibrium can, however, be displaced by reaction with a dienophile which removes the oxide from the equilibrium mixture. Thus, 2,7-dimethyloxepin readily forms an adduct with maleic anhydride. A 2-acetyl substituent also strongly stabilizes the oxepin form relative to the benzene oxide, but again 2-acetyloxepin (70; R = $COCH_3$, R′ = H) readily gives a maleic anhydride adduct [50].

2,3-Dihydro-oxepins are synthesized by a two-step process from dihydropyran, as illustrated by the following scheme.

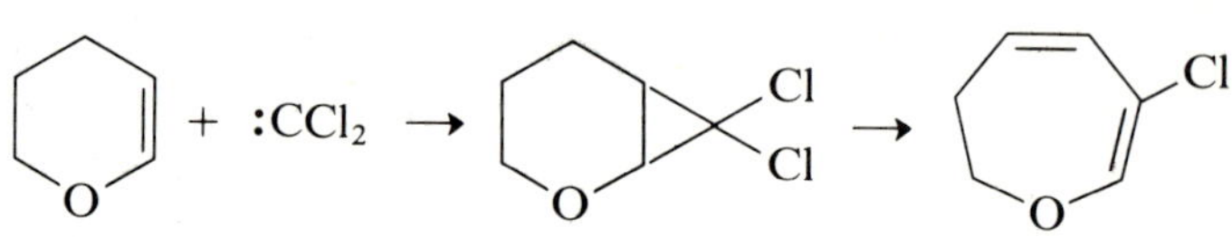

2,3-dihydro-6-chloro-oxepin

(75%) (83%)

The dichlorocarbene is generated from ethyl trichloroacetate and sodium methoxide and the adduct decomposed by distilling from quinoline under reduced pressure [51].

1,6-Oxido[10]annulene (72), prepared by dehydrobromination of the tetrabromo epoxide (71) with sodium methoxide, isomerizes in the presence of acid to α-naphthol (73) and rearranges on silica gel to benzo[*b*]oxepin (74) [52, 53].

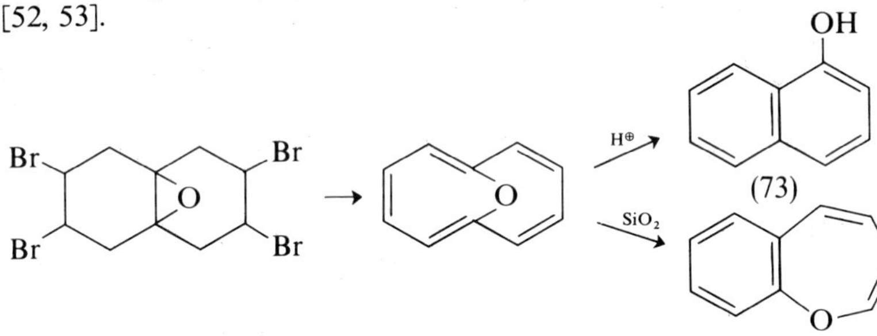

(71) (72) (73) (74)

An interesting ring expansion occurs when the diazoacetylpyrazoline (75) is heated in acetic acid at 85–90° [54]. The formation of 2,3–dihydro-5-methyl-6-phenyldiazepine-4-one (76) proceeds via (77) and (78); diazepines undergo many rearrangement reactions.

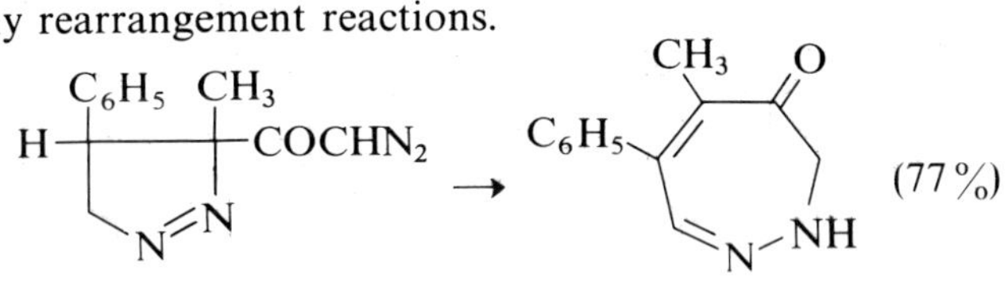

(75) (76)

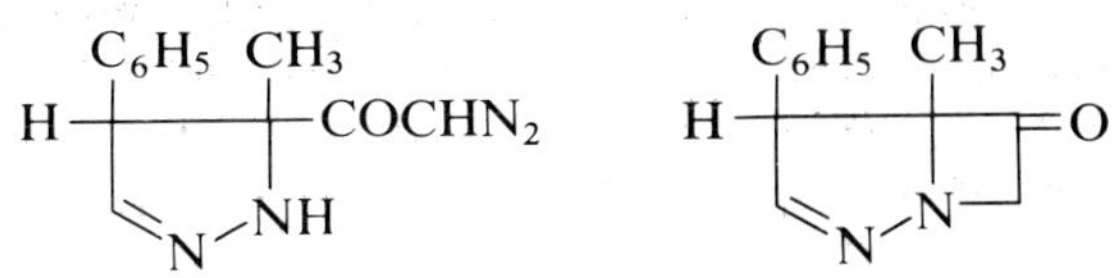

(77) (78)

Ring-expanded products are obtained when the quinazoline *N*-oxide (79) is treated with ammonia or primary amines, as a result of initial nucleophilic attack at C-2 of the quinazoline ring system. The product (80), obtained with methylamine, is a valuable tranquillizing agent and it is known as 'Librium' [55].

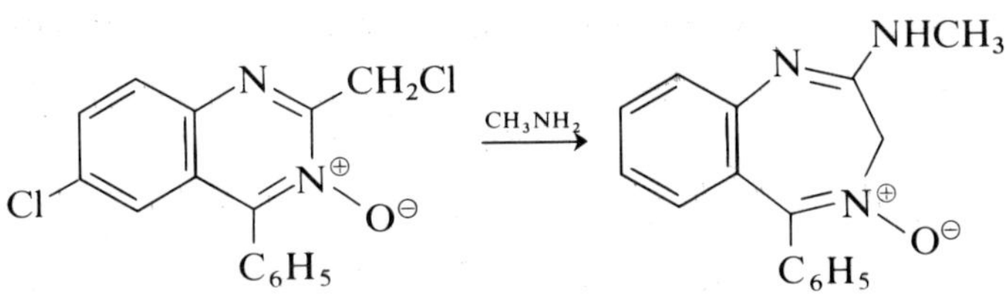

(79) (80)

SYNTHESES VIA BENZYNE INTERMEDIATES

This method of ring synthesis is illustrated by the following examples.

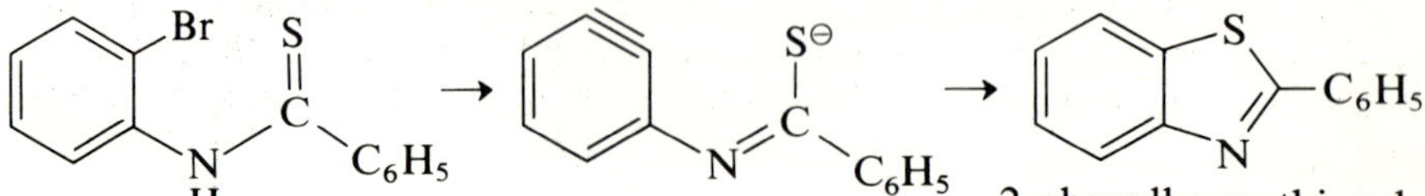

2-phenylbenzothiazole
(90%) [56, 57]

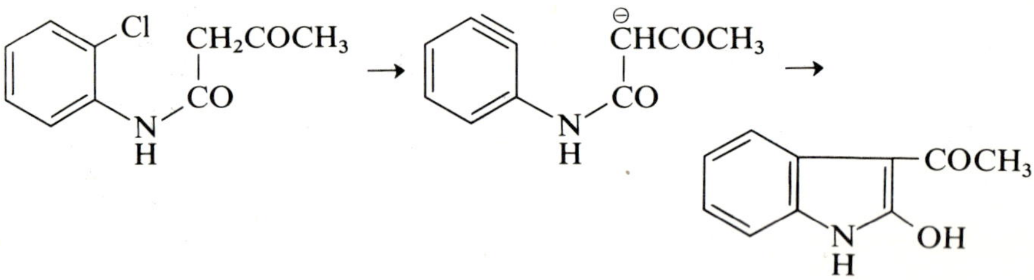

3-acetyloxindole
(74–78%) [56]

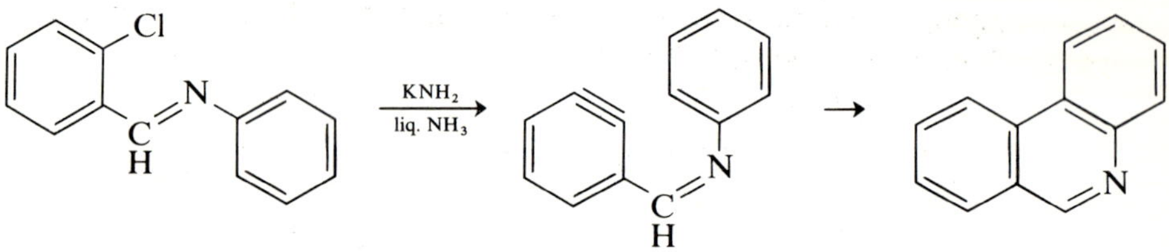

anil of *o*-chlorobenzaldehyde phenanthridine [58]

The intermediate benzynes undergo intramolecular nucleophilic attack with oxygen as well as with sulphur and carbon nucleophiles, and reactions of this type have been used to prepare a wide range of five- and six-membered heterocyclic and homocyclic systems.

SYNTHESES FROM *o*-AMINONITRILES

There have been many elegant syntheses of fused pyrimidine ring systems using *o*-aminonitriles particularly from Taylor and his co-workers [59]. For example, treatment of an *o*-aminonitrile with a thioamide gives a pyrimidine 4-thione. The *o*-aminonitrile is first condensed with the thioamide in ethanolic hydrogen chloride to give a 1,3-thiazine (81) which then undergoes rearrangement to the pyrimidine thione (83) on treatment with alkali.

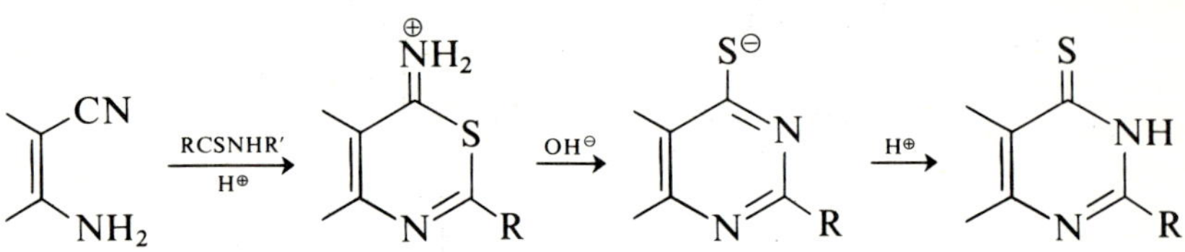

(81) (82) (83)

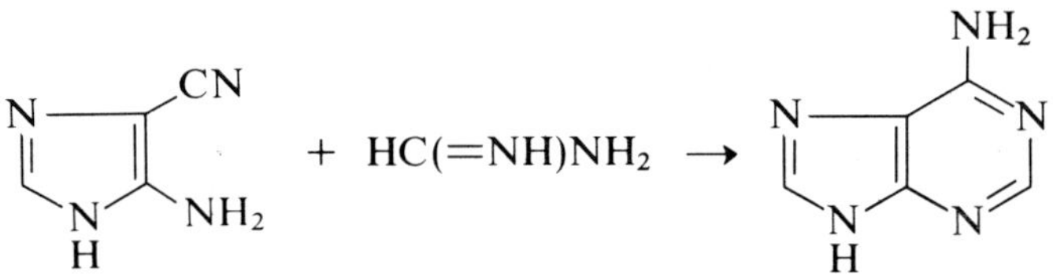

(68%)

(84) (85) (86)

Adenine (86) is obtained from 5-amino-4-cyanoimidazole (84) on boiling with formamidine acetate (85) in methoxyethanol [60].

SYNTHESES USING PYRIDINIUM SALTS

The Michael addition of the active methylene groups in pyridinium salts to $\alpha\beta$-unsaturated carbonyl compounds, or Mannich bases, e.g., ω-dimethylaminopropiophenone, which readily give $\alpha\beta$-unsaturated compounds, leads to an easy synthesis of pyridine derivatives [61].

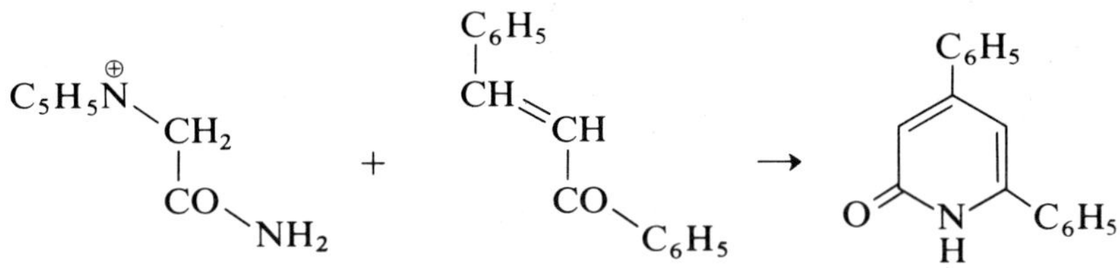

N-(aminoformylmethyl) pyridinium chloride — benzalacetophenone — 4,6-diphenyl-2-pyridone (82%)

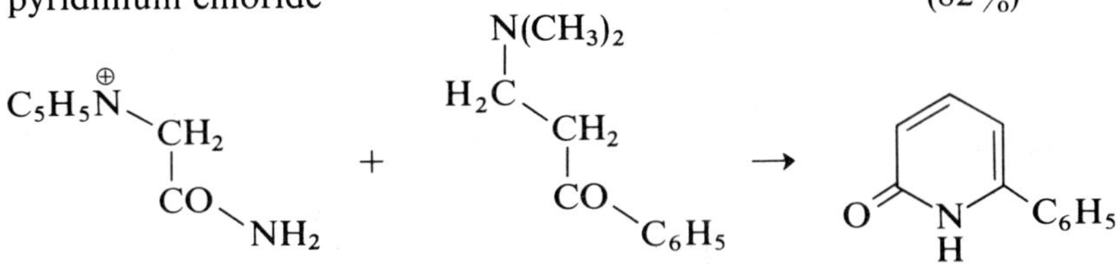

ω-dimethylaminopropiophenone — 6-phenyl-2-pyridone (56%)

Activation of the pyridinium methylene can be achieved by a variety of groups besides $CONH_2$; for example, $-COC_6H_5$, $-CN$, and $-CSSCH_3$; and internal Michael addition leads to pyrrolopyridinium salts.

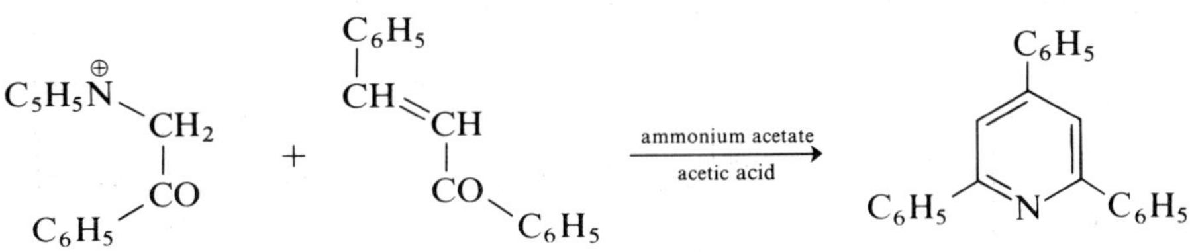

phenacylpyridinium bromide — 2,4,6-triphenylpyridine (92%)

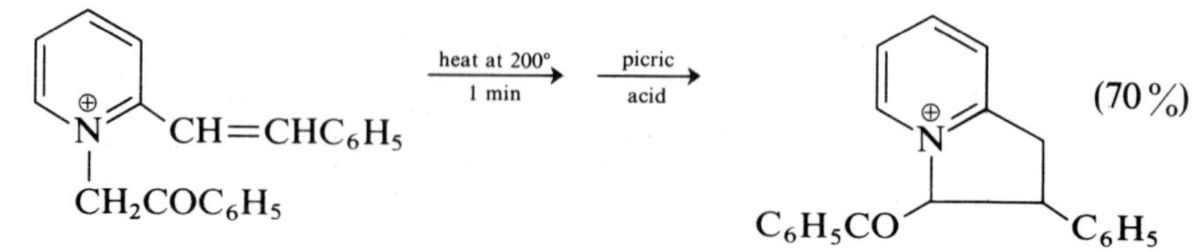

N-phenacyl-2-styrylpyridinium bromide

2-phenyl-3-benzoyl pyrrolino [1,2-*a*]pyridinium picrate

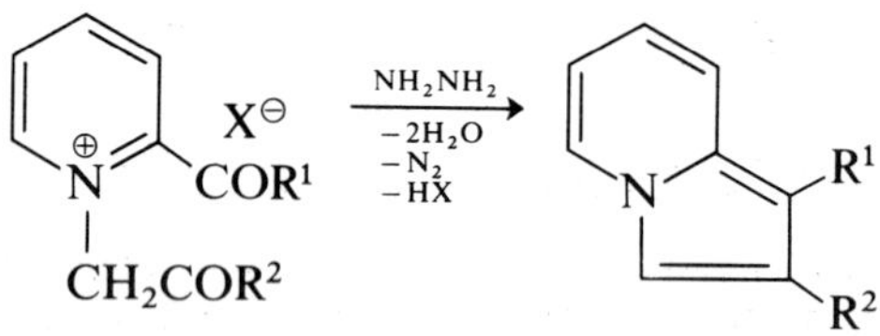

(87) (88)

Reaction of *N*-acylmethyl-2-acylpyridinium salts (87) with hydrazine provides a convenient synthesis of 1,2-disubstituted indolizines (88), presumably through an intermediate seven-membered ring ketazine [62].

SYNTHESES USING PYRYLIUM SALTS

Pyrylium salts are readiy prepared and of wide potential use in organic synthesis. For example, 2,4,6-trimethylpyrylium perchlorate (89) is prepared by either of the following methods [63, 64].

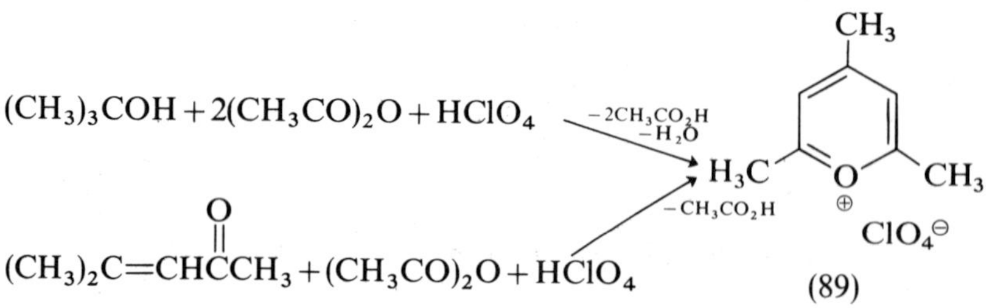

(89)

Reaction of 2,4,6-trimethylpyrylium perchlorate with hydrogen peroxide gives 2-acetyl-3,5-dimethylfuran in 50% yield [65]. Pyrylium salts are readily converted to a variety of pyridine derivatives [66, 67]; pyrylium iodides give phosphabenzenes on treatment with tris(trimethylsilyl)phosphine [68].

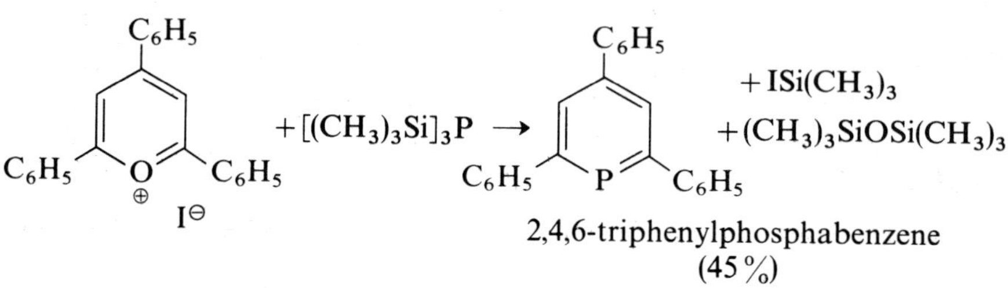

2,4,6-triphenylphosphabenzene (45%)

SYNTHESES USING PHOSPHONIUM SALTS

Acylmethylenephosphoranes (90; R = Me or Ar), prepared from the appropriate phosphonium salt and aqueous potassium carbonate, react with tosyl azide (91) to furnish 1,5-disubstituted 1,2,3-triazoles (92) in high yield. The 1-tosyl group is readily solvolysed in hot ethanol to give 5-substituted triazoles and ethyl tosylate [69].

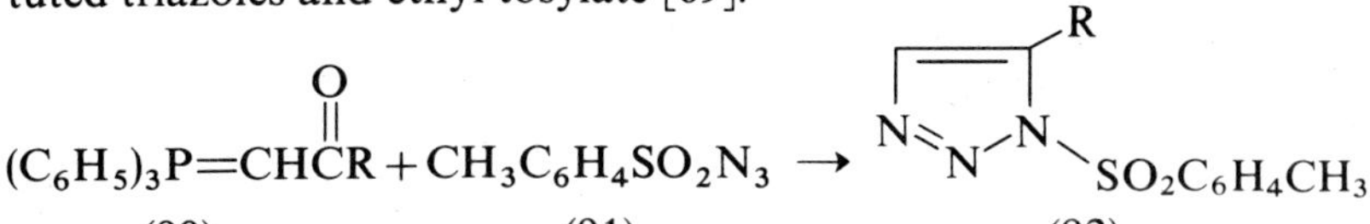

(90) (91) (92)

The sodium salts of α-hydroxyketones, *o*-hydroxybenzaldehyde and pyrrole-2-aldehyde cyclize as shown on treatment with vinyltriphenylphosphonium bromide (93). The reagent is conveniently prepared from 2-phenoxy-1-bromoethane [70].

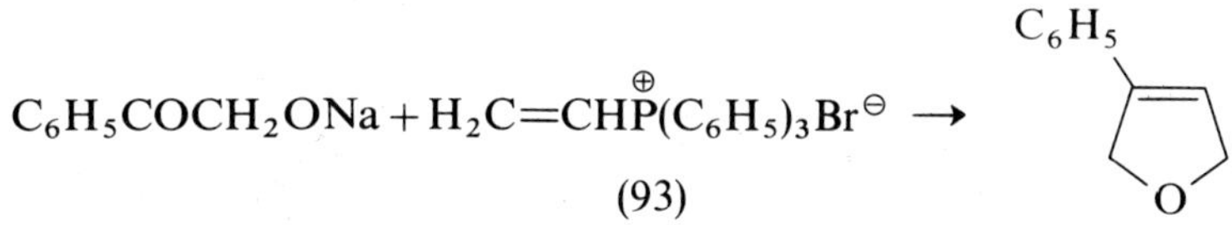

[71]

(93)

2,5-dihydro-3-phenylfuran

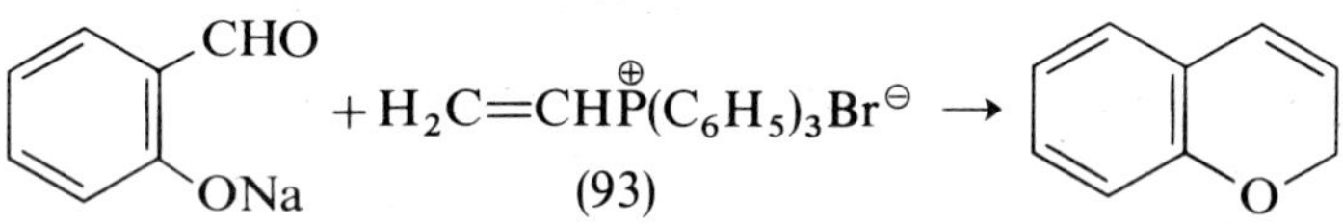

(93)

3,4-chromene

(62%) [72]

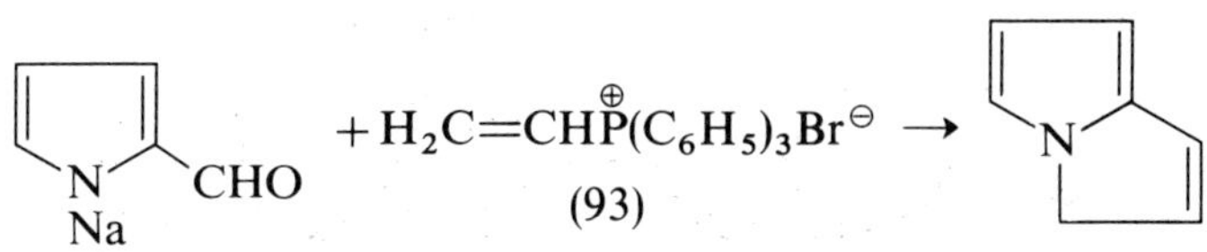

(93)

pyrrolizine

(87%) [73]

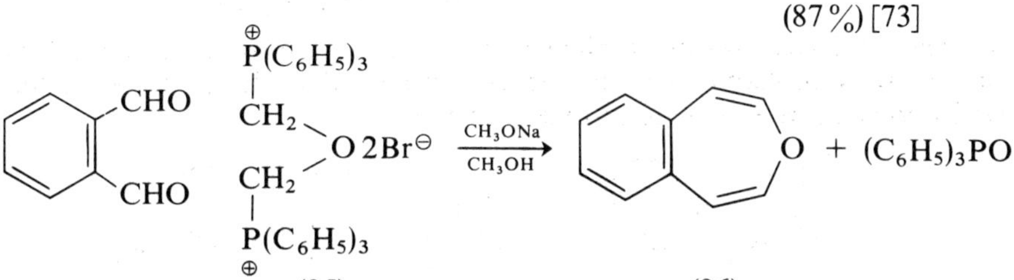

(94) (95) (96)

The bis-phosphonium salt (95) from αα′-dibromodimethyl ether reacts with *o*-phthalaldehyde (94) to give benzo[*d*]oxepin (96) [74].

REDUCTIVE CYCLIZATION WITH TRIETHYL PHOSPHITE

A number of heterocyclic systems have been prepared by reductive cyclization of nitro groups with triethyl phosphite.

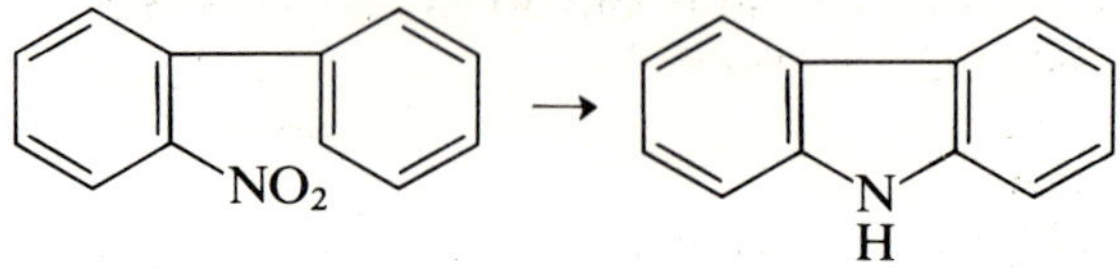

carbazole (83%) [75]

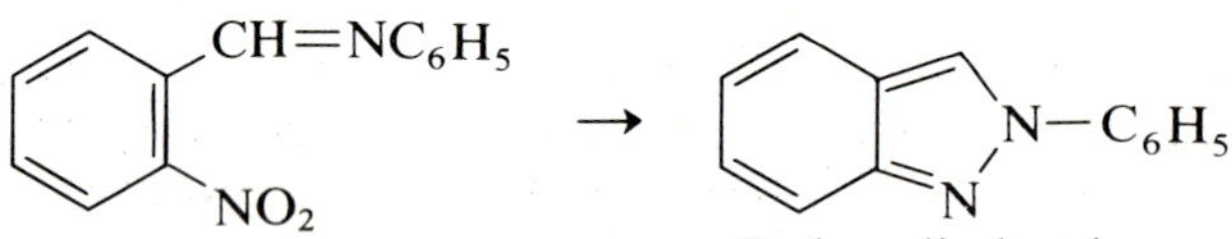

2-phenylindazole (67–78%) [76]

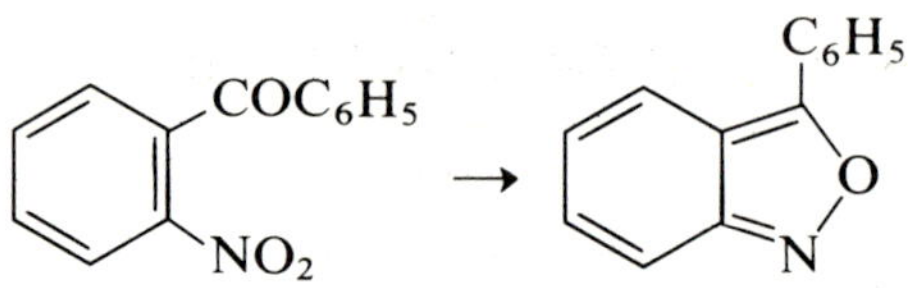

3-phenylanthranil (58%) [77]

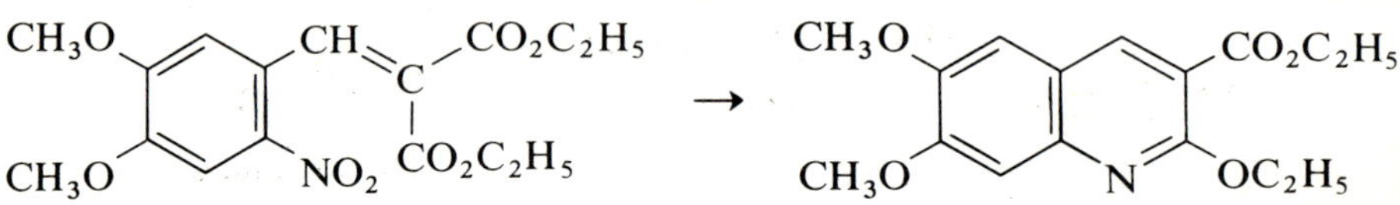

2-ethoxy-3-ethoxycarbonyl-6,7-dimethoxyquinoline (61%) [78]

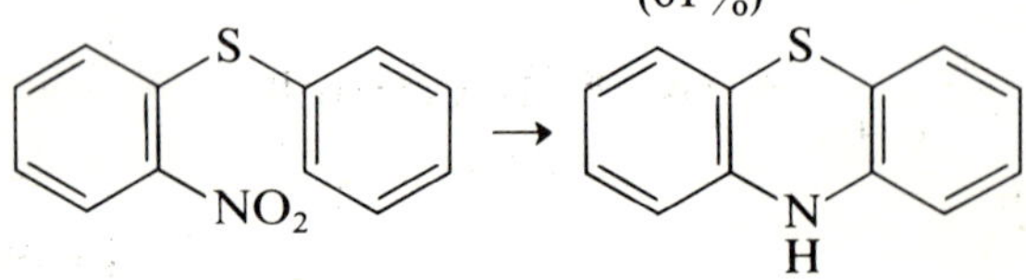

phenothiazine (54%) [77]

On heating 2-azido-4′-nitrobiphenyl (97) at 170°, 2-nitrocarbazole (98) is formed. It is probable that in this and the triethyl-phosphite-promoted cyclizations an intermediate nitrene is formed [79].

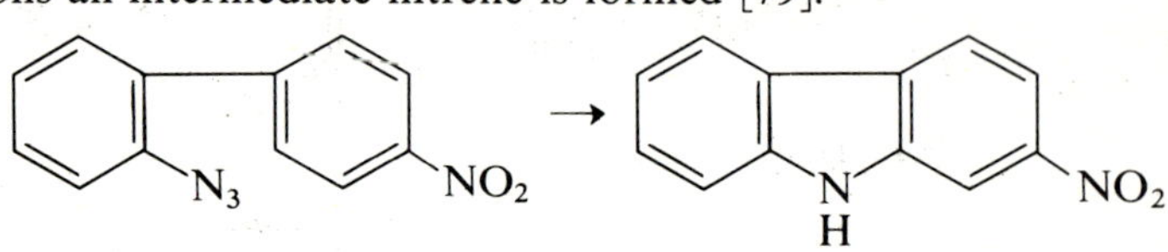

(77–82%)

(97) (98)

SYNTHESES USING SULPHUR YLIDES

An increasing number of heterocyclic syntheses has been carried out using sulphur ylides. Sulphonium ylides add to the double bond in carbonyl, imino and nitroso compounds [80]. For example, oxiranes (epoxides) are obtained by the addition of dimethylsulphonium methylide to ketones or $\alpha\beta$-unsaturated ketones.

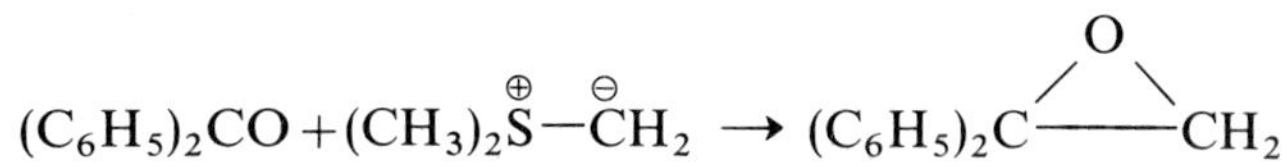

1,1-diphenylethylene oxide (100%) [81]

The following equations give further examples of this type of reagent.

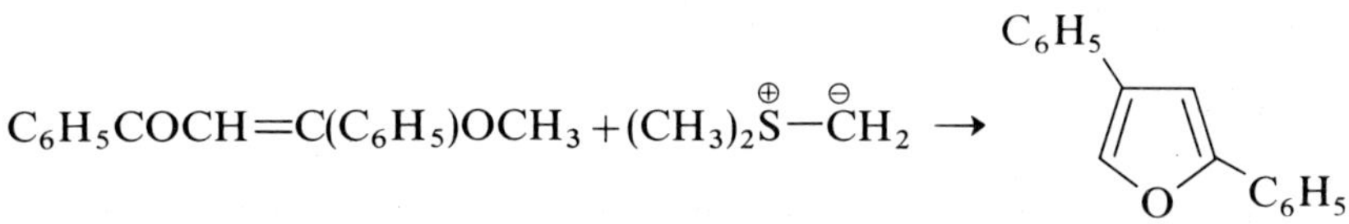

dibenzoylmethane enol ether — 2,4-diphenylfuran (42%) [82]

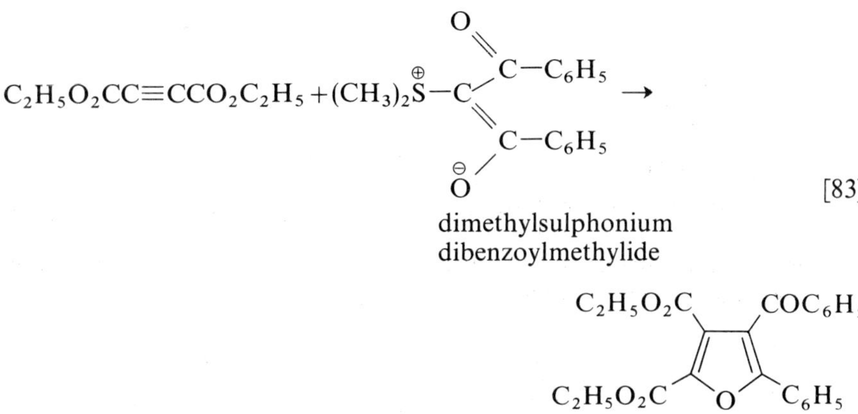

dimethylsulphonium dibenzoylmethylide [83]

3-benzoyl-4,5 bis-(ethoxycarbonyl)-2-phenylfuran (97%)

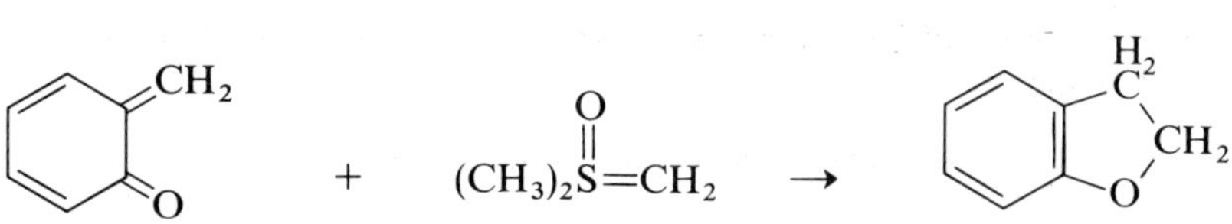

generated from *o*-hydroxybenzyl-trimethylammonium iodide and base — dimethylsulphoxonium methylide — 2,3-dihydrobenzofuran (47%)

[84]

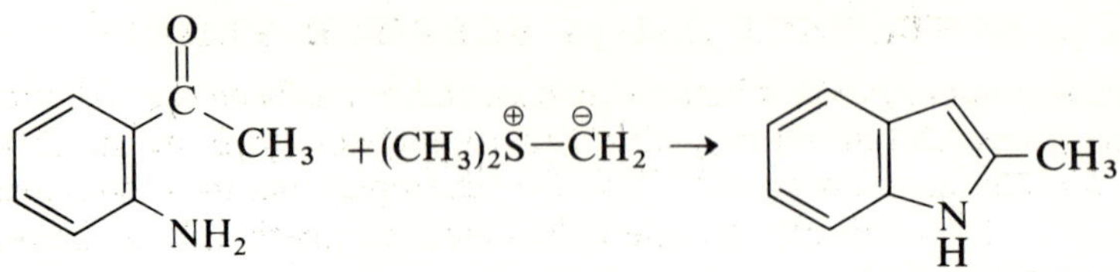

2-methylindole
(80%) [85]

MISCELLANEOUS RING SYNTHESES

The syntheses discussed in this section have been arranged in order of increasing size of the heterocyclic ring, and for a given ring size in order of increasing number of heteroatoms.

(a) Aziridines

The photodecomposition of triazolines has been found to be a general method for the preparation of aziridines [86]. Addition of iodine azide (IN_3) to substituted ethylenes gives β-iodoazides. The lithium aluminium hydride reduction of these adducts gives a stereospecific synthesis of aziridines.

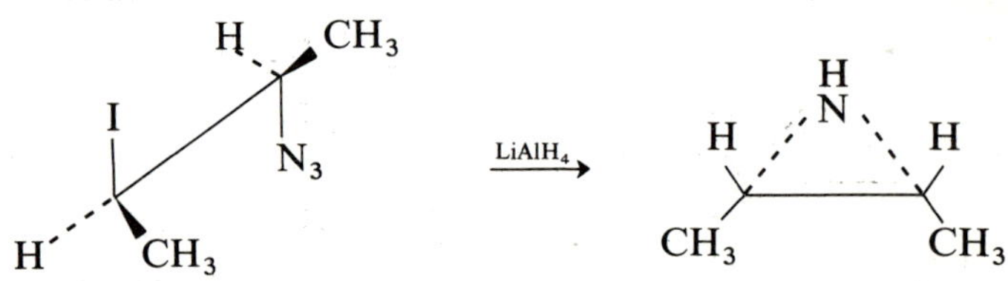

threo-2-azido-3-iodobutane   *cis*-2,3-dimethylaziridine [87]

Alternatively aziridines have been synthesized by lithium aluminium hydride reduction of ketoximes.

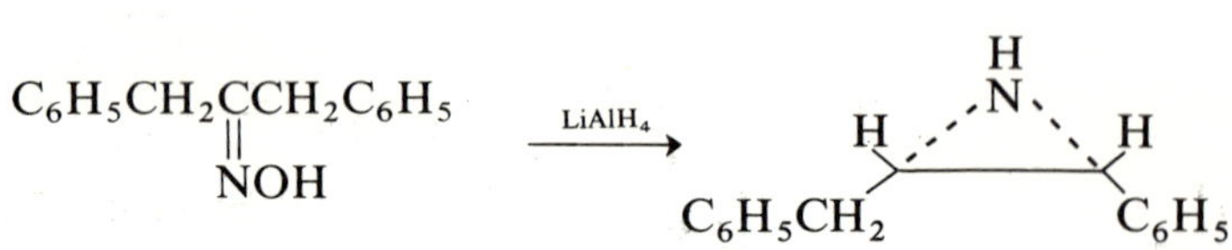

oxime of dibenzyl ketone   *cis*-2-benzyl-3-phenylaziridine [88]

A general method for preparing aziridines of type (99) is illustrated in the following equation.

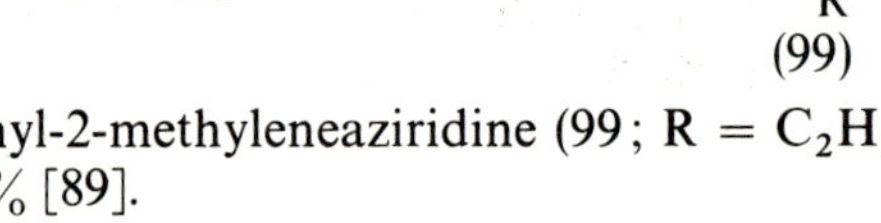

(99)

The yield of 1-ethyl-2-methyleneaziridine (99; R = C_2H_5) obtained by this method is 48–55% [89].

(b) Oxetanes

The following preparation of 2-methyloxetane illustrates a general method for the preparation of oxetanes from bromohydrins.

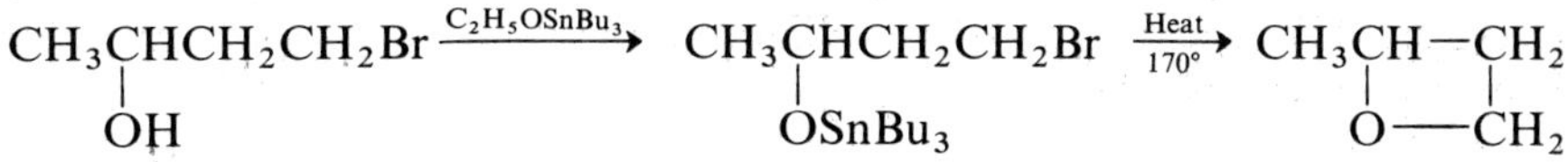

(80%) [90]

(c) Thietanes

Thietane-1,1-oxides are formed by the interaction of a sulphonyl chloride containing a hydrogen atom α to the sulphonyl group with ketene diethylacetal in the presence of triethylamine.

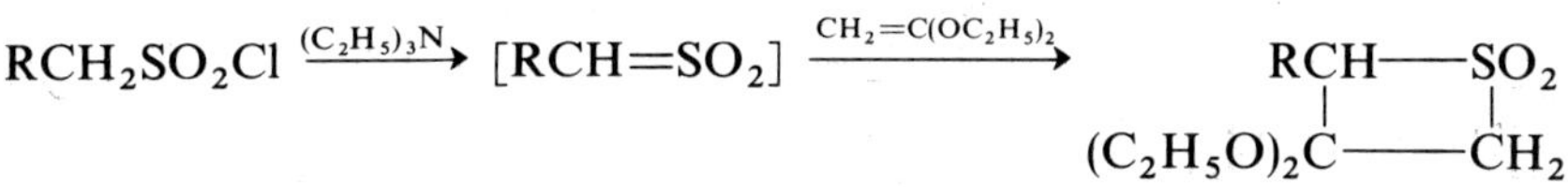

(100)

The yield of 3,3-diethoxythietane-1,1-dioxide (100; R = H) obtained from methanesulphonyl chloride and ketene diethylacetal is 65%, but the reaction fails if R is an alkyl group [91].

(d) Pyrroles

Esters of substituted pyrrole-2-carboxylic acids can be prepared from esters of toluene-*p*-sulphonylglycine in three steps: (1) base-catalysed addition to an $\alpha\beta$-unsaturated ketone; (2) dehydration of the resulting pyrrolidine with phosphoryl chloride and pyridine; and (3) elimination of *p*-toluenesulphinic acid to give the pyrrole with sodium ethoxide or potassium t-butoxide [92].

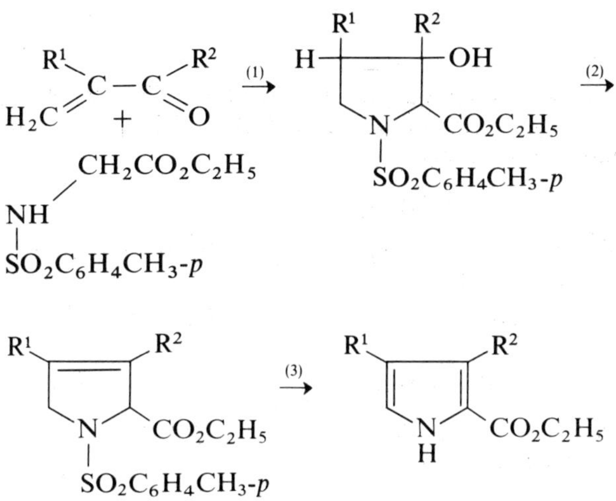

(e) Benzofurans

An improved preparation of benzofuran (coumarone) has recently been described [93]. A new general route to benzofurans by rearrangement of *O*-aryloximes, represents an extension of the well-known Fischer indole synthesis. The mechanisms of benzofuran and indole formation have been shown to be similar [94, 95].

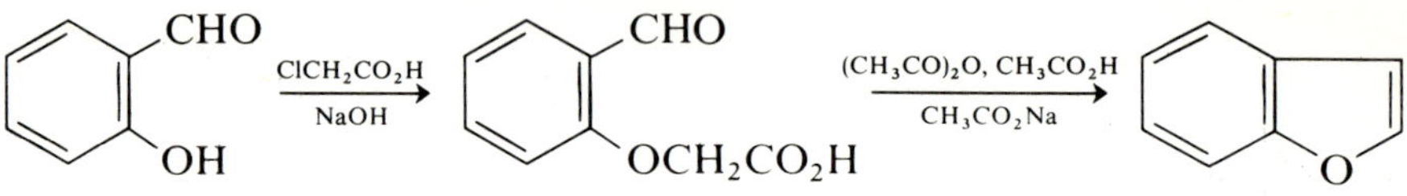

(52–66 % overall)

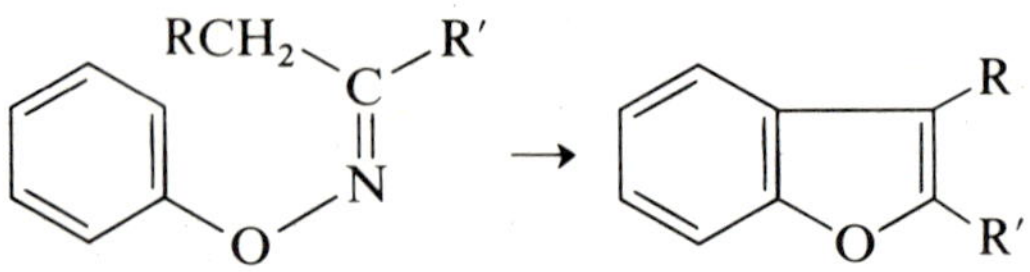

(f) Indazoles

The parent compound is conveniently prepared from *o*-toluidine by the following route [96].

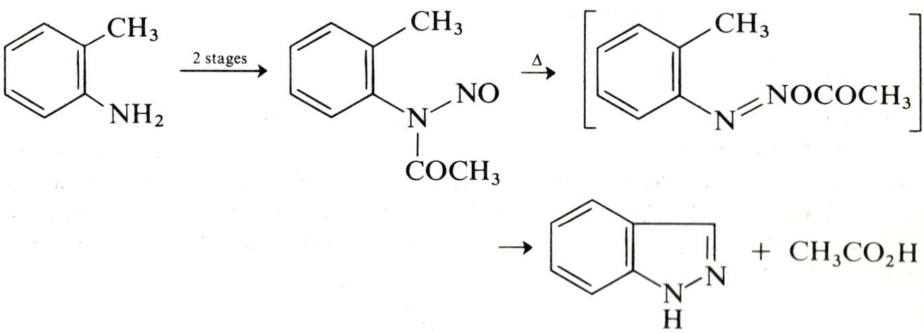

(36–47 %)

A general synthesis of indazoles involves the oxidation of aromatic ketone aryl hydrazones with lead tetra-acetate followed by treatment of the resulting azo-acetate with a Lewis acid [97].

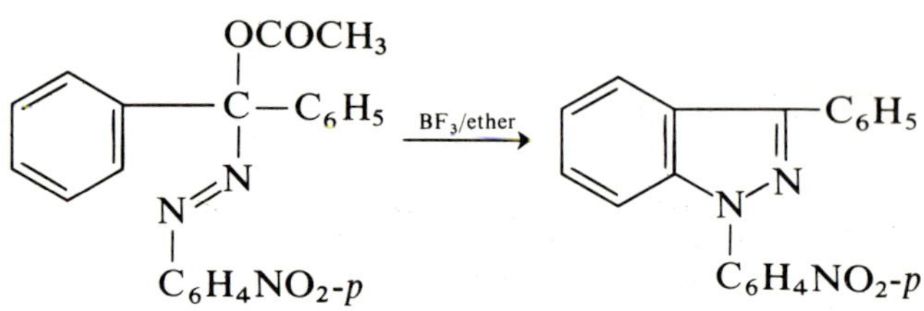

α-phenyl-α-*p*-nitrophenylazobenzyl acetate

1-*p*-nitrophenyl-3-phenylindazole (98 %)

(g) Imidazoles and Benzimidazoles

α-Ketol esters (101) give *C*-substituted imidazoles (102) on heating with formamide. α-Ketol esters are prepared from ketones by bromination followed by reaction with the potassium salt of the appropriate acid [98].

$$RCOCH_2R^1 \rightarrow RCOCHBrR^1 \rightarrow RCOCH(R^1)OCOR^2 \xrightarrow{HCONH_2}$$

(101)

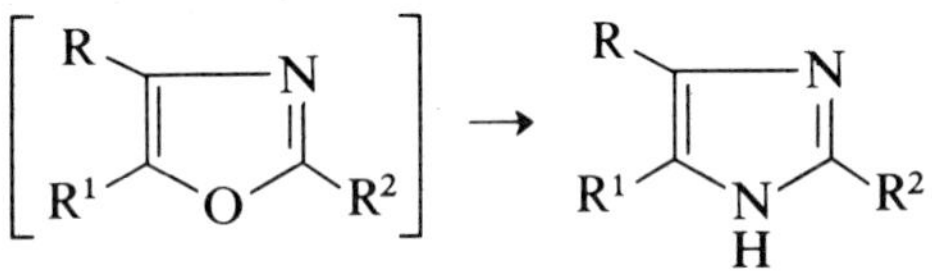

(102)

A variant on the usual methods of preparing benzimidazoles from *o*-phenylenediamines or *o*-nitroanilines involves the chlorination of an *N*-arylamidine with sodium hypochlorite, followed by treatment of the resulting *N*-chloro derivative with base [99, 100].

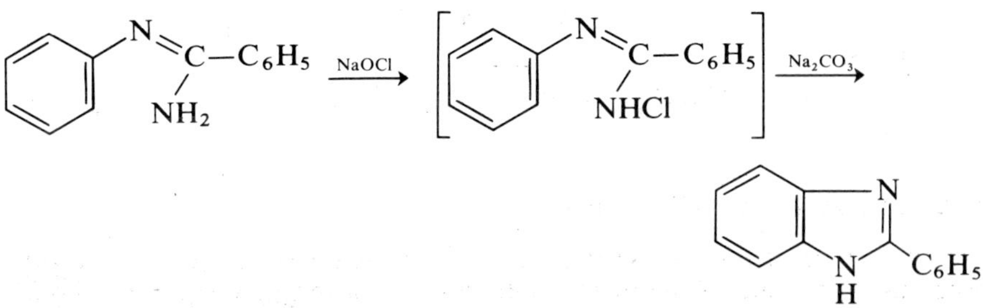

2-phenylbenzimidazole
(94%)

A novel approach to hydantoin synthesis involves intermediate 1,3,2-dioxophospholenes [101].

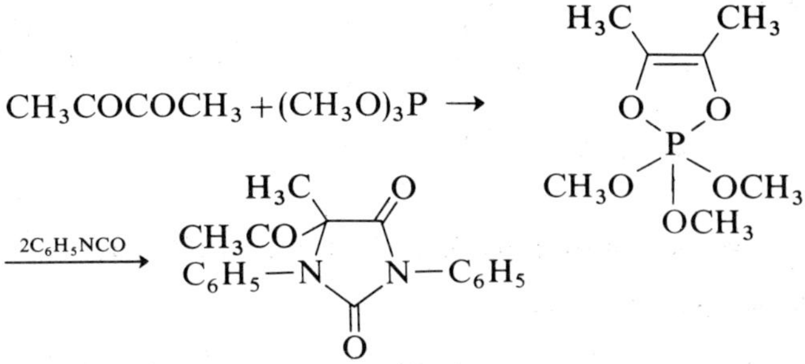

5-acetyl-5-methyl-
1,3-diphenylhydantoin
(70%)

(h) Oxazolidones

A new method for the synthesis of 2-oxazolidones is illustrated by the preparation of 3-phenyl-5-phenoxymethyl-2-oxazolidone (103) [102].

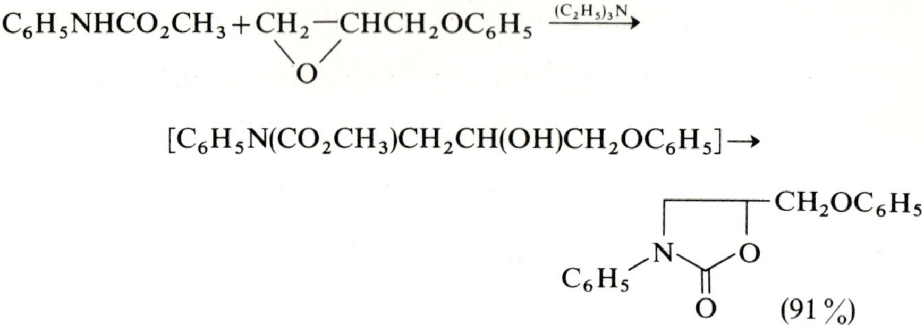

(103)

(i) Isothiazoles

The more readily available isoxazoles are converted into isothiazoles of similar substitution pattern by the following two-step synthesis [103].

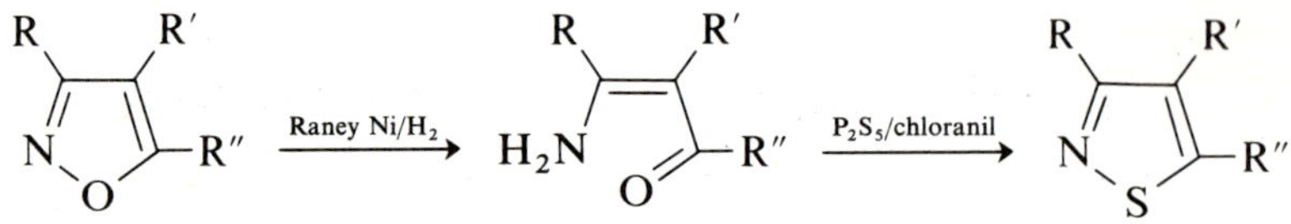

(j) β-Piperidones and β-Thianones

Compounds of both these types have been prepared from 2,3-dichloropropene. The plan of synthesis is to add an appropriate C—C—N or C—C—S fragment to the propene by nucleophilic displacement of its allylic chlorine and then to effect an intramolecular electrophilic ring closure of the resulting alkene [104].

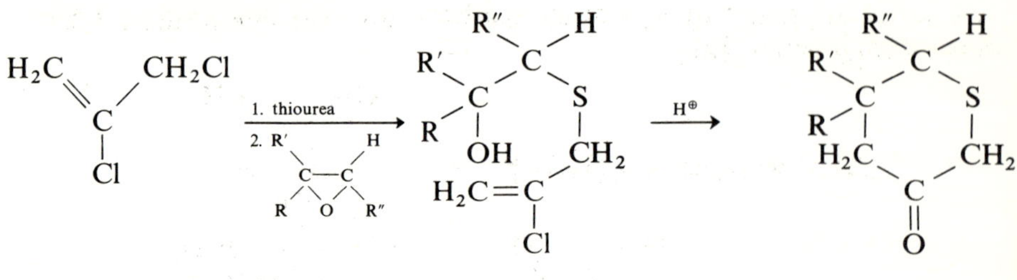

Where for example $R = C_6H_5$; $R' = R'' = H$

(k) α-Pyridones and α-Pyrones

These are readily prepared from diethyl 3-ethoxyallylidenemalonate (106). This intermediate is obtained by heating 1,1,3,3-tetraethoxypropane (104) with diethylmalonate (105) in acetic anhydride and in the presence of catalytic amounts of zinc chloride. Compound (106) cyclizes in acidic media to the α-pyrone (107) and is converted to the α-pyridine (108) by successive treatment with ammonia and alcoholic sodium ethoxide [105].

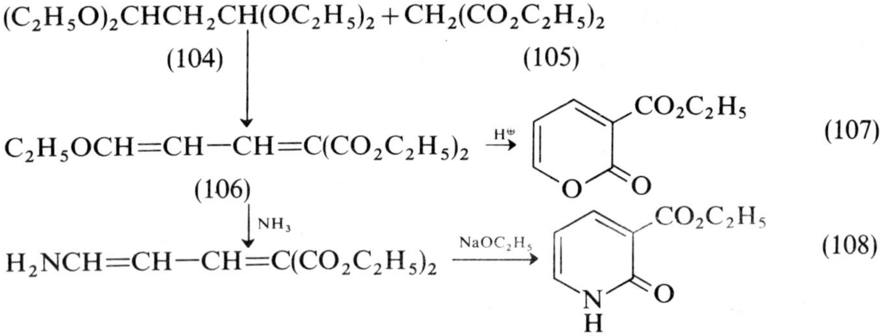

(104) (105) (106) (107) (108)

(l) Quinolines

β-chlorovinyl aldehydes (110) are readily available by treatment of α-methylene ketones (109) with dimethylformamide and phosphoryl chloride. They are converted into quinolines on reaction with primary aromatic amines, and to quinolinium salts on reaction with secondary aromatic amines [106].

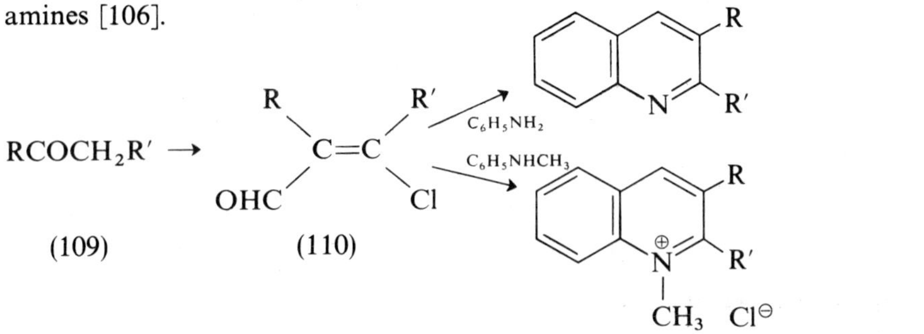

(109) (110)

(m) Isoquinolines

Cyclization of *o*-cyanobenzyl cyanides (111) with hydrogen bromide or hydrogen iodide affords 1-halo-3-aminoisoquinolines (112) in excellent yield [107].

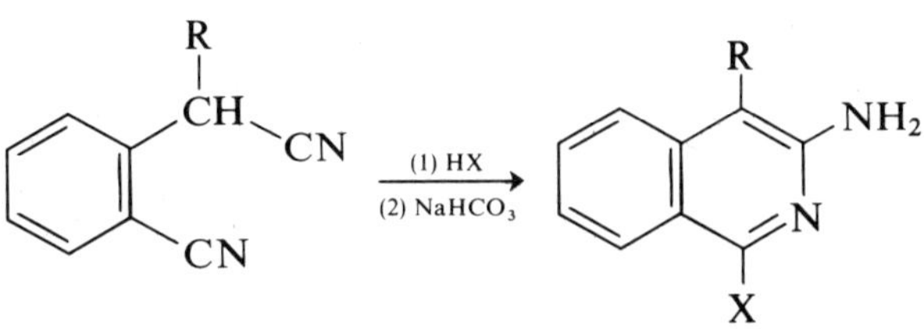

(111) (112)

3,4-Dihydroisoquinolines have been prepared from 2-methyl-allylbenzene as illustrated in the following equation. This method leads to 3,3-disubstituted derivatives which cannot be obtained by Bischler–Napieralski ring closure [108].

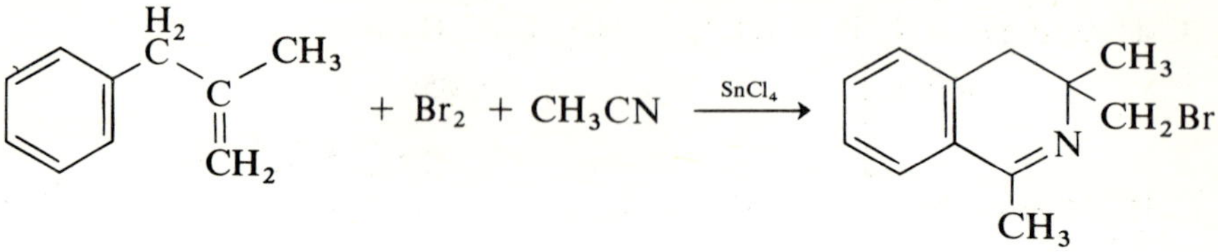

3-bromomethyl-1,3-dimethyl-
3,4-dihydroisoquinoline
(72%)

N-Substituted *o*-toluamides have been converted into 3,4-dihydro-isocarbostyrils by a route which involves the initial formation of their dilithio derivatives [109].

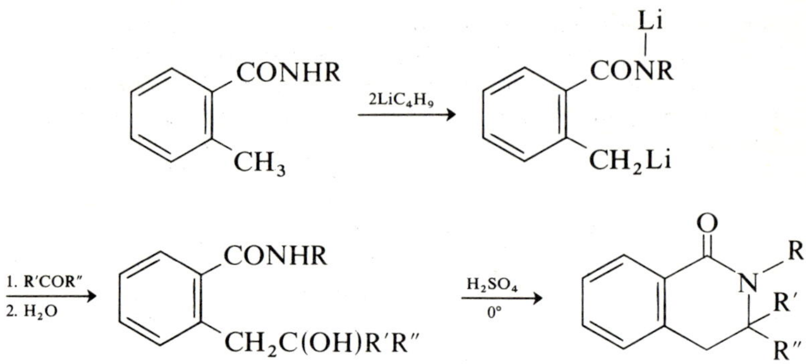

Where for example R = CH_3; R′ = R″ = C_6H_5

(n) Benzo[*a*]quinolizines

A novel approach to the synthesis of these compounds involves the addition of a Mannich base to a 3,4-dihydroisoquinoline [110].

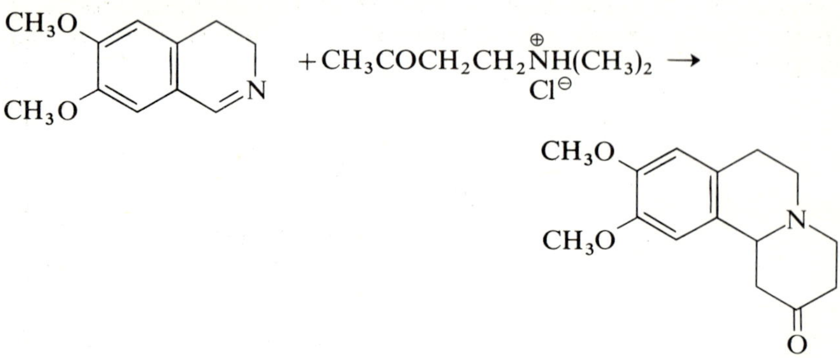

1,3,4,6,7,11b-hexahydro-
9,10-dimethoxyquinolizin-2-one
(67%)

(o) Pyrans

Condensation of α-olefins with paraformaldehyde and hydrogen halides in a non-aqueous medium yields *cis–trans* mixtures of 3-alkyl-4-halogeno-tetrahydropyrans [111].

$RCH{=}CH_2 + 2CH_2O + HX \rightarrow$ 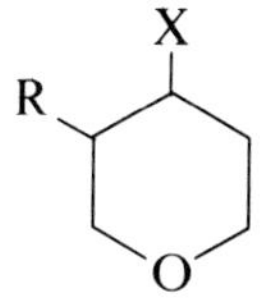 $+ H_2O$

Optimum yields are obtained with hydrogen chloride at $-60°$ to $-70°$.

(p) γ-Pyrones

The preparation of 2-(*p*-methoxyphenyl)-6-phenyl-4-pyrone (113) illustrates a general method for the synthesis of 2,6-disubstituted-4-pyrones [112].

$C_6H_5COCH_2COCH_2COC_6H_4OCH_3\text{-}p \rightarrow$ 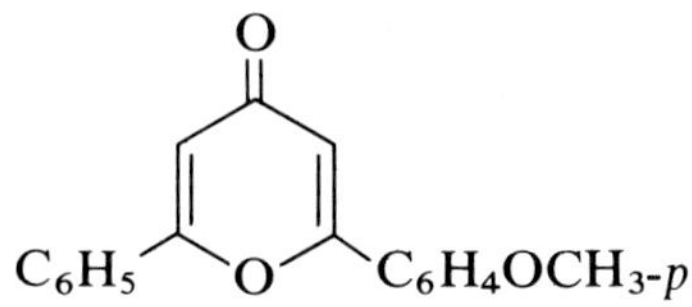

(88–99 %)

(113)

Alkyl substituted γ-pyrones have been prepared in good yield by heating aliphatic carboxylic acids or their anhydrides in polyphosphoric acid [113, 114].

$CH_3CH_2CO_2H \xrightarrow[3\ hr.]{PPA/200°}$ 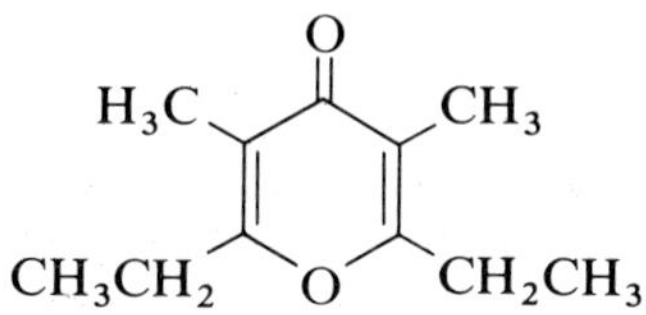

2,6-diethyl-3,5-dimethyl-γ-pyrone

(65 %)

(q) Coumarins

The enamine derivatives of β-ketonic acids condense smoothly with phenolic compounds. Coumarins which are only obtained in poor yield under normal Pechmann conditions have been successfully prepared by this procedure [115].

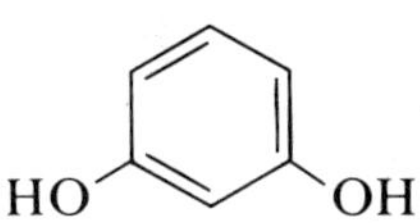 $+ CH_3C(NH_2){=}CHCO_2C_2H_5 \xrightarrow{180°}$ 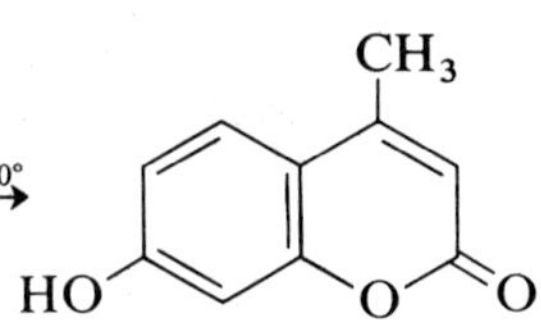

7-hydroxy-4-methylcoumarin

(85 %)

(r) Pyrimidines

The reaction of *s*-triazine with amidines or amidine salts having acidic α-methylene groups gives 5-substituted 4-aminopyrimidines in good yield [116].

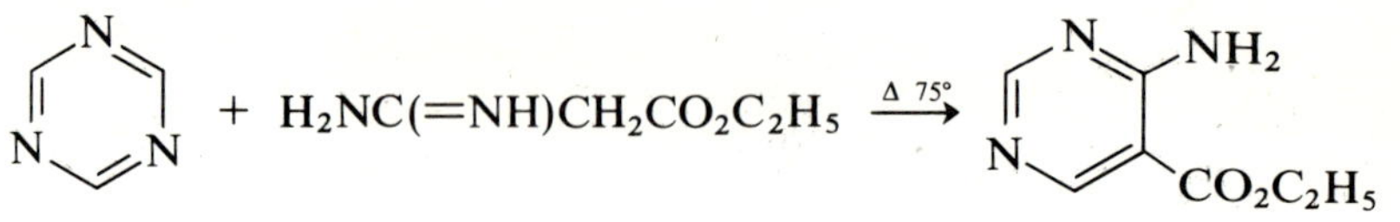

s-triazine — 2-carbethoxyacetamidine (as hydrochloride) — 4-amino-5-carbethoxypyridine (85%)

$$C_6H_5CON{=}CClC_6H_5 + CH_2(CN)_2 \xrightarrow{CH_3ONa\ CH_3OH} C_6H_5CONHC(C_6H_5){=}C(CN)_2 \quad (83\%)$$

(114) (115)

$$\xrightarrow{HCl/C_2H_5OH}$$

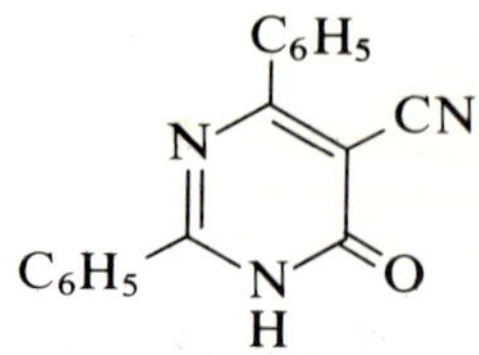

(100%)

(116)

Condensation of an *N*-acylimidochloride (e.g., 114) with malonitrile followed by cyclization of the resulting *N*-acylenamine (e.g., 115) gives 6-hydroxy-2,4-diaryl-5-cyanopyrimidines (e.g., 116) in excellent yield [117].

(s) Pyrazines

2-Aminopyrazine 1-oxides are obtained from the condensation of α-aminocyano compounds and α-oximinoketones according to the following equation. The amine oxides are important intermediates for unambiguous pteridine synthesis [118].

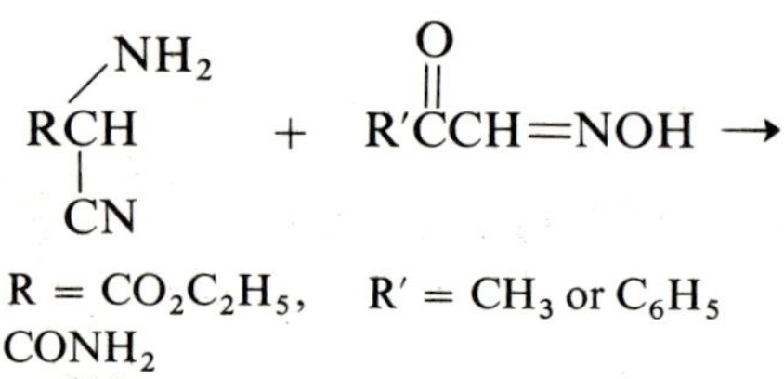

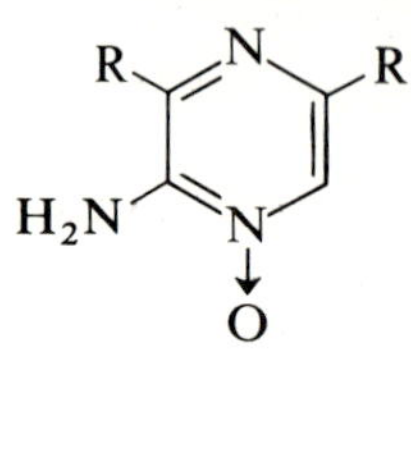

$R = CO_2C_2H_5$, $CONH_2$ or CN $\quad R' = CH_3$ or C_6H_5

(t) Quinoxalines

Quinoxaline di-*N*-oxides have been prepared in excellent yield from benzfuroxan. The latter compound undergoes ring opening with a variety of substrates which provide a two carbon fragment for quinoxaline ring formation [119].

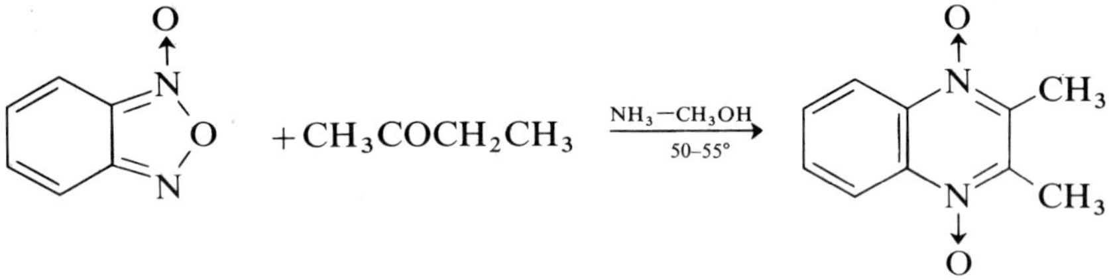

2,3-dimethylquinoxaline-1,4-dioxide (91 %)

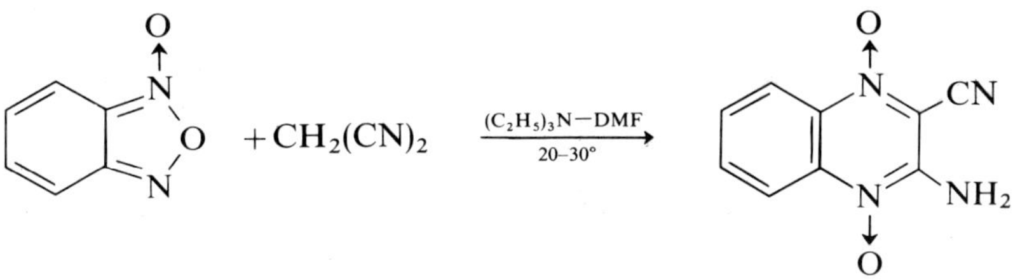

2-amino-3-cyanoquinoxaline-1,4-dioxide (83 %)

(u) Purines

In a new xanthine synthesis, the final step is the thermal cyclization of a 5-nitroso-4-aminouracil. For example, 1,3-dimethyl-6-methylamino-5-nitrosouracil (117) is converted on heating in boiling butanol to theophylline (118) in 98 % yield [120].

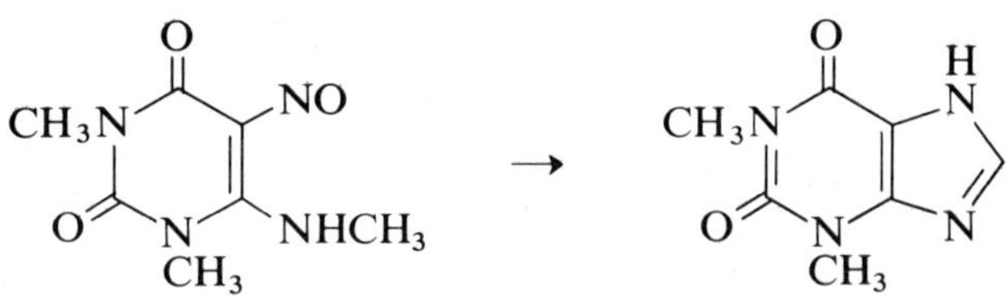

(117) (118)

Pyrimidines have also been converted into purines by a method which involves the introduction of a C_5-substituent into the pyrimidine ring with diethyl azodicarboxylate [121].

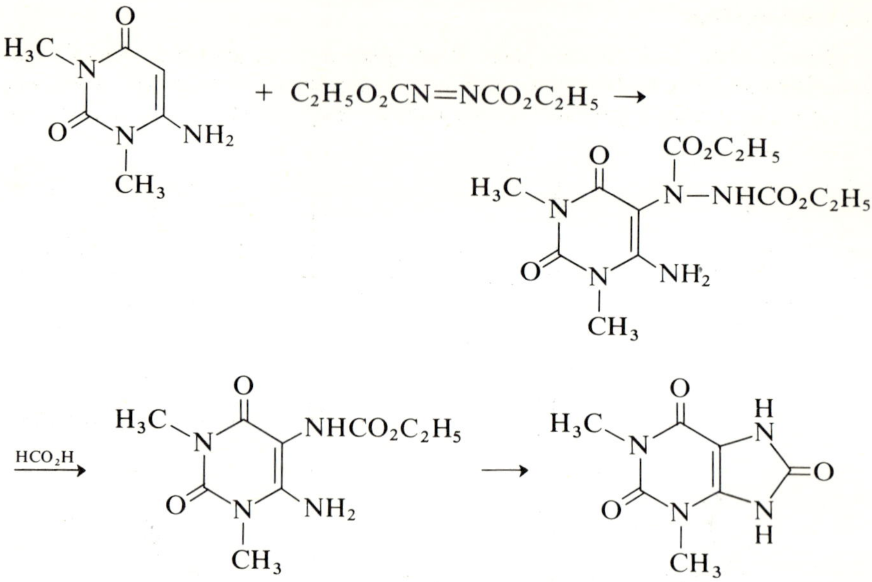

The pyrimidine-diethyl azodicarboxylate adducts have also been converted into azapteridines.

(v) 10,9-Borazarophenanthrenes

The following equations illustrate the preparation of the 10-methyl derivative (120) from *o*-aminobiphenyl (119) [122].

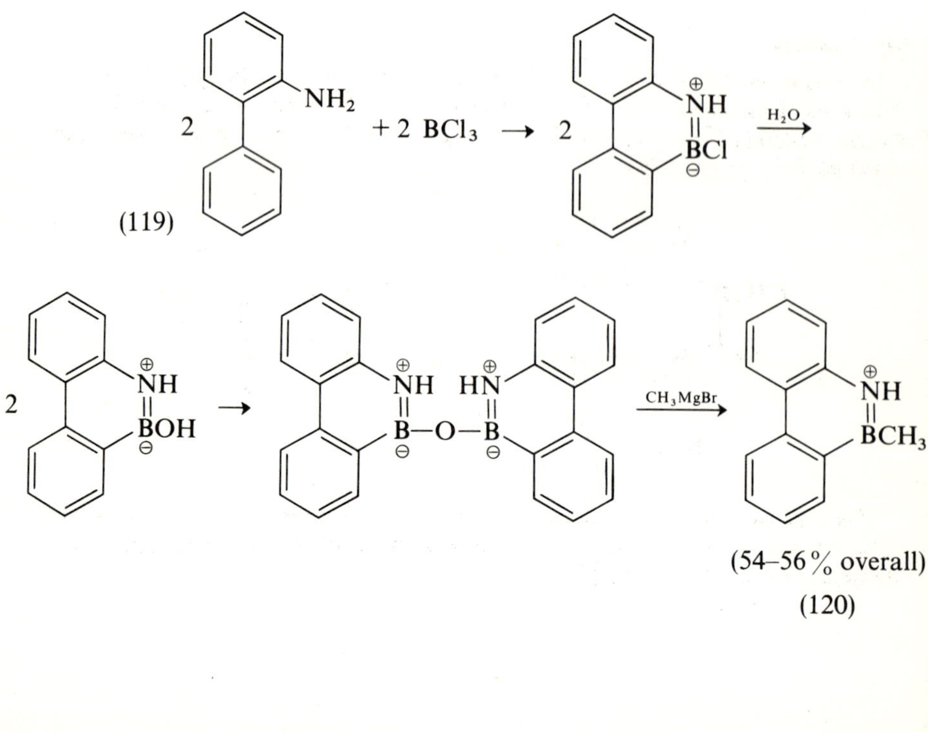

(54–56% overall)

(120)

REFERENCES

[1] Harris, R. L. N., Johnson, A. W. and Kay, I. T. *Q. Rev. chem. Soc.* **20**, 211 (1966).
[2] Winterfeldt, E. *Angew. Chem. int. Edn* **6**, 423 (1967).
[3] Hendrickson, J. B., Rees, R. and Templeton, J. F. *J. Am. chem. Soc.* **86**, 107 (1964).
[4] Courtin, A., Class, E. and Erlenmeyer, H. *Helv. chim. Acta* **47**, 1748 (1964).
[5] Iwanami, Y. *Bull. chem. Soc. Japan* **37**, 1740 (1964).
[6] Iwanami, Y. *et al. Bull. chem. Soc. Japan* **37**, 1745 (1964).
[7] Pachter, J. *J. org. Chem.* **26**, 4157 (1961).
[8] Acheson, R. M. *Adv. heterocycl. Chem.* **1**, 125 (1963).
[9] Acheson, R. M. and Plunkett, A. O. *J. chem. Soc.* 2676 (1965).
[10] Huisgen, R. in *Proceedings of the First International Congress of Heterocyclic Chemistry*, Albuquerque (1967).
[11] Winterfeldt, E. *Chem. Ber.* **98**, 3537 (1965).
[12] Boekelheide, V. *et al. J. Am. chem. Soc.* **83**, 453 (1961).
[13] Huisgen, R. and Herbig, K. *Annln Chem.* (*Liebigs*) **688**, 98 (1965); see also Gompper, R. and Elser, W. *Angew. Chem. int. Edn* **6**, 366 (1967).
[14] Acheson, R. M., Bailey, A. S. and Selby, I. A. *Chem. Commun.* 835 (1966).
[15] Krespan, C. G. *J. Am. chem. Soc.* **83**, 3434, 3438 (1961).
[16] Huisgen, R. *Angew. Chem. int. Edn* **2**, 565, 633 (1963).
[17] Huisgen, R. *et al. Org. Synth.* **46**, 127 (1966).
[18] Huisgen, R. and Mack, W. *Tetrahedron Lett.* 583 (1961).
[19] For review of 1,2,3-triazole synthesis see Boyer, J. H. in *Heterocyclic Compounds* (ed. R. C. Elderfield), Vol. 7, p. 387, Wiley, New York (1961).
[20] Hantzsch, A. and Vagt, A. *Annln Chem.* (*Liebigs*) **314**, 339 (1901).
[21] Seidl, H., Huisgen, R. and Knorr, R. *Chem. Ber.* **102**, 904 (1969).
[22] Huisgen, R. *Chem. Soc. sp. Publ.* **21**, 51 (1967).
[23] Hamer, J., (ed.) *1,4-Cycloaddition Reactions*, Academic Press, New York (1967).
[24] Sauer, J. *Angew. Chem. int. Edn* **5**, 211 (1966).
[25] Albrecht, R. and Kresze, G. *Chem. Ber.* **98**, 1431 (1965).
[26] Kealey, J. T. *J. Am. chem. Soc.* **84**, 966 (1962).
[27] Gillis, B. J. and Hagarty, J. D. *J. org. Chem.* **32**, 330 (1967).
[28] Firl, J. and Kresze, G. *Chem. Ber.* **99**, 3695 (1966).
[29] Kresze, G. and Braun, H. *Tetrahedron Lett.* 1743 (1969).
[30] Chumakov, Y. I. and Luzovskaya, L. P. *J. gen. Chem. U.S.S.R.* **34**, 3652 (1964); see also Chumakov, Y. I. and Sherstyuk, V. P. *Tetrahedron Lett.* 129 (1965).
[31] Harris, E. E. *et al. J. org. Chem.* **27**, 2705 (1962).
[32] Carruthers, W. *Chem. Commun.* 272 (1966).
[33] Chakraborty, D. P. *Tetrahedron Lett.* 661 (1966).
[34] Grellmann, K. H. and Tauer, E. *Tetrahedron Lett.* 1909 (1967).
[35] Thyagarajan, B. S., Kharasch, N., Lewis, H. B. and Wolf, W. *Chem. Commun.* 614 (1967).
[36] Kagan, J. *J. Am. chem. Soc.* **88**, 2617 (1966).
[37] Cleveland, P. G. and Chapman, O. L. *Chem. Commun.* 1064 (1967).
[38] Swenton, J. S. *Tetrahedron Lett.* 2855 (1967).
[39] Craigie, R. and Kreiger, M. *Chem. Ber.* **98**, 387 (1965).
[40] Helmkamp, G. K. *et al. J. Am. chem. Soc.* **88**, 1030 (1966).
[41] Heine, H. W. and Tomalia, D. A. *J. Am. chem. Soc.* **84**, 993 (1962).
[42] Dick, C. R. *J. org. Chem.* **32**, 72 (1967).
[43] Kemp, D. S. and Woodward, R. B. *Tetrahedron* **21**, 3019 (1965); Woodward, R. B. and Olofson, R. A. *Tetrahedron*, Supplement No. 7, p. 415 (1966).
[44] Moriconi, E. J. and Creegan F. J. *Org. Synth.* **48**, 90 (1968).

[45] HAFNER, K. and KÖNIG, C. *Angew. Chem. int. Edn* **3**, 165 (1964).
[46] MARSH, F. D. and SIMMONS, H. E. *J. Am. chem. Soc.* **87**, 3529 (1965).
[47] PAQUETTE, L. A. *J. Am. chem. Soc.* **85**, 3288 (1963); *Org. Synth.* **44**, 41 (1964).
[48] CHILDS, R. F. and JOHNSON, A. W. *J. chem. Soc.* (*C*) 1950 (1966).
[49] MOORE, H. W., SHELDON, H. R. and WEYLER W. *Tetrahedron Lett.* 1243 (1969).
[50] VOGEL, E. and GÜNTHER, H. *Angew. Chem. int. Edn* **6**, 385 (1967).
[51] SCHWEIZER, E. E. and PARHAM, W. E. *J. Am. chem. Soc.* **82**, 4085 (1960).
[52] SONDHEIMER, F. and SHANI, A. *J. Am. chem. Soc.* **86**, 3168 (1964).
[53] VOGEL, E. *et al. Angew. Chem. int. Edn* **3**, 642 (1964).
[54] MOORE, J. A. and MEDEIROS, R. W. *J. Am. chem. Soc.* **81**, 6026 (1959); MOORE, J. A. *Trans. N.Y. Acad. Sci.* **27**, 591 (1965).
[55] STERNBACH, L. H. and REEDER, E. *J. org. Chem.* **26**, 111 (1961).
[56] BUNNETT, J. F., HRUTFIORD, B. F. and WILLIAMSON, S. M. *Org. Synth.* **40**, 1 (1960).
[57] BUNNETT, J. F. *et al. J. org. Chem.* **28**, 1 (1963).
[58] KESSAR, S. V. and MANMOHAN SINGH. *Tetrahedron Lett.* 1155 (1969).
[59] TAYLOR, E. C., WARRENER, R. N. and MCKILLOP, A. *Angew. Chem. int. Edn* **5**, 309 (1966) and earlier papers.
[60] FERRIS, J. P. and ORGEL, L. E. *J. Amer. chem. Soc.* **87**, 4976 (1965).
[61] KRÖHNKE, F. and ZECHER, W. *Angew. Chem. int. Edn* **1**, 626 (1962).
[62] KRÖHNKE, F. and WEIS, W. *Annln Chem.* (*Liebigs*) **679**, 136 (1964).
[63] BALABAN, A. T. and NENITZESCU, C. D. *Org. Synth.* **44**, 98 (1964).
[64] HAFNER, K. and KAISER, H. *Org. Synth.* **44**, 101 (1964).
[65] BALABAN, A. T. *Organic Preparations and Procedures* **1**, 63 (1969).
[66] BALABAN, A. T. and NENITZESCU, C. D. *J. chem. Soc.* 3553 (1961).
[67] BOYD, G. V. and JACKMAN, L. M. *J. chem. Soc.* 548 (1963).
[68] MÄRKL, G., LIEB, F. and MERZ, A. *Angew. Chem. int. Edn* **6**, 458 (1967).
[69] HARVEY, G. R. *J. org. Chem.* **31**, 1587 (1966).
[70] SCHWEIZER, E. E. and BACH, R. D. *Org. Synth.* **48**, 129 (1968).
[71] SCHWEIZER, E. E. and LIEHR, J. G. *J. org. Chem.* **33**, 583 (1968).
[72] SCHWEIZER, E. E. *J. Am. chem. Soc.* **86**, 2744 (1964).
[73] SCHWEIZER, E. E. and LIGHT, K. K. *J. Am. chem. Soc.* **86**, 2963 (1964).
[74] DIMROTH, K., POHL, G. and FOLLMANN, H. *Chem. Ber.* **99**, 634 (1966).
[75] CADOGAN, J. I. G. *et al. J. chem. Soc.* 4831 (1965).
[76] CADOGAN, J. I. G. and MACKIE, R. B. *Org. Synth.* **48**, 113 (1968).
[77] CADOGAN, J. I. G., MACKIE, R. K. and TODD, M. J. *Chem. Commun.* 491 (1966).
[78] KAMETANI, T. *et al. Tetrahedron Lett.* 1027 (1969).
[79] MENDERHALL, G. D. and SMITH, P. A. S. *Org. Synth.* **46**, 101 (1966).
[80] JOHNSON, A. WILLIAM. *Ylid Chemistry*, p. 304, Academic Press, New York (1966).
[81] COREY, E. J. and CHAYKOVSKY, M. *J. Am. chem. Soc.* **87**, 1353 (1965).
[82] HARRIS, T. M., HARRIS, C. M. and CLEARY, J. C. *Tetrahedron Lett.* 1427 (1968).
[83] TAKAKU, M., HAYASI, Y. and NOZAKI, H. *Tetrahedron Lett.* 2053 (1969).
[84] BREUER, E. and MELUMAD, D. *Tetrahedron Lett.* 1875 (1969).
[85] BRAVO, P., GAUDIANO, G. and UMANI-RONCHI, A. *Tetrahedron Lett.* 679 (1969).
[86] SCHEINER, P. *Tetrahedron* **24**, 2757 (1968).
[87] HASSNER, A., MATTHEWS, G. J. and FOWLER, F. W. *J. Am. chem. Soc.* **91**, 5046 (1969).
[88] KOTERA, K. and KITAHONOKI, K. *Org. Synth.* **48**, 20 (1968).
[89] BOTTINI, A. T. and OLSEN, R. E. *Org. Synth.* **44**, 53 (1964).
[90] DELMOND, B., POMMIER, J. C. and VALADE, J. *Tetrahedron Lett.* 2089 (1969).
[91] TRUCE, W. E. and NORELL, J. R. *J. Am. chem. Soc.* **85**, 3231 (1963).
[92] TERRY, W. G., JACKSON, A. H., KENNER, G. W. and KORNIS, G. *J. chem. Soc.* 4389 (1965).
[93] BURGSTAHLER, W. and WORDEN, L. R. *Org. Synth.* **46**, 28 (1966).

[94] SHERADSKY, J. *Tetrahedron Lett.* 5225 (1966).
[95] MOORADIAN, A. and DUPONT, P. E. *Tetrahedron Lett.* 2867 (1967).
[96] HUISGEN, R. and BAST, H. *Org. Synth.* **42**, 69 (1962).
[97] GLADSTONE, W. A. F. and NORMAN, R. O. C. *J. chem. Soc.* (*C*) 1527 (1966).
[98] NOVELLI, A. and DE SANTIS, A. *Tetrahedron Lett.* 265 (1967).
[99] GRENDA, V. J. *et al. J. org. Chem.* **30**, 259 (1965).
[100] HARUKI, E., INAIKE, T. and IMOTA, E. *Bull. chem. Soc. Japan* **38**, 1805 (1965).
[101] RAMIREZ, F., BHATIA, S. B. and SMITH, C. P. *J. Am. chem. Soc.* **89**, 3030 (1967).
[102] IWAKURA, Y. and IZAWA, S. *J. org. Chem.* **29**, 379 (1964).
[103] MCGREGOR, D. N. *et al. Tetrahedron* **25**, 389 (1969).
[104] LANSBURY, P. T. and SCHARF, D. J. *J. Am. chem. Soc.* **90**, 536 (1968).
[105] WINDHOLZ, T. B., PETERSON, L. H. and KENT, G. J. *J. org. Chem.* **28**, 1443 (1963).
[106] GAGAN, J. M. F. and LLOYD, D. *Chem. Commun.* 1043 (1967).
[107] JOHNSON, F. and NASUTAVICUS, W. A. *J. org. Chem.* **27**, 3953 (1962).
[108] HASSNER, A. *et al. Tetrahedron Lett.* 1241 (1968).
[109] MAO, C.-L., BARNISH, I. T. and HAUSER, C. R. *J. heterocyclic Chem.* **6**, 83 (1969).
[110] WHITTAKER, N. *J. chem. Soc.* (*C*) 83 (1969).
[111] STAPP, P. R. *J. org. Chem.* **34**, 479 (1969).
[112] MILES, M. L. and HAUSER, C. R. *Org. Synth.* **46**, 60 (1966).
[113] MULLOCK, E. B. and SUSCHITZKY, H. *J. chem. Soc.* (*C*) 828 (1967).
[114] EMMICK, T. L. and LETSINGER, R. L. *Org. Synth.* **47**, 54 (1967).
[115] KAPPE, T. and ZIEGLER, E. *Org. Preparations and Procedures* **1**, 61 (1969).
[116] SCHAEFER, F. C., HUFFMAN, K. R. and PETERS, G. A. *J. org. Chem.* **27**, 548, 551 (1962).
[117] SCHMIDT, R. R. *Chem. Ber.* **98**, 346 (1965).
[118] TAYLOR, E. C. and LENARD, K. *J. Am. chem. Soc.* **90**, 2424 (1968).
[119] LEY, K. *et al. Angew. Chem. int. Edn.* **8**, 596 (1969).
[120] GOLDNER, H., DIETZ, G. and CARSTAIRS, E. *Annln Chem.* (*Liebigs*) **691**, 142 (1966); *J. prakt. Chem.* **32**, 43 (1966).
[121] TAYLOR, E. C. and SOWINSKI, F. *J. Am. chem. Soc.* **90**, 1374 (1968).
[122] DEWAR, M. J. S., DEWAR, R. B. K. and GAIBEL, Z. L. F. *Org. Synth.* **46**, 65 (1966).

Addendum

Excellent accounts of the use of enamines in heterocyclic synthesis appear in [123–124].

[123] PAQUETTE, L. A. *Principles of Modern Heterocyclic Chemistry*, Benjamin, New York (1968).
[124] COOK, A. G. (ed.). *Enamines: Synthesis, Structure and Reactions*, Dekker, New York (1969).

Index